Useful Statistical Functions in Excel	Description
AVERAGE(*data range*)	Computes the average value (arithmetic mean) of a set of data.
BETADIST(*x, alpha, beta, A, B*)	Returns the cumulative beta density function.
BINOMDIST(*number_s, trials, probability_s, cumulative*)	Returns the individual term binomial distribution.
CHITEST(*actual_range, expected_range*)	Returns the test for independence; the value of the chi-square distribution and the appropriate degrees of freedom.
CONFIDENCE(*alpha, standard_dev, size*)	Returns the confidence interval for a population mean.
CORREL(*arrayl, array2*)	Computes the correlation coefficient between two data sets.
EXPONDIST(*x, lambda, cumulative*)	Returns the exponential distribution.
FORECAST(*x, known_y's, known_x's*)	Calculates a future value along a linear trend.
GAMMADIST(*x, alpha, beta, cumulative*)	Returns the gamma distribution.
GROWTH(*known_y's, known_x's, new_x's, constant*)	Calculates predicted exponential growth.
LINEST(*known_y's, known_x's, new_x's, constant, stats*)	Returns an array that describes a straight line that best fits the data.
LOGNORMDIST(*x, mean, standard_deviation*)	Returns the cumulative lognormal distribution of x, where $\ln(x)$ is normally distributed with parameters mean and standard deviation.
MEDIAN(*data range*)	Computes the median (middle value) of a set of data.
MODE(*data range*)	Computes the mode (most frequently occurring) of a set of data.
NORMDIST(*x, mean, standard_deviation, cumulative*)	Returns the normal cumulative distribution for the specified mean and standard deviation.
NORMSDIST(*z*)	Returns the standard normal cumulative distribution (mean = 0, standard deviation = 1).
PERCENTILE(*array, k*)	Computes the kth percentile of data in a range.
POISSON(*x, mean, cumulative*)	Returns the Poisson distribution.
QUARTILE(*array, quart*)	Computes the quartile of a distribution.
SKEW(*data range*)	Computes the skewness, a measure of the degree to which a distribution is not symmetric around its mean.
STANDARDIZE(*x, mean, standard_deviation*)	Returns a normalized value for a distribution characterized by a mean and standard deviation.
STDEV(*data range*)	Computes the standard deviation of a set of data, assumed to be a sample.
STDEVP(*data range*)	Computes the standard deviation of a set of data, assumed to be an entire population.
TREND(*known_y's, known_x's, new_x's, constant*)	Returns values along a linear trend line.
TTEST(*arrayl, array2, tails, type*)	Returns the probability associated with a t-test.
VAR(*data range*)	Computes the variance of a set of data, assumed to be a sample.
VARP(*data range*)	Computes the variance of a set of data, assumed to be an entire population.
WEIBULL(*x, alpha, beta, cumulative*)	Returns the Weibull distribution.
ZTEST(*array, x, sigma*)	Returns the two-tailed p-value of a z-test.

Statistics, Data Analysis, and Decision Modeling

FOURTH EDITION

James R. Evans

University of Cincinnati

Prentice Hall

Boston   Columbus   Indianapolis   New York   San Francisco
Upper Saddle River   Amsterdam   Cape Town   Dubai   London   Madrid
Milan   Munich   Paris   Montreal   Toronto   Delhi   Mexico City   Sao Paulo
Sydney   Hong Kong   Seoul   Singapore   Taipei   Tokyo

To Beverly, Kristin, and Lauren, for all
their love and understanding.

—James R. Evans

Editor in Chief: Eric Svendsen
Acquisitions Editor: Mark Pfaltzgraff
Editorial Project Manager: Susan Abraham
Director of Marketing: Patrice Lumumba Jones
Senior Managing Editor: Judy Leale
Project Manager: Ann Pulido
Senior Operations Supervisor: Arnold Vila
Operations Specialist: Benjamin Smith
Creative Director: Jayne Conte
Cover Designer: Axell Designs
Manager, Rights and Permissions: Charles Morris
Manager, Cover Visual Research & Permissions: Karen Sanatar
Cover Art: Getty Images, Inc.
Media Project Manager/Production: Lisa Rinaldi
Media Project Manager/Editorial: Allison Longley
Full-Service Project Management: Andrea Shearer/GGS Higher Education Resources,
a Division of PreMedia Global, Inc.
Composition: GGS Higher Education Resources, a Division of PreMedia Global, Inc.
Printer/Binder: Hamilton Printing Company
Cover Printer: Hamilton Printing Company
Text Font: Palatino

Credits and acknowledgments borrowed from other sources and reproduced, with permission, in this textbook appear on appropriate page within text.

Microsoft® and Windows® are registered trademarks of the Microsoft Corporation in the U.S.A. and other countries. Screen shots and icons reprinted with permission from the Microsoft Corporation. This book is not sponsored or endorsed by or affiliated with the Microsoft Corporation.

Library of Congress Cataloging-in-Publication Data

Evans, James R. (James Robert)
    Statistics, data analysis, and decision modeling / James R. Evans.— 4th ed.
      p. cm.
    Includes bibliographical references and index.
    ISBN-13: 978-0-13-606600-2 (alk. paper)
    ISBN-10: 0-13-606600-3 (alk. paper)
    I. Industrial management—Statistical methods. 2. Statistical decision. I. Title.
    HD30.215.E93 2009
    658.4′033—dc22

                              2008054659

Prentice Hall
is an imprint of

www.pearsonhighered.com

10 9 8 7 6 5 4 3 2
ISBN: 13: 978-0-13-606600-2
ISBN: 10: 0-13-606600-3

BRIEF CONTENTS

CONTENTS

PART II: DECISION MODELING AND ANALYSIS   293

PREFACE

INTENDED AUDIENCE

Statistics, Data Analysis, and Decision Modeling was written to meet the need for an introductory text that provides a basic introduction to business statistics and decision models/optimization, focusing on practical applications of data analysis and decision modeling, all presented in a simple and straightforward fashion.

The text consists of 14 chapters in two distinct parts. The first eight chapters deal with statistical and data analysis topics, while the remaining chapters deal with decision models and applications. Thus, the text may be used for:

- MBA or undergraduate business programs that combine topics in business statistics and management science into a single, brief, quantitative methods course.
- Business programs that teach statistics and management science in short, modular courses.
- Executive MBA programs.
- Graduate refresher courses for business statistics and management science.

SUBSTANCE

The danger in using quantitative methods does not generally lie in the inability to perform the requisite calculations, but rather in the lack of a fundamental understanding of why to use a procedure, how to use it correctly, and how to properly interpret results. The principal focus of this text is conceptual understanding using simple and practical examples rather than a plug-and-chug or point-and-click mentality, as are often done in other texts, supplemented by appropriate theory. On the other hand, the text does not attempt to be an encyclopedia of detailed quantitative procedures, but focuses in on useful concepts and tools for today's managers.

To support the presentation of topics in business statistics and decision modeling, this text integrates fundamental theory and practical applications in a spreadsheet environment using *Microsoft Excel 2007* and various spreadsheet add-ins, specifically:

- *PHStat*, a collection of statistical tools that enhance the capabilities of Excel; published by Pearson Education.
- A time limited professional version of *Crystal Ball* (including *CBPredictor* for forecasting and *OptQuest* for optimization), the most popular commercial package for risk analysis.
- *TreePlan*, a decision analysis add-in.
- *SimQuick*, an Excel-based application for process simulation, published by Pearson Education.
- *Premium Solver*, a more powerful version of Excel's Solver.

These tools have been integrated throughout the text to simplify the presentations and implement tools and calculations so that more focus can be placed on interpretation and understanding the managerial implications of results. However, as not to disrupt the flow of the text discussion and distract from conceptual understanding, we have placed boxed "Notes" for Excel, *PHStat*, and other add-ins that provide procedural details of using specific functions, tools, or techniques where appropriate.

NEW TO THIS EDITION

The fourth edition of this text has been substantially re-written to improve clarity and pedagogical features. Many significant changes have been made in this edition. These changes include the following.

1. Spreadsheet-based tools and applications are now compatible with *Microsoft Excel 2007*, which is used throughout this edition.
2. Every chapter has been carefully revised to improve clarity of the material. Many explanations of critical concepts have been enhanced using new business examples and data sets.
3. Key decision modeling chapters in Part 2 of this book have been significantly revised and reorganized. These are Chapter 9–*Building and Using Decision Models*, Chapter 13–*Linear Optimization*, and Chapter 14–*Integer and Nonlinear Optimization*.
4. Theory and extensive computational formulas have been relegated to end of chapter Appendixes to provide better flexibility for instructors, and not impede learning essential concepts and skills.
5. End-of-chapter material has been enhanced and reorganized to include *Basic Concepts Review Questions* that focus on the understanding of fundamental terms concepts; *Skill-Building Exercises* that facilitate experiential learning and Excel-based skills, and *Problems and Applications*, which provide a wide variety of numerical exercises and practical applications to real and/or realistic data sets or problem scenarios. New cases are introduced in most chapters.

TO THE STUDENTS

The CD-ROM accompanying this text contains all the data and model files used throughout the text in examples, problems, and exercises. These are also available on the text's Web site, www.pearsonhighered.com/evans. Versions of a variety of software packages, including *PHStat*, *SimQuick*, *Crystal Ball*, and *Premium Solver*, are also available in connection with this text. For complete information on these, please also visit www.pearsonhighered.com/evans.

TO THE INSTRUCTORS

To access instructor solutions files please visit pearsonhighered.com/evans and choose the instructor resources option. A variety of instructor resources are available for instructors who register for our secure environment. The files for each chapter, including PowerPoint presentations, are available for download.

As a registered faculty member, you can login directly to download resource files, and receive immediate access and instructions for installing Course Management content to your campus server.

Need help? Our dedicated Technical Support team is ready to assist instructors with questions about the media supplements that accompany this text. Visit: http://247.pearsoned.com/ for answers to frequently asked questions and toll-free user support phone numbers.

ACKNOWLEDGEMENTS

I would like to thank the following individuals who have provided reviews and insightful suggestions for this edition:

Johannes Ledolter, University of Iowa
J. Morgan Jones, University of North Carolina
Ray Dacey, University of Idaho
Jim Frendewey, Jr., Michigan Technological University
Yasha Crnkovic, SUNY Albany

In addition, I thank the many students over the years who provided numerous suggestions, data sets and problem ideas, and insights into how to better present the material. Finally, appreciation goes to my editor Mark Pfaltzgraff, Ann Pulido, Susie Abraham, and the entire production staff at Pearson Education, for their dedication in developing and producing this text. If you have any suggestions or corrections, please contact me via email at james.evans@uc.edu.

James R. Evans
University of Cincinnati

Part

I

Statistics and Data Analysis

Chapter 1

Data and Business Decisions

INTRODUCTION

A phrase one often hears in many companies today is, "In God we trust; all others use data." Modern organizations truly manage by fact—they depend on complete and accurate data for performance evaluation, improvement, and decision making. However, many organizations ignore the most important data they need to make good decisions. This may occur for several reasons:

➤ *They may not fully understand what to measure or how to measure.*

➤ *They may be reluctant to spend the required time and effort.*

> ➤ *They may feel they can make decisions by instinct and do not need data.*

> ➤ *They may fear discovering problems or poor performance that data may uncover.*

Even if organizations do gather data, they may not interpret them properly.

Information derives from analysis of data. *Analysis* refers to extracting larger meaning from data to support evaluation and decision making. One of the most important tools for analyzing data in business is **statistics**, which is the science of *collecting, organizing, analyzing, interpreting,* and *presenting* data. Modern spreadsheet technology, such as Microsoft Excel, has made it quite easy to organize, analyze, and present data.

Data also provide key inputs to decision models. A **decision model** is a logical or mathematical representation of a problem or business situation. Decision models establish relationships between actions that decision makers might take and results that they might expect, thereby allowing the decision makers to predict what might happen based on the model assumptions. For instance, the manager of a grocery store might want to know how best to use price promotions, coupon programs, and advertising to increase sales. In the past, grocers have studied the relationship of sales volume to programs such as these by conducting controlled experiments to identify the relationship between actions and sales volumes.[1] That is, they implement different combinations of price promotions, coupon programs, and advertising (the decision variables) then observe the sales that result. Using the data from these experiments and a statistical technique known as *regression analysis* (which we cover in Chapter 6), we can develop a predictive model of sales as a function of the decision variables. Such a model might look like the following:

$$\text{Sales} = a + b \times \text{Price} + c \times \text{Coupons} + d \times \text{Advertising} + e \times \text{Price} \times \text{Advertising}$$

where *a, b, c, d,* and *e* are constants that are estimated from the data. By setting levels for price, coupons, and advertising, the model estimates a level of sales. The manager can use the model to identify effective pricing, promotion, and advertising strategies.

Statistics and data analysis have long been critical to business decisions, but they are becoming more important as an increasing amount of electronic information becomes available. These techniques help managers determine trends, projections, cause-and-effect relationships, and other significant meanings of data that might not be evident. The purpose of this book is to introduce you to practical approaches for analyzing data; ways of using data effectively to make informed decisions; and approaches for developing, analyzing, and solving models of decision problems. Part I of this book (Chapters 1–8) focuses on key issues of statistics and data analysis, and Part II (Chapters 9–14) introduces you to various types of decision models that rely on good data analysis.

In this chapter, we discuss the roles of data analysis in business, discuss how data are used in evaluating business performance, introduce some fundamental issues of statistics and measurement, and

---

[1] "Flanking in a Price War," *Interfaces*, Vol. 19, No. 2, 1989, 1–12.

introduce spreadsheets as a support tool for data analysis and decision modeling. The key concepts we will discuss are the following:

> *The importance of statistical thinking in business*
> *The importance of statistics in Six Sigma, which has become a widely accepted approach to business performance improvement*
> *The scope of business performance data and the concept of a "balanced scorecard," as well as the use of data outside the business environment*
> *The role of statistics in using sample data to understand and draw inferences about populations and monitor the effectiveness of business processes*
> *Classification of data and common types of measurement scales*
> *Basic Microsoft Excel skills and add-ins that are supplied with this book*
> *Using PivotTables to manipulate data*

STATISTICAL THINKING IN BUSINESS

The importance of applying statistical concepts to make good business decisions and improve performance cannot be overemphasized. **Statistical thinking** is a philosophy of learning and action for improvement that is based on the principles that

- all work occurs in a system of interconnected processes;
- variation exists in all processes; and
- understanding and reducing variation are keys to success.[2]

Work gets done in any organization through **processes**—systematic ways of doing things that achieve desired results. Understanding processes provides the context for determining the effects of variation and the proper type of action to be taken. Any process contains many sources of variation. In manufacturing, for example, different lots of material vary in strength, thickness, or moisture content. Cutting tools have inherent variation in their strength and composition. During manufacturing, tools experience wear, vibrations cause changes in machine settings, and electrical fluctuations cause variations in power. Workers may not position parts on fixtures consistently, and physical and emotional stress may affect workers' consistency. In addition, measurement gauges and human inspection capabilities are not uniform, resulting in variation in measurements even when the true value is constant. The complex interactions of these variations in materials, tools, equipment, people, and the environment are not easily understood and are often referred to as *common causes* of variation. Other variations, which we generally call *special causes*, arise from external sources that are not inherent in the process. Some factors that lead to special causes in manufacturing are a bad batch of material from a supplier, a poorly trained substitute machine operator, a broken or worn tool, or miscalibration of measuring instruments. These typically result in unusual variations that disrupt the statistical pattern of common causes. Similar phenomena occur in service processes because of variation in employee and customer behavior, application of technology, and so on.

---

[2] Galen Britz, Don Emerling, Lynne Hare, Roger Hoerl, and Janice Shade, "How to Teach Others to Apply Statistical Thinking," *Quality Progress*, June 1997, 67–79.

While variation exists everywhere, many business decisions do not often account for it, and managers frequently confuse common and special causes of variation and try to take action to eliminate a perceived special cause when it in fact is simply common cause variation. For example, if sales in some region fell from the previous year, the regional manager might quickly blame her sales staff for not working hard. If a new advertising campaign happens to coincide with a drop in sales, some managers would quickly drop the ad campaign without any further analysis. How often do managers make decisions based on a single data point or two, seeing trends when they don't exist, or manipulate financial figures they cannot truly control? Usually, it is simply a matter of ignorance of how to deal with data and information. A better approach would be to formulate a theory ("Certain ad campaigns positively affect sales") and test this theory in some way, either by collecting and analyzing some data ("Measure change in sales when advertising is adopted") and perhaps developing a model of the situation that will provide better insight ("When advertising is increased by 10%, sales increase by 15%"). Using statistical thinking in this fashion can provide much better insight into the facts and nature of relationships among the many factors that may have contributed to the event and enable managers to make better decisions.

The lack of broad and sustained use of statistical thinking in many organizations is due to two reasons.[3] First, statisticians historically have functioned as problem solvers in manufacturing, research, and development and, thereby, have focused on individual clients rather than on organizations. Second, statisticians have focused primarily on technical aspects of statistics rather than emphasizing process definition, measurement, control, and improvement—the key activities that will lead to bottom-line results. Today, many organizations, including General Electric (GE), Ford Motor Company, numerous healthcare organizations, and many others, are implementing "Six Sigma" initiatives and training all employees in statistical thinking and other problem-solving tools and techniques to improve organizational effectiveness and financial performance.

Six Sigma and Statistical Thinking[4]

Six Sigma can be best described as a business process improvement approach that seeks to find and eliminate causes of defects and errors, reduce cycle times and cost of operations, improve productivity, better meet customer expectations, and achieve higher asset use and returns on investment in manufacturing and service processes. It is based on a simple problem-solving methodology—**DMAIC**, which stands for Define, Measure, Analyze, Improve, and Control—that incorporates a wide variety of statistical and other types of process improvement tools.

Six Sigma is appealing to top executives because of its focus on measurable bottom-line results; a disciplined, fact-based approach to problem solving; and rapid project completion. Motorola pioneered the concept as an approach to measuring product and service quality, and it has garnered significant credibility over the past decade because of its acceptance at such major firms as Allied Signal (now part of Honeywell) and GE. The term *six sigma* is actually based on a statistical measure that equates to 3.4 or fewer errors or defects per million opportunities. An ultimate

---

[3] Ronald D. Snee, "Getting Better Business Results: Using Statistical Thinking and Methods to Shape the Bottom Line," *Quality Progress*, June 1998, 102–106.

[4] See James R. Evans and William M. Lindsay, *An Introduction to Six Sigma & Process Improvement* (Cincinnati: Thomson/South-Western), 2005, for an introduction to the tools and methodology of Six Sigma.

"stretch" goal of all organizations that adopt a Six Sigma philosophy is to have all critical processes, regardless of functional area, at a six-sigma level of capability.

Considerable evidence exists that Six Sigma initiatives positively impact bottom-line results. In the first year of Six Sigma implementation at GE, they trained 30,000 employees at a cost of $200 million and got back about $150 million in savings. From 1996 to 1997, GE increased the number of Six Sigma projects from 3,000 to 6,000 and achieved $320 million in productivity gains and profits. By 1998, the company had generated $750 million in Six Sigma savings over and above their investment, and would receive $1.5 billion in savings the next year.

GE had many early success stories. GE Capital, for example, fielded about 300,000 calls each year from mortgage customers who had to use voicemail or call back 24% of the time because employees were busy or unavailable. A Six Sigma team analyzed one branch that had a near perfect percentage of answered calls and applied their best practices to the other 41 branches, resulting in a 99.9% chance of customers' getting a representative on the first try. A team at GE Plastics improved the quality of a product used in CD-ROMs and audio CDs from a 3.8 sigma level to 5.7 level and captured a significant amount of new business from Sony.[5] GE credits Six Sigma with a tenfold increase in the life of CT scanner X-ray tubes, a 400% improvement in return on investment in its industrial diamond business, a 62% reduction in turnaround time at railcar repair shops, and $400 million in savings in its plastics business.[6]

Six Sigma has heightened the awareness of statistics among business professionals, and the material in this book will provide the foundation for more advanced topics commonly found in Six Sigma training courses in many organizations.

DATA IN THE BUSINESS ENVIRONMENT

An example from the Boeing Company shows the value of having good business data and analysis capabilities.[7] In the early 1990s, Boeing's assembly lines were morasses of inefficiency. A manual numbering system dating back to World War II bomber days was used to keep track of an airplane's four million parts and 170 miles of wiring; changing a part on a 737's landing gear meant renumbering 464 pages of drawings. Factory floors were covered with huge tubs of spare parts worth millions of dollars. In an attempt to grab market share from rival Airbus, the company discounted planes deeply and was buried by an onslaught of orders. The attempt to double production rates, coupled with implementation of a new production control system, resulted in Boeing being forced to shut down its 737 and 747 lines for 27 days in October 1997, leading to a $178 million loss and a shakeup of top management. Much of the blame was focused on Boeing's financial practices and lack of real-time financial data. With a new Chief Financial Officer and finance team, the company created a "control panel" of vital measures, such as materials costs, inventory turns, overtime, and defects, using a color-coded spreadsheet. For the first time, Boeing was able to generate a series of bar charts showing which of its programs were creating value and which were destroying it. The results were eye-opening and helped formulate a growth plan. As one manager noted, "The data will set you free."

---

[5] Jack Welch, *Jack: Straight from the Gut* (New York: Warner Books, 2001), 333–334.
[6] "GE Reports Record Earnings with Six Sigma," *Quality Digest*, December 1999, 14.
[7] Jerry Useem, "Boeing versus Boeing," *Fortune*, October 2, 2000, 148–160.

Metrics and Measurement

A **metric** is a unit of measurement that provides a way to objectively quantify performance. For example, senior managers might assess overall business performance using such metrics as net profit, return on investment, market share, and customer satisfaction. A supervisor in a manufacturing plant might monitor the quality of a production process for a brass sink fixture by counting surface defects. A useful metric would be the percentage of fixtures that have surface defects. For services, some common metrics are the percentage of orders filled accurately and the time taken to fill a customer's order. Metrics provide a means for objectively assessing performance and are necessary to ensure that decisions are made on the basis of facts. **Measurement** is the act of obtaining data. **Measures** refer to the numerical information that results from measurement; that is, measures and indicators are numerical values associated with a metric.

Metrics can be either discrete or continuous. A **discrete metric** is countable. For example, a dimension is either within tolerance or out of tolerance; an order is complete or incomplete; or an invoice can have one, two, three, or any number of errors. Some examples are determining whether the correct ZIP code was used in shipping an order or comparing a dimension to specifications, such as whether the diameter of a shaft falls within specification limits of 1.60 ± 0.01 inch. They are typically expressed as numerical counts or as proportions. **Continuous metrics**, such as length, time, or weight, are concerned with the degree of conformance to specifications. Thus, rather than determining whether the diameter of a shaft simply meets a specification of 1.60 ± 0.01 inch, a measure of the actual value of the diameter is taken.

A key performance dimension might be measured using either a continuous or a discrete metric. The differences are that discrete metrics are usually easier to capture, for example, by visual inspection, while continuous metrics usually require some type of measurement instrument such as a gauge or stopwatch. However, one generally must collect a larger amount of discrete data to draw appropriate statistical conclusions as compared to continuous data.

In Six Sigma projects, a common metric used to assess quality is the number of **defects per unit**. In services, each customer transaction provides an opportunity for many different types of errors; a measure of quality analogous to defects per unit is **errors per opportunity**. Regardless of the terminology, defects, nonconformities, or errors must be clearly defined in terms of failure to meet customer requirements or internal specifications. Defects per unit or errors per opportunity are often reported as rates per thousand or million. A common metric used in Six Sigma is **dpmo—defects per million opportunities**. Thus, a defect rate of 2 per 1,000 is equivalent to 2,000 dpmo.

The Balanced Scorecard

Data and analysis support a variety of company purposes, such as planning, reviewing company performance, improving operations, and comparing company performance with competitors' or "best practices" benchmarks. Data that organizations use should focus on critical success factors that lead to competitive advantage. Most organizations have traditionally focused on financial and market information, such as profit, sales volume, and market share. Today, however, many organizations create a "balanced scorecard" of measures that provide a comprehensive view of business performance. Such a scorecard is balanced by the interests of all stakeholders—customers, employees, stockholders, suppliers and partners, and the community—and allows organizations to focus on the critical success factors that lead to competitive advantage.

The term **balanced scorecard** was coined by Robert Kaplan and David Norton of the Harvard Business School in response to the limitations of traditional accounting measures.[8] Its purpose is "to translate strategy into measures that uniquely communicate your vision to the organization." Their version of the balanced scorecard consists of four perspectives:

1. *Financial Perspective.* Measures the ultimate results that the business provides to its shareholders. This includes profitability, revenue growth, return on investment, asset utilization, operating margins, earnings per share, economic value added (EVA), shareholder value, and other relevant measures.
2. *Internal Perspective.* Focuses attention on the performance of the key internal processes that drive the business. This includes such measures as quality levels, productivity, process yields, cycle time, cost, and legal compliance.
3. *Customer Perspective.* Focuses on customer needs and satisfaction as well as market share. This includes service levels, satisfaction ratings, repeat business, and other indicators such as complaints.
4. *Innovation and Learning Perspective.* Directs attention to the basis for future success—the organization's people and infrastructure. Key measures might include intellectual assets such as patent filings, employee satisfaction, turnover, market innovation, training effectiveness, skills development, environmental improvements, and supplier performance.

Other models for "balanced scorecards" exist in business. The Malcolm Baldrige National Quality Award Criteria for Performance Excellence, which many organizations use as a high-performance management framework, categorizes business results into six categories:

1. *Product and service outcomes*, such as quality measurements, field performance, defect levels, service errors, and response times.
2. *Customer-focused outcomes*, such as customer satisfaction and dissatisfaction, customer retention, complaint resolution, customer perceived value, and gains and losses of customers.
3. *Financial and market outcomes.* Financial outcomes might include revenue, profit and loss, net assets, cash-to-cash cycle time, earnings per share, financial operations efficiency (collections, billings, receivables). Market outcomes might include market share, business growth, and new products and service introductions.
4. *Workforce-focused outcomes*, such as workforce engagement and satisfaction, retention and turnover, safety, and training effectiveness.
5. *Process effectiveness outcomes* such as productivity, production flexibility, setup times, time to market, waste stream reductions, innovation, Six Sigma results, and supply chain effectiveness.
6. *Leadership outcomes*, such as strategic plan accomplishment, governance and accountability, environmental and regulatory compliance, ethical behavior, and organizational citizenship.

A good balanced scorecard contains both leading and lagging measures and indicators. **Lagging measures** (outcomes) tell what has happened; **leading measures** (performance drivers) predict what *will* happen. For example, customer survey results about recent transactions might be a leading indicator for customer retention

---

[8] Robert S. Kaplan and David P. Norton, *The Balanced Scorecard* (Boston: Harvard Business School Press, 1996).

(a lagging indicator), employee satisfaction might be a leading indicator for turnover, and so on.

Two examples illustrate how the analysis of leading and lagging measures can lead to useful decision models. IBM Rochester developed a model to quantify the causal relationships among the key measures in its balanced scorecard. This model suggests that improving internal capabilities such as people skills, product/service quality, and products and channels will lead to improved customer satisfaction and loyalty, which, in turn, lead to improved financial and market share performance. Another example is Sears, Roebuck and Company, which provided a consulting group with 13 financial measures, hundreds of thousands of employee satisfaction data points, and millions of data points on customer satisfaction. Using advanced statistical tools, the analysts discovered that employee attitudes about the job and the company are key factors that predict their behavior with customers, which, in turn, predicts the likelihood of customer retention and recommendations, which, in turn, predict financial performance. Sears can now predict that if a store increases its employee satisfaction score by five units, customer satisfaction scores will go up by two units and revenue growth will beat the stores' national average by 0.5%.[9] Such an analysis can help managers make decisions, for instance, on improved human resource policies.

Balanced scorecards are usually used at the executive management level to provide a broad picture of organizational performance. Down in the trenches, many organizations use the concept of a **dashboard** to monitor process performance. This term stems from the analogy to an automobile's dashboard—a collection of indicators (speed, RPM, oil pressure, temperature, etc.) that summarizes performance. Dashboards show how business processes are performing and provide real-time information for corrective action or maintenance purposes.

SOURCES AND TYPES OF DATA

Business data may come from a variety of sources: internal, external, and generated. Internal data are routinely collected by accounting, marketing, and operations functions of a business. Much of these data might be gathered using modern technology such as bar coding or automated transaction reporting. External data might include competitive performance acquired from annual reports, Standard & Poor's Compustat data sets, industry trade associations, or the government (see www.fedstats.gov for a comprehensive gateway to statistical data from more than 100 U.S. federal agencies). Other data must be generated through special efforts. For example, customer satisfaction data are often acquired by mail or telephone surveys, personal interviews, or focus groups.

The use of data for analysis and decision making certainly is not limited to business. Science, engineering, medicine, and sports, to name just a few, are examples of professions that rely heavily on data. Many of these types of data or statistical summaries pop up in the daily newspapers and magazines and provide readers with some basic notions (which are often flawed!) about statistics. Table 1.1 provides a list of data files that are available in the Statistics Data Files folder on the CD-ROM accompanying this book. All are saved in Microsoft Excel workbooks. These data files will be used throughout this book to illustrate various issues associated with

---

[9] "Bringing Sears into the New World," *Fortune*, October 13, 1997, 183–184.

Table 1.1
Data Files Available
on CD-ROM

Business and Economics

Accounting Professionals	Gasoline Sales
Atlanta Airline Data	Hatco
Automobile Quality	Hi-Definition Televisions
Baldrige	Home Market Value
Banking Data	House Sales
Beverage Sales	Housing Starts
California Census Data	Hypothetical Stock Data
Call Center Data	Microprocessor Data
Cereal Data	Mortgage Rates
China Trade Data	New Account Processing
Closing Stock Prices	New Car Sales
Coal Consumption	Nuclear Power
Coal Production	Prime Rate
Concert Sales	Quality Control Case Data
Consumer Price Index	Quality Measurements
Consumer Transportation Survey	Refrigerators
Credit Approval Decisions	Retirement Portfolio
Customer Support Survey	Room Inspection
Customer Survey	Salary Data
DJIA December Close	Sales Data
EEO Employment Report	Science and Engineering Jobs
Employee Salaries	State Unemployment Rates
Energy Production & Consumption	Syringe Samples
Federal Funds Rate	Treasury Yield Rates
Gas & Electric	Unions and Labor Law Data
Gasoline Prices	University Grant Proposals

Science and Engineering

Bicycle Wheels	Surface Finish
Pile Foundation	Washington, DC, Weather
Seattle Weather	

Social Sciences

Arizona Population	Graduate School Survey
Blood Pressure	Infant Mortality
Burglaries	MBA Student Survey
Census Education Data	Ohio Education Performance
Church Contributions	Ohio Prison Population
Colleges and Universities	Self Esteem
Death Cause Statistics	Social Networking
Demographics	Student Grades
Facebook Survey	Vacation Survey
Freshman College Data	

Sports

Baseball Attendance	National Football League
Major League Baseball	Olympic Track and Field Data
NASCAR Track Data	

Figure 1.1 Example of Cross-Sectional, Univariate Data (portion of *Automobile Quality*)

	A	B	C	D
1	2008 J.D. Power and Associates Initial Quality			
2				
3	Brand	Problems per 100 Vehicles		
4	Acura	119		
5	Audi	113		
6	BMW	126		
7	Buick	118		
8	Cadillac	113		
9	Chevrolet	113		

statistics and data analysis and also for many of the questions and problems at the end of the chapters. They show but a sample of the wide variety of applications for which statistics and data analysis techniques may be used.

Data Classification

When we deal with data sets, it is important to understand the nature of the data in order to select the appropriate statistical tool or procedure. One classification of data is the following:

1. Type of data
   - *Cross-sectional*—data that are collected over a single period of time
   - *Time series*—data collected over time
2. Number of variables
   - *Univariate*—data consisting of a single variable
   - *Multivariate*—data consisting of two or more (often related) variables

Figures 1.1 through 1.4 show examples of data sets from Table 1.1 representing each combination from this classification.

Figure 1.2 Example of Cross-Sectional, Multivariate Data (portion of *Atlanta Airline Data*)

	A	B	C	D	E	F
1	Delta Airline Flight Statistics, Atlanta Hartsfield International (ATL) December 24, 1997					
2						
3	Flight Number	Origin Airport	Scheduled Arrival Time	Actual Arrival Time	Time Difference (minutes)	Arrival Status
4	16	DFW	11:04	10:43	21	Early
5	52	PDX	19:55	20:24	29	Late
6	54	HNL	6:45	6:00	45	Early
7	105	CVG	17:50	18:33	43	Late
8	152	SLC	14:22	15:40	18	Late
9	160	SLC	18:20	19:16	56	Late

Figure 1.3 Example of Time-Series, Univariate Data (portion of *Federal Funds Rate*)

	A	B	C
1	**Federal Funds Rate: Percent**		
2			
3	**Month/Year**	**Rate**	
4	Jan-1960	3.99%	
5	Feb-1960	3.97%	
6	Mar-1960	3.84%	
7	Apr-1960	3.92%	
8	May-1960	3.85%	
9	Jun-1960	3.32%	
10	Jul-1960	3.23%	
11	Aug-1960	2.98%	

Another classification of data is by the type of measurement scale. Failure to understand the differences in measurement scales can easily result in erroneous or misleading analysis. Data may be classified into four groups:

1. **Categorical (nominal) data**, which are sorted into categories according to specified characteristics. For example, a firm's customers might be classified by their geographical region (North America, South America, Europe, and Pacific); employees might be classified as managers, supervisors, and associates. The categories bear no quantitative relationship to one another, but we usually assign an arbitrary number to each category to ease the process of managing the data and computing statistics. Categorical data are usually counted or expressed as percents; statistics such as averages are difficult to interpret and are usually meaningless.

2. **Ordinal data**, which are ordered or ranked according to some relationship to one another. For instance, J.D. Power and Associates' Initial Quality Study ranks new cars and trucks according to how many problems per 100 vehicles that owners report experiencing in the first 90 days from smallest to largest. Other examples include ranking regions according to sales levels each month and NCAA basketball rankings. Ordinal data are more meaningful than categorical data because data can be compared to one another. However, like categorical data, averages are generally meaningless, because ordinal data have no fixed units of measurement. In addition, meaningful numerical statements about differences between categories cannot be

Figure 1.4 Example of Time-Series, Multivariate Data (portion of *Treasury Yield Rates*)

	A	B	C	D	E	F	G	H	I	J	K	L
1	**Daily Treasury Yield Curve Rates**											
2												
3	**Date**	**1 mo**	**3 mo**	**6 mo**	**1 yr**	**2 yr**	**3 yr**	**5 yr**	**7 yr**	**10 yr**	**20 yr**	**30 yr**
4	1/2/2008	3.09	3.26	3.32	3.17	2.88	2.89	3.28	3.54	3.91	4.39	4.35
5	1/3/2008	3.19	3.24	3.29	3.13	2.83	2.85	3.26	3.54	3.91	4.41	4.37
6	1/4/2008	3.22	3.2	3.22	3.06	2.74	2.75	3.18	3.47	3.88	4.4	4.36
7	1/7/2008	3.27	3.27	3.29	3.11	2.76	2.76	3.16	3.46	3.86	4.37	4.34
8	1/8/2008	3.31	3.25	3.27	3.09	2.76	2.76	3.16	3.47	3.86	4.39	4.35
9	1/9/2008	3.34	3.22	3.22	3.04	2.69	2.69	3.1	3.4	3.82	4.35	4.32

made. For example, comparing the average rank of U.S. models versus imports does not meaningfully compare differences in reported problem rates.

3. **Interval data**, which are ordered, have a specified measure of the distance between observations but have no natural zero. A common example is temperature: Both the Fahrenheit and Celsius scales represent a specified measure of distance—degrees—but have no natural zero. Thus we cannot say that 50 degrees is twice as hot as 25 degrees. Nevertheless, in contrast to ordinal data, interval data allow meaningful comparison of ranges, averages, and other statistics.

In business, data from satisfaction and attribute rating scales (for example, 1 = poor, 2 = average, 3 = good, 4 = very good, 5 = excellent) are often considered to be interval data. Strictly speaking, this is not correct, as the numerical "distance" between categories on the measurement scale, such as between *good* and *very good*, is meaningless. If respondents select their response on the basis of the category description, the data are ordinal. However, if respondents clearly understand that response categories are associated with the numerical measurement scale, then the data are interval. Usually, this is a very tenuous assumption but, nevertheless, most users of survey data treat such data as interval when computing and analyzing statistics. You should remember, however, that care must be taken when collecting and interpreting survey data as interval data.

4. **Ratio data**, which have a natural zero. For example, sales dollars has an absolute zero (no sales activity at all). Knowing that the Seattle region sold $12 million in March while the Tampa region sold $6 million means that Seattle sold twice as much as Tampa. Most business and economic data fall into this category, and statistical methods are the most widely applicable to them.

This classification is hierarchical in that each level includes all of the information content of the one preceding it. For example, ratio information can be converted to any of the other types of data. Interval information can be converted to ordinal or categorical data but cannot be converted to ratio data without the knowledge of the absolute zero point. Thus, a ratio scale is the strongest form of measurement.

The managerial implications of this classification are in understanding the choice and validity of the statistical measures used. For example, consider the following statements:

- Sales occurred in March (categorical).
- Sales increased in March relative to February (ordinal).
- Sales increased by $50,000 in March over February (interval).
- Sales were $920,000 in March and $870,000 in February (ratio).

A higher level of measurement is more useful to a manager because more definitive information describes the data. Obtaining ratio data can be more expensive than categorical data, especially when surveying customers, but it may be needed for proper analysis. Thus, before data are collected, consideration must be given to the type of data needed.

POPULATIONS, SAMPLES, AND STATISTICS

Data that organizations collect can generally be classified into two categories: populations and samples. A **population** consists of all items of interest for a particular decision or investigation, for example, *all* married drivers over the age of 25 in the United States, *all* first-year MBA students at a college, or *all* stockholders of Google. It is important to understand that a population can be anything we define it to be, such as all customers who have purchased from Amazon over the past year or individuals who do not own a cell phone. A company like Amazon keeps extensive records on its customers, making it easy to retrieve data about the entire population of customers with prior purchases. However, it would probably be impossible to identify all individuals who do not own cell phones. A population may also be an existing collection of items (for instance, all teams in the National Football League) or the potential, but unknown, output of a process (such as automobile engines produced on an assembly line).

A **sample** is a subset of a population. For example, a list of individuals who purchased a CD from Amazon in the past year would be a sample from the population of all customers who purchased from the company. Whether this sample is representative of the population of customers—which depends on how the sample data are intended to be used—may be debatable; nevertheless, it is a sample. Sampling is desirable when complete information about a population is difficult or impossible to obtain. For example, it may be too expensive to send all previous customers a survey. In other situations, such as measuring the amount of stress needed to destroy an automotive tire, samples are necessary even though the entire population may be sitting in a warehouse. Most of the data files in Table 1.1 represent samples.

We use samples to provide information about populations. We are all familiar with survey samples of voters prior to and during elections. A small subset of potential voters, if properly chosen on a statistical basis, can provide accurate estimates of the behavior of the voting population. Thus, television network anchors can announce the winners of elections based on a small percentage of voters before all votes can be counted. Samples are routinely used for business and public opinion polls—magazines such as *Business Week* and *Fortune* often report the results of surveys of executive opinions on the economy and other issues. Many businesses rely heavily on sampling. Producers of consumer products conduct small-scale market research surveys to evaluate consumer response to new products before full-scale production, and auditors use sampling as an important part of audit procedures. In 2000, the U.S. Census began using statistical sampling for estimating population characteristics, which resulted in considerable opposition and debate.

You are undoubtedly familiar with the concept of statistics in daily life: baseball batting averages, the Dow Jones Industrial Average, the Consumer Price Index, and many more. Newspapers and magazines often use the term quite loosely and inappropriately. For example, baseball batting averages are summary measures of population data, and technically, are not statistics. **Statistics** are summary measures of population characteristics computed from samples. In business, statistical methods are used to present data in a concise and understandable fashion, to estimate population characteristics, to draw conclusions about populations from sample data, and to develop useful decision models for prediction and forecasting. For example, in the 2008 J.D. Power and Associates' Initial Quality Study, Porsche led the industry with a reported 87 problems per 100 vehicles. The number 87 is a statistic that summarizes the total number of problems reported per 100 vehicles and suggests that the entire population of Porsche owners (all of whom may not have responded to the survey) averaged less than one problem in their first 90 days of ownership. However, a particular automobile owner may have experienced zero, one, two, or perhaps more problems.

The process of collection, organization, and description of data is commonly called **descriptive statistics**. **Statistical inference** refers to the process of drawing conclusions about unknown characteristics of a population based on sample data. Finally, **predictive statistics**—developing predictions of future values based on historical data—is the third major component of statistical methodology. In subsequent chapters, we will cover each of these types of statistical methodology.

Spreadsheet software for personal computers has become an indispensable tool for business analysis, particularly for the manipulation of numerical data and the development and analysis of decision models. In this text, we will use Microsoft Excel 2007 to perform most calculations and analyses. Although Excel has some flaws and limitations, its widespread availability makes it the software of choice for many business professionals. We do wish to point out, however, that better and more powerful statistical software packages are available, and serious users of statistics should consult a professional statistician for advice on selecting the proper software.

Basic Excel Skills

To be able to apply the procedures and techniques we will study in this book, it is necessary for you to know many of the basic capabilities of Excel. We will assume that you are familiar with the most elementary spreadsheet concepts and procedures, such as:

- opening, saving, and printing files
- moving around a spreadsheet
- selecting ranges
- inserting/deleting rows and columns
- entering and editing text, numerical data, and formulas
- formatting data (number, currency, decimal places, etc.)
- working with text strings
- performing basic arithmetic calculations
- formatting data and text
- modifying the appearance of the spreadsheet

Excel has extensive online help, and many good manuals and training guides are available both in print and online, and we urge you to take advantage of these. However, to facilitate your understanding and ability, we will review some of the more important topics in Excel that you may not have used. Other tools and procedures in Excel, such as creating graphs and charts, will be introduced as we need them throughout this book.

Menus and commands in Excel 2007 reside in the "ribbon" shown in Figure 1.5. Menus and commands are arranged in logical *groups* under different *tabs*; small triangles pointing downward indicate *menus* of additional choices. The Office button contains other commands and options. We will often refer to certain commands or options and where they may be found in the ribbon.

Figure 1.5 Excel 2007 Ribbon

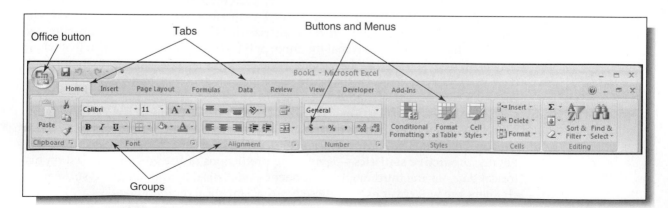

Copying Formulas and Cell References

Excel provides several ways of copying formulas to different cells. This is extremely useful in building decision models, because many models require replication of formulas for different periods of time, similar products, and so on. One way is to select the cell with the formula to be copied, click the *Copy* button from the *Clipboard* group under the *Home* tab (or simply press Ctrl-C on your keyboard), click on the cell you wish to copy to, and then click the *Paste* button (or press Ctrl-V). You may also enter a formula directly in a range of cells without copying and pasting by selecting the range, typing in the formula, and pressing Ctrl-Enter.

To copy a formula from a single cell or range of cells down a column or across a row, first select the cell or range, then click and hold the mouse on the small square in the lower right-hand corner of the cell (the "fill handle"), and drag the formula to the "target" cells you wish to copy to. To illustrate this technique, suppose we wish to compute the differences in projected employment for each occupation in the Excel file *Science and Engineering Jobs*. In Figure 1.6, we have added a column for the difference and entered the formula =C10-B10 in the first row. Highlight cell D4 and then simply drag the handle down the column. Figure 1.7 shows the results.

In any of these procedures, the *structure* of the formula is the same as in the original cell, but the cell references have been changed to reflect the *relative addresses* of the formula in the new cells. That is, the new cell references have the same relative relationship to the new formula cell(s) as they did in the original formula cell. Thus, if a formula is copied (or moved) one cell to the right, the relative cell addresses will have their column label increased by one; if we copy or move the formula two cells down, the row number is increased by 2. Figure 1.8 shows the formulas for the *Science and Engineering Jobs* spreadsheet example. For example, note that the formulas in each row are the same, except for the column reference.

Sometimes, however, you do not want to change the relative addressing because you would like all the copied formulas to point to a certain cell. We do this by using a $ before the column and/or row address of the cell. This is called an *absolute address*. For example, suppose we wish to compute the percent of the total for each occupation for 2010. In cell E4, enter the formula = C4/$C$12. Then, if we copy this formula down column E for other months, the numerator will change to reference each occupation,

Figure 1.6 Copying Formulas by Dragging

	A	B	C	D	E
1	Total science and engineering jobs in thousands: 2000 and projected 2010				
2					
3	Occupation	2000	2010	Difference	
4	Scientists	3,241	5,301	2,060	
5	Life scientists	184	218		
6	Mathematical/computer scientists	2,408	4,308		
7	Computer specialists	2,318	4,213		Click here and
8	Mathematical scientists	89	95		drag down
9	Physical scientists	239	283		
10	Social scientists	410	492		
11	Engineers	1,465	1,603		
12	All occupations	145,571	167,754		

Figure 1.7 Results of Dragging Formulas

	A	B	C	D	E
1	**Total science and engineering jobs in thousands: 2000 and projected 2010**				
2					
3	**Occupation**	**2000**	**2010**	**Difference**	
4	Scientists	3,241	5,301	2,060	
5	Life scientists	184	218	34	
6	Mathematical/computer scientists	2,408	4,308	1,900	
7	Computer specialists	2,318	4,213	1,895	
8	Mathematical scientists	89	95	6	
9	Physical scientists	239	283	44	
10	Social scientists	410	492	82	
11	Engineers	1,465	1,603	138	
12	All occupations	145,571	167,754		

but the denominator will still point to cell C12 (see Figure 1.9). You should be very careful to use relative and absolute addressing appropriately in your models.

Functions

Functions are used to perform special calculations in cells. Some of the more common functions that we will see are as follows:

MIN(*range*)—finds the smallest value in a range of cells
MAX(*range*)—finds the largest value in a range of cells
SUM(*range*)—finds the sum of values in a range of cells
AVERAGE(*range*)—finds the average of the values in a range of cells
AND(*condition 1, condition 2 . . .*)—a logical function that returns TRUE if all conditions are true, and FALSE if not
OR(*condition 1, condition 2 . . .*)—a logical function that returns TRUE if any condition is true, and FALSE if not

Figure 1.8 Formulas for *Science and Engineering Jobs* Spreadsheet

	A	B	C	D
3	**Occupation**	**2000**	**2010**	**Difference**
4	Scientists	3241	5301	=C4-B4
5	Life scientists	184	218	=C5-B5
6	Mathematical/computer scientists	2408	4308	=C6-B6
7	Computer specialists	2318	4213	=C7-B7
8	Mathematical scientists	89	95	=C8-B8
9	Physical scientists	239	283	=C9-B9
10	Social scientists	410	492	=C10-B10
11	Engineers	1465	1603	=C11-B11
12	All occupations	145571	167754	

Figure 1.9   Example of Absolute Address Referencing

	A	B	C	D	E
3	Occupation	2000	2010	Difference	2010 Percent of Total
4	Scientists	3241	5301	=C4-B4	=C4/$C$12
5	Life scientists	184	218	=C5-B5	=C5/$C$12
6	Mathematical/computer scientists	2408	4308	=C6-B6	=C6/$C$12
7	Computer specialists	2318	4213	=C7-B7	=C7/$C$12
8	Mathematical scientists	89	95	=C8-B8	=C8/$C$12
9	Physical scientists	239	283	=C9-B9	=C9/$C$12
10	Social scientists	410	492	=C10-B10	=C10/$C$12
11	Engineers	1465	1603	=C11-B11	=C11/$C$12
12	All occupations	145571	167754		

IF(*condition, value if true, value if false*)—a logical function that returns one
value if the condition is true and another if the condition is false
VLOOKUP(*value, table range, column number*)—looks up a value in a table

Excel has many other functions for statistical, financial, and other applications,
many of which we will use throughout the text. The easiest way to locate a particular
function is to select a cell and click on the *Insert function* button which can be found
under the ribbon. This is particularly useful even if you know what function to use
but you are not sure of what arguments to enter. Figure 1.10 shows the dialog box

Figure 1.10   *Insert Function* Dialog

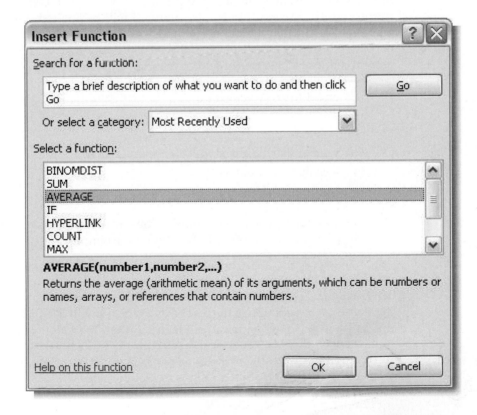

Figure 1.11 *Function Arguments* Dialog for Average

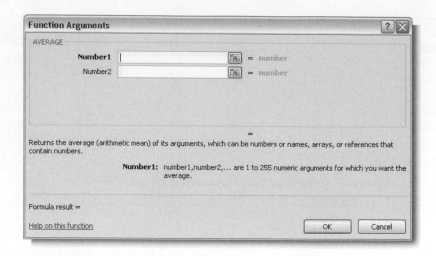

from which you may select the function you wish to use, in this case, the AVERAGE function. Once this is selected, the dialog box in Figure 1.11 appears. When you click in an input cell, a description of the argument is shown. Thus, if you were not sure what to enter for the argument Number 1, the explanation in Figure 1.11 will help you. For further information, you could click on the *Help* button in the lower left-hand corner.

The IF function, *IF(condition, value if true, value if false)*, allows you to choose one of two values to enter into a cell. If the specified *condition* is true, value A will be put in the cell. If the condition is false, value B will be entered. For example, if cell C2 contains the function =IF(A8=2,7,12), it states that if the value in cell A8 is 2, the number 7 will be assigned to cell C2; if the value in cell A8 is not 2, the number 12 will be assigned to cell C2. "Conditions" may include the following:

=	equal to
>	greater than
<	less than
>=	greater than or equal to
<=	less than or equal to
<>	not equal to

You may "nest" up to seven IF functions by replacing *value-if-true* or *value-if-false* in an IF function with another IF function, for example:

$$= IF(A8 = 2,(IF(B3 = 5,\text{"YES"},\text{""})),15)$$

This says that if cell A8 equals 2, then check the contents of cell B3. If cell B3 is 5, then the value of the function is the text string YES; if not, it is a blank space (a text string that is blank). However, if cell A8 is not 2, then the value of the function is 15 no matter what cell B3 is. You may use AND and OR functions as the *condition* within an IF function, for example: =IF(AND(B1=3,C1=5),12,22). Here, if cell B1 = 3 and cell C1 = 5, then the value of the function is 12, otherwise it is 22.

Other Useful Excel Tips

- *Split Screen.* You may split the worksheet horizontally and/or vertically to view different parts of the worksheet at the same time. The vertical splitter bar is just to the right of the bottom scroll bar, and the horizontal splitter bar is just above the right-hand scroll bar. Position your cursor over one of these until it changes shape, click, and drag the splitter bar to the left or down.
- *Paste Special.* When you normally copy (one or more) cells and paste them in a worksheet, Excel places an exact copy of the formulas or data in the cells (except for relative addressing). Often you simply want the *result* of formulas, so the data will remain constant even if other parameters used in the formulas change. To do this, use the *Paste Special* option found under the *Paste* menu in the *Clipboard* group under the *Home* tab instead of the *Paste* command. Choosing *Paste Values* will paste the result of the formulas from which the data were calculated.
- *Column and Row Widths.* Many times a cell contains a number that is too large to display properly because the column width is too small. You may change the column width to fit the largest value or text string anywhere in the column by positioning the cursor to the right of the column label so that it changes to a cross with horizontal arrows then double-click. You may also move the arrow to the left or right to manually change the column width. You may change the row heights in a similar fashion by moving the cursor below the row number label. This can be especially useful if you have a very long formula to display. To break a formula within a cell, position the cursor at the break point in the formula bar and press Alt-Enter.
- *Displaying Formulas in Worksheets.* Choose *Show Formulas* in the *Formula Auditing* group under the *Formulas* tab. You will probably need to change the column width to display the formulas properly.
- *Displaying Grid Lines and Row and Column Headers for Printing.* Check the *Print* boxes for gridlines and headings in the *Sheet Options* group under the *Page Layout* tab. Note that the *Print* command can be found by clicking on the *Office* button.
- *Filling a Range with a Series of Numbers.* Suppose you want to build a worksheet for entering 100 data values. It would be tedious to have to enter the numbers from 1 to 100 one at a time. Simply fill in the first few values in the series and highlight them. Now click and drag the small square (fill handle) in the lower right-hand corner down (Excel will show a small pop-up window that tells you the last value in the range) until you have filled in the column to 100; then release the mouse.

Excel Add-Ins

Microsoft Excel will provide most of the computational support required for the material in this book. Excel provides an add-in called the *Analysis Toolpak,* which contains a variety of tools for statistical computation, and Solver, which is used for optimization. These add-ins are not included in a standard Excel installation. To install them, click the *Office* button at top left of the toolbar. Click on *Excel Options* button at the bottom of the pop-up and choose *Add-Ins* from the left column. At the bottom of the dialog, make sure *Excel Add-ins* is selected in the *Manage:* box and click *Go.* In the *Add-Ins* dialog, if *Analysis Toolpak, Analysis Toolpak VBA,* and *Solver Add-in* are not checked, simply check the boxes and click OK. You will not have to repeat this procedure every time you run Excel in the future.

Four other add-ins available with this book provide additional capabilities and features not found in Excel and will be used in various chapters in this book. You should install these add-ins now on your computer. Prentice-Hall's *PHStat2* (which we will simply refer to as *PHStat*) add-in provides useful statistical support that extends the capabilities of Excel.[10] Refer to the installation procedures on the CD-ROM in the back of this book. The student version of *Crystal Ball* provides a comprehensive set of tools for performing risk analysis simulations. *TreePlan* provides Excel support for decision trees. Finally, Frontline Systems' *Premium Solver for Education* is included as a replacement for the default Solver in Excel. This add-in provides a more stable and accurate solution algorithm for optimization problems than Excel's default tool. The CD-ROM also includes an Excel workbook, SimQuick-v2.xls, which will be used for process simulation in Chapter 12.

Throughout this book we will provide many notes that describe how to use specific features of Microsoft Excel, *PHStat,* or other add-ins. These are introduced as needed to supplement examples and discussions of applications. It is important to read these notes and apply the procedures described in them in order to gain a working knowledge of the software features to which they refer.

WORKING WITH DATA IN EXCEL

In many cases, data on Excel worksheets may not be in the proper form to use a statistical tool. Figure 1.12, for instance, shows the worksheet *Process Capability* from the Excel file *Quality Measurements*, which we use for a case problem later in this book. Some tools in the *Analysis Toolpak* require that the data be listed in a single column in the worksheet. As a user, you have two choices. You can manually move the data within the worksheet, or you can use a utility from the *Data Preparation* menu in *PHStat* called *Stack Data* (see *PHStat Note: Using the Stack Data and Unstack Data Tools*).

The tool creates a new worksheet called "Stacked" in your Excel workbook, a portion of which is shown in Figure 1.14. If the original data columns have group labels (headers), then the column labeled "Group" will show them; otherwise, as in this example, the columns are simply labeled as Group1, Group2, and so on. If you apply the *Unstack Data* tool to the data in Figure 1.14, you will put the data in its original form.

PivotTables

Excel provides a powerful tool for distilling a complex data set into meaningful information: PivotTables. PivotTables allows you to create custom summaries and charts (see Chapter 2) of key information in the data. To apply PivotTables, you need a data set with column labels in the first row. The data set in the Excel file *Accounting Professionals,* shown in Figure 1.15, which provides the results of a survey of 27 employees in a tax division of a Fortune 100 company, satisfies this condition. Select any cell in the data set and choose *PivotTable* and *PivotChart Report* from the *Data* tab and follow the steps of the wizard (see *Excel Note: Creating PivotTables*).

---

[10] The latest version of *PHStat, PHStat2,* is included on the CD-ROM. Enhanced versions and updates may be published on the *PHStat* Web site at www.prenhall.com/phstat. To date, *PHStat* is not available for Mac.

Figure 1.12 Excel Worksheet *Process Capability*

	A	B	C	D	E	F	G	H
1	**Process Capability**							
2								
3	5.21	5.87	4.85	4.95	5.07	4.96	4.96	5.11
4	5.02	5.33	4.82	4.86	4.82	4.96	5.06	5.11
5	4.90	5.11	5.02	5.13	5.03	4.94	4.86	5.08
6	5.00	5.07	4.90	4.95	4.85	5.19	4.96	5.03
7	5.16	4.93	4.73	5.22	4.89	4.91	4.99	4.94
8	5.03	4.99	5.04	4.81	4.82	5.01	4.94	4.88
9	4.96	5.04	5.07	4.91	5.18	4.93	5.06	4.91
10	5.04	5.14	4.81	4.95	5.02	5.05	4.95	4.86
11	4.98	5.09	5.04	4.94	5.05	4.96	5.02	4.89
12	5.07	5.06	5.03	4.81	4.88	4.92	5.01	4.91
13	5.02	4.85	5.01	5.11	5.08	4.95	5.04	4.87
14	5.08	4.93	5.14	4.81	4.98	5.08	5.01	4.93
15	4.85	5.04	5.12	4.97	5.02	4.97	5.02	5.14
16	4.90	5.09	4.89	5.07	4.99	5.04	5.03	4.87
17	4.97	5.07	4.91	5.03	5.02	4.94	5.18	4.98
18	5.09	4.99	4.97	4.81	5.03	4.98	5.08	4.88
19	4.89	5.01	4.98	4.95	5.02	5.03	5.14	4.88
20	4.87	4.88	5.01	4.89	5.07	5.05	4.92	5.01
21	5.01	4.93	5.01	5.08	4.95	4.91	4.97	4.93
22	4.97	5.10	5.09	4.93	4.95	5.09	4.92	4.93
23	4.76	4.94	4.93	4.99	4.94	5.21	5.14	4.99
24	4.94	4.88	5.04	4.94	5.12	4.87	4.92	4.91
25	4.92	4.89	5.11	5.13	5.08	5.02	5.03	4.96
26	4.91	4.89	5.07	5.02	4.91	4.81	4.98	4.78
27	4.96	5.02	5.13	5.13	4.92	4.98	4.89	4.88

You should first decide what types of tables you wish to create. For example, in the accounting department survey data, suppose you wish to count the average number of years of service for males and females with and without a graduate degree. If you drag the variable *Gender* from the *PivotTable Field List* in Figure 1.17 to the *Row Labels* area, the variable *Graduate Degree?* into the *Column Labels* area, and the variable *Years of Service* into the *Values* area, you will have created the PivotTable shown in Figure 1.18. However, the sum of years of service (default) is probably not what you would want. Click on the *Options* tab under *PivotTable Tools* in the menu bar. In the *Active Field* group, click on *Value Field Settings* and you will be able to change the summarization method in the PivotTable (see Figure 1.19). Selecting *Average* results in the PivotTable shown in Figure 1.20, we see that average years is not much different for holders of graduate degrees, but that females have many fewer years of service than males.

The beauty of PivotTables is that if you wish to change the analysis, you can simply uncheck the boxes in the *PivotTable Field List* or drag the variable names to different field areas. You may easily add multiple variables in the fields to create different views of the data. Figure 1.21 shows a count of the number of employees by gender

PHSTAT NOTE
Using the *Stack Data* and *Unstack Data* Tools

From the *PHStat* menu, select *Data Preparation* then either *Stack Data* (to create a single column from multiple columns) or *Unstack Data* (to split a single column into multiple according to a grouping label). Figure 1.13 shows the dialog boxes that appear. To stack data in columns (with optional column labels), enter the range of the data in the *Unstacked Data Cell Range*. If the first row of the range contains a label, check the box *First cells contain group labels*. These labels will appear in the first column of the stacked data to help you identify the data if appropriate.

To unstack data in a single column and group them according to a set of labels in another column, enter the range of the column that contains the labels for the grouping variable in the *Grouping variable cell range* box and the range of the data in the *Stacked data cell range* box. If the top row contains descriptive labels, check the *First cells contain labels* box. This tool is useful when you wish to sort data into different groups.

Figure 1.13 *PHStat* Dialogs for *Stack Data* and *Unstack Data*

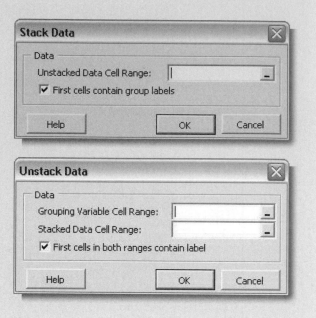

and CPA in the rows, and graduate degree in the columns. For example, we see that 9 employees are female and do not have a graduate degree, while 8 employees are male and do not have a graduate degree. We also see that 4 employees are female and do not have a CPA or a graduate degree, while 3 females have both a CPA and a

	A	B
1	Group	Value
2	Group1	5.21
3	Group1	5.02
4	Group1	4.90
5	Group1	5.00
6	Group1	5.16
7	Group1	5.03
8	Group1	4.96
9	Group1	5.04
10	Group1	4.98
11	Group1	5.07

Figure 1.14 Portion of *Stacked* Worksheet

Figure 1.15 Portion of Excel File *Accounting Professionals*

	A	B	C	D	E	F	G
1	Accounting Department Survey Data						
2							
3	Employee	Gender	Years of Service	Years Undergraduate Study	Graduate Degree?	CPA?	Age Group
4	1	F	17	4	N	Y	5
5	2	F	6	2	N	N	2
6	3	M	8	4	Y	Y	3
7	4	F	8	4	Y	N	3
8	5	M	16	4	Y	Y	4
9	6	F	21	1	N	Y	7
10	7	M	27	4	N	N	7

graduate degree. The best way to learn about PivotTables is simply to experiment with them!

PHStat includes two procedures that facilitate the process of creating certain types of PivotTables by simplifying the input process and automatically creating charts (see *PHStat Note: One- and Two-Way Tables and Charts*).

EXCEL NOTE
Creating PivotTables

Start by clicking on any cell within the data matrix. Choose *PivotTable* from the *Tables* group under the *Insert* tab. The *Create PivotTable* dialog (Figure 1.16) asks you for the range of the data. If your data are in organized rows and columns, Excel will generally default to the complete range of your list. You may either put the PivotTable into a new worksheet or in a blank range of the existing worksheet. The result is shown in Figure 1.17

You may create other PivotTables without repeating all the steps in the Wizard. Simply copy and paste the first table.

Figure 1.16 *Create PivotTable* Dialog

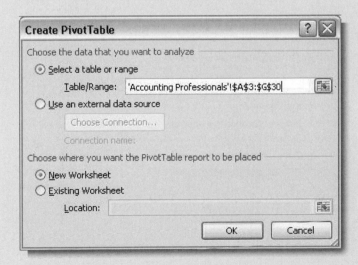

Figure 1.17 Blank PivotTable

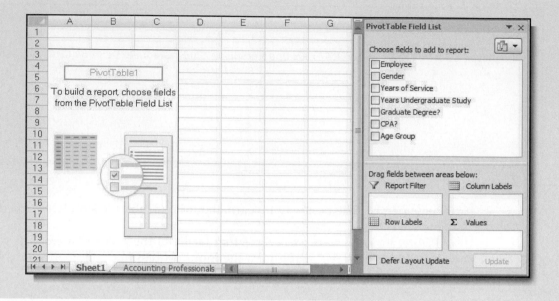

Figure 1.18 PivotTable for the Sum of Years of Service

	A	B	C	D
1				
2				
3	Sum of Years of Service	Column Labels ▾		
4	Row Labels ▾	N	Y	Grand Total
5	F	95	46	141
6	M	168	88	256
7	Grand Total	263	134	397

Figure 1.19 *Value Field Settings* Dialog

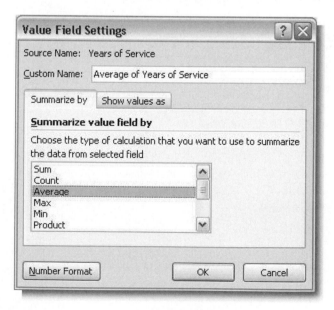

Figure 1.20 PivotTable for Average Years of Service

	A	B	C	D
1				
2				
3	Average of Years of Service	Column Labels		
4	Row Labels	N	Y	Grand Total
5	F	10.55555556	9.2	10.07142857
6	M	21	17.6	19.69230769
7	Grand Total	15.47058824	13.4	14.7037037

Figure 1.21 Count of Employees by Gender and CPA versus Graduate Degree

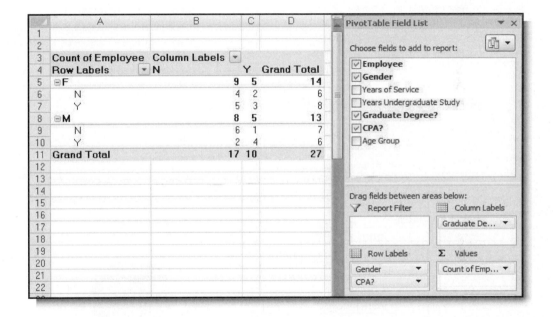

PHSTAT NOTE
One- and Two-Way Tables and Charts

To generate a one-way PivotTable for a set of categorical data, select *Descriptive Statistics* from the *PHStat* menu then *One-Way Tables & Charts*. The dialog box in Figure 1.22 prompts you for the type and location of the data and optional charts that you would like to create

ate (which we discuss further in Chapter 2). The type of data may be either:

- *Raw Categorical Data.* If selected, the single-column cell range entered in the dialog box

Figure 1.22 *PHStat One-Way Tables & Charts Dialog*

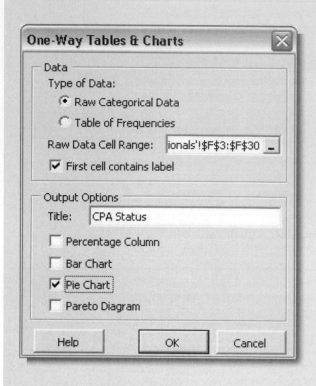

contains raw, unsummarized data. Selected by default.

- *Table of Frequencies.* If selected, the two-column cell range entered in the dialog box contains a frequency table containing categories and frequency counts for each category.

Enter the single-column (raw data) or two-column (frequency table) cell range containing the categorical data to be summarized in the *Raw Data Cell Range* field, and check *First cell contains label* if the contents of the first cell (or first row) are treated as a descriptive label and not as a data value. Figure 1.23 shows the table and pie chart generated for the CPA status in the Excel file *Accounting Professionals*.

A two-way table may also be constructed by selecting *Two-Way Tables & Charts* from the menu. However, we note that the PivotTables created by this tool include only the variables selected and not all the variables in the data set, thus, limiting the flexibility of creating different views of PivotTables as can be done using the Excel PivotTable tool.

Figure 1.23 One-Way Table and Chart for CPA Status

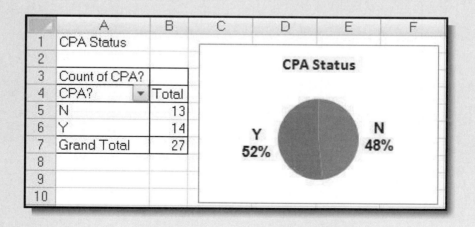

BASIC CONCEPTS REVIEW QUESTIONS

1. Define statistics and decision models. Why are they important in business?
2. Explain the principles of statistical thinking. Why is statistical thinking an important managerial skill?
3. What is Six Sigma? Why are statistics and data analysis important in Six Sigma studies?
4. What is a metric, and how does it differ from a measure?
5. Explain the difference between a discrete and a continuous metric.
6. Describe some of the common types of metrics used in Six Sigma analyses.
7. What is a balanced scorecard, and what value does it have for managers?
8. Explain the difference between leading and lagging measures.
9. Explain the difference between data and information.
10. Explain the differences between categorical, ordinal, interval, and ratio data.
11. Explain the difference between cross-sectional and time-series data.
12. What is the difference between a population and a sample?

SKILL-BUILDING EXERCISES

1. Sort the data in the Excel file *Automobile Quality* from lowest to highest number of problems per 100 vehicles using the sort capability in Excel.
2. Modify the Excel file *Science and Engineering Jobs* to compute the percent increase in the number of jobs for each occupational category.
3. Modify the Excel file *Federal Funds Rate* to apply the *PHStat Unstack* tool to group the data into columns by year, and then compute the minimum, maximum, and average rate for each year using Excel functions.
4. Modify the Excel file *Automobile Quality* to label each car brand as either Foreign or Domestic, use the *PHStat Unstack* tool to group them, and compute the average number of problems per 100 vehicles for each group.
5. Create a PivotTable to find the average number of years of undergraduate study for each age group

with and without a CPA in the Excel file *Accounting Professionals*.
6. Create a PivotTable to find the number of employees in each age group with and without a CPA in the Excel file *Accounting Professionals*.
7. Develop a one-way table using *PHStat* to count the number of early and late flights in the Excel file *Atlanta Airline Data*.
8. Develop a two-way table using *PHStat* to count the number of early and late flights by origin airport in the Excel file *Atlanta Airline Data*.
9. Develop a two-way table using *PHStat* to count the number of employees with different CPA status and graduate degrees in the Excel file *Accounting Professionals*.

PROBLEMS AND APPLICATIONS

1. Call centers have high turnover rates because of the stressful environment. The national average is approximately 50%. The director of human resources for a large bank has compiled data about 70 former employees at one of the bank's call centers (see the Excel file *Call Center Data*). Use PivotTables to find these items:
   a. the average length of service for males and females in the sample
   b. the average length of service for individuals with and without a college degree
   c. the average length of service for males and females with and without prior call center experience

   What conclusions might you reach from this information?
2. A national homebuilder builds single-family homes and condominium-style townhouses. The Excel file *House Sales* provides information on the selling price, lot cost, type of home, and region of the country (M = Midwest, S = South) for closings during one month. Use PivotTables to find the average selling price and lot cost for each type of home in each region of the market. What conclusions might you reach from this information?
3. The Excel file *Cereal Data* provides a variety of nutritional information about 67 cereals and their shelf location in a supermarket. Use PivotTables to

determine if there appear to be differences in nutritional values among cereal manufacturers.

4. The Excel file *MBA Student Survey* provides data on a sample of students' social and study habits. Use PivotTables to find the average age, number of nights out per week, and study hours per week by gender, whether the student is international or not, and undergraduate concentration.

5. A mental health agency measured the self-esteem score for randomly selected individuals with disabilities who were involved in some work activity within the past year. The Excel file *Self Esteem* provides the data, including the individuals' marital status, length of work, type of support received (direct support includes job-related services such as job coaching and counseling), education, and age. Use PivotTables to find the average length of work and self-esteem score for individuals in each classification of marital status and support level. What conclusions might you reach from this information?

6. The Excel file *Unions and Labor Law Data* reports the percent of public and private sector employees in unions in 1982 for each state, along with indicators whether the states had a bargaining law that covered public employees or right-to-work laws. Use PivotTables to find the following:
   a. the average percent of public and private sector employees in unions for states with and without bargaining laws
   b. the average percent of public and private sector employees in unions for states with and without right-to-work laws
   c. the average percent of public and private sector employees in unions for states that have both bargaining laws and right-to-work laws, and those that have neither
   What conclusions might you reach from this information?

7. The Excel file *Freshman College Data* shows data for four years at a large urban university. Use PivotTables to examine differences in student high school performance and first-year retention among different colleges at this university. What conclusions do you reach?

8. The Excel file *Bicycle Wheels* provides data on different types of bicycle wheels. Use PivotTables to examine differences in weight and rotational drag, first by type of wheel and second by size. Do there seem to be any significant differences?

9. The Excel file *University Grant Proposals* provides data on the dollar amount of proposals, gender of the researcher, and whether the proposal was funded or not. Construct a PivotTable to find the average amount of proposals by gender and outcome.

10. The Excel file *Credit Approval Decisions* provides data about credit decisions. Construct a PivotTable to find the average credit score, years of credit history, revolving balance, and revolving utilization based on whether the applicant was a homeowner and whether the application was approved.

CASE

A Data Collection and Analysis Project

Develop a simple questionnaire to gather data that include a set of both categorical variables and ratio variables. In developing the questionnaire, think about some meaningful questions that you would like to address using the data. Obtain a sample of at least 20 responses from fellow students or coworkers. The questionnaire should pertain to any subject of interest to you, for example, customer satisfaction with products or school-related issues, investments, hobbies, leisure activities, and so on—be creative! Record the data on an Excel worksheet.

a. Apply PivotTables to gain information and insight about the nature of your data and write a formal report that clearly summarizes the data you have gathered and the information it conveys.

b. As you learn new material in Chapters 2 through 7, apply the statistical tools as appropriate to analyze your data and write a comprehensive report that describes how you drew statistical insights and conclusions.

Chapter

2

Displaying and Summarizing Data

INTRODUCTION

$\mathcal{I}$n Chapter 1, we discussed the role of data in modern organizations, ways of extracting meaningful information from data using PivotTables, and introduced you to rudimentary concepts of statistics. In this chapter we discuss how to effectively display and summarize data quantitatively for useful managerial information and insight. Because of the ease with which data can be generated and transmitted today, managers, supervisors, and even front-line workers can be overwhelmed. Hence, it is vital that critical data be displayed, aggregated, and summarized in as succinct a fashion as possible. Data visualization through various charts and plots provides simple communication vehicles that all employees can easily understand. Statistical summaries, such as measures of central tendency, dispersion, and relationships among variables, provide more precise quantitative information on which to base decisions. Spreadsheet software makes it possible to easily create visual displays and compute statistical measures. Our focus is on learning how to understand and incorporate these tools to make better decisions, as well as becoming proficient with the capabilities of Microsoft Excel.

The key concepts that we will discuss in this chapter are the following:

➤ *Visual data display: Using Excel-based charts such as line charts, bar charts, pie charts, area charts, and scatter plots; and other visual methods for displaying data, including box-and-whisker plots, stem-and-leaf displays, and dot-scale diagrams*

➤ *Descriptive statistics for numerical data: Computing and understanding measures of central tendency (mean, median, mode), dispersion (range, standard deviation, and variance), shape (skewness and kurtosis), data profiles (quartiles, deciles, and percentiles), the coefficient of variation, and correlation*

➤ *Descriptive statistics for categorical data: Proportions and cross-tabulations*

DISPLAYING DATA WITH CHARTS AND GRAPHS

The Excel file *EEO Employment Report* provides data on the employment in the state of Alabama for 2006. Figure 2.1 shows a portion of this data set. Raw data such as these are often difficult to understand and interpret. Graphs and charts provide a convenient way to visualize data and provide information and insight for making better decisions. Microsoft Excel offers a variety of charts to express data visually. These include vertical and horizontal bar charts, line charts, pie charts, area charts, scatter plots, three-dimensional charts, and many other special types of charts.

Column and Bar Charts

Excel distinguishes between vertical and horizontal bar charts, calling the former *column charts* and the latter *bar charts*. A *clustered column chart* compares values across categories using vertical rectangles; a *stacked column chart* displays the contribution of

Figure 2.1   Portion of *EEO Commission Employment Report*

	A	B	C	D	E	F
1	Equal Employment Opportunity Commission Report - Number Employed in State of Alabama, 2006					
2						
3	Racial/Ethnic Group and Gender	Total Employment	Officials & Managers	Professionals	Technicians	Sales Workers
4	ALL EMPLOYEES	632,329	60,258	80,733	39,868	62,019
5	Men	349,353	41,777	39,792	19,848	23,727
6	Women	282,976	18,481	40,941	20,020	38,292
7						
8	WHITE	407,545	51,252	67,622	28,830	41,091
9	Men	237,516	36,536	34,842	16,004	17,756
10	Women	170,029	14,716	32,780	12,826	23,335
11						
12	MINORITY	224,784	9,006	13,111	11,038	20,928
13	Men	111,837	5,241	4,950	3,844	5,971
14	Women	112,947	3,765	8,161	7,194	14,957

each value to the total by stacking the rectangles; and a *100% stacked column chart* compares the percentage that each value contributes to a total. These charts are shown in Figure 2.2. Bar charts present information in a similar fashion, only horizontally instead of vertically.

Excel provides an easy way to create charts within your spreadsheet (see *Excel Note: Creating Charts in Excel 2007*). We generally will not guide you through every application but will provide some guidance for new procedures as appropriate.

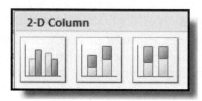

Figure 2.2   Types of Two-Dimensional Column Charts

EXCEL NOTE
Creating Charts in Excel 2007

In this and other Excel Notes in this book, we provide some basic information about key features in Excel. More detail can be found in Excel's Help files and we encourage you to use them as needed. Excel provides a very comprehensive charting capability with many features. With a little experimentation, you can create very professional charts for business presentations.

It is best to first highlight the range of the data you wish to chart. The Excel Help files provide guidance on formatting your data for a particular type of chart. Click the *Insert* tab in the Excel ribbon (Figure 2.3). From the *Charts* group, click the chart type, and then click a chart subtype that you want to use. Once a basic chart is created, you may use the options within the *Chart Tools* tabs to customize your chart (Figure 2.4). In the *Design* tab, you can

change the type of chart, data included in the chart, chart layout, and styles. From the *Layout* tab, you can modify the layout of titles and labels, axes and gridlines, and other features. The *Format* tab provides various formatting options. Many of these options can also be invoked by right-clicking on elements of the chart.

We will illustrate a simple bar chart for the various employment categories for all employees in the EEO data. First, highlight the range C3:K6, which includes the headings and data for each category. Click on the *Column Chart* button and then on the first chart type in the list (a clustered column chart). To add a title, click on the first icon in the *Chart Layouts* group. Click on "Chart Title" in the chart and change it to "EEO Employment Report – Alabama."

The names of the data series can be changed by clicking on the *Select Data* button in the *Data* group of the *Design* tab. In the *Select Data Source* dialog (see Figure 2.5), click on "Series1" and then the *Edit* button. Enter the name of the data series, in this case "All Employees." Change the names of the other data series to "Men" and "Women" in a similar fashion. You can also change the order in which the data series are displayed on the chart using the up and down buttons. The final chart is shown in Figure 2.6.

Figure 2.3 Excel *Insert* Tab

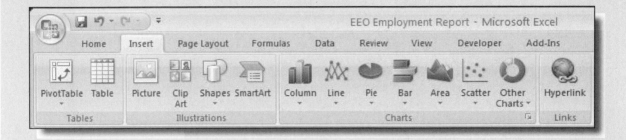

Figure 2.4 Excel *Chart Tools*

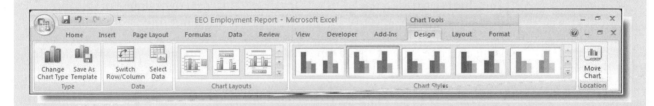

Figure 2.5 *Select Data Source* Dialog

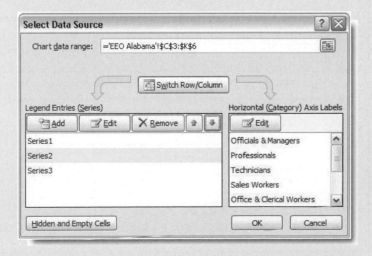

Figure 2.6 Final Column Chart for EEO Data

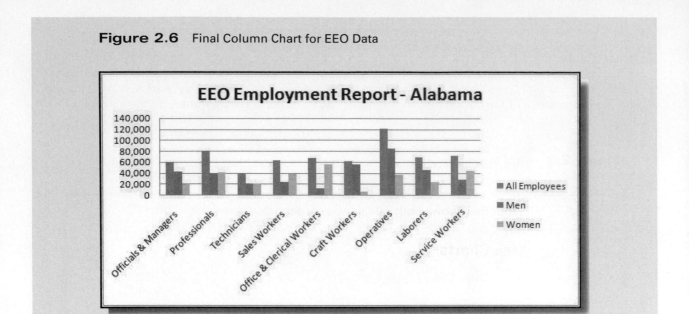

Line Charts

Line charts provide a useful means for displaying data over time. For instance, a line chart showing the amount of U.S. exports to China in billions of dollars from the Excel file *China Trade Data* is shown in Figure 2.7. The chart clearly shows a trend of rising amounts of exports starting in the year 2000. You may plot multiple data series

Figure 2.7 Line Chart for U.S. to China Exports

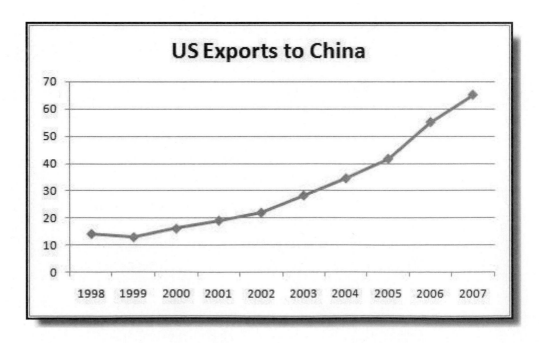

in line charts; however, the information can be difficult to interpret if the magnitude of the data values differs greatly. In this case, it would be advisable to create separate charts for each data series.

Pie Charts

For many types of data, we are interested in understanding the relative proportion of each data source to the total. For example, consider the marital status of individuals in the U.S. population in the Excel file *Census Education Data*, a portion of which is shown in Figure 2.8. To show the relative proportion in each category, we can use a **pie chart**, as shown in Figure 2.9. This chart uses a layout option that shows the labels associated with the data, but not the actual values or proportions. A different layout that shows both can also be chosen.

Area Charts

An **area chart** combines the features of a pie chart with those of line charts. For example, Figure 2.10 displays total energy consumption (billion Btu) and consumption of fossil fuels from the Excel file *Energy Production & Consumption*. This chart shows that while total energy consumption has grown since 1949, the relative proportion of fossil fuel consumption has remained generally consistent at about half of the total, indicating that alternative energy sources have not replaced a significant portion of fossil fuel consumption. Area charts present more information than pie or line charts alone but may clutter the observer's mind with too many details if too many data series are used; thus, they should be used with care.

Scatter Diagrams

Scatter diagrams show the relationship between two variables. Figure 2.11 shows a scatter diagram of the earned run average versus number of games won from the Excel file *Major League Baseball*, which provides data for the 2007 baseball season. Earned run average (ERA) is a measure of a pitching effectiveness, calculated as the average number of earned runs allowed by the pitcher for every nine innings. Each point on the chart represents one team; thus, a scatter diagram plots pairs of observations and displays the value of one variable against the corresponding value of the other. We see that a clear relationship appears to exist between these variables— a smaller ERA corresponds to a larger number of wins. Later in this chapter, we shall see how to describe such a relationship numerically.

Figure 2.8   Portion of *Census Education Data*

	A	B
18	**Marital Status**	
19	Never Married	25,752,000
20	Married, spouse present	107,008,000
21	Married, spouse absent	6,844,000
22	Separated	4,605,000
23	Widowed	13,577,000
24	Divorced	19,030,000

Figure 2.9 Pie Chart for Marital Status

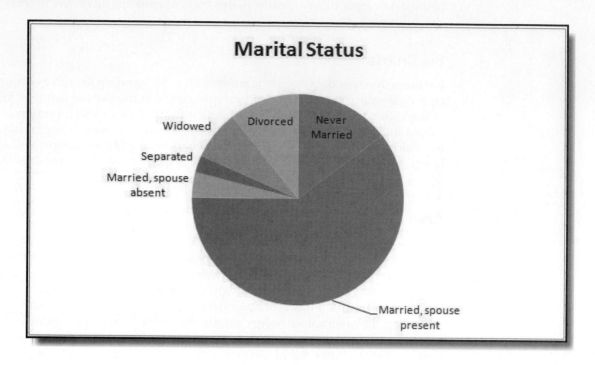

Miscellaneous Excel Charts

Excel provides several additional charts for special applications (see Figure 2.12). A **stock chart** allows you to plot stock prices, such as the daily high, low, and close. It may also be used for scientific data such as temperature changes. A **surface chart**

Figure 2.10 Area Chart for Energy Consumption

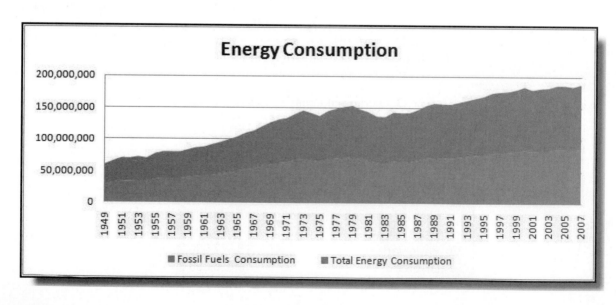

Figure 2.11 Scatter Diagram of *Major League Baseball* Data

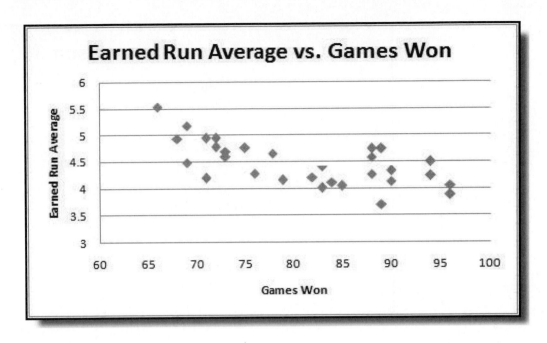

Figure 2.12 Other Excel Charts

Figure 2.13 Example of a Radar Chart

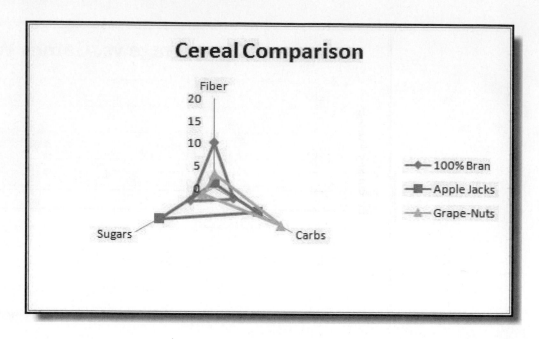

shows three-dimensional data. A **doughnut chart** is similar to a pie chart but can contain more than one data series. A **bubble chart** is a type of scatter chart in which the size of the data marker corresponds to the value of a third variable; consequently, it is a way to plot three variables in two dimensions. Finally, a **radar chart** allows you to plot multiple dimensions of several data series. An example is shown in Figure 2.13. Using the Excel file *Cereal Data*, we plotted the values of fiber, carbohydrates, and sugars for three cereals in the database. This chart clearly shows the differences in nutritional attributes. We encourage you to experiment with different charts in displaying and interpreting data.

Summary of Graphical Display Methods

In summary, tables of numbers often hide more than they inform. Graphical displays clearly make it easier to gain insights about the data. Thus, graphs and charts are a means of converting raw data into useful managerial information. However, it can be easy to distort data by manipulating the scale on the chart. For example, Figure 2.14 shows the U.S. exports to China in Figure 2.7 displayed on a different scale. The pattern looks much flatter and suggests that the rate of exports is not increasing as fast as it really is. It is not unusual to see distorted graphs in newspapers and magazines that are intended to support the author's conclusions. Creators of statistical displays have an ethical obligation to report data honestly and without attempts to distort the truth. Another drawback of visual displays is that they provide no *quantitative* summaries of the data. We will discuss statistical measures useful to convey information for decision making later in this chapter.

Figure 2.14 An Alternate View of U.S. Exports to China

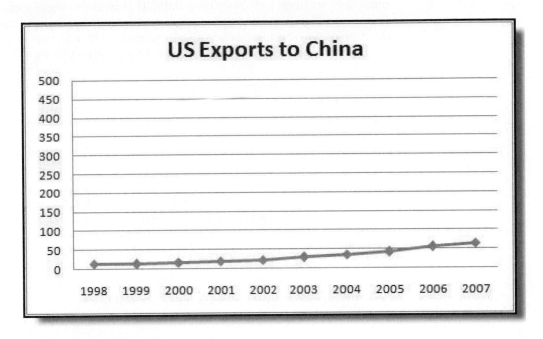

DESCRIPTIVE STATISTICS:
CONCEPTS AND APPLICATIONS

The Excel file *Facebook Survey* provides data from a sample of 33 students (see Figure 2.15). What can we learn from these data? We need to be able to get a "big picture" view of what the data tell us. This is particularly important in such applications as market research, when sample data are used for important business decisions. Statistical measures provide an effective and efficient way of obtaining meaningful

Figure 2.15 Portion of *Facebook Survey*

	A	B	C	D	E	F
1	Facebook Survey					
2						
3	Sample	Hours online/week	Log-ins/ day	Friends	Student	Gender
4	1	4	3	68	yes	female
5	2	2	4	61	yes	female
6	3	7	5	124	yes	male
7	4	3	5	405	yes	male
8	5	3	5	260	yes	female
9	6	2	1	74	yes	female
10	7	7	1	90	yes	female
11	8	1	1	250	yes	male
12	9	2	1	115	yes	female
13	10	2	5	19	yes	female

information from data. **Descriptive statistics** refers to a collection of quantitative measures and ways of describing data. This includes measures of central tendency (mean, median, mode, proportion), measures of dispersion (range, variance, standard deviation) and frequency distributions and histograms.

Statistical support within Microsoft Excel can be accomplished in three ways:

1. Using statistical functions that are entered in worksheet cells directly or embedded in formulas.
2. Using the Excel *Analysis Toolpak* add-in to perform more complex statistical computations.
3. Using the *Prentice-Hall Statistics* add-in, *PHStat,* available with this book to perform analyses not designed into Excel.

Table 2.1 summarizes the descriptive statistics functions and tools available. One important point to note about the use of the tools in the *Analysis Toolpak* versus Excel functions is that while functions dynamically change as the data in the spreadsheet are changed, the results of the tools do not. For example, if you compute the mean directly using the functions AVERAGE(*range*), then changing the data in the range will automatically update the result. However, you would have to rerun the *Descriptive Statistics* tool after changing the data.

Table 2.1
Excel Statistical Functions and Tools

EXCEL FUNCTIONS	DESCRIPTION
AVERAGE(*data range*)	Computes the average value (arithmetic mean) of a set of data
MEDIAN(*data range*)	Computes the median (middle value) of a set of data
MODE(*data range*)	Computes the mode (most frequently occurring) of a set of data
VAR(*data range*)	Computes the variance of a set of data, assumed to be a sample
VARP(*data range*)	Computes the variance of a set of data, assumed to be an entire population
STDEV(*data range*)	Computes the standard deviation of a set of data, assumed to be a sample
STDEVP(*data range*)	Computes the standard deviation of a set of data, assumed to be an entire population
SKEW(*data range*)	Computes the skewness, a measure of the degree to which a distribution is not symmetric around its mean
PERCENTILE(*array, k*)	Computes the *k*th percentile of data in a range
QUARTILE(*array, quart*)	Computes the quartile of a distribution
COVAR(*array1, array2*)	Computes the covariance, assuming population data
CORREL(*array1, array2*)	Computes the correlation coefficient between two data sets
ANALYSIS TOOLPAK TOOLS	DESCRIPTION
Descriptive Statistics	Provides a summary of a variety of basic statistical measures
Histogram	Creates a frequency distribution and graphical histogram for a set of data
Rank and Percentile	Computes the ordinal and percentage rank of each value in a data set
Correlation	Computes the correlation coefficient between two data sets
PHSTAT ADD-IN	DESCRIPTION
Box-and-Whisker Plot	Creates a box-and-whisker plot of a data set
Stem-and-Leaf Display	Creates a stem-and-leaf display of a data set
Dot-Scale Diagram	Creates a dot-scale diagram of a data set
Frequency Distribution	Creates a table of frequency counts and percentage frequency values
Histogram & Polygons	Creates a frequency table and histogram and optional frequency polygons

EXCEL NOTE
Using the Descriptive Statistics Tool

Click on *Data Analysis* in the *Analysis* group under the *Data* tab in the Excel menu bar. Select *Descriptive Statistics* from the list of tools. The *Descriptive Statistics* dialog shown in Figure 2.16 will appear. You need only enter the range of the data, which must be in a *single row or column*. If the data are in multiple columns, the tool treats each row or column as a separate data set, depending on which you specify. In the Facebook example, we have specified the *Input Range* as the three columns of data. Thus, the tool will treat these as three separate data sets. This means that if you have a single data set arranged in a matrix format, you would have to stack the data (for example, using the *PHStat Stack Data* tool described in Chapter 1) in a single column before applying the *Descriptive Statistics* tool. Check the box *Labels in First Row* if labels are included in the input range. You may choose to save the results in the current worksheet or in a new one. For basic summary statistics, check the box *Summary statistics*; you need not check any others.

Figure 2.16 *Descriptive Statistics* Dialog

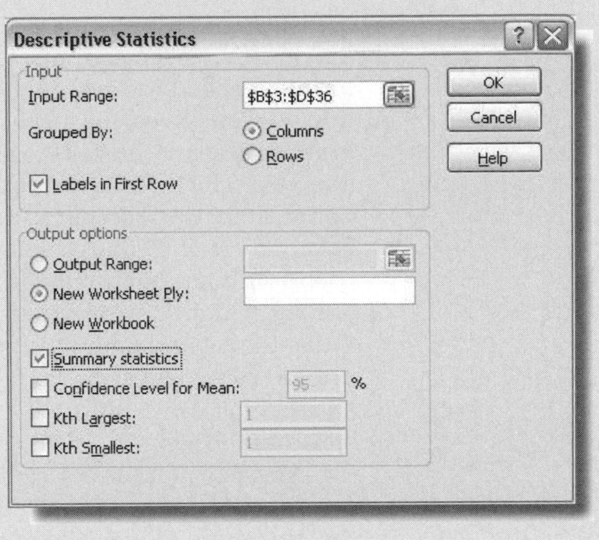

Excel Descriptive Statistics Tool

Excel provides a useful tool for basic data analysis, *Descriptive Statistics* (see *Excel Note: Using the Descriptive Statistics Tool*), which provides a variety of statistical measures that describe central tendency, dispersion, and shape of the distribution of sample data.

To illustrate the use of this tool, we will apply it to the data in the *Facebook Survey* file The results are shown in Figure 2.17. We will describe the information provided in

Figure 2.17 Facebook Data Descriptive Statistics

	A	B	C	D	E	F
1	Hours online/week		Log-ins/day		Friends	
2						
3	Mean	2.727272727	Mean	2.818181818	Mean	113.4242424
4	Standard Error	0.346171505	Standard Error	0.378029544	Standard Error	16.37660646
5	Median	2	Median	2	Median	90
6	Mode	2	Mode	1	Mode	90
7	Standard Deviation	1.988603896	Standard Deviation	2.171614397	Standard Deviation	94.07644176
8	Sample Variance	3.954545455	Sample Variance	4.715909091	Sample Variance	8850.376894
9	Kurtosis	0.07769041	Kurtosis	2.059631596	Kurtosis	2.240398535
10	Skewness	1.062645667	Skewness	1.203529872	Skewness	1.572124585
11	Range	7	Range	10	Range	402
12	Minimum	0	Minimum	0	Minimum	3
13	Maximum	7	Maximum	10	Maximum	405
14	Sum	90	Sum	93	Sum	3743
15	Count	33	Count	33	Count	33

the results; however, we will discuss the standard error and confidence level in Chapters 3 and 4, respectively, so you may ignore them for now.

Measures of Central Tendency

Measures of central tendency provide estimates of a single value that in some fashion represents "centering" of the entire set of data. The most common is the *average,* formally called the **arithmetic mean** (or simply the *mean*). We all use averages routinely in our lives, for example, to measure student accomplishment in a curriculum, to measure the scoring ability of sports figures, and to measure performance in business. The mean is the sum of the observations divided by the number of observations. In the Facebook data, the mean hours per week that students used Facebook is 2.73 hours, mean log-ins per day is 2.82, and the mean number of friends in a student's network is about 113. Note that the sum and number of observations (Count) are provided in the last two lines of the *Descriptive Statistics* results. Thus, the mean hours used per week was calculated as 90/33 = 2.73. The mean is unique for every set of data and is meaningful for both interval and ratio data. However, it can be affected by **outliers**—rare observations that are radically different from the rest.

Another measure of central tendency is the **median**, the middle value when the data are arranged from smallest to largest. For an odd number of observations, the median is the middle of the sorted numbers. For an even number of observations, the median is the arithmetic mean of the two middle numbers. We could have used the *Sort* option in Excel to rank order the values in the columns of the Facebook data and then determine the median. Since we have 33 observations, the median would be the 17th observation. The Excel function MEDIAN(*data range*) would also provide this result. Half the data are below the median, and half the data are above it. For example, half the students are on Facebook at most 2 hours per week, log in at most 2 times per day, and have at least 90 friends. The median is meaningful for ratio, interval, and ordinal data. As opposed to the mean, the median is *not* affected by outliers. In this case, the median is very close in value to the mean.

A third measure of central tendency is the **mode**. The mode is the observation that occurs most frequently. For instance, in the Facebook data, 11 of the 33 values for Hours online/week are 2, and 10 of the 33 values for Log-ins/day are 1. This can easily be counted or you may use the Excel function MODE(*data range*) to compute it. The mode is not affected by order or differences in scale, but only by the count for each observation. The mode is more useful for data sets that consist of a relatively small number of unique values. The mode for the Friends is 90. While this value occurs the most frequently, only 2 of the values are 90; the rest of the data values are all unique. Thus, when a data set has few repeating values, the mode is not very useful.

Another measure of central tendency that some people use, which is not computed by the Excel tool, is the **midrange**. This is simply the average of the largest and smallest values in the data set. For the Facebook data, the maximum value for Hours online/week is 7 and the minimum value is 0. Thus, the midrange is $(7 - 0)/2 = 3.5$. Caution must be exercised when using this statistic because extreme values easily distort the result. Note that the midrange uses only two pieces of data, while the mean uses *all* the data; thus, it is usually a much rougher estimate than the mean and is often used for only small sample sizes.

Measures of Dispersion

Dispersion refers to the degree of variation in the data, that is, the numerical spread (or compactness) of the data. For instance, a cursory examination of the Facebook data clearly shows more variation in Friends than in the other two variables. Several statistical measures characterize dispersion: the *range, variance,* and *standard deviation.* The **range** is the simplest and is computed as the difference between the maximum value and the minimum value in the data set. Although Excel does not provide a function for the range, it can be computed easily by the formula = MAX(*data range*) – MIN(*data range*). The range is provided in the *Descriptive Statistics* results. Like the midrange, the range is affected by outliers.

A more commonly used measure of dispersion is the **variance**, whose computation depends on *all* the data. The formula used for calculating the variance is different for populations and samples (we will discuss the theory behind this later in the chapter). The Excel function VAR(*data range*) may be used to compute the sample variance, while the Excel function VARP(*data range*) is used to compute the variance of a population. Note that the *Descriptive Statistics* tool computes the sample variance (and implicitly assumes that the data represents a sample and not a population).

A related measure, which is perhaps the most popular and useful measure of dispersion, is the **standard deviation**, which is defined as the square root of the variance. The Excel function STDEVP(*data range*) calculates the standard deviation for a population; the function STDEV(*data range*) calculates it for a sample. The value shown in the *Descriptive Statistics* output is the sample standard deviation. The larger the standard deviation, the more the data are "spread out" from the mean, and the more variability one can expect in the observations. Observe that the standard deviations of Hours online/week and Log-ins/day are quite small compared to the standard deviation of Friends, whose values cover a much larger range.

The standard deviation is generally easier to interpret than the variance because its units of measure are the same as the units of the data. Thus, it can be more easily related to the mean or other statistics measured in the same units. One of the most important results in statistics is **Chebyshev's theorem**, which states that for *any set of data,* the proportion of values that lie within k standard deviations ($k > 1$) of the mean is at least $1 - 1/k^2$. Thus, for $k = 2$, at least three-fourths of the data lie within two standard deviations of the mean; for $k = 3$, at least 8/9, or 89% of the data lie within three standard deviations of the mean. For the Facebook Friends data, a three standard deviation range is $113.42 \pm 3(94.08)$ or $[-168.82, 395.66]$, and you can easily see that all data fall within this interval.

For many data sets encountered in practice, the percentages are generally much higher than Chebyshev's theorem specifies. For example, a two standard deviation spread might include close to 95% of the data, and a three standard deviation spread might include as much as 98% or 99% of the data. Thus, two or three standard deviations around the mean are commonly used to describe the variability of most practical sets of data. For example, the capability of a manufacturing process, which is characterized by the expected variation of output, is generally quantified as the mean plus or minus three standard deviations. This range is used in many quality control and Six Sigma applications.

The standard deviation is also a useful measure of risk, particularly in financial analysis. For example, the Excel file *Closing Stock Prices* (see Figure 2.18) lists weekly closing prices for four stocks, Google, Yahoo, Cisco, and Apple, over the first six months of 2008. The average closing price for Yahoo and Cisco are quite similar, $26.21 and $24.83, respectively. However, the standard deviation of Yahoo's price is

Figure 2.18 Portion of Excel File *Closing Stock Prices*

	A	B	C	D	E
1	**Closing Stock Prices**				
2					
3	Date	GOOG	YHOO	CSCO	AAPL
4	1/2/2008	$657.00	$23.16	$26.12	$180.05
5	1/7/2008	$638.25	$23.36	$25.87	$172.69
6	1/14/2008	$600.25	$20.78	$24.30	$161.36
7	1/22/2008	$566.40	$21.94	$24.20	$130.01
8	1/28/2008	$515.90	$28.38	$24.94	$133.75
9	2/4/2008	$516.69	$29.20	$23.54	$125.48
10	2/11/2008	$529.64	$29.66	$23.30	$124.63
11	2/19/2008	$507.80	$28.42	$23.60	$119.46
12	2/25/2008	$471.18	$27.78	$24.39	$125.02

$2.92 while Cisco's is $1.13. This means that over this time frame, the returns on Yahoo have generally varied between $26.21 − 3($2.92) = $17.45 and $26.21 + 3($2.92) = $34.97. Cisco's stock price, however, varied approximately between $24.83 − 3($1.13) = $21.44 and $24.83 + 3($1.13) = $28.22. Cisco has the smaller standard deviation and, therefore, more stable and predictable returns. The larger standard deviation for Yahoo implies that the variability is much higher; while a greater potential exists of a higher return, there is also greater risk of realizing a significantly lower return. Many investment publications and Web sites provide standard deviations of stocks and mutual funds to help investors assess risk in this fashion. We will learn more about risk in Part II of this book.

The **coefficient of variation (CV)** provides a relative measure of the dispersion in data relative to the mean and is defined as:

$$CV = \text{Standard Deviation/Mean} \tag{2.1}$$

Sometimes the coefficient of variation is multiplied by 100 to express it as a percent. This statistic is useful when comparing the variability of two or more data sets when their scales differ. One practical application of the coefficient of variation is in comparing stock prices. For example, by examining only the standard deviations in the *Closing Stock Prices* worksheet, we would conclude that the variation in Google is much larger than the other stocks. However, the mean stock price of Google is much larger than the other stocks. Thus, comparing standard deviations directly provides little information. The coefficient of variation provides a more comparable measure. Using the data in Figure 2.18 for the stock price data, we calculate the coefficients of variation as:

	GOOG	YHOO	CSCO	AAPL
Mean	$ 531.94	$ 26.21	$ 24.83	$ 156.35
Std Dev	$ 60.97	$ 2.92	$ 1.13	$ 2.13
CV	0.115	0.111	0.046	0.014

We see that the coefficients of variation of Google and Yahoo are about equal, while those of Cisco and Apple are lower, indicating a lower relative level of risk. Another way to see this is as follows. Suppose that each stock's price falls by one standard deviation. What would be the percentage loss to the investor? For Google and Yahoo, the percentage loss would be 11%, while the percentage losses for Cisco and Apple are only 4.6% and 1.4%, respectively.

Frequency Distributions and Histograms

A **frequency distribution** is a tabular summary showing the frequency of observations in each of several non-overlapping groups, or cells (called "bins" in Excel). A graphical depiction of a frequency distribution in the form of a column chart is called a **histogram**. Frequency distributions and histograms can be created using the *Analysis Toolpak* in Excel (see *Excel Note: Creating a Frequency Distribution and Histogram*). *PHStat* also provides tools for creating frequency distributions and histograms in the *Descriptive Statistics* menu option.

The bins can be discrete values or intervals. Discrete values should be used when there are only a small number of unique data values, each having multiple observations. When the number of data values is large, intervals should be used. This is usually the case for most sample data in business applications.

EXCEL NOTE
Creating a Frequency Distribution and Histogram

Click the *Data Analysis* tools button in the *Analysis* group under the *Data* tab in the Excel menu bar and select *Histogram* from the list. In the dialog box (see Figure 2.19), specify the *Input Range* corresponding to the data. If you include the column header, then also check the *Labels* box so Excel knows that the range contains a label. If you do not specify a *Bin Range*, Excel will automatically determine bin values for the frequency distribution and histogram, which often results in a rather poor histogram. We recommend that you define your own bin values by specifying the upper cell limits of each interval in a column in your worksheet after examining the range of the data. Generally, you should choose between 5 to 15 cells, and bin ranges should be of equal width. The more data you have, the more cells you should generally use. Note that with fewer cells, the cell widths will be wider. Wider cell widths provide a "coarse" histogram. Sometimes you need to experiment to find the best number of cells that provide a useful visualization of the data. Choose the width by calculating (Max value – Min value)/Number of cells, and round the result to a reasonable value. (If you check the *Data Labels* box, be sure you include a column label such as "Bin" or "Upper Cell Limit" in the bin range column or else Excel will treat your first value as a label.) If you have a small number of unique values, use discrete values for the bin range. Check the *Chart Output* box to display a histogram in addition to the frequency distribution.

Figure 2.19 *Histogram* Dialog

Figure 2.20 *Facebook Survey* Worksheet with Bin Range

	A	B	C	D	E	F	G	H
1	Facebook Survey							
2								
3	Sample	Hours online/week	Log-ins/day	Friends	Student	Gender		Friends
4	1	4	3	68	yes	female		Upper Cell Limit
5	2	2	4	61	yes	female		50
6	3	7	5	124	yes	male		100
7	4	3	5	405	yes	male		150
8	5	3	5	260	yes	female		200
9	6	2	1	74	yes	female		250
10	7	7	1	90	yes	female		300
11	8	1	1	250	yes	male		350
12	9	2	1	115	yes	female		400
13	10	2	5	19	yes	female		

Figure 2.20 shows a portion of the *Facebook Survey* worksheet to which we added a bin range for Friends. From the descriptive statistics results, we noted that the minimum value is 3 and the maximum value is 405; we chose 8 cells and selected the cell width as $(400 - 0)/8 = 50$. (Figure 2.19 shows the completed *Histogram* dialog for this example.) Figure 2.21 shows the results. The left column of the frequency distribution shows the upper limit of each cell in which the data fall. For example, the first cell includes all data below or equal to 50; we see that 8 observations fall in this cell. The second cell includes all observations greater than 50 and less than or equal to 100; we see that 12 values fell in this range, and so on. The histogram shows that the distribution of Friends begins to fall off rapidly after about 150. While some students have large numbers of friends, most have 150 or less.

To illustrate an example with discrete bin values, note that the number of Hours online/week in the Facebook data consists only of the values 0 through 7. Thus, we create a column of values from 0 to 7 in the worksheet and use this as the bin range. The frequency distribution and histogram are shown in Figure 2.22.

Figure 2.21 Frequency Distribution and Histogram for Friends

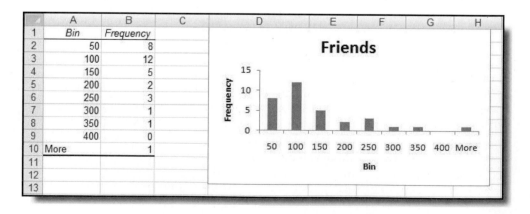

Figure 2.22 Frequency Distribution and Histogram of Hours Online/Week

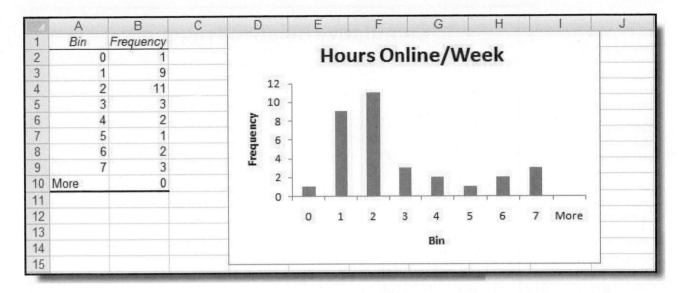

You may easily modify the spreadsheet to count the total number of observations and compute the fraction that falls within each interval—called the **relative frequency**. By summing all the relative frequencies at or below each upper limit, we obtain the **cumulative relative frequency**. The cumulative frequency represents the proportion of the sample that falls at or below the upper limit value. Note that relative frequencies must be between zero and one and must add up to one over all cells. This also means that the cumulative frequency for the last cell must equal one. We have done these computations below for Friends:

UPPER CELL LIMIT	FREQUENCY	RELATIVE FREQUENCY	CUMULATIVE RELATIVE FREQUENCY
50	8	0.242	0.242
100	12	0.364	0.606
150	5	0.152	0.758
200	2	0.061	0.818
250	3	0.091	0.909
300	1	0.030	0.939
350	1	0.030	0.970
400	0	0.000	0.970
More	1	0.030	1.000
Sum	33	1.000	

Figure 2.23 shows a chart for the cumulative relative frequency. From this chart, you can easily estimate the proportion of observations that fall below a certain value. For example, the second data point tells us that about 60% of the data fall at or below 100; the fourth data point shows that about 80% of the data falls at or below 200, and so on.

A limitation of the Excel *Histogram* tool is that the frequency distribution and histogram are not linked to the data; thus, if you change any of the data, you must repeat the entire procedure to construct a new frequency distribution and histogram. An alternative is to use Excel's FREQUENCY function and the *Chart Wizard*. First, define

Figure 2.23 Cumulative Relative Frequency Chart

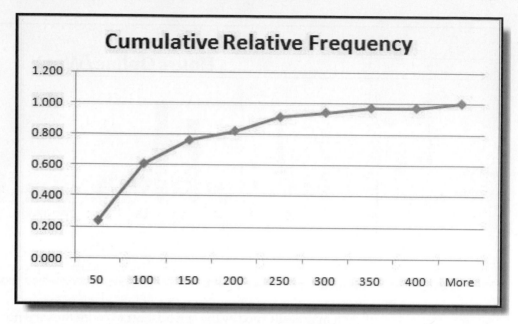

the bins as you would by using the *Histogram* tool. Select the range of cells adjacent to the bin range and add one additional empty cell below it (this provides an overflow cell). Then enter the formula =FREQUENCY(*range of data, range of bins*) and press Ctrl-Shift-Enter simultaneously. This is necessary because FREQUENCY is an array function in Excel. This will create the frequency distribution. You may then construct a histogram using a column chart, customizing it as appropriate. Now, if the data are changed, the frequency distribution and histogram will be updated automatically. An exercise at the end of the chapter will ask you to try this.

Measures of Shape

Histograms of sample data can take on a variety of different shapes. Figure 2.24 shows a histogram for carbohydrates per serving from the Excel file *Cereal Data*. Compare this to the distribution of the Facebook Friends data in Figure 2.21. This carbohydrate distribution is relatively symmetric, having its modal value in the middle and falling away from the center in roughly the same fashion on either side. The Facebook histogram is asymmetrical, or *skewed*; that is, more of the mass is concentrated on one side and the distribution "tails off" to the other. Distributions that tail off to the right, like the Facebook example, are called *positively skewed*; those that tail off to the left are said to be *negatively skewed*.

The **coefficient of skewness (CS)**, which can be found using an Excel function in Table 2.1 or from the *Descriptive Statistics* tool results, measures the degree of asymmetry of a distribution around its mean. If CS is positive, the distribution is positively skewed; if negative, it is negatively skewed. The closer CS is to zero, the less the degree of skewness in the distribution. A distribution whose coefficient of skewness is greater than 1 or less than −1 is highly skewed. A value between 0.5 and 1 or between −0.5 and −1 is moderately skewed. Coefficients between 0.5 and −0.5 indicate relative symmetry. The coefficient of skewness for the Facebook Friends data is 1.57, indicating a relatively high positive skewness.

Figure 2.24 Histogram of Carbohydrates per Serving for Cereals

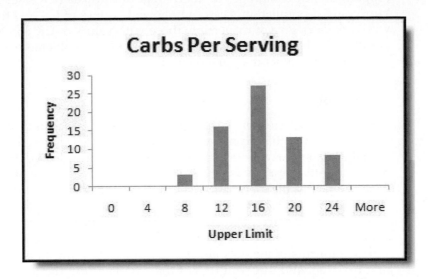

Distributions that have only one "peak," such as Carbs Per Serving, are called **unimodal**. (If the distribution has exactly two peaks, we call it **bimodal**. This often signifies a mixture of samples from different populations.) For unimodal distributions that are relatively symmetric, the mode is a fairly good estimate of the mean. For example, in Figure 2.24 we see that the mode occurs in the cell (12, 16); thus, the midpoint, 14, would be a good estimate of the mean (the true mean is 14.77). On the other hand, for the Facebook data in Figure 2.21, the mode occurs in the cell (50, 100). The midpoint estimate, 75, is not very close to the true mean of 113. Skewness pulls the mean away from the mode.

Comparing measures of central tendency can sometimes reveal information about the shape of a distribution. For example, in a perfectly symmetrical unimodal distribution, the mean, median, and mode would all be the same. For a highly negatively skewed unimodal distribution, we would generally find that mean < median < mode, while for a highly positively skewed distribution, mode < median < mean (see Figure 2.25).

Figure 2.25 Characteristics of Skewed Distributions

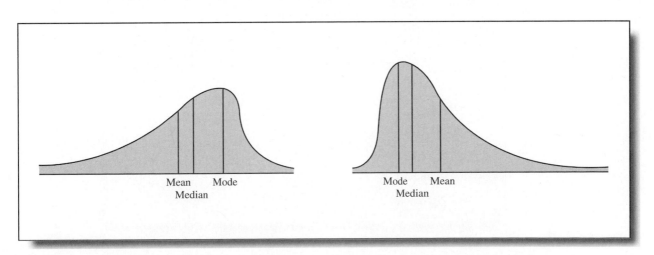

Kurtosis refers to the peakedness (i.e., high, narrow) or flatness (i.e., short, flat-topped) of a distribution. The **coefficient of kurtosis (CK)** measures the degree of kurtosis of a population. Distributions with values of CK less than 3 are more flat with a wide degree of dispersion; those with values of CK greater than 3 are more peaked with less dispersion. The higher the kurtosis, the more area the distribution has in the tails rather than in the middle.

Data Profiles

Many statistical measures are expressed as percentiles and quartiles. You are no doubt familiar with percentiles from standardized tests used for college or graduate school entrance examinations (SAT, ACT, GMAT, GRE, etc.). Percentiles specify the percent of other test takers who scored at or below the score of a particular individual. Specifically, the *kth percentile* is a value at or below which at least *k* percent of the observations lie. *Quartiles* divide the sorted data into four sets, representing the 25th percentile, 50th percentile, 75th percentile, and 100th percentile. Note that the median is actually the second quartile, with two-fourths of the data below the median and two-fourths of the data above the median. One-fourth of the data is below the first quartile, and three-fourths of the data is below the third quartile. Similarly, *deciles* divide the data into 10 sets: the 10th percentile, 20th percentile, and so on. Such measures are called **data profiles** (or **fractiles**).

Excel has a Data Analysis tool called *Rank and Percentile* for computing percentiles. Figure 2.26 shows the dialog box for this tool, using the Facebook Friends data; results are shown in Figure 2.27. While the *Rank and Percentile* tool sorts the data into percentiles, the column labeled *Percent* does not correctly display the percent at or below the given value. For example, while 100% of the data fall at or below 405, the correct percentage of the data falling at or below 99 is 20/33 = 60.61%, not 59.30%, and clearly 0% of the data do not fall at or below 3. This is a flaw in Excel, and we caution you not to use this tool to compute percentiles.

Figure 2.26 *Rank and Percentile* Dialog

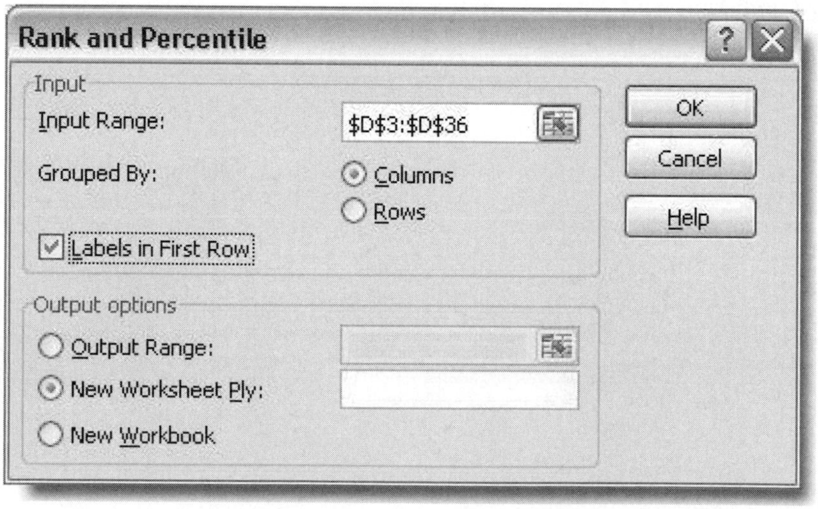

	A	B	C	D
1	Point	Friends	Rank	Percent
2	4	405	1	100.00%
3	26	340	2	96.80%
4	5	260	3	93.70%
5	8	250	4	90.60%
6	18	231	5	87.50%
7	25	224	6	84.30%
8	28	160	7	81.20%
9	15	155	8	78.10%
10	3	124	9	75.00%
11	20	118	10	71.80%
12	9	115	11	68.70%
13	33	108	12	65.60%
14	13	103	13	62.50%
15	22	99	14	59.30%
16	24	96	15	56.20%
17	7	90	16	50.00%
18	16	90	16	50.00%
19	11	79	18	46.80%
20	6	74	19	43.70%
21	32	73	20	40.60%
22	1	68	21	37.50%
23	17	66	22	34.30%
24	2	61	23	31.20%
25	14	60	24	28.10%
26	19	51	25	25.00%
27	31	48	26	21.80%
28	21	45	27	18.70%
29	23	36	28	12.50%
30	27	36	28	12.50%
31	29	33	30	9.30%
32	30	23	31	6.20%
33	10	19	32	3.10%
34	12	3	33	0.00%

Figure 2.27 Rank and Percentile Results for Friends

Correlation

Two variables have a strong statistical relationship with one another if they appear to move together. We see many examples on a daily basis; for instance, attendance at baseball games is often closely related to the win percentage of the team, and ice cream sales likely have a strong relationship with daily temperature. Figure 2.11 showed a scatter diagram that suggested that the number of games won by a major league baseball team increases as the earned run average of its pitchers decreases. In these cases, you might suspect a cause-and-effect relationship. Sometimes, however, statistical relationships exist even though a change in one variable is not caused by a change in the other. For example, the *New York Times* reported a strong relationship between the golf handicaps of corporate CEOs and their companies' stock market performance over three years. CEOs who were better-than-average golfers were likely to deliver above-average returns to shareholders![1] Therefore, you must be cautious in drawing inferences about causal relationships based solely on statistical relationships. (On the other hand, you might want to spend more time out on the course!)

Understanding the relationships between variables is extremely important in making good business decisions, particularly when cause-and-effect relationships can be justified. When a company understands how internal factors such as product quality, employee training, and pricing factors affect such external measures as profitability and customer satisfaction, it can make better decisions. Thus, it is helpful to have statistical tools for measuring these relationships.

Correlation is a measure of a linear relationship between two variables, X and Y, and is measured by the **correlation coefficient**. The Excel file *Colleges and Universities,* a portion of which is shown in Figure 2.28, contains data from 49 top liberal arts and research universities across the United States. Several questions might be raised about statistical relationships among these variables. For instance, does a higher percentage of students in the top 10% of their high school class suggest a higher graduation rate? Is acceptance rate related to the amount spent per student? Do schools with lower acceptance rates tend to accept students with higher SAT scores? Questions such as these can be addressed by computing the correlation between the variables.

Excel's CORREL function (see Table 2.1) computes the correlation coefficient of two data arrays, and the *Data Analysis Correlation* tool computes correlation coefficients for more than two arrays (see *Excel Note: Using the Correlation Tool*). The correlation coefficient is a number between -1 to $+1$. A correlation of 0 indicates that the two variables have no linear relationship to each other. Thus, if one changes, we cannot reasonably predict what the other variable might do. A positive correlation coefficient indicates a linear relationship for which one variable increases as the other also increases. A negative correlation coefficient indicates a linear relationship for one variable that increases while the other decreases. In economics, for instance, a price-elastic product has a negative correlation between price and sales; as price increases, sales decrease, and vice versa. These relationships are illustrated in Figure 2.29. Note that although Figure 2.29 (d) has a clear relationship between the variables, the relationship is not linear and the correlation is zero.

Figure 2.28   Portion of Excel File *Colleges and Universities*

	A	B	C	D	E	F	G
1	**Colleges and Universities**						
2							
3	**School**	**Type**	**Median SAT**	**Acceptance Rate**	**Expenditures/Student**	**Top 10% HS**	**Graduation %**
4	Amherst	Lib Arts	1315	22%	$ 26,636	85	93
5	Barnard	Lib Arts	1220	53%	$ 17,653	69	80
6	Bates	Lib Arts	1240	36%	$ 17,554	58	88
7	Berkeley	University	1176	37%	$ 23,665	95	68
8	Bowdoin	Lib Arts	1300	24%	$ 25,703	78	90
9	Brown	University	1281	24%	$ 24,201	80	90
10	Bryn Mawr	Lib Arts	1255	56%	$ 18,847	70	84
11	Cal Tech	University	1400	31%	$ 102,262	98	75

---

[1] Adam Bryant, "CEOs' Golf Games Linked to Companies' Performance," *Cincinnati Enquirer*, June 7, 1998, El.

Figure 2.29 Examples of Correlation

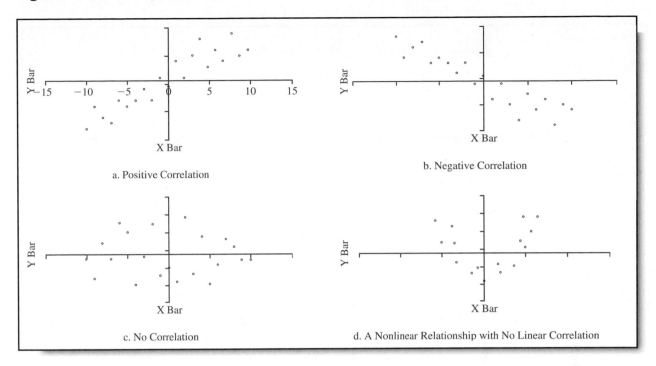

a. Positive Correlation

b. Negative Correlation

c. No Correlation

d. A Nonlinear Relationship with No Linear Correlation

EXCEL NOTE
Using the Correlation Tool

Select *Correlation* from the *Data Analysis* tool list. The dialog box is shown in Figure 2.30. You need only input the range of the data (which must be in contiguous columns; if not, you must move them in your worksheet), specify whether the data are grouped by rows or columns (most applications will be grouped by columns), and indicate whether the first row contains data labels. The output of this tool is a matrix giving the correlation between each pair of variables. This tool provides the same output as the CORREL function for each pair of variables.

Figure 2.30 *Correlation* Tool Dialog

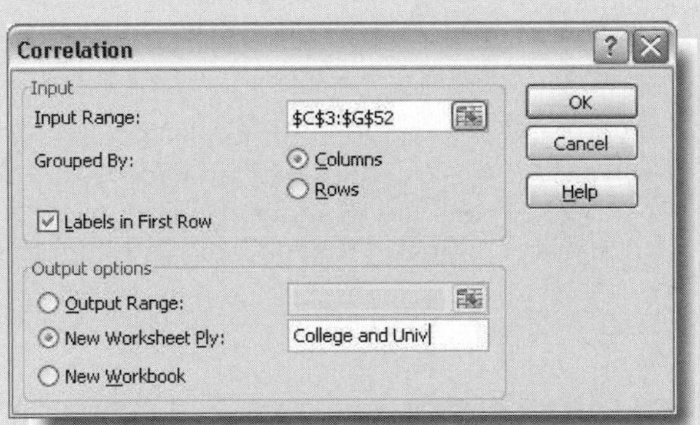

The correlation matrix among all the variables in the *Colleges and Universities* worksheet is shown in Figure 2.31. None of the correlations are very high; however, we see a moderate positive correlation between the graduation rate and SAT score (see Figure 2.32), indicating that schools with higher median SATs have higher graduation rates, and a moderate negative correlation between acceptance rate and graduation rate, indicating that schools with lower acceptance rates have higher

Figure 2.31 Correlation Results for *Colleges and Universities* Data

	A	B	C	D	E	F
1		*Median SAT*	*Acceptance Rate*	*Expenditures/Student*	*Top 10% HS*	*Graduation %*
2	Median SAT	1				
3	Acceptance Rate	-0.601901959	1			
4	Expenditures/Student	0.572741729	-0.284254415	1		
5	Top 10% HS	0.503467995	-0.609720972	0.505782049	1	
6	Graduation %	0.564146827	-0.55037751	0.042503514	0.138612667	1

Figure 2.32 Scatter Chart of Graduation Rate vs. Median SAT

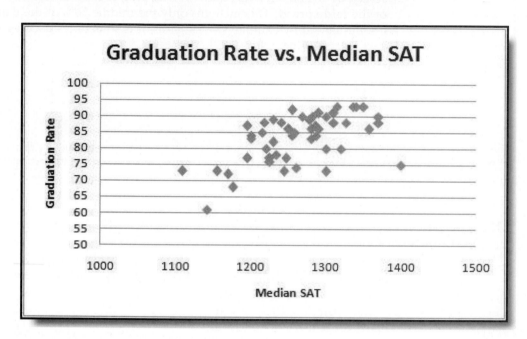

graduation rates. We also see that acceptance rate is negatively correlated with the median SAT and Top 10% HS, suggesting that schools with lower acceptance rates have higher student profiles. The correlations with Expenditures/Student also suggest that schools with higher student profiles spend more money per student.

VISUAL DISPLAY OF STATISTICAL MEASURES

Statisticians use other types of graphs to visually display statistical measures. Three useful tools are *box-and-whisker plots, stem-and-leaf displays,* and *dot-scale diagrams,* all of which are available in the *PHStat* Excel add-in.

Box-and-Whisker Plots

Box-and-whisker plots graphically display five key statistics of a data set—the minimum, first quartile, median, third quartile, and maximum—and are very useful in identifying the shape of a distribution and outliers in the data. Box-and-whisker plots can be created in Excel using *PHStat* (see *PHStat Note: Creating Box-and-Whisker Plots*).

A box-and-whisker plot for Facebook Friends along with the five-number summary is shown in Figure 2.34. Box-and-whisker plots use dashed lines to represent the minimum and maximum values in a data set, and a box encloses the first and third quartiles, with a line representing the median inside the box. Box-and-whisker plots that have very long whiskers help identify potential outliers, which appear to be evident in this example. Since the box is somewhat off center to the left, and the median line is also slightly off center within the box, the distribution is positively skewed (from Figure 2.21, the coefficient of skewness is 1.572).

The difference between the first and third quartiles, $Q_3 - Q_1$, which is represented by the box in a box-and-whisker plot, is often called the **interquartile range**, or the **mid-spread**. This includes only the middle 50% of the data and, therefore, is not influenced by extreme values. Thus, it is sometimes used as an alternative measure of dispersion instead of the standard deviation.

Stem-and-Leaf Displays

Another useful tool for visually displaying data is a **stem-and-leaf display** (see *PHStat Note: Creating Stem-and-Leaf Displays*). The concept behind the stem-and-leaf display is to classify the data into cells, similar to a histogram, but at different levels of aggregation as defined by the *stem unit*. Each observation is represented by two

PHSTAT NOTE
Creating Box-and-Whisker Plots

From the *PHStat* menu, select *Descriptive Statistics* then *Box-and-Whisker Plot*. The dialog box is shown in Figure 2.33. In the *Raw Data Cell Range* box, enter the range of the data; if the first cell contains a label, check the box below. For a single data set, check the *Single Group Variable* radio button. For multiple groups of data, check the appropriate button (see the *PHStat* note on stacked and unstacked data in Chapter 1). In the *Output Options* section, you may enter a title for the chart. Checking the *Five-Number Summary* box will provide a worksheet with the minimum, first quartile, median, third quartile, and maximum values of the data set(s).

Figure 2.33   *Box-and-Whisker Plot* Dialog

Figure 2.34 Box-and-Whisker Plot for Facebook Friends

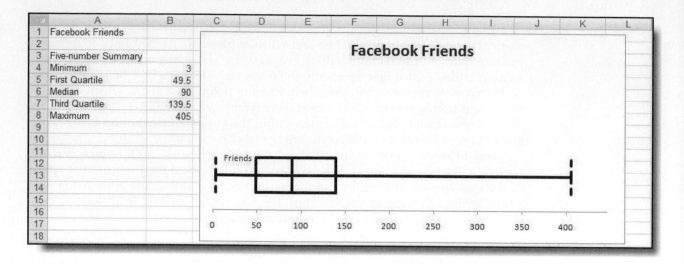

numbers: $x \mid y$, where x is the stem and y is the leaf. The stem indicates the cell, while the leaf indicates the value within the cell. For example, consider the numbers 117, 113, 124, and 126. If we define the stem to be the first two digits (i.e., a stem unit of 10), then each number can be represented as:

NUMBER	STEM | LEAF
117	11 | 7
113	11 | 3
124	12 | 4
125	12 | 6

A stem-and-leaf display aggregates and sorts all leaves within the same stem, for example:

11 | 37
12 | 46

PHSTAT NOTE
Creating Stem-and-Leaf Displays

From the *PHStat* menu, select *Descriptive Statistics* and then *Stem-and-Leaf Display*. The dialog box is shown in Figure 2.35. Enter the range of the data in the first box, checking the *First cell contains label* box if appropriate. You may have the tool automatically calculate the stem unit, or you may specify it as a power of 10. For example, if the numbers are large, say in the range of 800 to 900, then you might wish to specify the stem unit as 100 so the stem values are 800, 900, and so on, and leaves represent values after the first digit. The stem-and-leaf routine also provides summary statistics if the box is checked.

Figure 2.35 *PHStat Stem-and-Leaf Display* Dialog

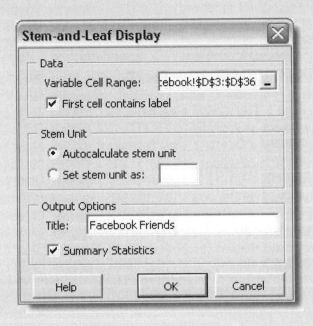

The stem unit is a power of 10; the higher the stem unit, the higher the degree of aggregation of the data. A stem unit of 1 (that is, 10^0) creates essentially a histogram of the individual observations. Zeros are used to indicate the number of observations (leaves) for each value (stem). Turned sideways, this looks like an ordinary histogram. A stem unit of 10 (as illustrated earlier) defines the last digit as the leaf and sets all other leading digits as the stem.

Figure 2.36 shows a stem-and-leaf display for the Facebook Friends data using a stem unit of 100. Note that each leaf corresponds to the last two digits of the

Figure 2.36 Stem-and-Leaf Display Results

	A	B	C	D	E
1				**Facebook Friends**	
2					
3				Stem unit 100	
4					
5	**Statistics**			0	0 2 2 3 4 4 5 5 5 6 6 7 7 7 7 8 9 9
6	**Sample Size**	33		1	0 0 0 1 2 2 2 6 6
7	**Mean**	113.4242		2	2 3 5 6
8	**Median**	90		3	4
9	**Std. Deviation**	94.07644		4	0
10	**Minimum**	3			
11	**Maximum**	405			

observations. For example, there are four observations in the 200 range: 224, 231, 250, and 260. Each leaf value rounds up the last two digits to the 10s value: 24 becomes 2, 31 becomes 3, 50 becomes 5, and 60 becomes 6. Thus, in the stem-and-leaf display at this level of aggregation, we do not know the exact values of the observations. To estimate the values, multiply the stems by the stem unit then add the leaves multiplied by 1 if the stem is positive or −1 if the stem is negative to the next 10s digit. Thus,

$$2 \mid 2\ 3\ 5\ 6$$

provides estimates of 220, 230, 250, and 260.

Dot-Scale Diagrams

A **dot-scale diagram** is another visual display that shows a histogram of data values as dots corresponding to individual data points, along with the mean, median, first and third quartiles, and ±1, 2, and 3 standard deviation ranges from the mean. The mean essentially acts as a fulcrum as if the data were balanced along an axis. Figure 2.37 shows a dot-scale diagram for Facebook Friends data generated by *PHStat* from the *Descriptive Statistics* menu item. Dot-scale diagrams provide a better visual picture and understanding of the data than either box-and-whisker plots or stem-and-leaf displays.

Visual displays such as box-and-whisker plots, stem-and-leaf displays, and dot-scale diagrams give more complete pictures of data sets. They are highly useful tools in exploring the characteristics of data before computing other statistical measures.

Figure 2.37   Dot-Scale Diagram

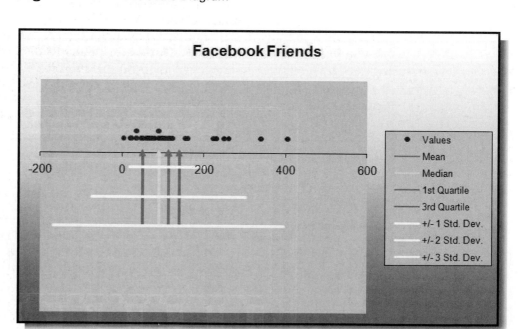

DESCRIPTIVE STATISTICS FOR CATEGORICAL DATA

Statistics such as means and variances are not appropriate for categorical data. Instead, we are generally interested in the fraction of data that have a certain characteristic. The formal statistical measure is called the **proportion**. Proportions are key descriptive statistics for categorical data, such as defects or errors in quality control applications or consumer preferences in market research. For example, in the *Facebook Survey* Excel file, column F lists the gender of each respondent. The proportion of females is $20/33 = .606$. The Excel function =COUNTIF (*data range, criteria*) is useful in determining how many observations meet specified characteristics. For instance, to find the number of females, we used the function =COUNTIF (F4:F36, "female"). The criterion field can also be numerical, such as ">15" or "=0" and so on.

One of the most basic statistical tools used to summarize categorical data and examine the relationship between two categorical variables is cross-tabulation. A **cross-tabulation** is a tabular method that displays the number of observations in a data set for different subcategories of two categorical variables. A cross-tabulation table is often called a **contingency table**. The subcategories of the variables must be mutually exclusive and exhaustive, meaning that each observation can be classified into only one subcategory and, taken together over all subcategories, they must constitute the complete data set. Excel PivotTables, which are described in Chapter 1, provide an easy method of constructing cross-tabulations. The *PHStat* tool *Two-Way Tables & Charts*, described in Chapter 1, provides a quick method of creating cross-tabulations in the form of Excel PivotTables (see *PHStat Note: Two-Way Tables & Charts*).

PHSTAT NOTE
Two-Way Tables & Charts

Click on *PHStat* from the *Menu Commands* group under the *Add-Ins* tab. In the *Descriptive Statistics* menu, select *Two-Way Tables & Charts*. In the dialog (see Figure 2.38) enter the ranges of the categorical variables for the contingency table. If you want a bar chart to display the results, check the box.

Figure 2.38   *Two-Way Tables & Charts* Dialog

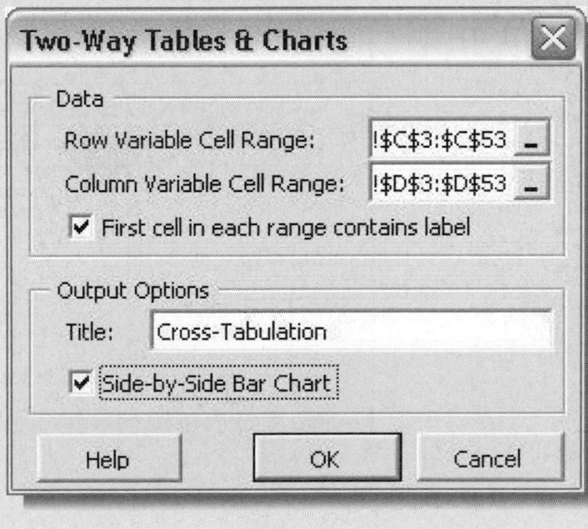

Figure 2.39 Portion of the Excel File *Social Networking*

	A	B	C	D	E	F	G
1	Social Networking						
2							
3	Gender	Age	Year	Facebook	Myspace	Hrs. checking E-mail/week	Hours online/week
4	F	18	Fr	Y	Y	5	10
5	F	19	Fr	Y	Y	8	15
6	F	20	Jr	Y	Y	14	30
7	F	23	Sr	Y	Y	5	7
8	F	21	So	Y	Y	10	20
9	F	22	Sr	Y	Y	20	30
10	F	21	Jr	Y	Y	14	20

To illustrate, consider the Excel file *Social Networking*, shown in Figure 2.39. This provides another set of data from a sample of students about their social networking habits. A contingency table created with *PHStat* for the variables Year and Facebook is shown in Figure 2.40. This clearly suggests that younger students, such as freshmen and sophomores, have greater interest in Facebook than upperclassmen and graduate students. A bar chart visualizing these results from the *PHStat* tool is shown in Figure 2.41. Cross-tabulations are commonly used in marketing research to provide insight into characteristics of different market segments using categorical variables such as gender, educational level, marital status, and so on.

Figure 2.40 Cross-Tabulation of Year versus Facebook

	A	B	C	D
1	Cross-Tabulation			
2				
3	Count of Year	Facebook		
4	Year	N	Y	Grand Total
5	Fr	1	10	11
6	G	7	1	8
7	Jr	5	3	8
8	So	0	13	13
9	Sr	5	5	10
10	Grand Total	18	32	50

Figure 2.41 Bar Chart of Cross-Tabulation

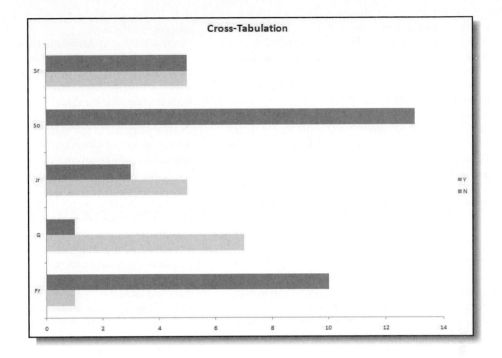

Basic Concepts Review Questions[2]

1. List the different types of charts available in Excel, and explain characteristics of data sets that make each chart most appropriate to use.
2. What types of chart would be best for displaying the data in each of the following data sets on the CD-ROM? If several charts are appropriate, state this, but justify your best choice.
   a. *Mortgage Rates*
   b. *Census Education Data*
   c. *Consumer Transportation Survey*
   d. *MBA Student Survey*
   e. *Vacation Survey*
   f. *Washington, DC, Weather*
3. Explain the principal types of descriptive statistics measures that are used for describing data.
4. Explain the difference between the mean, median, mode, and midrange. In what situations might one be more useful than the others?
5. What statistical measures are used for describing dispersion in data? How do they differ from one another?
6. Explain the importance of the standard deviation in interpreting and drawing conclusions about risk.
7. What does Chebyshev's theorem state and how can it be used in practice?
8. Explain the value of the coefficient of variation.
9. What are frequency distributions and histograms? What information do they provide?
10. Explain how to compute the relative frequency and cumulative relative frequency.
11. Explain the concepts of skewness and kurtosis and what they tell about the distribution of data.
12. Explain the concept of correlation and how to interpret correlation coefficients of 0.3, 0, and –0.95.
13. Explain how to compute the mean and variance of a sample and a population. How would you explain the formulas in simple English?
14. How can one estimate the mean and variance of data that are summarized in a grouped frequency distribution? Why are these only estimates?
15. Explain the concept of covariance. How is covariance used in computing the correlation coefficient?
16. Explain the information contained in box-and-whisker plots, stem-and-leaf displays, and dot-scale diagrams.
17. What is a proportion? Provide some practical examples where proportions are used in business.
18. What is a cross-tabulation? How can it be used by managers to provide insight about data, particularly for marketing purposes?

---

[2] Questions 13–15 require information presented in the appendix to this chapter.

SKILL-BUILDING EXERCISES

1. Create the column chart shown in Figure 2.1 for the *EEO Employment Report* data.
2. Create line charts for U.S. Imports from China and Chinese Exports and Imports in the Excel file *China Trade Data*.
3. Create a pie chart showing the breakdown of students in the *Social Networking* Excel file that are freshmen, sophomores, juniors, seniors, and graduate students.
4. Create a scatter diagram showing the relationship between Hours online/week and Log-ins/day in the *Facebook Survey* data.
5. Create a bubble chart for the data in the Excel file *Sales Data* for which the *x*-axis is the Industry Code, *y*-axis is Competitive Rating, and bubbles represent the average profit associated with each combination of Industry Code and Competitive Rating. (Hint: Use PivotTables to process the data.)
6. Use the *Descriptive Statistics* tool to summarize the numerical data in the Excel file *Social Networking*.
7. Compute the coefficients of variation for the numerical data in the Excel file *Facebook Survey*.
8. Construct frequency distributions and histograms for the numerical data in the Excel file *Social Networking*. Also, compute the relative frequencies and cumulative relative frequencies. Plot the cumulative relative frequencies on a line chart similar to Figure 2.23.
9. Use the FREQUENCY function to construct a frequency distribution for the data in the *Facebook Survey* Excel file.
10. Use the *Rank and Percentile* tool to generate the results shown in Figure 2.27. Then add a new column to the spreadsheet that computes the correct percentiles for the data.
11. Compute the correlations between all pairs of numerical variables in the Excel file *Major League Baseball*. Drawing upon your knowledge of the game, explain why the results make sense.
12. Recognizing that the data in the Excel file *Major League Baseball* represents population data, develop a spreadsheet to calculate the mean, variance, and standard deviation using formulas 2A.2, 2A.4, and 2A.6 in the appendix to this chapter.
13. Using the frequency distributions developed for Hours online/week and Friends data in the *Facebook Survey* Excel file, develop a spreadsheet to calculate the sample variance using formula (2A.11) in the Appendix to this chapter.
14. Construct a frequency distribution for the number of vacations per year in the Excel file *Vacation Survey*. Use the frequency distribution to calculate the mean and standard deviation and verify your calculations using appropriate Excel functions.
15. Construct box-and-whisker plots, stem-and-leaf diagrams, and dot-scale diagrams for the numerical data in the Excel file *Social Networking*.
16. Use the COUNTIF function as described in the text to find the proportions of freshmen, sophomores, juniors, seniors, and graduate students in the sample data in the *Social Networking* Excel file.
17. Develop a cross-tabulation and chart of the categorical variables Student and Gender in the *Facebook Survey* Excel file using the *PHStat Two-Way Tables & Charts* tool.

PROBLEMS AND APPLICATIONS

1. Construct a column chart for the data in the Excel file *State Unemployment Rates* to allow comparison of the June rate with the historical highs and lows. Would any other charts be better to visually convey this information? Why or why not?
2. Data from the 2000 U.S. Census show the following distribution of ages for residents of Ohio:

TOTAL HOUSEHOLDS	4,445,773
Family households (families)	2,993,023
With own children under 18 years	1,409,912
Married-couple family	2,285,798
With own children under 18 years	996,042
Female householder, no husband present	536,878
With own children under 18 years	323,095
Nonfamily households	1,452,750
Householder living alone	1,215,614
Householder 65 years and over	446,396

a. Construct a column chart to visually represent these data.

b. Construct a stacked bar chart to display the subcategories where relevant. (Note that you will have to compute additional subcategories, for instance, under Family households, the number of families without children under 18, so that the total of the subcategories equals the major category total. The sum of all categories does not equal the total.)

c. Construct a pie chart showing the proportion of households in each category.

3. The Excel file *Energy Production & Consumption* provides various energy data from 1949 through 2007.

a. Construct an area chart showing the fossil fuel production as a proportion of total energy production.

b. Construct line charts for each of the variables.

c. Construct a line chart showing both the total energy production and consumption during these years.

d. Construct a scatter diagram for total energy exports and total energy production.

e. Discuss what information the charts convey.

4. The Excel file *Colleges and Universities* provides data on top liberal arts colleges and research universities.

a. Construct appropriate charts that will allow you to compare any differences between liberal arts colleges and research universities (hint: first use PivotTables).

b. What conclusions can you draw from these charts?

5. The Excel file *Freshman College Data* provides data from different colleges and branch campuses within one university over four years.

a. Construct appropriate charts that allow you to contrast the differences among the colleges and branch campuses.

b. Write a report to the academic vice president explaining the information.

6. Construct whatever charts you deem appropriate to convey comparative information on the two categories of televisions in the Excel file *Hi-Definition Televisions*. What conclusions can you draw from these?

7. Construct whatever charts you deem appropriate to convey comparative information on deaths by major causes in the Excel file *Death Cause Statistics*. What conclusions can you draw from these?

8. Construct an appropriate chart to show the proportion of funds in each investment category in the Excel file *Retirement Portfolio*.

9. The Excel file *Baseball Attendance* shows the attendance in thousands at San Francisco Giants baseball games for the 10 years before the Oakland A's moved to the Bay Area in 1968, as well as the combined attendance for both teams for the next 11 years. What is the mean and standard deviation of the number of baseball fans attending before and after the A's move to the San Francisco area? What conclusions might you draw?

10. The Excel file *Seattle Weather* contains weather data for Seattle, Oregon. Apply the *Descriptive Statistics* tool to these data. Show that Chebyshev's theorem holds the average temperature and rainfall.

11. For the Excel file *University Grant Proposals*, compute descriptive statistics for all proposals and also for the proposals that were funded and those that were rejected. Are any differences apparent?

12. Compute descriptive statistics for liberal arts colleges and research universities in the Excel file *Colleges and Universities*. Compare the two types of colleges. What can you conclude?

13. Compute descriptive statistics for all colleges and branch campuses for each year in the Excel file *Freshman College Data*. Are any differences apparent from year to year?

14. Find the first and third quartiles for each of the performance statistics in the Excel file *Ohio Education Performance.*
15. Find the 10th and 90th percentile of home prices in the Excel file *Home Market Value.*
16. A deep-foundation engineering contractor has bid on a foundation system for a new world headquarters building for a Fortune 500 company. A part of the project consists of installing 311 auger cast piles. The contractor was given bid information for cost-estimating purposes, which consisted of the estimated depth of each pile; however, actual drill footage of each pile could not be determined exactly until construction was performed. The Excel file *Pile Foundation* contains the estimates and actual pile lengths after the project was completed. Compute the correlation coefficient between the estimated and actual pile lengths. What does this tell you?
17. Call centers have high turnover rates because of the stressful environment. The national average is approximately 50%. The director of human resources for a large bank has compiled data from about 70 former employees at one of the bank's call centers (see the Excel file *Call Center Data*). For this sample, how strongly is length of service correlated with starting age?
18. A national homebuilder builds single-family homes and condominium-style townhouses. The Excel file *House Sales* provides information on the selling price, lot cost, type of home, and region of the country (M = Midwest, S = South) for closings during one month.
    a. Construct a scatter diagram showing the relationship between sales price and lot cost. Does there appear to be a linear relationship? Compute the correlation coefficient.
    b. Construct scatter diagrams showing the relationship between sales price and lot cost *for each region.* Do linear relationships appear to exist? Compute the correlation coefficients.
    c. Construct scatter diagrams showing the relationship between sales price and lot cost for each type of house. Do linear relationships appear to exist? Compute the correlation coefficients.
19. The Excel file *Salary Data* provides information on current salary, beginning salary, previous experience in months when hired, and total years of education for a sample of 100 employees in a firm. Find the correlation matrix for these data. What conclusions can you draw?
20. The Excel file *Infant Mortality* provides data on infant mortality rate (deaths per 1,000 births), female literacy (percent who read), and population density (people per square kilometer) for 85 countries. Compute the correlation matrix for these three variables. What conclusions can you draw?
21. The Excel file *Refrigerators* provides data on various brands and models. Compute the correlation matrix for the variables. What conclusions can you draw?
22. Compute the mean, variance, and standard deviation for all the variables in the Excel file *National Football League.* Note that the data represent a population. If you had applied the *Descriptive Statistics* tool, what differences would you have encountered?
23. Data obtained from a county auditor (see the Excel file *Home Market Value*) provides information about the age, square footage, and current market value of houses along one street in a particular subdivision.
    a. Considering these data as describing the population of homeowners on this street, compute the mean, median, variance, and standard deviation for each of these variables using the formulas presented in the appendix to this chapter.
    b. Compute the coefficient of variation for each variable. Which has the least and greatest relative dispersion?

24. A community health status survey obtained the following demographic information from the respondents:

AGE	FREQUENCY
18 to 29	297
30 to 45	661
46 to 64	634
65+	369

Compute the relative frequency and cumulative relative frequency of the age groups. Also, estimate the average age of the sample of respondents. What assumptions do you have to make to do this?

25. A marketing study of 800 adults in the 18–34 age group reported the following information:
  - Spent less than $100 on children's clothing per year: 9.3%
  - Spent $100–$499 on children's clothing per year: 24.6%
  - Spent $500–$999 on children's clothing per year: 3%
  - The remainder reported spending nothing. Estimate the sample mean and sample variance of spending on children's clothing for this age group.

26. Data from the 2000 U.S. Census in the Excel file *California Census Data* show the distribution of ages for residents of California. Estimate the mean age and standard deviation of age for California residents, assuming these data represent a sample of current residents.

27. The data in the Excel file *Church Contributions* were reported on annual giving for a church. Estimate the mean and standard deviation of the annual contributions, assuming these data represent the entire population of parishioners.

28. Construct box-and-whisker plots and dot-scale diagrams for each of the variables in the data set *Ohio Education Performance*. What conclusions can you draw from them?

29. A producer of computer-aided design software for the aerospace industry receives numerous calls for technical support. Tracking software is used to monitor response and resolution times. In addition, the company surveys customers who request support using the following scale:
  - 0—Did not exceed expectations
  - 1—Marginally met expectations
  - 2—Met expectations
  - 3—Exceeded expectations
  - 4—Greatly exceeded expectations

The questions are as follows:
  - Q1: Did the support representative explain the process for resolving your problem?
  - Q2: Did the support representative keep you informed about the status of progress in resolving your problem?
  - Q3: Was the support representative courteous and professional?
  - Q4: Was your problem resolved?
  - Q5: Was your problem resolved in an acceptable amount of time?
  - Q6: Overall, how did you find the service provided by our technical support department?

A final question asks the customer to rate the overall quality of the product using this scale:

0—Very poor
1—Poor
2—Good
3—Very good
4—Excellent

A sample of survey responses and associated resolution and response data are provided in the Excel file *Customer Support Survey.* Use box-and-whisker plots, stem-and-leaf displays, and dot-scale diagrams as you deem appropriate to visually convey these sample data and write a report to the manager explaining your findings and conclusions.

30. The Excel file *EEO Employment Report* shows the number of people employed in different professions for various racial and ethnic groups. Find the proportion of men and women in each ethnic group for the total employment and in each profession.

31. The Excel file *Unions and Labor Law Data* reports the percent of public and private sector employees in unions in 1982 for each state, along with indicators of whether the states had a bargaining law that covered public employees or right-to-work laws.

    a. Compute the proportion of employees in unions in each of the four categories: public sector with bargaining laws, public sector without bargaining laws, private sector with bargaining laws, and private sector without bargaining laws.

    b. Compute the proportion of employees in unions in each of the four categories: public sector with right-to-work laws, public sector without right-to-work laws, private sector with right-to-work laws, and private sector without right-to-work laws.

    c. Construct a cross-tabulation of the number of states within each classification of having or not having bargaining laws and right-to-work laws.

32. A mental health agency measured the self-esteem score for randomly selected individuals with disabilities who were involved in some work activity within the past year. The Excel file *Self Esteem* provides the data, including the individuals' marital status, length of work, type of support received (direct support includes job-related services such as job coaching and counseling), education, and age. Construct a cross-tabulation of the number of individuals within each classification of marital status and support level.

CASE

The Malcolm Baldrige National Quality Award

The Malcolm Baldrige National Quality Award recognizes U.S. companies that excel in high-performance management practice and have achieved outstanding business results. The award is a public–private partnership, funded primarily through a private foundation and administered through the National Institute of Standards and Technology (NIST) in cooperation with the American Society for Quality (ASQ). It was created to increase the awareness of American business for quality and good business practices and has become a worldwide standard for business excellence. See the National Quality Program Web site at www.baldrige.nist.gov for more information.

The award examination is based on a rigorous set of criteria, called the *Criteria for Performance Excellence*, which consists of seven major categories: Leadership; Strategic Planning; Customer and Market Focus;

Measurement, Analysis, and Knowledge Management; Workforce Focus; Process Management; and Business Results. Each category consists of several *items* that focus on major requirements on which businesses should focus. For example, the two items in the Leadership category are Senior Leadership and Governance and Social Responsibilities. Each item, in turn, consists of a small number of *areas to address*, which seek specific information on approaches used to ensure and improve competitive performance, the deployment of these approaches, or results obtained from such deployment. The current year's criteria may be downloaded from the Web site.

Applicants submit a 50-page document that describes their management practices and business results that respond to the criteria. The evaluation of applicants for the award is conducted by a volunteer board of examiners selected by NIST. In the first stage, each application is reviewed by a team of examiners. They evaluate the applicant's response to each criteria item, listing major strengths and opportunities for improvement relative to the criteria. Based on these comments, a score from 0 to 100 in increments of 10 is given to each item. Scores for each examination item are computed by multiplying the examiner's score by the maximum point value that can be earned for that item, which varies by item. These point values weight the importance of each item in the criteria. Then the examiners share information on a secure Web site and discuss

issues via telephone conferencing to arrive at consensus comments and scores. The consensus stage is an extremely important step of the process. It is designed to smooth out variations in examiners' scores, which inevitably arise because of different perceptions of the applicants' responses relative to the criteria, and provide useful feedback to the applicants. In many cases, the insights of one or two judges may sway opinions, so consensus scores are not simple averages. A national panel of judges then reviews the scores and selects the highest-scoring applicants for site visits. At this point, a team of examiners visits the company for the greater part of a week to verify information contained in the written application and resolve issues that are unclear or about which the team needs to learn more. The results are written up and sent to the judges who use the site visit reports and discussions with the team leaders to recommend award recipients to the Secretary of Commerce.

Statistics and data analysis tools can be used to provide a summary of the examiners' scoring profiles and to help the judges review the scores. Figure 2.42 illustrates a hypothetical example (Excel file *Baldrige*).[3] Your task is to apply the concepts and tools discussed in this chapter to analyze the data and provide the judges with appropriate statistical measures and visual information to facilitate their decision process regarding a site visit recommendation.

Figure 2.42 Baldrige Examination Scores

	A	B	C	D	E	F	G	H	I	J	K
1	Baldrige Examination Scores										
2											
3					Individual Assessment Percentage Scores						
4	Item	Maximum	Examiner	Examiner	Examiner	Examiner	Examiner	Examiner	Examiner	Examiner	Consensus
5		Points	1	2	3	4	5	6	7	8	Score
6	1.1	70	80	80	50	60	60	70	70	50	75
7	1.2	50	30	50	30	40	40	60	60	50	50
8	2.1	40	50	70	50	50	40	60	70	40	65
9	2.2	45	30	40	50	50	60	40	30	50	55
10	3.1	40	30	60	40	60	50	30	50	30	45
11	3.2	45	30	50	60	60	60	50	30	60	50
12	4.1	45	40	70	50	60	40	30	20	50	50
13	4.2	45	30	20	40	40	30	30	10	30	40
14	5.1	45	70	50	60	40	40	60	60	50	60
15	5.2	40	50	20	40	40	70	40	40	20	40
16	6.1	35	50	60	50	50	50	40	30	40	45
17	6.2	50	40	40	60	50	40	30	60	50	50
18	7.1	100	60	70	70	70	80	70	70	70	75
19	7.2	70	50	60	70	50	70	50	70	70	70
20	7.3	70	50	40	50	50	70	30	30	50	50
21	7.4	70	40	50	50	50	50	40	20	60	45
22	7.5	70	70	70	60	70	50	60	80	50	75
23	7.6	70	60	80	70	60	70	40	60	70	70
24	Weighted score		499.5	565.5	546.5	543	564	478.5	503	523	585

---

[3] The criteria undergo periodic revision, so the items and maximum points will not necessarily coincide with the current year's criteria.

Appendix

Descriptive Statistics: Theory and Computation

In this appendix, we discuss some basic theory and mathematical basis for descriptive statistics calculations. While most business applications deal with sample data, there are times when you will need to calculate statistical information for population data. Calculations of some statistical measures differ between samples and populations, and it is important to understand the differences.

Mean, Variance, and Standard Deviation

It is common practice in statistics to use Greek letters to represent population measures and Roman letters to represent sample statistics. We will use N to represent the number of items in a population, and n to represent the number of observations in a sample. The mean of a sample of n observations, $x_1, \ldots, x_n$, denoted by "x-bar" is calculated as:

$$\bar{x} = \frac{\sum_{i=1}^{n} x_i}{n} \tag{2A.1}$$

If a population consists of N observations $x_1, \ldots, x_N$, population mean, μ, is calculated as:

$$\mu = \frac{\sum_{i=1}^{N} x_i}{N} \tag{2A.2}$$

Note that the calculations for the mean are the same; only the notation differs between a sample and a population. One property of the mean is that the sum of the deviations of each observation from the mean is zero:

$$\sum_{i}(x_i - \bar{x}) = 0 \tag{2A.3}$$

The formula for the variance of a population is:

$$\sigma^2 = \frac{\sum_{i=1}^{N}(x_i - \mu)^2}{N} \tag{2A.4}$$

where x_i is the value of the ith item, N is the number of items in the population, and μ is the population mean. Essentially, the variance is the average of the squared deviations of the observations from the mean.

A major difference exists between the variance of a population and that of a sample. The variance of a sample is calculated using the formula:

$$s^2 = \frac{\sum_{i=1}^{n}(x_i - \bar{x})^2}{n - 1} \tag{2A.5}$$

where n is the number of items in the sample, and $\bar{x}$ is the sample mean. It may seem peculiar to use a different denominator to "average" the squared deviations from the mean for populations and samples, but statisticians have shown that the formula for the sample variance provides a more accurate representation of the true population variance. We will discuss this more formally in Chapter 4. For now, simply understand that the proper calculations of the population and sample variance use different denominators based on the number of observations in the data.

The standard deviation is the square root of the variance. For a population, the standard deviation is computed as:

$$\sigma = \sqrt{\frac{\sum_{i=1}^{N}(x_i - \mu)^2}{N}} \tag{2A.6}$$

and for samples, it is:

$$s = \sqrt{\frac{\sum_{i=1}^{n}(x_i - \bar{x})^2}{n - 1}} \tag{2A.7}$$

Statistical Measures for Grouped Data

When sample data are summarized in a frequency distribution, the mean may be computed using the formula:

$$\bar{x} = \frac{\sum_{i=1}^{n} f_i x_i}{n} \tag{2A.8}$$

where f_i is the frequency of observation x_i. For populations, the formula is similar:

$$\mu = \frac{\sum_{i=1}^{N} f_i x_i}{N} \tag{2A.9}$$

To illustrate this, consider the Hours online/week in the Facebook data. The calculations are shown below:

OBSERVATION	FREQUENCY	OBSERVATION × FREQUENCY
0	1	0
1	9	9
2	11	22
3	3	9
4	3	12
5	1	5
6	2	12
7	3	21
Sum	33	90

Mean = 90/33 = 2.727273

If the data are grouped into k cells in a frequency distribution, we can use modified versions of these formulas to estimate the mean by replacing x_i with a representative value (such as the midpoint) for all the observations in each cell. Thus, using the Facebook Friends data, we would have:

UPPER LIMIT	MIDPOINT	FREQUENCY	MIDPOINT × FREQUENCY
50	25	8	200
100	75	12	900
150	125	5	625
200	175	2	350
250	225	3	675
300	275	1	275
350	325	1	325
400	375	0	0
450	425	1	425
	Sum	33	3775

Estimate of mean = 3375/33 = 114.394

Note that this is not identical, but is very close, to the true mean of 113.424. This is because we have not used all the original data, but only representative values for each cell. Although most statistics are simple concepts, they must be applied correctly, and we need to understand how to interpret them properly.

We may use similar formulas to compute the population variance for grouped data:

$$\sigma^2 = \frac{\sum_{i=1}^{N} f_i(x_i - \mu)^2}{N} \qquad (2A.10)$$

and sample variance:

$$s^2 = \frac{\sum_{i=1}^{n} f_i(x_i - \bar{x})^2}{n - 1} \qquad (2A.11)$$

An exercise at the end of this chapter will ask you to apply these formulas.

Skewness and Kurtosis

The coefficient of skewness is computed as:

$$CS = \frac{\frac{1}{N}\sum_{i=1}^{N}(x_i - \mu)^3}{\sigma^3} \qquad (2A.12)$$

For sample data, replace the population mean and standard deviation with the corresponding sample statistics.

The **coefficient of kurtosis** is computed as:

$$CK = \frac{\frac{1}{N}\sum_{i=1}^{N}(x_i - \mu)^4}{\sigma^4} \qquad (2A.13)$$

Again, for sample data, use the sample statistics instead of the population measures.

Correlation

The correlation coefficient for a population is computed as:

$$\rho_{xy} = \frac{\text{cov}(X, Y)}{\sigma_x \sigma_y} \qquad (2A.14)$$

The numerator is called the **covariance** and is the average of the products of deviations of each observation from its respective mean:

$$\text{cov}(X, Y) = \frac{\sum_{i=1}^{N}(x_i - \mu_x)(y_i - \mu_y)}{N} \qquad (2A.15)$$

To understand this, examine the formula for the covariance. This is the average of the product of the deviations of each pair of observations from their respective means. Suppose that large (small) values of X are generally associated with large (small) values of Y. Then in most cases, both x_i and y_i are either above or below their respective means. If so, the product of the deviations from the means will be a positive number and when added together and averaged will give a positive value for the covariance. On the other hand, if small (large) values of X are associated with large (small) values of Y, then one of the deviations from the mean will generally be negative while the other is positive. When multiplied together, a negative value results, and the value of the covariance will be negative. The Excel function COVAR computes the covariance of a population.

In a similar fashion, the **sample correlation coefficient** is computed as:

$$r_{xy} = \frac{\text{cov}(X, Y)}{s_x s_y} \qquad (2A.16)$$

However, the sample covariance is computed as:

$$\text{cov}(X, Y) = \frac{\sum_{i=1}^{n}(x_i - \bar{x})(y_i - \bar{y})}{n - 1} \qquad (2A.17)$$

Like the sample variance, note the use of $n - 1$ in the denominator. Unfortunately, Excel does not have parallel functions to COVAR for sample data. Thus, to compute the covariance for sample data, one must multiply the results obtained using the COVAR function by $n/(n - 1)$ to obtain the correct value.

Chapter 3

Probability Distributions and Applications

INTRODUCTION

*M*ost business decisions involve some elements of uncertainty and randomness. For example, in models of manufacturing operations, times of job arrivals, job types, processing times, times between machine breakdowns, and repair times all involve uncertainty. Similarly, a model to predict the future return of an investment portfolio requires a variety of assumptions about uncertain economic conditions and market behavior. Specifying the nature of such assumptions is a key modeling task that relies on fundamental knowledge of random variables and probability distributions—the subject of this chapter. Random variables and probability distributions are also important in applying statistics to analyze sample data from business processes because sample data are usually assumed to stem from some underlying probability distribution. Thus, we will also examine characteristics of sampling distributions in this chapter and discuss errors associated with sampling. The key concepts and tools that we will study are as follows:

➤ *The notion of probability and a random variable*

➤ *Basic rules for computing probabilities*

➤ *Useful families of discrete and continuous probability distributions*

➤ *Joint, marginal, and conditional probability distributions*

➤ *The concept of random numbers and Monte Carlo simulation methods for random sampling from probability distributions*

➤ *Sampling distributions and sampling error*

PROBABILITY: CONCEPTS AND APPLICATIONS

The notion of probability is used every day, from the World Series of Poker, to weather forecasts, to market research, to stock market predictions. Probability quantifies the variation and uncertainty that we encounter in business and in our daily lives, and is an important element of analytical modeling and decision making.

An **experiment** is the act of gathering data through observation. An experiment might be as simple as rolling dice, observing and recording weather conditions, conducting a market research study, or watching the stock market. The **outcome** of an experiment is a result that we observe; it might be the sum of two dice, a qualitative characterization of the weather, the proportion of consumers who favor a new product, or the change in the Dow Jones Industrial Average (DJIA) at the end of a week. The collection of all possible outcomes of an experiment is called the **sample space**. For instance, if we roll two fair dice, the possible outcomes are the numbers 2 through 12; if we observe the weather, the outcome might be clear, partly cloudy, or cloudy; the outcomes for customer reaction to a new product in a market research study would be favorable or unfavorable; and the weekly change in the DJIA can theoretically be any positive or negative real number. Note that a sample space may consist of a small number of discrete outcomes or an infinite number of outcomes. Probability is the likelihood that an outcome occurs. Two basic facts govern probability:

1. The probability associated with any outcome must be between 0 and 1.
2. The sum of the probabilities over all possible outcomes must be 1.0.

Probability may be defined from one of three perspectives. First, if the process that generates the outcomes is known, probabilities can be deduced from theoretical arguments; this is the *classical definition* of probability. For example, if we count the possible outcomes associated with rolling two dice, we can easily determine that out of 36 possible outcomes, one outcome will be the number 2, two outcomes will be the number 3 (you can roll a 1 on the first die and 2 on the second, and vice versa), six outcomes will be the number 7, and so on. Thus, the probability of rolling any number is the ratio of the number of ways of rolling that number to the total number of possible outcomes. For instance, the probability of rolling a 2 is $1/36$, the probability of rolling a 3 is $2/36 = 1/18$, and the probability of rolling a 7 is $6/36 = 1/6$. Similarly, if two consumers are asked their opinion about a new product, there could be four possible outcomes:

1. (Favorable, Favorable)
2. (Favorable, Not Favorable)
3. (Not Favorable, Favorable)
4. (Not Favorable, Not Favorable)

If these are assumed to be equally likely, the probability that at least one consumer would respond unfavorably is $3/4$.

The second approach to probability, called the *relative frequency definition*, is based on empirical data. For example, a sample of weather in the Excel file *Seattle Weather* shows that on average in January in Seattle, 3 days were clear, 5 were partly cloudy, and 23 were cloudy (see Figure 3.1). Thus, the probability of a clear day in Seattle in January would be computed as $3/31 = 0.097$. As more data become available (and if global weather changes), the distribution of outcomes and hence, the probability may change.

Finally, the *subjective definition* of probability is based on judgment, as financial analysts might do in predicting a 75% chance that the DJIA will increase 10% over the next year, or as sports experts might predict a one-in-five chance (0.20 probability) of a certain team making it to the Super Bowl at the start of the football season.

Which definition to use depends on the specific application. Throughout this book we will see various examples that draw upon each of these perspectives.

Basic Probability Rules

An **event** is a collection of one or more outcomes from a sample space. An event might be rolling a 7 or an 11 with two dice, having a clear or partly cloudy day, or obtaining a positive weekly change in the DJIA.

Figure 3.1 Portion of Excel File *Seattle Weather*

	A	B	C	D	E	F
1	Seattle Weather					
2						
3		Average	Average			
4		Temperature	Rainfall	Clear	Partly Cloudy	Cloudy
5	January	41.3	5.4	3	5	23
6	February	44.3	4	3	6	19
7	March	46.6	3.8	4	8	19
8	April	50.4	2.5	5	9	16
9	May	56.1	1.8	7	10	14
10	June	61.4	1.6	7	8	15
11	July	65.3	0.9	12	10	9
12	August	65.7	1.2	10	10	11
13	September	60.8	1.9	9	8	13
14	October	53.5	3.3	5	8	18
15	November	46.3	5.7	3	6	21
16	December	41.6	6	3	5	23

This leads to the following rule:

Rule 1. The probability of any event is the sum of the probabilities of the outcomes that compose that event.

For example, consider the event of rolling a 7 or 11 on two dice. The probability of rolling a 7 is 6/36 and the probability of rolling an 11 is 3/36; thus, the probability of rolling a 7 or 11 is 6/36 + 3/36 = 9/36. Similarly, the probability of a clear or partly cloudy day in January in Seattle is 3/31 + 5/31 = 8/31.

Two events are **mutually exclusive** if both cannot occur at the same time. Clearly, any two individual outcomes are mutually exclusive.

Rule 2. If events A and B are mutually exclusive, then $P(A \text{ or } B)$ = $P(A) + P(B)$.

To illustrate this, let A be the event "roll a 7 or 11" and B be the event "roll a 2, 3, or 12." These events are mutually exclusive because if we roll a 7 or 11, then clearly we cannot have rolled either a 2, 3, or 12. As we have seen, the probability of event A is $P(A) = 9/36$. The probability of event B is $P(B) = 4/36$. Therefore, the probability that either event A or B occurs, that is, the roll of the dice is either 2, 3, 7, 11, or 12, is 9/36 + 4/36 = 13/36.

If two events are not mutually exclusive, then adding their probabilities would result in double counting some outcomes, so an adjustment is necessary. This leads to the following rule:

Rule 3. If two events A and B are *not* mutually exclusive, then $P(A \text{ or } B)$ = $P(A) + P(B) - P(A \text{ and } B)$.

For example, suppose that A is the event "clear or partly cloudy day in Seattle in January" and B is the event "partly cloudy or cloudy day in Seattle in January." The probability of A is 8/31 and the probability of B is 28/31. A and B are not mutually exclusive because partly cloudy belongs to both events A and B. Thus, the probability that both *A* and *B* occur together, $P(A \text{ and } B)$ is 5/31. Applying Rule 3 yields $P(A \text{ or } B) = 8/31 + 28/31 - 5/31 = 1$. Because the outcome must be either clear, partly cloudy, or cloudy, then obviously one of these two events must occur. Similarly in the dice example, if $A = \{2, 3 \text{ or } 12\}$ and $B = \{\text{odd number}\}$, then $P(A \text{ and } B)$ is the probability of rolling a 3 because 3 is the only element in common to *A* and *B*. "*A* or *B*" must consist of the outcomes $\{2, 3, 5, 7, 9, 11, 12\}$. Thus, $P(A \text{ or } B) = 4/36 + 18/36 - 2/36 = 5/9$.

Random Variables

Some experiments naturally have numerical outcomes, as rolls of dice or the weekly change in the DJIA. For other experiments, such as observing the weather, the sample space is categorical, for example, {clear, partly cloudy, cloudy}. To have a consistent mathematical basis for dealing with probability, we would like the outcomes of all experiments to be numerical. A **random variable** is a numerical description of the outcome of an experiment. Formally, a random variable is a function that assigns a real number to each element of a sample space. If we have categorical outcomes, we can associate an arbitrary numerical value to them, such as 0 = clear, 1 = partly cloudy, and 2 = cloudy, but there is no physical or natural meaning to this scheme. Similarly, a favorable product reaction in a market research study might be assigned a value of 1, and an unfavorable reaction a value of 0. Random variables are usually denoted by capital Roman letters, such as *X* or *Y*.

Random variables may be discrete or continuous. A **discrete random variable** is one for which the number of possible outcomes can be counted. For example, the outcomes of rolling dice, the type of weather for the next day, and customer reactions to a product are discrete random variables. The number of outcomes may be theoretically infinite, such as the number of hits on a Web site link during some period of time—we cannot place a guaranteed upper limit on this number—nevertheless, the outcomes can be counted. A **continuous random variable** has outcomes over a continuous range of real numbers, such as the weekly change in the DJIA. Other examples of continuous random variables include the daily temperature, the time to complete a task, the time to repair a failed machine, and the return on an investment.

PROBABILITY DISTRIBUTIONS

A **probability distribution** is a characterization of the possible values that a random variable may assume along with the probability of assuming these values. We may develop probability distributions using any one of the three perspectives of probability. First, if we can quantify the probabilities associated with the values of a random variable from theoretical arguments, then we can easily define the probability distribution. For example, the probabilities of the outcomes for rolling two dice, calculated by counting the number of ways to roll each number divided by the total number of possible outcomes, along with an Excel column chart depicting the probability distribution are shown in Figure 3.2.

Second, we can calculate the relative frequencies from a sample of empirical data to develop a probability distribution. Figure 3.3 shows the distribution of weather in Seattle in January based on the data in the Excel file *Seattle Weather*. Because this is based on sample data, we usually call this an **empirical probability distribution**. An empirical probability distribution is an approximation of the probability distribution of the associated random variable, whereas the probability distribution of a random variable, such as one derived from counting arguments, is a theoretical model of the random variable.

Figure 3.2 Probability Distribution of Rolls of Two Dice

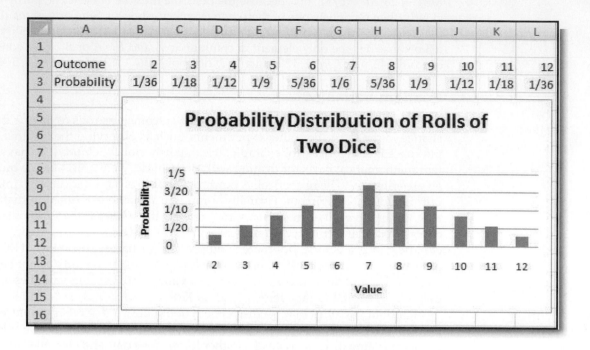

	A	B	C	D	E	F	G	H	I	J	K	L
1												
2	Outcome	2	3	4	5	6	7	8	9	10	11	12
3	Probability	1/36	1/18	1/12	1/9	5/36	1/6	5/36	1/9	1/12	1/18	1/36

Figure 3.3 Empirical Probability Distribution of Seattle Weather

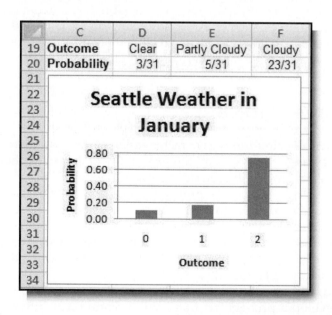

	C	D	E	F
19	Outcome	Clear	Partly Cloudy	Cloudy
20	Probability	3/31	5/31	23/31

Finally, we could simply specify a probability distribution using subjective values and expert judgment. This is often done in creating decision models for phenomena for which we have no historical data. We will see many examples of this in Part II of this book. Figure 3.4 shows a hypothetical example of the distribution of one expert's assessment of the how the DJIA might change in the next year.

Figure 3.4   Subjective Probability Distribution of DJIA Change

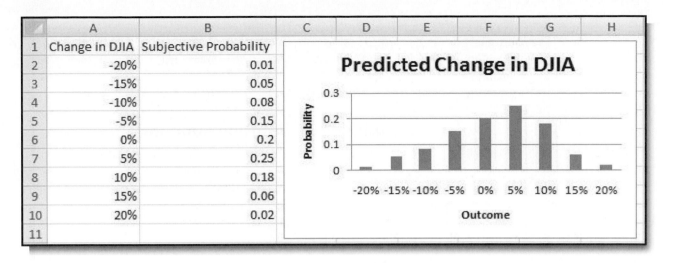

	A	B	C	D	E	F	G	H
1	Change in DJIA	Subjective Probability						
2	-20%	0.01						
3	-15%	0.05						
4	-10%	0.08						
5	-5%	0.15						
6	0%	0.2						
7	5%	0.25						
8	10%	0.18						
9	15%	0.06						
10	20%	0.02						
11								

Discrete Probability Distributions

In each of these examples, the random variable is discrete. For a discrete random variable X, the probability distribution of the discrete outcomes is called a **probability mass function**, and is denoted by a mathematical function $f(x)$. The symbol x_i represents the ith value of the random variable X and $f(x_i)$ is the probability. For instance, in Figure 3.2 for the dice example, $x_1 = 2$ and $f(x_1) = 1/36; x_2 = 3$ and $f(x_2) = 1/18$, and so on. A probability mass function has the properties that (1) the probability of each outcome must be between 0 and 1 and (2) the sum of all probabilities must add to 1; that is:

$$0 \le f(x_i) \le 1 \tag{3.1}$$

$$\sum_i f(x_i) = 1 \tag{3.2}$$

You may easily verify that this holds in each of the examples.

A **cumulative distribution function**, $F(x)$, specifies the probability that the random variable X will assume a value *less than or equal to* a specified value, x. This is also denoted as $P(X \le x)$ and read as "the probability that the random variable X is less than or equal to x." For example, the cumulative distribution function for rolling two dice is shown in Figure 3.5 along with an Excel line chart. To use this, suppose we want to know the probability of rolling a 6 or less. We simply look up the cumulative probability for 6, which is 5/12. Alternatively, we could locate the point for $x = 6$ in the chart and estimate the probability from the graph. Also note that if the probability of rolling a 6 or less is 5/12, then the probability of rolling a 7 or more must be $1 - 5/12 = 7/12$. We can also use the cumulative distribution function to find probabilities over intervals. For example, to find the probability of rolling a number between 4 and 8, $P(4 \le X \le 8)$, we can find $P(X \le 8)$ and subtract $P(X \le 3)$, that is:

$$P(4 \le X \le 8) = P(X \le 8) - P(X \le 3) = 13/18 - 1/12 = 23/36$$

Be careful with the endpoints; because 4 is included in the interval we wish to compute, we need to subtract $P(X \le 3)$, not $P(X \le 4)$!

Figure 3.5 Cumulative Distribution Function for Rolling Dice

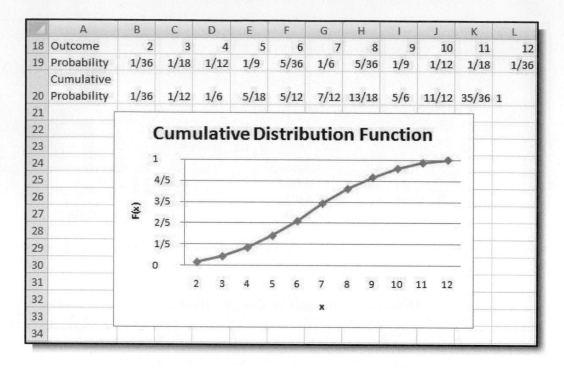

	A	B	C	D	E	F	G	H	I	J	K	L
18	Outcome	2	3	4	5	6	7	8	9	10	11	12
19	Probability	1/36	1/18	1/12	1/9	5/36	1/6	5/36	1/9	1/12	1/18	1/36
20	Cumulative Probability	1/36	1/12	1/6	5/18	5/12	7/12	13/18	5/6	11/12	35/36	1
21												

Continuous Probability Distributions

A continuous random variable has an infinite number of possible outcomes. Suppose that the expert who predicted the probabilities associated with next year's change in the DJIA kept refining the estimates over larger and larger ranges of values. Figure 3.6 shows what such a probability distribution might look like using 2.5% increments rather than 5%. Notice that the distribution is similar in shape to the one in Figure 3.4 but simply has more outcomes. If this refinement process continues, then the distribution will approach the shape of a smooth curve as shown in the figure. Such a curve that characterizes outcomes of a continuous random variable is called a **probability density function**, and is described by a mathematical function $f(x)$. A probability density function has the properties that (1) $f(x) \geq 0$ for all values of x, and (2) the total area under the function above the x-axis is 1.0.

For continuous random variables, it does not make mathematical sense to attempt to define a probability for a specific value of x because there are an infinite number of values; thus, $P(X = x) = 0$. For continuous random variables, probabilities are only defined over intervals, such as $P(a \leq X \leq b)$ or $P(X > c)$. The probability that a random variable will assume a value between a and b is given by the area under the density function between a and b. We shall use this fact in examples later in the chapter.

The cumulative distribution function for a continuous random variable is denoted the same way as for discrete random variables, $F(x)$, and represents the probability that the random variable X is less than or equal to x, $P(X \leq x)$. Intuitively, $F(x)$ represents the area under the density function to the left of x. $F(x)$ can often be derived mathematically from $f(x)$ using techniques of calculus.

Figure 3.6 Refined Probability Distribution of DJIA Change

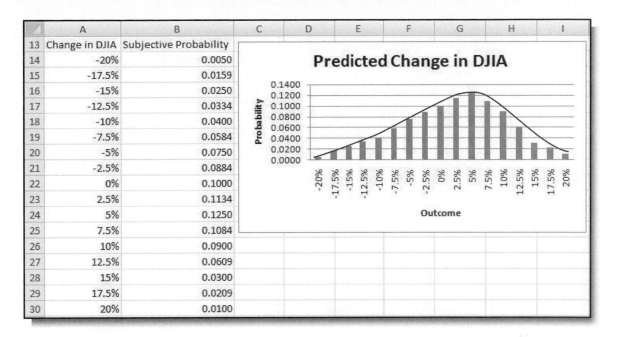

	A	B	C	D	E	F	G	H	I
13	Change in DJIA	Subjective Probability							
14	-20%	0.0050							
15	-17.5%	0.0159							
16	-15%	0.0250							
17	-12.5%	0.0334							
18	-10%	0.0400							
19	-7.5%	0.0584							
20	-5%	0.0750							
21	-2.5%	0.0884							
22	0%	0.1000							
23	2.5%	0.1134							
24	5%	0.1250							
25	7.5%	0.1084							
26	10%	0.0900							
27	12.5%	0.0609							
28	15%	0.0300							
29	17.5%	0.0209							
30	20%	0.0100							

Knowing $F(x)$ makes it easy to compute probabilities over intervals for continuous distributions. The probability that X is between a and b is equal to the difference of the cumulative distribution function evaluated at these two points; that is:

$$P(a \leq X \leq b) = P(X \leq b) - P(X \leq a) - F(b) - F(a) \tag{3.3}$$

For continuous distributions we need not be concerned about the endpoints as we were with discrete distributions because $P(a \leq X \leq b)$ is the same as $P(a < X < b)$.

Continuous probability distributions depend on one or more parameters. Many continuous distributions can assume different shapes and sizes, depending on the value of the parameters, similar to what we will see for the binomial and Poisson. There are three basic types of parameters. A **shape parameter** controls the basic shape of the distribution. For certain distributions, changing the shape parameter will cause major changes in the form of the distribution. For others, the changes will be less severe. A **scale parameter** controls the unit of measurement within the range of the distribution. Changing the scale parameter either contracts or expands the distribution along the horizontal axis. Finally, a **location parameter** specifies the location of the distribution relative to zero on the horizontal axis. The location parameter may be the midpoint or the lower endpoint of the range of the distribution. Not all distributions will have all three parameters; some may have more than one shape parameter. Understanding the effects of these parameters is important in selecting distributions as inputs to decision models.

Expected Value and Variance of a Random Variable

The **expected value** of a random variable corresponds to the notion of the mean, or average, for a sample. For a discrete random variable X, the expected value is the weighted average of all possible outcomes, where the weights are the probabilities. For example, suppose that you play a lottery in which you buy a ticket for $50 and are told you have a 1 in 1,000 chance of winning $25,000. The random variable X is your net winnings, and its probability distribution is:

x	$f(x)$
$-\$50$	0.999
$\$24{,}950$	0.001

The expected value, $E[X]$, is $-\$50(0.999) + \$24{,}950(0.001) = -\$25.00$. This means that if you played this game repeatedly over the long run, you would lose an average of $25.00 each time you play. Of course, for any *one* game you would either lose $50 or win $24,950.

We may also compute the variance of a discrete random variable X as a weighted average of the squared deviations from the expected value. For the lottery example, the variance is calculated as:

$$\text{Var}[X] = (-50 - [-25.00])^2(0.999) + (24{,}950 - [-25.00])^2(0.001) = 624{,}375$$

Similar to our discussion in Chapter 2, the variance measures the uncertainty of the random variable; the higher the variance, the higher the uncertainty of the outcome. Although variances are easier to work with mathematically, we usually measure the variability of a random variable by its standard deviation, which is simply the square root of the variance. Thus, the standard deviation for the lottery example is $\sqrt{624{,}375} = 790.17$.

The formal definitions of expected value and variance for a continuous random variable are similar; however, to understand them, we must rely on notions of calculus, so we will not discuss them in this book. It is important to understand that the expected value, variance, and standard deviation of random variables are not sample statistics like the mean, sample variance, and sample standard deviation we introduced in Chapter 2. Rather, they are measures associated with the set of *all* possible outcomes of the random variable.

COMMON PROBABILITY DISTRIBUTIONS

A working knowledge of common families of probability distributions is important for several reasons. First, it can help you to understand the underlying process that generates sample data. We will investigate the relationship between distributions and samples later in this chapter. Second, many phenomena in business and nature follow some theoretical distribution and, therefore, are useful in building decision models. In Chapter 9, we will discuss how to fit sample data to the best theoretical distribution. Finally, working with distributions is essential in computing probabilities of occurrence of outcomes to assess risk and make decisions.

Excel and *PHStat* have a variety of functions and tools for working with many of the distributions that we will introduce. These are summarized in Table 3.1. First, we will discuss some useful discrete distributions: the Bernoulli, binomial, and Poisson. Then we will review some of the more common types of probability distributions that are used in decision modeling; discuss how shape, scale, and location parameters affect the distributions; and describe typical situations for which each distribution often applies. The distributions we have chosen are incorporated into the *Crystal Ball* software that we will use in Part II of this book when discussing decision modeling and risk analysis.

Table 3.1   Probability Distribution Support in Excel

EXCEL FUNCTION	DESCRIPTION
BINOMDIST(*number_s, trials, probability_s, cumulative*)	Returns the individual term binomial distribution
POISSON(*x, mean, cumulative*)	Returns the Poisson distribution
NORMDIST(*x, mean, standard_cumulative*)	Returns the normal cumulative distribution for *deviation,* the specified mean and standard deviation
NORMSDIST(*z*)	Returns the standard normal cumulative distribution (mean = 0, standard deviation = 1)
STANDARDIZE(*x, mean, standard_deviation*)	Returns a normalized value for a distribution characterized by a mean and standard deviation
EXPONDIST(*x, lambda, cumulative*)	Returns the exponential distribution
LOGNORMDIST(*x, mean, standard_deviation*)	Returns the cumulative lognormal distribution of *x,* where ln(*x*) is normally distributed with parameters mean and standard deviation
BETADIST(*x, alpha, beta, A, B*)	Returns the cumulative beta density function
GAMMADIST(*x, alpha, beta, cumulative*)	Returns the gamma distribution
WEIBULL(*x, alpha, beta, cumulative*)	Returns the Weibull distribution
PHSTAT ADD-IN	DESCRIPTION
Binomial Probabilities	Computes binomial probabilities and histogram
Poisson Probabilities	Computes Poisson probabilities and histogram
Normal Probabilities	Computes normal probabilities
Exponential Probabilities	Computes exponential probabilities
Hypergeometric Probabilities	Computes hypergeometric probabilities
Simple and Joint Probabilities	Computes simple and joint probabilities for a 2 × 2 cross tabulation
Sampling Distribution Simulation	Generates a simulated sampling distribution from a uniform, standardized normal, or discrete population

Bernoulli Distribution

The *Bernoulli distribution* characterizes a random variable with two possible outcomes with constant probabilities of occurrence. Typically, these outcomes represent "success" ($x = 1$) or "failure" ($x = 0$). A "success" can be any outcome you define. For example, in attempting to boot a new computer just off the assembly line, we might define a "success" as "does not boot up" in defining a Bernoulli random variable to characterize the probability distribution of failing to boot. Thus, "success" need not be a positive result in the traditional sense.

The probability mass function of the Bernoulli distribution is:

$$f(x) = \begin{cases} p & if \ x = 1 \\ 1 - p & if \ x = 0 \end{cases} \qquad (3.4)$$

where p represents the probability of success. The expected value is p, and the variance is $p(1-p)$.

A Bernoulli distribution might be used to model whether an individual responds positively ($x = 1$) or negatively ($x = 0$) to a telemarketing promotion. For example, if you estimate that 20% of customers contacted will make a purchase, the probability distribution that describes whether or not a particular individual makes a purchase is Bernoulli with $p = 0.2$. Think of the following experiment. Suppose that you have a box with 100 marbles, 20 red and 80 white. For each customer, select one marble at random (and then replace it). The outcome will have a Bernoulli distribution. If a red marble is chosen, then that customer makes a purchase; if it is white, the customer does not make a purchase.

Binomial Distribution

The *binomial distribution* models n independent replications of a Bernoulli experiment, each with a probability p of success. The random variable X represents the number of successes in these n experiments. In the telemarketing example, suppose that we call $n = 10$ customers. Then the probability distribution of the number of positive responses is binomial. Using the binomial distribution, we can calculate the probability that exactly x customers out of the 10 will make a purchase.

Binomial probabilities are tedious to compute by hand (see the appendix at the end of this chapter) but can be computed in Excel easily using the function:

$$\text{BINOMDIST}(number\_s, trials, probability\_s, cumulative)$$

In this function, *number_s* plays the role of x, and *probability_s* is the same as p. If *cumulative* is set to TRUE, then this function will provide cumulative probabilities; otherwise the default is FALSE, and it provides values of $f(x)$.

Figure 3.7 shows the results of using this function to compute the distribution for this example. For instance, the probability that exactly 4 individuals will make a purchase is 0.088080, and the probability that 4 or less individuals will make a purchase is 0.967207. Correspondingly, the probability that more than 4 out of 10 individuals will make a purchase is $1 - F(4) = 1 - .967207 = 0.032793$. A binomial distribution might also be used to model the results of sampling inspection in a production operation or the effects of drug research on a sample of patients.

Figure 3.7 Computing Binomial Probabilities in Excel

	A	B	C	D	E	F
1	Binomial Probabilities					
2				=BINOMDIST(A7,$B$3,$B$4,FALSE)		
3		n	10			
4		p	0.2	=BINOMDIST(A7,$B$3,$B$4,TRUE)		
5						
6	x	f(x)	F(x)			
7	0	0.107374	0.107374			
8	1	0.268435	0.375810			
9	2	0.301990	0.677800			
10	3	0.201327	0.879126			
11	4	0.088080	0.967207			
12	5	0.026424	0.993631			
13	6	0.005505	0.999136			
14	7	0.000786	0.999922			
15	8	0.000074	0.999996			
16	9	0.000004	1.000000			
17	10	0.000000	1.000000			

Figure 3.8 Two Examples of Binomial Distributions

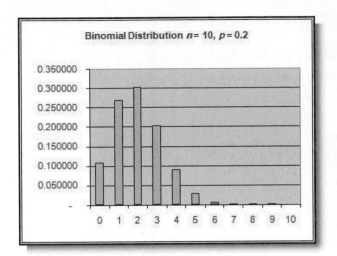

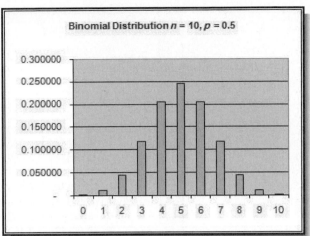

The expected value of the binomial distribution is np, and the variance is $np(1 - p)$. The binomial distribution can assume different shapes and amounts of skewness, depending on the parameters. Figure 3.8 shows two examples. When $p = 0.5$, the distribution is symmetric. For larger values of p, the binomial distribution is negatively skewed; for smaller values, it is positively skewed.

Poisson Distribution

The Poisson distribution is a discrete distribution used to model the number of occurrences in some unit of measure, for example, the number of events occurring in an interval of time, number of items demanded per customer from an inventory, or the number of errors per line of software code. The Poisson distribution assumes no limit on the number of occurrences (meaning that the random variable X may assume any nonnegative integer value), that occurrences are independent, and that the average number of occurrences per unit is a constant, λ (Greek lowercase lambda). The expected value of the Poisson distribution is λ, and the variance also is equal to λ.

Like the binomial, Poisson probabilities are cumbersome to compute by hand. Probabilities can easily be computed in Excel using the function POISSON(x, *mean, cumulative*). For example, suppose that the average number of customers arriving at an ATM during lunch hour is $\lambda = 12$ customers per hour. The probability that exactly x customers will arrive during the hour is given by a Poisson distribution with a mean of 12.

Figure 3.9 shows the results of using this function to compute the distribution for this example. Thus, the probability of exactly 1 arrival during the lunch hour is 0.000074, the probability of 2 arrivals is 0.000442, and so on. Because the possible values of a Poisson random variable are infinite, we have not shown the complete distribution in Figure 3.9. As x gets large, the probabilities become quite small. Figure 3.10 shows this Poisson distribution. Like the binomial, the specific shape depends on the value of the parameter λ; the distribution is more skewed for smaller values.

Figure 3.9   Computing Poisson Probabilities in Excel

	A	B	C	D	E	F
1	**Poisson Probabilities**					
2				=POISSON(A7,$B$3,FALSE)		
3	Mean	12				
4				=POISSON(A7,$B$3,TRUE)		
5						
6	x	f(x)	F(x)			
7	0	0.000006	0.000006			
8	1	0.000074	0.000080			
9	2	0.000442	0.000522			
10	3	0.001770	0.002292			
11	4	0.005309	0.007600			
12	5	0.012741	0.020341			
13	6	0.025481	0.045822			
14	7	0.043682	0.089504			
15	8	0.065523	0.155028			
16	9	0.087364	0.242392			
17	10	0.104837	0.347229			
18	11	0.114368	0.461597			
19	12	0.114368	0.575965			
20	13	0.105570	0.681536			
21	14	0.090489	0.772025			
22	15	0.072391	0.844416			
23	16	0.054293	0.898709			
24	17	0.038325	0.937034			
25	18	0.025550	0.962584			
26	19	0.016137	0.978720			
27	20	0.009682	0.988402			

Figure 3.10   Poisson Distribution for $\lambda = 12$

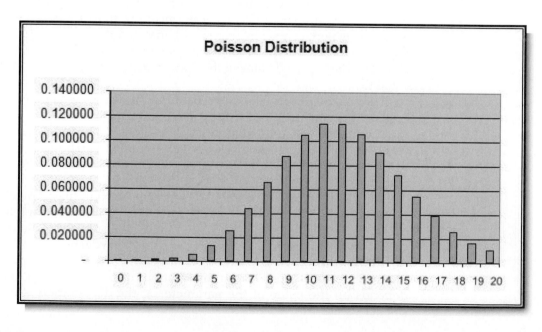

Uniform Distribution

The uniform distribution characterizes a continuous random variable for which all outcomes between some minimum value a and maximum value b are equally likely. The density function for the uniform distribution is shown in Figure 3.11. You can easily verify that the area under the density function is 1 using simple geometry. The expected value of the uniform distribution is $(a + b)/2$, and the variance is $(b - a)^2/12$. Note that a can be considered to be a location parameter since it controls the location of the distribution along the horizontal axis. If a is fixed, the value of b plays the role of a scale parameter. Increasing b elongates the distribution; decreasing b compresses it. There is no shape parameter since any uniform distribution is flat. A variation of the uniform distribution is one for which the random variable is restricted to be integer values between a and b (also integers); this is called a **discrete uniform distribution**.

The uniform distribution is often used when little knowledge about a random variable is available; the parameters a and b are chosen judgmentally to reflect a modeler's best guess about the range of the random variable. Although Excel does not provide a function to compute uniform probabilities, the formula is simple enough to incorporate into a spreadsheet. See the Theory and Computation appendix later in this chapter for an example.

Normal Distribution

The *normal distribution* is a continuous distribution that is described by the familiar bell-shaped curve and is perhaps the most important distribution used in statistics. The normal distribution is observed in many natural phenomena. Errors of various types, such as deviations from specifications of machined items, often are normally distributed. Thus, the normal distribution finds extensive applications in quality control. Processing times in some service systems also follow a normal distribution. Another useful application is that the distribution of the averages of random variables having any distribution tends to be normal as the number of random variables increases.

The normal distribution is characterized by two parameters: the mean, μ (the location parameter), and the variance, σ^2 (the scale parameter). Thus, as μ changes, the location of the distribution on the x-axis also changes, and as σ^2 is decreased or increased, the distribution becomes narrower or wider, respectively. Figure 3.12 provides a sketch of a special case of the normal distribution called the **standard normal distribution**—the normal distribution with $\mu = 0$ and $\sigma^2 = 1$. This distribution is important in performing many probability calculations. A standard normal random variable is usually denoted by Z, and its density function by $f(z)$. The scale along the z-axis represents the number of standard deviations from the mean of zero.

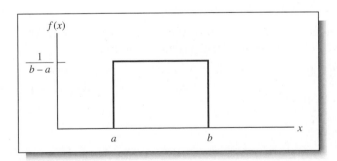

Figure 3.11 Uniform Probability Density Function

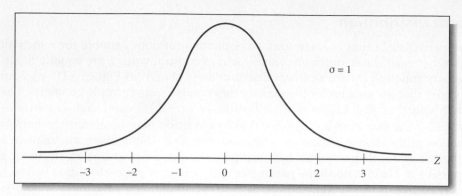

Figure 3.12 Standard Normal Distribution

The normal distribution is symmetric and has the property that the median equals the mean. Thus, half the area falls above the mean and half below it. Although the range of x is unbounded, meaning that the tails of the distribution extend to negative and positive infinity, most of the density is close to the mean; in fact, over 99% of the area is within (that is, plus or minus) three standard deviations of the mean.

Two Excel functions are used to compute normal probabilities: NORMDIST(x, *mean, standard_deviation, cumulative*) and NORMSDIST(z). NORMDIST(x, *mean, standard_deviation, TRUE*) calculates the cumulative probability $F(x) = P(X \leq x)$ for a specified mean and standard deviation. (If *cumulative* is set to *FALSE*, the function simply calculates the value of the density function $f(x)$, which has little practical application.) NORMSDIST(z) generates the cumulative probability for a standard normal distribution.

To illustrate the application of the normal distribution, suppose that a company has determined that the distribution of customer demand (X) is normal with a mean of 750 units/month and a standard deviation of 100 units/month and would like to know the following:

1. What is the probability that demand will be at most 900 units?
2. What is the probability that demand will exceed 700 units?
3. What is the probability that demand will be between 700 and 900 units?
4. What level of demand would be exceeded at most 10% of the time?

Figure 3.13 shows some cumulative probabilities calculated with the NORMDIST function. To answer the questions, first draw a picture. Figure 3.14(a) shows the probability that demand will be at most 900 units, or $P(X < 900)$. This is simply the cumulative probability for $x = 900$, or 0.9332. Figure 3.14(b) shows the probability that demand will exceed 700 units, $P(X > 700)$. Using the principles we have previously discussed, this can be found by subtracting $P(X < 700)$ from 1, or:

$$P(X > 700) = 1 - P(X < 700) = 1 - F(700) = 1 - 0.3085 = 0.6915$$

The probability that demand will be between 700 and 900, $P(700 < X < 900)$ is illustrated in Figure 3.14(c). This is calculated by:

$$P(700 < X < 900) = P(X < 900) - P(X < 700)$$
$$= F(900) - F(700) = 0.9332 - 0.3085 = 0.6247$$

	A	B	C	D	E	F
1	Normal Probabilities					
2						
3	Mean	750				
4	Standard Deviation	100				
5			=NORMDIST(A7,$B$3,$B$4,TRUE)			
6	**x**	**F(x)**				
7	500	0.0062				
8	550	0.0228				
9	600	0.0668				
10	650	0.1587				
11	700	0.3085				
12	750	0.5000				
13	800	0.6915				
14	850	0.8413				
15	900	0.9332				
16	950	0.9772				
17	1000	0.9938				

Figure 3.13
Normal Probability
Calculations

Figure 3.14 Computing Normal Probabilities

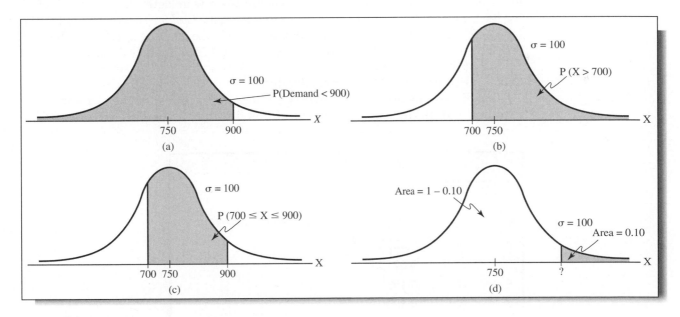

EXCEL NOTE
Using *Goal Seek*

On the *Data* tab, in the *Data Tools* group, click *What-If Analysis*, and then click *Goal Seek*. The dialog box shown in Figure 3.15 will be shown. In the *Set cell* box, enter the reference for the cell that contains the formula that you want to resolve. In the *To value* box, type the formula result that you want. In the *By changing cell* box, enter the reference for the cell that contains the value that you want to adjust.

Figure 3.15 *Goal Seek* Dialog Box

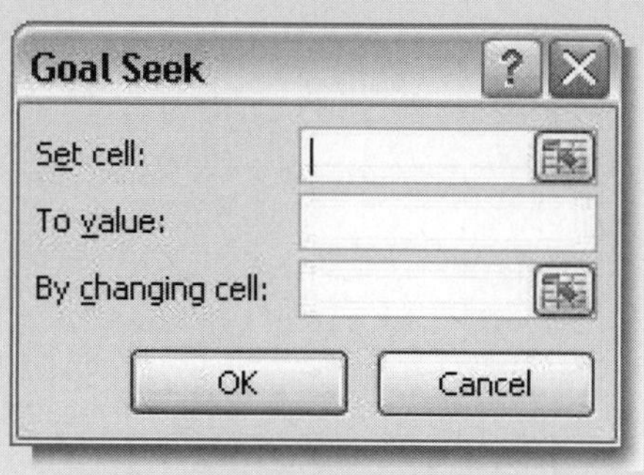

The third question is a bit tricky. We wish to find the level of demand that will be exceeded only 10% of the time; that is, find the value of x so that $P(X > x) = 0.10$. This is illustrated in Figure 3.14(d). An upper tail probability of 0.10 is equivalent to a cumulative probability of 0.90. From Figure 3.13, we can see that the correct value must be somewhere between 850 and 900 because $F(850) = 0.8413$ and $F(900) = 0.9332$. One way to do this is to simply calculate $F(x)$ for values between 850 and 900. A more efficient and precise way is to use Excel's *Goal Seek* tool (see *Excel Note: Using Goal Seek*). If you know the result that you want from a formula, but are not sure what input value the formula needs to get that result, use *Goal Seek*. In this case, we seek to find the value of x that yields a value of 0.9 for $F(x)$. We can choose any row in Figure 3.13 to use in defining the inputs to *Goal Seek*. We will choose row 14. In the *Goal Seek* dialog box, enter B14 for the *Set cell*, enter 0.9 in the *To value* box, and A14 in the *By changing cell* box. The *Goal Seek* tool determines that a demand of approximately 878 will satisfy the criterion.

All of these questions can also be answered using the *PHStat Normal* probability tool. See *PHStat Note: Normal Probability Tools*.

The Excel function NORMSDIST(z) finds probabilities for the standard normal distribution. To illustrate the use of this function, let us find the areas under

	A	B	C	D	E
1	Standard Normal Probabilities				
2					
3	z	F(z)		=NORMSDIST(A4)	
4	-3	0.0013			
5	-2	0.0228			
6	-1	0.1587			
7	0	0.5000			
8	1	0.8413			
9	2	0.9772			
10	3	0.9987			

Figure 3.16
Standard Normal Probabilities Using the NORMSDIST Function

PHSTAT NOTE
Normal Probability Tools

PHStat has a useful tool for computing probabilities for normal distributions. From the *Probability & Prob. Distributions* menu, select *Normal*. The Normal Probability Distribution dialog, shown in Figure 3.17, allows you to compute probabilities for any interval and also find the value of *X* for a given cumulative percentage similar to what the Goal Seek tool did. The dialog box is filled out to answer the same questions as we did in the normal probability example. The results are shown in Figure 3.18. The tool also calculates *z*-values for the standard normal distribution as discussed in the appendix to this chapter.

Figure 3.17 *Normal Probability Distribution* Dialog

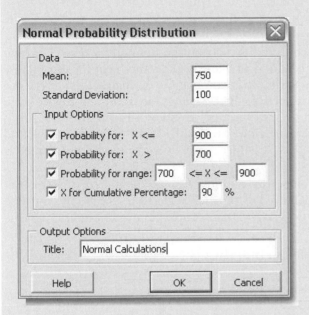

Figure 3.18 *PHStat* Normal Calculations

	A	B	C	D	E
1	**Normal Calculations**				
2					
3	**Common Data**				
4	Mean	750			
5	Standard Deviation	100			
6				**Probability for a Range**	
7	**Probability for X <=**			From X Value	700
8	X Value	900		To X Value	900
9	Z Value	1.5		Z Value for 700	-0.5
10	P(X<=900)	0.9331928		Z Value for 900	1.5
11				P(X<=700)	0.3085
12	**Probability for X >**			P(X<=900)	0.9332
13	X Value	700		P(700<=X<=900)	0.6247
14	Z Value	-0.5			
15	P(X>700)	0.6915		**Find X and Z Given Cum. Pctage.**	
16				Cumulative Percentage	90.00%
17	**Probability for X<900 or X >700**			Z Value	1.281552
18	P(X<900 or X >700)	1.6247		X Value	878.1552

the standard normal distribution within one, two, and three standard deviations of the mean. Figure 3.16 shows calculations using NORMSDIST. Therefore, $P(-1 < Z < 1) = 0.8413 - 0.1587 = 0.6826$. In a similar fashion, you should verify that the area within two standard deviations of the mean is 0.9544 and the area within three standard deviations of the mean is 0.9973. Notice that these values are much larger than specified by Chebyshev's theorem discussed in Chapter 2. These are important characteristics of the normal distribution.

As a final note, we can use the normal distribution with $\mu = np$ and $\sigma^2 = np(1 - p)$ to approximate the binomial distribution. This approximation holds well when $np \geq 5$ and $n(1 - p) \geq 5$.

Triangular Distribution

The triangular distribution is a continuous distribution defined by three parameters: the minimum, a; maximum, b; and most likely, c. Outcomes near the most likely value have a higher chance of occurring than those at the extremes. By varying the position of the most likely value relative to the extremes, the triangular distribution can be symmetric or skewed in either direction, as shown in Figure 3.19. From Figure 3.19, you can see that a is the location parameter, b is the scale parameter, and c is the shape parameter. The expected value is $(a + b + c)/3$ and the variance is $(a^2 + b^2 + c^2 - ab - ac - bc)/18$.

The triangular distribution is often used as a rough approximation of other distributions, such as the normal, or when no data are available and a distribution must be assumed judgmentally. Because it depends on three simple parameters and can assume

Figure 3.19 Examples of Triangular Distributions

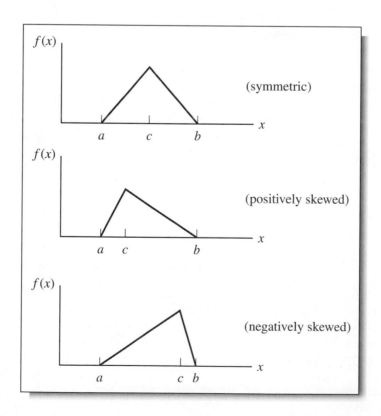

a variety of shapes—for instance, it can be skewed in either direction by changing the value of c—it is very flexible in modeling a wide variety of assumptions. One drawback, however, is that it is bounded, thereby eliminating the possibility of extreme values that might possibly occur. Excel does not have a function to compute triangular probabilities.

Exponential Distribution

The exponential distribution is a continuous distribution that models the time between randomly occurring events. Thus, it is often used in such applications as modeling the time between customer arrivals to a service system, or the time to failure of machines, lightbulbs, and other mechanical or electrical components. A key property of the exponential distribution is that it is *memoryless*; that is, the current time has no effect on future outcomes. For example, the length of time until a machine failure has the same distribution no matter how long the machine has been running.

Similar to the Poisson distribution, the exponential distribution has one parameter λ. In fact, the exponential distribution is closely related to the Poisson; if the number of events occurring during an interval of time has a Poisson distribution, then the time between events is exponentially distributed. For instance, if the number of arrivals at a bank is Poisson distributed, say with mean 12/hour, then the time between arrivals is exponential, with mean 1/12 hour, or 5 minutes.

The expected value of the exponential distribution = $1/\lambda$ and the variance = $(1/\lambda)^2$. The exponential distribution has no shape or location parameters; λ is the scale parameter. Figure 3.20 provides a sketch of the exponential distribution. The exponential distribution has the properties that it is bounded below by 0, it has its greatest density at 0, and the density declines as x increases.

The Excel function EXPONDIST(x, *lambda*, *cumulative*) can be used to compute exponential probabilities. To illustrate the exponential distribution, suppose that the mean time to failure of a critical component of an engine is $1/\lambda = 8,000$ hours. Figure 3.21 shows a portion of the cumulative distribution function. Note that we

Figure 3.20 Example of an Exponential Distribution ($\lambda = 1$)

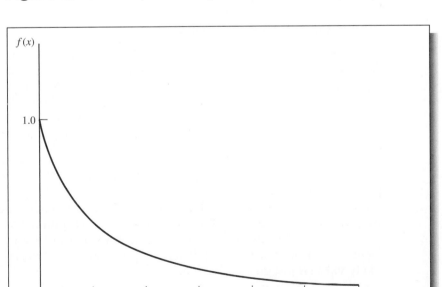

Figure 3.21 Computing Exponential Probabilities

	A	B	C	D	E	F
1	**Exponential Probabilities**					
2						
3	Mean	8000				
4						
5			=EXPONDIST(A7,1/$B$3,TRUE)			
6	x	F(x)				
7	1000	0.117503				
8	2000	0.221199				
9	3000	0.312711				
10	4000	0.393469				
11	5000	0.464739				
12	6000	0.527633				
13	7000	0.583138				
14	8000	0.632121				
15	9000	0.675348				
16	10000	0.713495				
17	11000	0.747160				
18	12000	0.776870				
19	13000	0.803088				
20	14000	0.826226				
21	15000	0.846645				

used the mean in cell B3 as an input in the worksheet; thus, we entered λ as 1/$B$3 in the EXPONDIST function. The probability that the component will fail before x hours is given by the cumulative distribution function $F(x)$. For example, the probability of failing before 5,000 hours is $F(5000) = 0.465$.

Other Useful Distributions

Many other probability distributions, especially those distributions that assume a wide variety of shapes, find application in decision modeling for characterizing a wide variety of phenomena. Such distributions provide a great amount of flexibility in representing both empirical data or when judgement is needed to define an appropriate distribution. We provide a brief description of these distributions; further details may be found in more advanced texts on probability and statistics.

- *Lognormal Distribution.* If the natural logarithm of a random variable X is normal, then X has a lognormal distribution. Because the lognormal distribution is positively skewed and bounded below by zero, it finds applications in modeling phenomena that have low probabilities of large values and cannot have negative values, such as the time to complete a task. Other common examples include stock prices and real estate prices. The lognormal distribution is also often used for "spiked" service times, that is, when the probability of zero is very low but the most likely value is just greater than zero.

- *Gamma Distribution.* The gamma distribution is a family of distributions defined by a shape parameter α, a scale parameter β, and a location parameter L. L is the lower limit of the random variable X; that is, the gamma distribution is defined for $X > L$. Gamma distributions are often used to model the time to complete a task, such as customer service or machine repair. It is used to measure the time between the occurrence of events when the event process is not completely random. It also finds application in inventory control and insurance risk theory.

A special case of the gamma distribution when $\alpha = 1$ and $L = 0$ is called the *Erlang distribution*. The Erlang distribution can also be viewed as the sum of k independent and identically distributed exponential random variables. The mean is k/λ, and the variance is k/λ^2. When $k = 1$, the Erlang is identical to the exponential distribution. For $k = 2$, the distribution is highly skewed to the right. For larger values of k, this skewness decreases, until for $k = 20$, the Erlang distribution looks similar to a normal distribution. One common application of the Erlang distribution is for modeling the time to complete a task when it can be broken down into independent tasks, each of which has an exponential distribution.

- **Weibull Distribution.** The Weibull distribution is another probability distribution capable of taking on a number of different shapes defined by a scale parameter α and a shape parameter β. Both α and β must be greater than zero. When the location parameter $L = 0$ and $\beta = 1$, the Weibull distribution is the same as the exponential distribution with $\lambda = 1/\alpha$. By choosing the scale parameter L different from 0, you can model an exponential distribution that has a lower bound different from zero. When $\beta = 3.25$, the Weibull approximates the normal distribution. Weibull distributions are often used to model results from life and fatigue tests, equipment failure times, and times to complete a task.

- **Beta Distribution.** One of the most flexible distributions for modeling variation over a fixed interval from 0 to a positive value s is the beta. The beta distribution is a function of two shape parameters, α and β, both of which must be positive. The parameter s is the scale parameter. Note that s defines the upper limit of the distribution range. If α and β are equal, the distribution is symmetric. If either parameter is 1.0 and the other is greater than 1.0, the distribution is in the shape of a J. If α is less than β, the distribution is positively skewed; otherwise, it is negatively skewed. These properties can help you to select appropriate values for the shape parameters.

- **Geometric Distribution.** This distribution describes the number of trials until the first success where the probability of a success is the same from trial to trial. An example would be the number of parts manufactured until a defect occurs, assuming that the probability of a defect is constant for each part.

- **Negative Binomial Distribution.** Like the geometric distribution, the negative binomial distribution models the distribution of the number of trials until the rth success, for example, the number of sales calls needed to sell 10 orders.

- **Hypergeometric Distribution.** This is similar to the binomial, except that it applies to sampling without replacement. The hypergeometric distribution is often used in quality control inspection applications.

- **Logistic Distribution.** This is commonly used to describe growth of a population over time.

- **Pareto Distribution.** This describes phenomena in which a small proportion of items accounts for a large proportion of some characteristic. For example, a small number of cities constitutes a large proportion of the population. Other examples include the size of companies, personal incomes, and stock price fluctuations.

- **Extreme Value Distribution.** This describes the largest value of a response over a period of time, such as rainfall, earthquakes, and breaking strengths of materials.

Figure 3.22   Shapes of Some Useful Probability Distributions

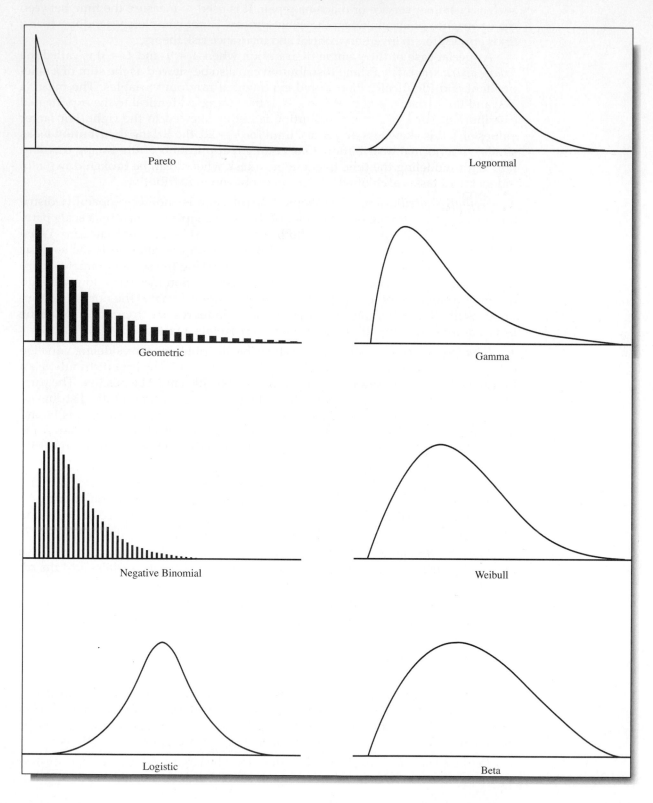

Figure 3.22 is a graphical summary of some of the distributions we have discussed. Understanding the shapes of these distributions makes it easier to select an appropriate distribution for a decision modeling application. Several other probability distributions, notably the t, F, and chi-square (χ^2) distributions, are used in statistical inference applications but have limited application in modeling. We will introduce these as needed in the next several chapters.

Probability Distributions in *PHStat*

PHStat has several routines for generating probabilities of distributions we have discussed. This allows you to compute probabilities without requiring you to develop a detailed worksheet in Excel. The following distributions available are:

- normal
- binomial
- exponential
- Poisson
- hypergeometric

PHSTAT NOTE
Generating Binomial Probabilities

From the *PHStat* menu, select *Probability & Prob. Distributions* and then *Binomial*. The dialog box in Figure 3.23 prompts you for the distribution's parameters and range of outputs desired. By checking the boxes for *Cumulative Probabilities*, the routine provides additional information as shown in Figure 3.24, specifically columns D through G. Checking the *Histogram* box provides a graphical chart of the distribution over the range of outcomes specified.

Figure 3.23 *Binomial Probability Distribution* Dialog in *PHStat*

Figure 3.24 Output from *PHStat* Binomial Probability Distribution Option

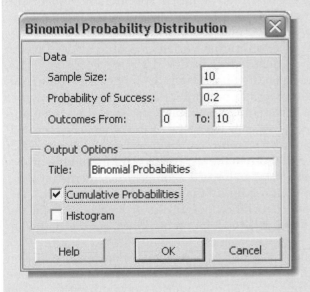

	A	B	C	D	E	F	G
1	Binomial Probabilities						
2							
3		Data					
4	Sample size	10					
5	Probability of success	0.2					
6							
7		Statistics					
8	Mean	2					
9	Variance	1.6					
10	Standard deviation	1.264911					
11							
12	Binomial Probabilities Table						
13		X	P(X)	P(<=X)	P(<X)	P(>X)	P(>=X)
14		0	0.107374	0.107374	0	0.892626	1
15		1	0.268435	0.37581	0.107374	0.62419	0.892626
16		2	0.30199	0.6778	0.37581	0.3222	0.62419
17		3	0.201327	0.879126	0.6778	0.120874	0.3222
18		4	0.08808	0.967207	0.879126	0.032793	0.120874
19		5	0.026424	0.993631	0.967207	0.006369	0.032793
20		6	0.005505	0.999136	0.993631	0.000864	0.006369
21		7	0.000786	0.999922	0.999136	7.79E-05	0.000864
22		8	7.37E-05	0.999996	0.999922	4.2E-06	7.79E-05
23		9	4.1E-06	1	0.999996	1.02E-07	4.2E-06
24		10	1.02E-07	1	1	0	1.02E-07

JOINT, MARGINAL, AND CONDITIONAL PROBABILITY DISTRIBUTIONS

Many applications involve the simultaneous consideration of two or more random variables. For example, we might observe the movement of both the Dow Jones and NASDAQ stock indexes on the same day. In a marketing study, we might record both the gender and brand preference of respondents. Let us suppose that a sample of 100 individuals who were asked to evaluate their preference for three new proposed energy drinks in a blind taste test yielded the following cross-tabulation:

Cross-tabulation	Brand 1	Brand 2	Brand 3	Total
Male	25	17	21	63
Female	9	6	22	37
Total	34	23	43	100

We can define gender as the random variable X, where $x = 0$ corresponds to Male, and $x = 1$ corresponds to Female. Similarly, define the random variable Y to be the brand preference, where $y = 1$ corresponds to Brand 1, $y = 2$ to Brand 2, and $y = 3$ to Brand 3. We may convert these counts to probabilities by dividing each by the total of 100:

Joint and Marginal Probabilities	Brand 1	Brand 2	Brand 3	Total
Male	0.25	0.17	0.21	0.63
Female	0.09	0.06	0.22	0.37
Total	0.34	0.23	0.43	1

This is called a joint probability distribution because it specifies the probabilities of outcomes of two different variables—one associated with gender and the other associated with brand preference—that occur at the same time, or jointly. Formally, a **joint probability distribution** of two random variables X and Y specifies the probability that $X = x$ and $Y = y$, denoted as $P(X = x, Y = y) = f(x, y)$. Thus, $f(0, 1) = 0.25$, $f(1, 1) = 0.09$, and so on. If we add the probabilities across the rows and down the columns, we obtain marginal probability distributions as shown in the table. **Marginal probabilities** represent the probability associated with the outcomes of each random variable regardless of the value of the other. The marginal probabilities in the right hand column, $f(x)$, represent the probability distribution of X independent of the value of Y; and those in the last row, $f(y)$, represent the probability distribution of Y independent of the value of X. For example, no matter what the gender of the respondent is, the probability that Brand 1 is preferred is 0.34. Note that the sum of the marginal probabilities for either X or Y add to 1.0. *PHStat* has a tool for computing simple (marginal) and joint probabilities, but it is limited to 2×2 cross-tabulations.

Conditional probability is the probability of occurrence of one event A, given that another event B is known to have occurred. We will restrict our discussion here to events having only one outcome (that is, outcomes associated with random variables) to simplify the discussion. In general, the conditional probability of A given B is:

$$P(A \mid B) = \frac{P(A \text{ and } B)}{P(B)} \qquad (3.5)$$

To illustrate this, let A = respondent prefers Brand 1 and B = respondent is a male. For example, $P(\text{Brand 1} \mid \text{Male})$ is the probability that a respondent who is male will prefer Brand 1. We can use the cross-tabulation counts as an alternative to formula (3.5) to easily calculate this. We know that we have 63 males in our sample. Of these 63 males, 25 preferred Brand 1; therefore, $P(\text{Brand 1} \mid \text{Male}) = 25/63 = 0.397$. Similarly, $P(\text{Brand 1} \mid \text{Female}) = 9/37 = 0.243$.

We may also use the joint and marginal probabilities to calculate conditional probabilities. Note that $P(A \text{ and } B)$ is simply the joint probability of A and B, and $P(B)$ is the marginal probability. Then

$$P(\text{Brand 1} \mid \text{Male}) = P(\text{Brand 1 and Male})/P(\text{Male}) = 0.25/0.63 = 0.397$$

and

$$P(\text{Brand 1} \mid \text{Female}) = P(\text{Brand 1 and Female})/P(\text{Female}) = 0.09/0.37 = 0.243$$

All conditional probabilities for this example are shown in the table below:

Conditional Probabilities P (Brand | Gender)	Brand 1	Brand 2	Brand 3	Total
Male	0.397	0.270	0.333	1.00
Female	0.243	0.162	0.595	1.00

Such information can be important in marketing efforts. Knowing that there is a difference in preference by gender can help focus advertising. For example, we see that among the male subgroup, Brand 1 is preferred the most, whereas among the female subgroup, Brand 3 seems to be the clear favorite. Suppose that the company decides to market both brands 1 and 3. The data suggest that it would make more sense to focus advertising efforts more on brand 3 in women's magazines and women's oriented television programs.

We may formalize this concept by defining the notion of statistical independence. Two random variables are **statistically independent** if $f(x, y) = f(x) f(y)$, or equivalent if $f(x, y) = f(x \mid y)f(y) = f(y \mid x)f(x)$ for all values of x and y. This means that the value of one random variable does not depend on the value of the other. By checking whether or not these conditions hold, we can determine if the random variables are independent or not. In this example, we can easily determine that they are not. If they were, then it would not be beneficial to focus advertising of specific brands toward different demographic groups.

MONTE CARLO METHODS IN STATISTICS

Monte Carlo methods involve sampling experiments whose purpose is to estimate the distribution of an outcome variable that depends on several input random variables. Many decision models contain variables of interest that are functions of random variables. For example, in a financial model, we might be interested in the distribution of the cumulative discounted cash flow over several years when sales, sales growth rate, operating expenses, and inflation factors are all uncertain and are described by probability distributions. One approach is to input many different values from the probability distributions of these random variables into a spreadsheet model and record the value of the cash flow for each combination of inputs. If we use many different combinations of inputs in a random fashion, we will have created a distribution of possible values of the cumulative discounted cash flow that provides an indication of the likelihood of what we might expect. This process is often called *Monte Carlo simulation*. The term *Monte Carlo simulation* was first used during the development of the atom bomb as a code name for computer simulations of nuclear fission. Researchers coined this term because of the similarity to random sampling in games of chance such as roulette in the famous casino in Monte Carlo. Although we will discuss the use of Monte Carlo simulation for analyzing risk in decision problems in Part II of this book, Monte Carlo methods are also important in statistics. We will use them to gain insight into important statistical issues involving probability distributions and sample data.

To apply Monte Carlo simulation, we will need to generate outcomes from many different types of probability distributions, such as a discrete distribution or a normal, exponential, or Poisson distribution. Sometimes physical processes can be used to generate random outcomes from specific distributions. For example, rolling a die is a physical experiment that randomly generates a number from a discrete uniform probability distribution between 1 and 6. Other experiments that can generate values from different types of distributions would include drawing from a deck of shuffled cards, or selecting numbered balls drawn from a cage as is done for state lotteries. While such experiments are highly intuitive, these approaches are not very practical. We will describe the fundamental concepts of sampling from probability distributions so that you will understand how this is accomplished in Excel and use them to gain insight about sampling distributions.

Random Numbers

The basis for generating samples from probability distributions is the concept of a random number. In simulation, a **random number** is defined as one that is uniformly distributed between 0 and 1. Technically speaking, computers cannot generate truly random numbers since they must use a predictable algorithm. However, the algorithms are designed to generate a stream of numbers that appear to be random. In Excel we may generate a random number within any cell using the function RAND(). This function has no arguments; therefore, nothing should be placed within the parentheses (but the parentheses are required). Table 3.2 shows a table of 100 random numbers generated

Table 3.2   One Hundred Random Numbers

0.007120	0.215576	0.386009	0.201736	0.457990	0.127602	0.387275	0.639298	0.757161	0.285388
0.714281	0.165519	0.768911	0.687736	0.466579	0.481117	0.260391	0.508433	0.528617	0.755016
0.226987	0.454259	0.487024	0.269659	0.531411	0.197874	0.527788	0.613126	0.716988	0.747900
0.339398	0.434496	0.398474	0.622505	0.829964	0.288727	0.801157	0.373983	0.095900	0.041084
0.692488	0.137445	0.054401	0.483937	0.954835	0.643596	0.970131	0.864186	0.384474	0.134890
0.962794	0.808060	0.169243	0.347993	0.848285	0.216635	0.779147	0.216837	0.768370	0.371613
0.824428	0.919011	0.820195	0.345563	0.989111	0.269649	0.433170	0.369070	0.845632	0.158662
0.428903	0.470202	0.064646	0.100007	0.379286	0.183176	0.180715	0.008793	0.569902	0.218078
0.951334	0.258192	0.916104	0.271980	0.330697	0.989264	0.770787	0.107717	0.102653	0.366096
0.635494	0.395185	0.320618	0.003049	0.153551	0.231191	0.737850	0.633932	0.056315	0.281744

in Excel. You should be aware that unless the automatic recalculation feature is suppressed, whenever any cell in the spreadsheet is modified, the values in any cell containing the RAND() function will change. Automatic recalculation can be changed to manual by choosing *Calculation Options* in the *Calculation* group under the *Formulas* tab. Under manual recalculation mode, the worksheet is recalculated only when the F9 key is pressed.

Random Sampling from Probability Distributions

Sampling from discrete probability distributions using random numbers is quite easy. We will illustrate this process using the discrete probability distribution for rolling two dice. The probability mass function and cumulative distribution are shown below.

x	$f(x)$	$F(x)$
2	0.028	0.028
3	0.056	0.083
4	0.083	0.167
5	0.111	0.278
6	0.139	0.417
7	0.167	0.583
8	0.139	0.722
9	0.111	0.833
10	0.083	0.917
11	0.056	0.972
12	0.028	1.000

Two properties of discrete probability distributions that allow us to use random numbers to generate samples are (1) the probability of any outcome is always between 0 and 1, and (2) the sum of the probabilities of all outcomes adds up to 1. We can, therefore, break up the range from 0 to 1 into smaller intervals that correspond to the probabilities of the outcomes in the probability mass function. Any random number, then, must fall within one of these intervals. For instance, the interval from 0 up to but not including 0.028 would correspond to the outcome $x = 2$; the interval from 0.028 up to but not including 0.083 corresponds to $x = 3$; and so on. (To prevent overlap, we do not include the upper limit of an interval in the interval itself; this is also necessary because a random number will never equal 1.0 precisely.) This is summarized as follows:

Interval	Outcome
0 to 0.028	2
0.028 to 0.083	3
0.083 to 0.167	4
0.167 to 0.278	5
0.278 to 0.417	6
0.417 to 0.583	7
0.583 to 0.722	8
0.722 to 0.833	9
0.833 to 0.917	10
0.917 to 0.972	11
0.972 to 1.000	12

To generate an outcome from this distribution, all we need to do is to select a random number and determine the interval into which it falls. Suppose we use the first column in Table 3.2. The first random number is 0.007120. This falls in the first interval; thus, the first sample outcome is $x = 2$. The second random number is 0.714281. This number falls in the seventh interval, generating a sample outcome $x = 8$. Essentially, we have developed a technique to roll dice on a computer! If this is done repeatedly, the frequency of occurrence of each outcome should be proportional to the size of the random number range (that is, the probability associated with the outcome) because random numbers are uniformly distributed. We can easily use this approach to generate outcomes from any discrete distribution; the VLOOKUP function in Excel can be used to implement this on a spreadsheet (see *Excel Note: Using the VLOOKUP Function*).

This approach of generating random numbers and transforming them into outcomes from a probability distribution may be used to sample from most any distribution. A value randomly generated from a specified probability distribution is called a **random variate**. For example, it is quite easy to transform a random number into a random variate from a uniform distribution between a and b. Consider the formula:

$$U = a + (b - a) \times \text{RAND}() \tag{3.6}$$

Note that when $\text{RAND} = 0$, $U = a$, and when RAND approaches 1, U approaches b. For any other value of RAND between 0 and 1, $(b - a) \times \text{RAND}$ represents the same proportion of the interval (a,b) as RAND does of the interval $(0, 1)$. Thus, all real numbers between a and b can occur. Since RAND is uniformly distributed, so also is U.

EXCEL NOTE
Using the *VLOOKUP* Function

This function allows you to look up a value in a table. It is similar in concept to the IF function, but it allows you to pick an answer from an entire table of values. VLOOKUP stands for a vertical lookup table; (a similar function, HLOOKUP—for horizontal lookup table, is also available). The function *VLOOKUP(A,X:Y,B)* uses three arguments:

1. The value to look up (*A*)
2. The table range (*X:Y*)
3. The number of the column whose value we want (*B*)

To illustrate this, suppose we want to place in cell G14 a number 1, 2, or 3, depending on the contents of cell B9. If the value in cell B9 is .55 or less, then G14 should be 1; if it is greater than .55 but .85 or less, then G14 should be 2; and if it is greater than .85, then cell G14 should be 3. We must first put the data in a table in cells B4 through C6:

	B	C
4	0	1
5	0.55	2
6	0.85	3

Now consider the formula in cell G14:

$$= VLOOKUP(B9,B4:C6,2)$$

The VLOOKUP function takes the value in cell B9 and searches for a corresponding value in the first column of the range B4:C6. The search ends when the first value greater than the value in cell B9 is found. The function then returns to the previous row in column B, picks the number found in the cell in the second column of that row in the table range, and enters it in cell G14. Suppose the number in cell B9 is 0.624. The function would search column B until it finds the first number larger than 0.624. This is 0.85 in cell B6. Then it returns to row 5 and picks the number in column C as the value of the function. Thus, the value placed in cell G14 is 2.

While this is quite easy, it is certainly not obvious how to generate random variates from other distributions such as a normal or exponential. We will not describe the technical details of how this is done, but rather just describe the capabilities available in Excel.

Generating Random Variates in Excel

Excel allows you to generate random variates from discrete distributions and certain others using the *Random Number Generation* option in the *Analysis Toolpak* (see *Excel Note: Sampling from Probability Distributions*). However, one disadvantage with using the *Random Number Generation* tool is that you must repeat the process to generate a new set of sample values; pressing the recalculation (F9) key will not change the values. This can make it difficult to use this tool to analyze decision models.

Excel also has several functions that may be used to generate random variates. The most common ones are the following:

- NORMINV(*probability, mean, standard_deviation*)—normal distribution
- NORMSINV(*probability*)—standard normal distribution
- LOGINV(*probability, mean, standard_deviation*)—lognormal distribution, where $\ln(x)$ has the specified mean and standard deviation

And for some advanced distributions, you might see:

- BETAINV(*probability, alpha, beta, A, B*)—beta distribution
- GAMMAINV(*probability, alpha, beta*)—gamma distribution

To use these functions, simply enter RAND() in place of *probability* in the function. For example, NORMINV(RAND(), 5, 2) will generate random variates from a normal distribution with mean 5 and standard deviation 2. Each time the worksheet is recalculated, a new random number and, hence, a new random variate are generated. These functions may be embedded in cell formulas and will generate new values whenever the worksheet is recalculated.

PHStat also includes the ability to generate samples from a uniform (0, 1) distribution, standard normal distribution, and an arbitrary discrete distribution (see *PHStat Note: Sampling Distributions Simulation*). As with the Excel *Random Number Generation* tool, this *PHStat* tool generates the samples "off line"; that is, they cannot be embedded directly into other cell formulas.

EXCEL NOTE
Sampling from Probability Distributions

From the *Data* tab in Excel 2007, select *Data Analysis* in the *Analysis* group and then *Random Number Generation*. The *Random Number Generation* dialog box, shown in Figure 3.25, will appear. From the *Random Number Generation* dialog box, you may select from seven distributions: uniform, normal, Bernoulli, binomial, Poisson, and patterned, as well as discrete. (The patterned distribution is characterized by a lower and upper bound, a step, a repetition rate for values, and a repetition rate for the sequence.) You are asked to specify the upper-left cell reference of the output table that will store the outcomes, the number of variables (columns of values you want generated), number of random numbers (the number of data points you want generated for each variable), and the type of distribution. The default distribution is the discrete distribution, which we illustrate. To use the discrete distribution, the spreadsheet must contain a table with two columns: the left column containing the outcomes and the right column containing the probabilities associated with the outcomes (which must sum to 1.0). Figure 3.26 shows an example of simulating dice rolls.

The dialog box in Figure 3.25 also allows you the option of specifying a random number seed. A **random number seed** is a value from which a stream of random numbers is generated. By specifying the same seed, you can produce the same random numbers at a later time. This is desirable when we wish to reproduce an identical sequence of "random" events in a simulation in order to test the effects of different policies or decision variables under the same circumstances.

Figure 3.25 Excel *Random Number Generation* Dialog

Figure 3.26 Simulating Dice Rolls using the Random Number Generation Tool

	A	B	C
1	x	f(x)	Simulated Dice Roll
2	2	0.028	7
3	3	0.056	5
4	4	0.083	9
5	5	0.111	4
6	6	0.139	9
7	7	0.167	8
8	8	0.139	7
9	9	0.111	6
10	10	0.083	6
11	11	0.056	8
12	12	0.028	

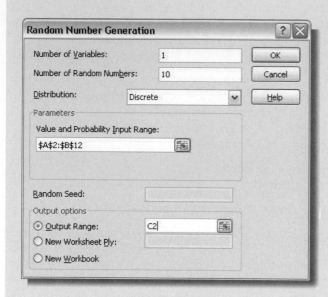

The capabilities of Excel and *PHStat* for sampling from probability distributions are rather limited. However, this book also provides a timelimited download of *Crystal Ball*, an Excel add-in that was developed and is published by Decisioneering, Inc., and is designed to facilitate the process of Monte Carlo simulation. *Crystal Ball* allows you to define cells in spreadsheets as random variables with specified distributions, draw samples from these distributions, evaluate the spreadsheet formulas using the sample data, and collect extensive statistical information about the

PHSTAT NOTE
Sampling Distributions Simulation

From the *PHStat* menu, select *Sampling*, then *Sampling Distributions Simulation*. The dialog box shown in Figure 3.27 appears. You must enter the number of samples to be generated, the sample size, and the type of distribution (uniform, standard normal, or discrete). If you select the discrete distribution, you also need to enter the range in a worksheet that contains the probability mass function. You may also opt for a histogram as part of the output. The procedure creates a new worksheet with the sample output in columns, along with the mean of each sample, overall mean, and standard error (to be discussed shortly). An example for the standard normal distribution is shown in Figure 3.28.

Figure 3.27 *PHStat Sampling Distributions Simulation* Dialog

Figure 3.28 *PHStat* Output for Normal Sampling

	A	B	C	D	E
1	Normal Samples				
2					
3	-0.71438	0.903342	1.256008	-0.60087	0.262189
4	0.826285	-0.12513	1.938579	1.271824	-0.6131
5	-0.446	-0.49869	-0.30303	-0.92132	-1.50879
6	-1.25265	0.711325	-0.43218	1.302294	-1.17316
7	0.971727	0.367784	0.611437	-0.43588	-0.68807
8	1.570602	-0.70818	0.204325	-0.92554	-1.60764
9	-0.41153	-0.02054	-0.74635	-0.57513	-0.44136
10	-1.59088	-0.60858	1.683766	1.172402	0.850308
11	0.197534	-0.89921	-0.35342	1.677486	1.942599
12	0.062807	0.099728	-0.07531	-0.46177	0.356592
13	Sample Means:				
14	-0.07865	-0.07781	0.378383	0.15035	-0.26204
15	Overall Average:				
16	0.022045				
17	Standard Error of the Mean:				
18	0.247111				

distribution of one or more output cells of interest. *Crystal Ball* provides functions for all of the distributions described in this chapter that can be entered directly into cell formulas in a spreadsheet: uniform, normal, triangular, binomial, Poisson, geometric, hypergeometric, lognormal, exponential, Weibull, beta, gamma, logistic, Pareto, extreme value, and negative binomial, as well as user-defined custom distributions. *Crystal Ball* is used extensively in analyzing risks associated with decisions, as we will explore further in Part II.

Applications of Monte Carlo Methods in Statistics

Monte Carlo methods are often used when a problem cannot apparently be solved using mathematics. For example, we know that dispersion of sample data can be measured using either the range or the standard deviation. If we use the range, which is a much simpler calculation if done manually, how might we estimate the standard deviation of the data, if we assume the data are normally distributed? (This is actually an important issue in quality control.) Suppose that we conduct an experiment in which samples of size n are generated from a normal distribution having a known standard deviation σ. If we compute the range, R, of each sample, we can estimate the distribution of the statistic R/σ. The expected value of this statistic is a factor that statisticians have labeled as d_2. If we know this value, then we can estimate σ by R/d_2.

Figure 3.29 Monte Carlo Experiment for Estimating the Statistical Factor d_2

	A	B	C	D	E	F	G	H	I
1	**Simulated Standard Normal Random Variates**								
2				Sample					
3	Experiment	1	2	3	4	5		R/σ	Mean
4	1	0.854561	-1.03003	-0.64778	-0.95043	-0.57717		1.884591	2.329859
5	2	-0.20255	-0.17015	-0.5465	-0.01631	0.408602		0.955103	
6	3	1.873159	-1.41493	-0.30729	-0.23232	1.740569		3.288087	
7	4	0.539949	-0.32848	-0.92189	-1.02624	1.470603		2.496845	
8	5	1.73728	-0.95969	-0.64225	2.415582	-0.49173		3.375272	
9	6	0.159216	0.051848	-0.78048	-1.01611	-0.18376		1.175329	
10	7	0.156621	-0.53387	0.966172	0.129511	-1.77263		2.738798	
11	8	-0.04529	-2.6699	1.982951	0.992474	0.229182		4.652852	
12	9	0.031989	0.399565	0.964993	1.148798	-0.11804		1.266834	
13	10	-0.41537	0.3036	-0.01131	-0.03853	0.997583		1.412953	

Figure 3.29 shows a portion of a simple spreadsheet for performing this experiment. Each row generates five random variates from a standard normal distribution with a mean of 0 and standard deviation of 1 and computes the value of R/σ. The entire spreadsheet performs 1,000 experiments. The value in cell I4 is the average of these 1,000 experiments; that is, the estimate of d_2, which is 2.327. Published statistical tables give this value as 2.326, so we see that the average simulated value is indeed very close to the value identified by statisticians.

Here is another example of using Monte Carlo simulation. In finance, one way of evaluating capital budgeting projects is to compute a profitability index (*PI*), which is defined as the ratio of the present value of future cash flows (*PV*) to the initial investment (*I*):

$$PI = PV/I$$

Due to uncertainty in cash flow and initial investment that may be required for a particular project, these values may not be constant, resulting in uncertainty of the profitability index. For example, suppose that *PV* is estimated to be normally distributed with a mean of $12 million and a standard deviation of $2.5 million, and the initial investment is also estimated to be normal with a mean of $3.0 million and standard deviation of $0.8 million. Intuitively, one might believe that the mean value of the profitability index is 12/3 = 4; however, as we shall see, this is not the case. We can use Monte Carlo simulation to identify the probability distribution of *PI* for these assumptions.

Figure 3.30 shows a simple model. The values of *PV* and *I* are simulated using the NORMINV function with the appropriate means and standard deviations. The value in cell E8 is the average *PI* for 1,000 experiments. We see that this is not equal to 4 as previously suspected. The histogram in Figure 3.31 demonstrates that the ratio of two normal distributions is not normally distributed. When the initial investment is low, the *PI* increases significantly more than when it is high. Thus, the distribution is skewed to the right, and the mean is larger than 4. This Monte Carlo simulation was able to provide some insights that were not readily apparent.

	A	B	C	D	E
1	**Profitability Index Analysis**				
2					
3		Mean	Standard Deviation		
4	PV	12	2.5		
5	I	3	0.8		
6					
7	Experiment	PV	I	PI	Mean
8	1	13.69538555	3.041646632	4.502622167	4.564967
9	2	8.537313995	3.271118185	2.609906923	
10	3	11.10957016	3.494903314	3.178791847	
11	4	15.35942722	1.264141592	12.15008455	
12	5	11.93490023	4.156000464	2.871727358	
13	6	13.85675231	3.008288775	4.606190879	
14	7	9.382066543	2.913704148	3.219979129	
15	8	10.95787015	4.138994624	2.647471462	
16	9	14.328562	2.652381236	5.402150266	
17	10	5.871110242	2.898073827	2.025866349	

Figure 3.30 Monte Carlo Simulation for Understanding Profitability in Capital Budgeting

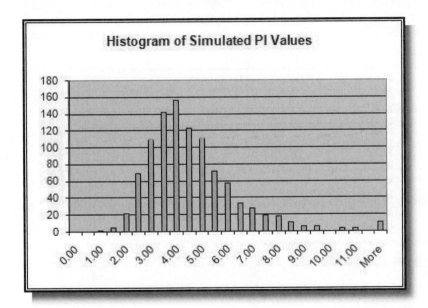

Figure 3.31 Histogram of Simulated PI Values

SAMPLING DISTRIBUTIONS AND SAMPLING ERROR

Whenever we collect data, we are essentially taking a sample from some generally unknown probability distribution. Usually the goal is to estimate a population parameter, such as the mean. An important statistical question is: How good is the estimate obtained from the sample? We could gain some insight into this question by performing a Monte Carlo experiment.

Let us assume that a random variable is uniformly distributed between 0 and 10. First, we will compute the characteristics of this random variable. Using the formulas introduced earlier in this chapter, we find that the expected value is $(0 + 10)/2 = 5$, the variance is $(10 - 0)^2/12 = 8.33$, and the standard deviation is 2.89. Suppose we generate a sample of size $n = 25$ for this random variable using the Excel function $10 \times \text{RAND}()$ and compute the sample mean. Note that each of the 25 values is uniformly distributed between 0 and 10. When averaged together, we would expect the sample mean to be close to 5, but probably not exactly equal to 5 because of the randomness in the simulated values.

Now suppose that we repeat this experiment several more times, say 20, and obtain a set of 20 sample means. Figure 3.32 shows a portion of a spreadsheet for this experiment. (The Excel file is available on the CD-ROM as *Sampling Error Experiment* if you wish to experiment further.) Figure 3.33 shows a histogram of 20 sample means generated in this fashion. For a sample size of 25, the sample means seem to be rather uniformly spread out.

Figure 3.32   Portion of Spreadsheet for Sampling Experiment

	A	B	C	D	E	F	G	H	I	J	K	L	M	N	O	P	Q	R	S	T	U
1	Sampling Error Experiment																				
2	Instructions: The worksheet is designed for 20 samples with sample sizes of up to 500. To change the sample size, simply change the																				
3			range in the formulas in row 6 for computing the sample mean to include the appropriate number of observations.																		
4			Pressing the F9 key will recalculate all values.																		
5	Experiment	1	2	3	4	5	6	7	8	9	10	11	12	13	14	15	16	17	18	19	20
6	Sample Mean	5.029	4.949	5.007	4.823	4.739	5.066	5.016	4.850	5.270	4.916	5.126	5.032	4.992	5.089	4.919	4.903	4.850	4.706	4.943	5.068
7																					
8	Sample																				
9	1	5.845	8.400	0.095	5.414	2.568	7.209	5.338	3.513	3.990	8.568	1.734	9.065	5.520	0.952	7.485	7.944	3.631	1.226	1.468	5.609
10	2	4.574	8.951	4.769	4.148	2.964	2.416	0.659	4.541	7.477	4.325	7.236	8.207	2.788	7.185	4.607	4.376	1.074	1.807	0.150	5.098
11	3	4.123	4.006	1.507	8.680	4.780	1.563	0.812	4.019	9.156	3.411	3.828	1.743	5.993	6.857	5.607	0.647	0.878	6.728	2.790	5.483
12	4	2.185	7.242	3.038	2.665	6.094	4.578	9.730	9.507	1.932	1.909	2.368	2.079	8.238	7.787	6.482	7.585	8.339	6.676	9.444	0.068
13	5	7.583	4.564	2.507	7.847	4.495	9.317	2.689	6.201	2.038	6.117	5.157	5.017	4.756	5.926	6.445	8.685	3.605	7.376	3.367	2.775
14	6	1.528	4.534	5.769	6.263	3.079	4.982	8.817	1.417	8.160	5.726	3.741	2.390	0.554	2.779	5.740	1.338	5.013	2.346	0.033	1.109
15	7	5.152	9.212	6.076	9.673	5.300	9.991	2.060	5.904	5.394	1.193	8.452	6.445	7.796	2.148	6.068	2.102	9.230	9.656	6.734	8.136
16	8	5.272	9.569	6.300	8.841	7.292	8.774	9.163	4.160	6.966	8.415	8.107	6.359	8.252	1.964	4.884	0.386	2.923	2.208	1.167	8.518
17	9	0.245	6.387	1.792	6.082	9.844	5.001	2.511	6.680	7.342	3.167	3.950	2.188	2.777	8.187	6.301	7.665	1.451	5.798	6.110	7.636
18	10	1.136	0.561	1.975	4.979	4.217	5.431	0.058	7.563	3.517	2.128	2.188	8.889	4.832	9.524	2.694	4.629	0.616	3.322	5.312	7.804

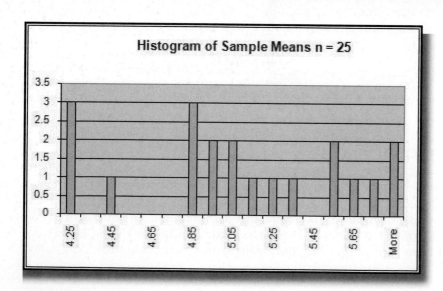

Figure 3.33   Histogram of 20 Sample Means for $n = 25$

Table 3.3 Results from Sampling Error Experiment

Sample Size	Average of 20 Sample Means	Standard Deviation of 20 Sample Means
25	5.166	0.639449
50	4.933	0.392353
100	4.948	0.212670
250	5.028	0.167521
500	4.997	0.115949

Now let us repeat this experiment for larger sample sizes. Table 3.3 shows some results. Notice that as the sample size gets larger, the average of the 20 sample means seems to be getting closer to the expected value of 5, and also, the standard deviation of the 20 sample means becomes smaller, meaning that the means of these 20 samples are clustered closer together around the true expected value. Figure 3.34 shows histograms of the sample means for each of these cases. You can clearly see that the variability becomes smaller as the sample size becomes larger. This suggests that the estimates we obtain from larger sample sizes provide greater accuracy in estimating the true population mean. In other words, larger sample sizes have *less sampling error*. Perhaps more surprisingly, the distribution of the sample means appears to assume the shape of a normal distribution for larger sample sizes!

The means of multiple samples of a fixed size n from some population will form a distribution, which we call the **sampling distribution of the mean**. The sampling

Figure 3.34 Histograms of Sample Means for Increasing Sample Sizes

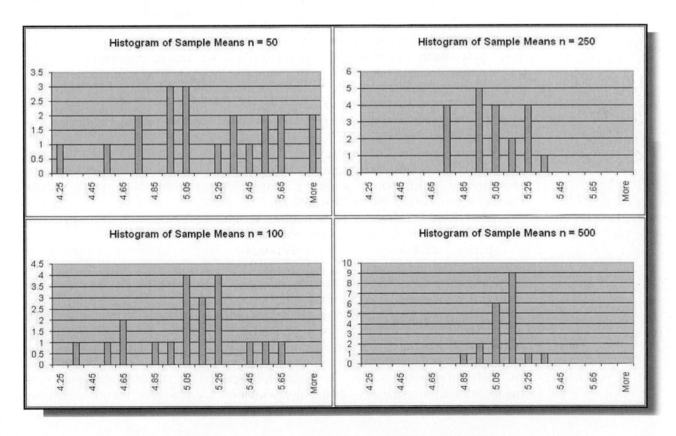

distribution of the mean characterizes the population of means of *all possible* samples of a given size. (In Figures 3.32 and 3.33, we have 20 samples from the sampling distribution of the mean for each sample size.) In practice, we base our estimate of the true population mean based on the sample mean of only *one* sample. However, if we understand the sampling distribution that characterizes the sample, we can draw conclusions about the accuracy of our estimate. We will discuss this further in Chapter 4.

Standard Error of the Mean

The standard deviation of the sampling distribution of the mean is called the **standard error of the mean** and is computed as:

$$\text{Standard Error of the Mean} = \sigma/\sqrt{n} \qquad (3.7)$$

where σ is the standard deviation of the population from which the individual observations are drawn, and n is the sample size. From this formula, we see that as n increases, the standard error decreases, just as our experiment demonstrated.

For our experiment, we know that the variance of the population is 8.33 (because the values were uniformly distributed). Therefore, the standard deviation of the population is $\sigma = 2.89$. We may compute the standard error of the mean for each of the sample sizes in our experiment using formula (3.7). This yields:

Sample Size, n	Standard Error of the Mean
25	0.577
50	0.408
100	0.289
250	0.183
500	0.129

The standard deviations in Table 3.3 are *estimates* of the standard error of the mean based on the 20 samples that we generated. If we compare these estimates with the theoretical values above, we see that they are close but not exactly the same. This is simply because we used only 20 sample means in our experiment; the true standard error is based on all possible sample means in the sampling distribution. If you repeat the experiment with a larger number of samples, the observed values of the standard error would be closer to these theoretical values.

In practice, we will never know the true population standard deviation, and generally only take a limited sample of n observations. However, we may estimate the standard error of the mean using the sample data by simply dividing the sample standard deviation by the square root of n.

What about the shape of the sampling distribution of the mean? Our experiment seemed to suggest that it becomes normal as the sample size increases. Indeed, statisticians have shown that if the population is exactly normal, then the sampling distribution of the mean will also be normal for *any* sample size and that the mean of the sampling distribution will be the same as that of the population. Furthermore, the **central limit theorem**, one of the most important practical results in statistics, states that if the sample size is large enough, the sampling distribution of the mean can be approximated by a normal distribution, *regardless* of the shape of the population distribution. This is exactly what we observed in our experiment. The distribution of the population was uniform, yet the sampling distribution of the mean converges to a normal distribution as the sample size increases. The central limit theorem allows us to use the theory we learned about calculating probabilities for normal distributions to make conclusions about sample means. This will be quite important in the next chapter.

Applying Sampling Distributions

Suppose that the size of individual customer orders (in dollars), X, from a major discount book publisher Web site is normally distributed with a mean of $36 and standard deviation of $8. The probability that the next individual who places an order at the Web site will purchase more than $40 can be found by calculating:

$$1 - \text{NORMDIST}(40, 36, 8, \text{TRUE}) = 1 - 0.6915 = 0.3085$$

Now suppose that a sample of 16 customers is chosen. What is the probability that the *mean purchase* for these 16 customers will exceed $40? To find this, we must realize that we must use the sampling distribution of the mean to carry out the appropriate calculations. The sampling distribution of the mean will have a mean of $36, but a standard error of $8/$\sqrt{16}$ = $2. Then the probability that the mean purchase exceeds $40 for a sample size of $n = 16$ is:

$$1 - \text{NORMDIST}(40, 36, 2, \text{TRUE}) = 1 - 0.9772 = 0.0228$$

While about 30% of individuals will make purchases exceeding $40, the chance that 16 customers will collectively average more than $40 is much smaller. It would be very unlikely for 16 customers to all make high-volume purchases, as some individual purchases would as likely be less than $36 as more, making the variability of the mean purchase amount for the sample of 16 much smaller than for individuals.

The key to applying sampling distribution correctly is to understand whether the probability that you wish to compute relates to an individual observation or to the mean of a sample. If it relates to the mean of a sample, then you must use the sampling distribution of the mean, whose standard deviation is the standard error, $\sigma/\sqrt{n}$. Understanding the standard error of the mean and characteristics of the sampling distribution is also important for designing sampling experiments and performing various statistical tests. We address these issues in other chapters.

BASIC CONCEPTS REVIEW QUESTIONS

1. Explain the concepts of an experiment, outcome, sample space, and event. Provide some examples different from those in the chapter.
2. Define probability and explain its three perspectives. Provide an example of each.
3. Explain the concept of mutually exclusive events. How do you compute the probability $P(A \text{ or } B)$ when A and B are, and are not, mutually exclusive?
4. What is a random variable? Explain the difference between discrete and continuous random variables.
5. What is a probability distribution? How does a probability mass function differ from a probability density function?
6. What information does a cumulative distribution function provide? How is it used to compute probabilities over intervals?
7. Explain how to compute the expected value of a discrete random variable.
8. Briefly summarize the important characteristics and applications of the Bernoulli, binomial, and Poisson distributions.

9. Explain the role of shape, scale, and location parameters in continuous probability distributions.
10. Explain practical situations in which the uniform, normal, triangular, and exponential distributions might be used.
11. What is a joint probability distribution? How are marginal probabilities calculated?
12. Explain the notion of conditional probability.
13. Explain the concept of statistical independence. How can it be used in marketing decisions?
14. What are Monte Carlo methods, and how are they used in statistics?
15. Define a random number. How is it different from a random variate?
16. Explain the sampling distribution of the mean. What properties does it have?
17. What is the standard error of the mean? How does it relate to the standard deviation of the population from which a sample is taken?
18. Explain the Central Limit theorem. Why is it important in statistical analysis?

SKILL-BUILDING EXERCISES

1. Develop a spreadsheet for computing the expected value and variance for the probability distribution of the rolls of two dice (see Figure 3.2).
2. Use the *PHStat Probability & Prob. Distributions* tool to generate 50 samples from a Poisson distribution with a mean of 12. Compare your results with Figure 3.10.
3. Develop a spreadsheet for computing the cumulative distribution of the triangular distribution (see the appendix to this chapter) for any input values of a, b, and c. Apply your work to generate a chart of the cumulative distribution function of a triangular distribution with $a = 3$, $b = 15$, and $c = 12$.
4. Develop a spreadsheet for the cross-tabulation example of the evaluation of energy drinks. Use the spreadsheet to compute the joint and marginal probabilities, and all conditional probabilities $P(\text{Brand} \mid \text{Gender})$ and $P(\text{Gender} \mid \text{Brand})$.
5. Develop a spreadsheet using the VLOOKUP function to simulate 50 rolls of a pair of dice using the approach described in this chapter. Construct a histogram of the outcomes and compare it to the theoretical probability distribution.
6. Construct charts for the probability mass function and cumulative distribution function for a binomial distribution with $n = 10$ and $p = 0.1$. Compare this with Figure 3.7. What conclusion can you reach?
7. Construct charts for the probability mass function for a Poisson distribution with means of 1, 5, and 20. Compare these to Figure 3.10. What conclusion can you reach?
8. Use Formula 3.6 to generate 100 uniform random variates between 1 and 100. Construct a histogram with bins having width 10 to examine the distribution of the results.
9. Use the NORMSINV function to generate samples of 50, 100, 500, and 1000 normal random variates. Construct histograms using a bin range $\{-3, -2.5, -2, -1.5, -1, -.5, 0, .5, 1, 1.5, 2, 2.5, 3\}$ and comment on how closely the data reflect a true normal distribution. What does this tell you about sample sizes?

10. Replicate the sampling error experiment described in this chapter. How closely do your results coincide with those in Table 3.3 and Figure 3.34?
11. Generate a set of 200 random numbers and place these in cells A1 through A200 of an Excel worksheet. Convert these to random variates by entering the following formula in the next column to the right of the random numbers. For instance, in cell B1, enter: $= -50 * LN(1 - A1)$. Copy this formula down through all 200 cells through cell B200. Compute the mean and variance of these values and construct a histogram. From what distribution do these random variates appear to be? Why?
12. Construct a spreadsheet similar to Figure 3.29 to estimate the factor d_2 for sample sizes of $n = 2$ through 10. Compare your results to published factors shown here:

n	d_2
2	1.128
3	1.693
4	2.059
5	2.326
6	2.534
7	2.704
8	2.847
9	2.970
10	3.078

13. Generate 20 groups of 10 uniformly distributed numbers, and calculate the mean of each group. Compute the mean and variance of all 200 values, as well as the mean and variance of the 20 means. Compute the standard error of the mean using the 20 sample means and compare this to $s/\sqrt{n}$ for the entire sample. Explain your results.
14. Generate three data sets of normally distributed random variates, all with a mean of 80 and a standard deviation of 5. Let data set 1 be 50 groups of 5, data set 2 be 50 groups of 10, and data set 3 be 50 groups of 30. Calculate the mean of each group. Compare the average of the means for each set, as well as their variances. What can you conclude?

PROBLEMS AND APPLICATIONS

1. Construct the probability distribution for the value of a 2-card hand dealt from a standard deck of 52 cards (all face cards have a value of 10 and an ace has a value of 11).
   a. What is the probability of being dealt 21?
   b. What is the probability of being dealt 20?
   c. Construct a chart for the cumulative distribution function. What is the probability of being dealt a 16 or less? Between 12 and 16? Between 17 and 20?
   d. Find the expected value and standard deviation of a two-card hand.

2. Three coins are dropped on a table.
   a. List all possible outcomes in the sample space.
   b. Find the probability associated with each outcome.
   c. Let A be the event "exactly 2 heads." Find $P(A)$.
   d. Let B be the event "at most 1 head." Find $P(B)$.
   e. Let C be the event "at least 2 heads." Find $P(C)$.
   f. Are the events A and B mutually exclusive? Find $P(A$ or $B)$.
   g. Are the events A and C mutually exclusive? Find $P(A$ or $C)$.
3. Based on the data in the Excel file *Consumer Transportation Survey*, develop a probability mass function and cumulative distribution function (both tabular and as charts) for the random variable Number of Children. What is the probability that an individual in this survey has less than two children? At least two children? Five or more?
4. Using the data in the Excel file *Cereal Data*, develop a probability mass function and cumulative distribution function (both tabular and as charts) for the random variable Calories. What is the probability that a box of cereal from this population contains less than 100 calories/serving? Between 90 and 110? More than 120?
5. An airline tracks data on its flight arrivals. Over the past six months, 65 flights on one route arrived early, 273 arrived on time, 218 were late, and 44 were cancelled.
   a. What is the probability that a flight is early? On time? Late? Cancelled?
   b. Are these outcomes mutually exclusive?
   c. What is the probability that a flight is either early or on time?
6. A survey of 100 MBA students found that 60 owned mutual funds, 40 owned stocks, and 20 owned both.
   a. What is the probability that a student owns a stock? A mutual fund?
   b. What is the probability that a student owns neither stocks nor mutual funds?
   c. What is the probability that a student owns either a stock or a mutual fund?
7. Roulette is played at a table similar to the one in Figure 3.35. A wheel with the numbers 1 through 36 (evenly distributed with the colors red and black) and two green numbers 0 and 00 rotates in a shallow bowl with a curved wall. A small ball is spun on the inside of the wall and drops into a pocket corresponding to one of the numbers. Players may make 11 different types of bets by placing chips on different areas of the table. These include bets on a single number, two adjacent numbers, a row of three numbers, a block of four numbers, two adjacent rows of six numbers, and the five number combinations of 0, 00, 1, 2, and 3; bets

Figure 3.35 Layout of a Typical Roulette Table

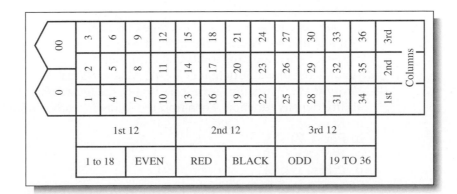

on the numbers 1–18 or 19–36; the first, second, or third group of 12 numbers; a column of 12 numbers; even or odd; and red or black. Payoffs differ by bet. For instance, a single-number bet pays 35 to 1 if it wins; a three-number bet pays 11 to 1; a column bet pays 2 to 1; and a color bet pays even money. Define the following events: $C1$ = column 1 number, $C2$ = column 2 number, $C3$ = column 3 number, O = odd number, E = even number, G = green number, $F12$ = first 12 numbers, $S12$ = second 12 numbers, and $T12$ = third 12 numbers.

a. Find the probability of each of these events.

b. Find $P(G \text{ or } O)$, $P(O \text{ or } F12)$, $P(C1 \text{ or } C3)$, $P(E \text{ and } F12)$, $P(E \text{ or } F12)$, $P(S12 \text{ and } T12)$, $P(O \text{ or } C2)$.

8. The weekly demand of a slow-moving product has the probability mass function:

Demand, x	Probability, $f(x)$
0	0.1
1	0.4
2	0.3
3	0.2
4 or more	0

Find the expected value, variance, and standard deviation of weekly demand.

9. A major application of data mining in marketing is determining the attrition of customers. Suppose that the probability of a long-distance carrier's customer leaving for another carrier from one month to the next is 0.15. What distribution models the retention of an individual customer? What is the expected value and standard deviation?

10. What type of distribution models the random variable Prior Call Center Experience in the file *Call Center Data?* Define the parameter(s) for this distribution based on the data.

11. If a cell phone company conducted a telemarketing campaign to generate new clients, and the probability of successfully gaining a new customer was 0.08, what are the probabilities that contacting 20 potential customers would result in 0, 1, 2, 3, or 4 new customers?

12. A telephone call center where people place marketing calls to customers has a probability of success of 0.05. The manager is very harsh on those who do not get a sufficient number of successful calls each hour. Find the number of calls needed to ensure that there is a probability of 0.80 of obtaining 4 or more successful calls.

13. A financial consultant has an average of 6 customers arrive each day. The consultant's overhead requires that at least 5 customers arrive in order that fees cover expenses. Find the probabilities of 0 through 4 customers arriving in a given day. What is the probability that at least 5 customers arrive?

14. Verify that the function corresponding to the figure on next page is a valid probability density function. Then find the following probabilities:

a. $P(x < 8)$

b. $P(x > 7)$

c. $P(6 < x < 10)$

d. $P(8 < x < 11)$

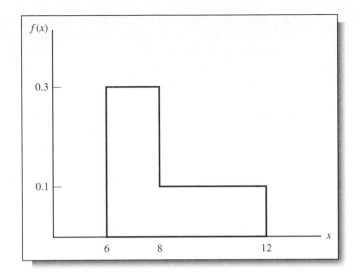

15. The time required to play a game of Battleship is uniformly distributed between 20 and 60 minutes.
    a. Find the expected value and variance of the time to complete the game.
    b. What is the probability of finishing within 30 minutes?
    c. What is the probability that the game would take longer than 45 minutes?
16. In determining automobile mileage ratings, it was found that the mpg in the city (x) for a certain model is normally distributed, with a mean of 24 mpg and a standard deviation of 1.0 mpg. Find the following:
    a. $P(X < 25)$
    b. $P(22 < X < 24)$
    c. $P(X > 26)$
    d. $P(X < 23)$
    e. The mileage rating that the upper 5% of cars achieve
17. The distribution of an MBA class's GMAT scores has a mean of 650 and standard deviation of 25. Assume that the scores are normally distributed. Find the probability that an individual's GMAT score is:
    a. less than 600
    b. between 600 and 700
    c. greater than 700
18. A popular soft drink is sold in 2-liter (2,000 mL) bottles. Because of variation in the filling process, bottles have a mean of 2,000 mL and a standard deviation of 20, normally distributed.
    a. If the process fills the bottle by more than 60 mL, the overflow will cause a machine malfunction. What is the probability of this occurring?
    b. What is the probability of underfilling the bottles by at least 30 mL?
19. A supplier contract calls for a key dimension of a part to be between 1.95 and 2.05 cm. The supplier has determined that the standard deviation of its process, which is normally distributed, is 0.10 cm.
    a. If the mean is 1.98, what fraction of parts will meet specifications?
    b. If the mean is adjusted to 2.00, what fraction of parts will meet specifications?
    c. How small must the standard deviation be to ensure that no more than 2% of parts are nonconforming, assuming the mean is 2.00?

20. Find the probability density function and cumulative distribution function, and compute the mean and variance for each of the following parameter sets for a triangular distribution.
    a. $a = 0, b = 8, c = 4$
    b. $a = 0, b = 8, c = 2$
    c. $a = 0, b = 8, c = 6$
21. A lightbulb is warranted to last for 3,000 hours. If the time to failure is exponentially distributed with a true mean of 2,750 hours, what is the probability that it will last at least 3,000 hours?
22. Using the data in the Excel file *Consumer Transportation Survey*, develop a contingency table for Gender and Vehicle Driven then convert this table to a joint probability distribution and compute marginal probabilities.
    a. What is the probability that a respondent is male and drives a minivan?
    b. What is the probability that a female respondent drives either a truck or an SUV?
    c. If it is known that an individual drives a car, what is the probability that the individual is male?
    d. If it is known that an individual is female, what is the probability that she drives an SUV?
    e. Determine whether the random variables "gender" and the event "vehicle driven" are statistically independent. What would this mean for advertisers?
23. Compute the joint probability distribution of age and educational status in the Excel file *Census Education* worksheet, and determine if any of the events associated with age and educational status are statistically independent.
24. Using the data in the Excel file *Census Education Data,* construct a joint probability distribution for marital status and educational status.
    a. What is the probability that a person is divorced and has a bachelor's degree?
    b. What is the probability that someone known to have an advanced degree is married with spouse present?
    c. What is the probability that an individual who is known to have never married has at least an associate's degree?
25. Suppose that we observed the outcomes associated with the stock market for 100 days and recorded the number of days for each of the following outcomes:
    1. DJIA up; NASDAQ up: 26
    2. DJIA down; NASDAQ up: 5
    3. DJIA unchanged; NASDAQ up: 3
    4. DJIA up; NASDAQ down: 10
    5. DJIA down; NASDAQ down: 42
    6. DJIA unchanged; NASDAQ down: 5
    7. DJIA up; NASDAQ unchanged: 4
    8. DJIA down; NASDAQ unchanged: 3
    9. DJIA unchanged; NASDAQ unchanged: 2
    a. Develop a joint probability distribution for the movement of the DJIA and NASDAQ indexes.
    b. What is the probability P(DJIA up or DJIA unchanged)?
    c. What is the probability P(NASDAQ unchanged)?
    d. What is the probability P(DJIA unchanged or NASDAQ unchanged)?
    e. If an investor heard on the news that the DJIA was down, what is the probability that the NASDAQ was also down?
26. Devise and implement a Monte Carlo experiment for estimating the standard deviation of the range, σ_R, of a sample of normally-distributed random variables as a function of the standard deviation of a normal distribution, σ. That is, determine

the distribution of σ_R/σ and estimate the expected value, d_3, for sample size 5. Compare your results with the published value 0.864.

27. A government agency is putting a large project out for low bid. Bids are expected from 10 different contractors and will have a normal distribution with a mean of $3.5 million and a standard deviation of $0.25 million. Devise and implement a Monte Carlo experiment for estimating the distribution of the minimum bid and the expected value of the minimum bid.

28. Consider the situation described in Problem 16. Suppose that the car manufacturer samples 9 cars from its assembly line and tests them for mileage ratings.
   a. What is the distribution of the mean mpg for the sample?
   b. What is the probability that the sample mean will be greater than 25 mpg?
   c. What is the probability that the sample mean will be less than 24.5 mpg?

29. Consider the situation described in Problem 18.
   a. If the manufacturer samples 100 bottles, what is the probability that the mean is less than 1,950 mL?
   b. What mean overfill or more will occur only 10% of the time for the sample of 100 bottles?

Questions 30–32 rely on optional material found in the appendix to this chapter.

30. A home pregnancy test is not always accurate. Suppose the probability that the test indicates that a woman is pregnant when she actually is not is 0.02, and the probability the test indicates that a woman is not pregnant when she really is, is 0.03. Assume that the probability that a woman who takes the test is actually pregnant is 0.7. What is the probability that a woman is pregnant if the test yields a not pregnant result?

31. Hahn Engineering is planning on bidding on a job and often competes against a major competitor, Sweigart and Associates (S&A), as well as other firms. Historically, S&A has bid for the same jobs 80% of the time. If S&A bids on a job, the probability that Hahn Engineering will win it is 0.30. If S&A does not bid on a job, the probability that Hahn will win the bid is 0.60. What is the probability that Hahn Engineering will win the bid? If they do, what is the probability that S&A did bid on it?

32. A consumer products company found that 48% of successful products also received favorable results from test market research, while 12% had unfavorable results but nevertheless were successful. They also found that 28% of unsuccessful products had unfavorable research results, while 12% of them had favorable research results. Construct the joint probability distribution for the two random variables: product success and test market research results.

CASE

Probability Analysis for Quality Measurements

A manufacturer of home and industrial lawn and garden equipment collects a variety of data from special studies, many of which are related to quality control. The company routinely collects data about functional test performance of its mowers after assembly; results from the past 30 days are given in the worksheet *Mower Test* in the Excel file *Quality Measurements*. In addition, many in-process measurements are taken to ensure that manufacturing processes remain in control and can produce according to design specifications. The worksheet *Process*

Capability provides the results of 200 samples of blade weights taken from the manufacturing process that produces mower blades. You have been asked you to evaluate these data. Specifically,

1. What fraction of mowers fails for each of the 30 samples in the worksheet *Mower Test*? What distribution might be appropriate to model the failure of an individual mower? Using these data, estimate the sampling distribution of the mean, the overall fraction of failures, and the standard error of the mean. Is a normal distribution an appropriate assumption for the sampling distribution of the mean?

2. What fraction of mowers fails the functional performance test using all the data in the worksheet *Mower Test*? Using this result, what is the probability of having *x* failures in the next 100 mowers tested, for *x* from 0 to 20?

3. Do the data in the worksheet *Process Capability* appear to be normally distributed? (Construct a frequency distribution and histogram and use these to draw a conclusion.) If not, based on the histogram, what distribution might better represent the data?

4. Estimate the mean and standard deviation for the data in the worksheet *Process Capability*. Using these values, and assuming that the process capability data are normal, find the probability that blade weights from this process will exceed 5.20. What is the probability that weights will be less than 4.80? What is the actual percent of weights that exceed 5.20 or are less than 4.80 from the data in the worksheet? How do the normal probability calculations compare? What do you conclude?

Summarize all your findings to these questions in a well-written report.

Appendix

Probability: Theory and Computation

In this appendix, we expand upon the basic concepts developed earlier in this chapter and provide details regarding the formulas used to perform many of the calculations involving probabilities and probability distributions.

Expected Value and Variance of a Random Variable
As noted earlier in the chapter, the expected value of a discrete random variable, $E[X]$, is the sum of the product of the values of the random variable and their probabilities. $E[X]$ is computed using the following formula:

$$E[X] = \sum_{i=1}^{\infty} x_i f(x_i) \qquad (3A.1)$$

Note the similarity to computing the population mean using Formula (2A.9) in Chapter 2:

$$\mu = \frac{\sum_{i=1}^{N} f_i x_i}{N}$$

If we write this as the sum of x_i times (f_i/N), then we can think of f_i/N as the probability of x_i. Then this expression for the mean has the same basic form as the expected

value formula. Similarly, the variance of a random variable is computed as:

$$\text{Var}[X] = \sum_{j=1}^{\infty} (x_j - E[X])^2 f(x_j) \qquad (3A.2)$$

Note the similarity between this and Formula (2A.10) in Chapter 2.

Binomial Distribution
The probability mass function for the binomial distribution is:

$$f(x) = \begin{cases} \binom{n}{x} p^x (1-p)^{n-x}, & \text{for } x = 0, 1, 2, \ldots, n \\ 0, & \text{otherwise} \end{cases} \qquad (3A.3)$$

The notation $\binom{n}{x}$ represents the number of ways of choosing x distinct items from a group of n items and is computed as:

$$\binom{n}{x} = \frac{n!}{x!(n-x)!} \qquad (3A.4)$$

where $n!$ (n factorial) $= n(n-1)(n-2)\ldots(2)(1)$, and $0!$ is defined to be 1.

For example, if the probability that any individual will react positively to a new drug is 0.8, then the probability distribution that x individuals will react positively out of a sample of 10 is:

$$f(x) = \begin{cases} \binom{10}{x}(0.8)^x(0.2)^{10-x}, & \text{for } x = 0, 1, 2, \ldots, 10 \\ 0, & \text{otherwise} \end{cases}$$

If $x = 4$, for example, we have:

$$f(4) = \binom{10}{4}(0.8)^4(0.2)^{10-4} = \frac{10!}{4!6!}(0.4096)(0.000064)$$
$$= 0.005505$$

Poisson Distribution

The probability mass function for the Poisson distribution is:

$$f(x) = \begin{cases} \dfrac{e^{-\lambda}\lambda^x}{x!}, & \text{for } x = 0, 1, 2, \ldots \\ 0, & \text{otherwise} \end{cases} \tag{3A.5}$$

where the mean number of occurrences in the defined unit of measure is λ.

Suppose that on average, 12 customers arrive at an ATM during lunch hour. The probability that exactly x customers will arrive during the hour would be calculated using the formula:

$$f(x) = \begin{cases} \dfrac{e^{-12}12^x}{x!}, & \text{for } x = 0, 1, 2, \ldots \\ 0, & \text{otherwise} \end{cases} \tag{3A.5}$$

For example, the probability that exactly 5 customers will arrive is $e^{-12}(12^5)/5!$ or 0.12741.

Uniform Distribution

For a uniform distribution with a minimum value a and a maximum value b, the density function is:

$$f(x) = \frac{1}{b-a} \quad \text{if } a \le x \le b \tag{3A.6}$$

and the cumulative distribution function is:

$$f(x) = \begin{cases} 0, & \text{if } x < a \\ \dfrac{x-a}{b-a}, & \text{if } a \le x \le b \\ 1, & \text{if } b < x \end{cases} \tag{3A.7}$$

For example, suppose that sales revenue for a product varies uniformly each week between $a = \$1,000$ and $b = \$2,000$. Then the probability that sales revenue will be less than $x = \$1,300$ is:

$$F(1300) = \frac{1300 - 1000}{2000 - 1000} = 0.3$$

Similarly, the probability that revenue will be between \$1,500 and \$1,700 is $F(1700) - F(1500) = 0.7 - 0.5 = 0.2$.

Normal Distribution

The probability density function for the normal distribution is quite complex:

$$f(x) = \frac{e^{\frac{-(x-\mu)^2}{2\sigma^2}}}{\sqrt{2\pi\sigma^2}} \tag{3A.8}$$

In addition, the cumulative distribution function cannot be expressed mathematically, only numerically. A numerical tabulation of the standard normal distribution is provided in the appendix at the end of this book. Table A.1 presents a table of the cumulative probabilities for values of z from -3.9 to 3.99. To illustrate the use of this table, let us find the area under the normal distribution within one standard deviation of the mean. Note that $P(-1 < Z < 1) = F(1) - F(-1)$. Using Table A.1, we find $F(1) = 0.8413$ and $F(-1) = 0.1587$. Therefore, $P(-1 < Z < 1) = 0.8413 - 0.1587 = 0.6826$.

To simplify the calculation of normal probabilities for any random variable X with an arbitrary mean μ and standard deviation σ, we may transform any value of the random variable X to a random variable Z having a standard normal distribution by applying the following formula:

$$z = \frac{x - \mu}{\sigma} \tag{3A.9}$$

These are called **standardized normal values**, or sometimes simply **z-values**. Standardized z-values are expressed in units of standard deviations of X. Thus, a z-value of 1.5 means that the associated x-value is 1.5 standard deviations above the mean μ; similarly, a z-value of -2.0 corresponds to an x-value that is two standard deviations below the mean μ.

Standardized z-values are particularly useful in solving problems involving arbitrary normal distributions and allow us to find probabilities using the standard normal table in Table A.1. For example, suppose that we determine that the distribution of customer demand (X) is normal with a mean of 750 units/month and a standard deviation of 100 units/month. To find this probability using Table A.1, we transform this into a standard normal distribution by finding the z-value that corresponds to $x = 900$:

$$z = \frac{900 - 750}{100} = 1.5$$

This means that $x = 900$ is 1.5 standard deviations above the mean of 750. From Table A.1, $F(1.5)$ is 0.9332.

Therefore, the probability that Z *exceeds* 1.5 (equivalently, the probability that X exceeds 900) is $1 - 0.9332 = 0.0668$. To summarize,

$$P(X > 900) = P(Z > 1.5) = 1 - P(Z < 1.5)$$
$$= 1 - 0.9332$$
$$= 0.0668$$

Another common calculation involving the normal distribution is to find the value of x corresponding to a specified probability, or area. For this example, suppose that we wish to find the level of demand that will be exceeded only 10% of the time; that is, find the value of x so that $P(X > x) = 0.10$. An upper tail probability of 0.10 is equivalent to a cumulative probability of 0.90. Using Table A.1, we search for 0.90 as close as possible in the *body* of the table and obtain $z = 1.28$, corresponding to 0.8997. This means that a value of z that is 1.28 standard deviations above the mean has an upper tail area of approximately 0.10. Using the standard normal transformation,

$$z = \frac{x - 750}{100} = 1.28$$

and solving for x, we find $x = 878$.

Triangular Distribution

The probability density function for the triangular distribution is given by:

$$f(x) = \begin{cases} \dfrac{2(x - a)}{(b - a)(c - a)}, & \text{if } a \le x \le c \\ \dfrac{2(b - x)}{(b - a)(b - c)}, & \text{if } c < x \le b \\ 0, & \text{otherwise} \end{cases} \quad \text{(3A.10)}$$

The cumulative distribution function is:

$$F(x) = \begin{cases} 0, & \text{if } x < a \\ \dfrac{(x - a)^2}{(b - a)(c - a)}, & \text{if } a \le x \le c \\ 1 - \dfrac{(b - x)^2}{(b - a)(b - c)}, & \text{if } c < x \le b \\ 1, & \text{if } b < x \end{cases} \quad \text{(3A.11)}$$

For example, suppose that $a = 4, b = 5$, and $c = 4.5$. Then the probability that X is greater than 4.7 would be computed as $1 - F(4.7)$. Because 4.7 is between c and b, we compute $F(4.7)$ using the formula:

$$1 - \frac{(b - x)^2}{(b - a)(b - c)} = 1 - \frac{(5 - 4.7)^2}{(5 - 4)(5 - 4.5)}$$
$$= 0.18$$

See Figure 3A.1.

Figure 3A.1 Triangular Distribution

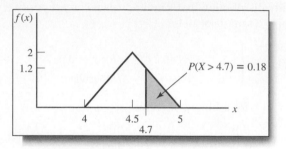

Exponential Distribution

The exponential distribution has the density function:

$$f(x) = \lambda e^{-\lambda x}, \quad x \ge 0 \quad \text{(3A.12)}$$

and its cumulative distribution function is:

$$F(x) = 1 - e^{-\lambda x}, \quad x \ge 0 \quad \text{(3A.13)}$$

Suppose that the mean time to failure of a critical component of an engine is $1/\lambda = 8,000$ hours. The probability that the component will fail before x hours is given by the cumulative distribution function. Thus, the probability of failing before 5,000 hours is:

$$F(5000) = 1 - e^{-(1/8000)(5000)} = 1 - e^{-5/8} = 0.465$$

Conditional Probability

We can use the conditional probability formula in many ways. One is to compute joint probabilities. By rearranging Formula (3.5) and exchanging the roles of A and B, we obtain:

$$P(A \text{ and } B) = P(A|B) \times P(B) = P(B|A) \times P(A)$$
$$\text{(3A.14)}$$

For example, suppose that in a game of Texas Hold 'Em, a player receives an ace on his first card. The probability that he will end up with "pocket aces" (two aces in the hand) is $P(ace\ on\ first\ card$ and $ace\ on\ second\ card) = P(ace\ on\ second\ card \mid ace\ on\ first\ card) \times P(ace\ on\ first\ card)$. Since the probability of an ace on the first card is 4/52, and the probability of an ace on the second card if he has already drawn an ace is 3/51, we have:

$$P(ace\ on\ first\ card \text{ and } ace\ on\ second\ card)$$
$$= P(ace\ on\ second\ card \mid ace\ on\ first\ card)$$
$$\times P(ace\ on\ first\ card)$$
$$= (3/51) \times (4/52) = 0.004525$$

We can present a general formula for computing marginal probabilities based on conditional probabilities:

$$P(A) = P(A \mid B_1) \times P(B_1) + P(A \mid B_2)$$
$$\times P(B_2) + \ldots + P(A \mid B_k) \times P(B_k) \quad \text{(3A.15)}$$

where $B_1, B_2, \ldots, B_k$ are k mutually exclusive and exhaustive outcomes. For example, in the brand preference example,

$$P(\text{Brand 1} \mid \text{male}) = 0.397$$
$$P(\text{Brand 1} \mid \text{female}) = 0.243$$
$$P(\text{male}) = 0.63$$
$$P(\text{female}) = 0.37$$

Then applying the general formula, we have:

$$\begin{aligned}P(\text{Brand 1}) &= P(\text{Brand 1} \mid \text{male}) \times P(\text{male}) \\ &+ P(\text{Brand 1} \mid \text{female}) \times P(\text{female}) \\ &= (0.397)(0.63) + (0.243)(0.37) \\ &= 0.25 + 0.09 = 0.34\end{aligned}$$

Bayes's Theorem

Bayes's theorem extends the concept of conditional probability to revise historical probabilities based on other information. Suppose that a digital camera manufacturer is developing a new model. Historically, 70% of cameras that have been introduced have been successful in terms of meeting a required return on investment, while 30% have been unsuccessful. Analysis of past market research studies, conducted prior to worldwide market introduction, has found that 90% of all successful product introductions had received favorable consumer response, while 60% of all unsuccessful products had also received favorable consumer response. If the new model receives a favorable response from a market research study, what is the probability that it will actually be successful?

Define the events:

A_1 = new camera successful
A_2 = new camera unsuccessful
B_1 = marketing response favorable
B_2 = marketing response unfavorable

Using these definitions and the information presented, we have:

$$P(A_1) = 0.7$$
$$P(A_2) = 0.3$$
$$P(B_1 \mid A_1) = 0.9$$
$$P(B_1 \mid A_2) = 0.6$$

This implies that $P(B_2 \mid A_1) = 0.1$ and $P(B_2 \mid A_2) = 0.4$, because the probabilities $P(B_1 \mid A_i) + P(B_2 \mid A_i)$ must add up to one for each A_i.

It is important to carefully distinguish between $P(A \mid B)$ and $P(B \mid A)$. As stated, *among all successful product introductions*, 90% received a favorable market research response. Thus, the probability of a favorable response *given* a successful introduction is 0.90, and not the other way around.

Suppose that $A_1, A_2, \ldots, A_k$ is a set of mutually exclusive and collectively exhaustive events, and we seek the probability that some event A_i occurs given that another event B has occurred. We will use the conditional probability formula:

$$P(A_i \mid B) = P(A_i \text{ and } B)/P(B)$$

to establish Bayes's theorem. Previously, we observed that $P(A_i \text{ and } B) = P(B \mid A_i) \times P(A_i)$. Substituting this into the numerator of the conditional probability formula, we obtain:

$$P(A_i \mid B) = \frac{P(B \mid A_i) \times P(A_i)}{P(B)}$$

If we substitute the general formula for computing marginal probabilities that we introduced earlier for $P(B)$ in the denominator, we may find the probability of event A_i given event B as:

$$P(A_i|B) =$$
$$\frac{P(B|A_i) \times P(A_i)}{P(B|A_1) \times P(A_1) + P(B|A_2) \times P(A_2) + \cdots + P(B|A_k) \times P(A_k)}$$
(3A.16)

This is a statement of Bayes's theorem. Bayes's theorem applies to situations in which we know the conditional probability of some event B given each of a set of mutually exclusive and exhaustive events, A_i, and seek the probability that event A_i will occur, knowing that event B has occurred.

We may apply it to our example as follows. The probability that the new camera will be successful given that the marketing response is favorable is:

$$P(A_1|B_1) = \frac{P(B_1|A_1) \times P(A_1)}{P(B_1|A_1) \times P(A_1) + P(B_1|A_2) \times P(A_2)}$$

$$P(A_1|B_1) = \frac{(0.9)(0.7)}{(0.9)(0.7) + (0.6)(0.3)} = 0.778$$

Although 70% of all previous new models have been successful, knowing that the marketing report is favorable increases the likelihood to 77.8%. A favorable market research response is not a guarantee of success because the research is occasionally wrong; however, it does increase the likelihood of success as a result of Bayes's theorem.

Chapter 4

Sampling and Estimation

INTRODUCTION

$\mathcal{J}$n Chapters 1 through 3, we discussed the use of data for managing and decision making, introduced methods for data visualization and descriptive statistics, and gained an understanding of some important probability distributions used in statistics and decision modeling. These topics were focused on how to view data to gain better understanding and insight. However, managers need to go further than simply understanding what data tell; they need to be able to *draw conclusions* about population characteristics from the data to make effective decisions. Sampling methods and estimation, as well as hypothesis testing and statistical inference, which we focus on in the next chapter, provide the means for doing this.

In this chapter, we focus on using sample data to estimate and make inferences about key population parameters. The key concepts and tools that we will present are as follows:

➤ *Statistical sampling, including sample design, simple random sampling, and other sampling schemes*

➤ *Understanding sampling error*

➤ *Estimation using point estimates and interval estimates*

➤ *Construction and interpretation of confidence intervals*

➤ *Sample size determination*

STATISTICAL SAMPLING

Sampling approaches play an important role in providing information for making business decisions. Market researchers, for example, need to sample from a large base of customers or potential customers; auditors must sample among large numbers of transactions; and quality control analysts need to sample production output to verify quality levels. Most populations, even if they are finite, are generally too large to deal with effectively or practically. For instance, it would be impractical as well as too expensive to survey the entire population of TV viewers in the United States. Sampling is also clearly necessary when data must be obtained from destructive testing or from a continuous production process. Thus, the purpose of sampling is to obtain sufficient information to draw a valid inference about a population.

Sample Design

The first step in sampling is to design an effective sampling plan that will yield representative samples of the populations under study. A **sampling plan** is a description of the approach that will be used to obtain samples from a population prior to any data collection activity. A sampling plan states the objectives of the sampling activity, the target population, the population **frame** (the list from which the sample is selected), the method of sampling, the operational procedures for collecting the data, and the statistical tools that will be used to analyze the data.

The objective of a sampling study might be to estimate key parameters of a population, such as a mean, proportion, or standard deviation. For example, *USA Today* reported on May 19, 2000, that the U.S. Census Bureau began a statistical sampling procedure to estimate the number and characteristics of people who might have been missed in the traditional head count. Another application of sampling is to determine if significant differences exist between two populations. For instance, the Excel file *Burglaries* provides data about monthly burglaries in an urban neighborhood before and after a citizen-police program was instituted. You might wish to determine whether the program was successful in reducing the number of burglaries.

The ideal frame is a complete list of all members of the target population. However, for practical reasons, a frame may not be the same as the target population. For example, a company's target population might be all golfers in America, which might be impossible to identify, whereas a practical frame might be a list of golfers who have registered handicaps with the U.S. Golf Association. Understanding how well the frame represents the target population helps us to understand how representative of the target population the actual sample is and, hence, the validity of any statistical conclusions drawn from the sample. In a classic example, *Literary Digest* polled individuals from telephone lists and membership rolls of country clubs for the 1936 presidential election and predicted that Alf Landon would defeat Franklin D. Roosevelt. The problem was that the frame—individuals who owned telephones and belonged to country clubs—was heavily biased toward Republicans and did not represent the population at large.

Sampling Methods

Sampling methods can be *subjective* or *probabilistic*. Subjective methods include **judgment sampling**, in which expert judgment is used to select the sample (survey the "best" customers), and **convenience sampling**, in which samples are selected based on the ease with which the data can be collected (survey all customers I happen to visit this month). Probabilistic sampling involves selecting the items in the sample using some random procedure. Probabilistic sampling is necessary to draw valid statistical conclusions.

The most common probabilistic sampling approach is simple random sampling. **Simple random sampling** involves selecting items from a population so that every subset of a given size has an equal chance of being selected. If the population data are stored in a database, simple random samples can generally be obtained easily by generating random numbers, as we discussed in Chapter 3. For example, consider the Excel file *Cereal Data*, which contains nutritional and marketing information for 67 cereals, a portion of which is shown in Figure 4.1. Suppose that we wish to sample 20

Figure 4.1   Portion of the Excel File *Cereal Data*

	A	B	C	D	E	F	G	H	I
1	Cereal Data								
2									
3	Product	Cereal Name	Manufacturer	Calories	Sodium	Fiber	Carbs	Sugars	Shelf
4	1	100% Bran	Nabisco	70	130	10	5	6	3
5	2	All-Bran	Kellogg	70	260	9	7	5	3
6	3	All-Bran w/Extra Fiber	Kellogg	50	140	14	8	0	3
7	4	Almond Delight	Ralston Purina	110	200	1	14	8	3
8	5	Apple Cinn Cheerios	General Mills	110	180	1.50	10.50	10	1
9	6	Apple Jacks	Kellogg	110	125	1	11	14	2
10	7	Basic 4	General Mills	130	210	2	18	8	3
11	8	Bran Chex	Ralston Purina	90	200	4	15	6	1
12	9	Bran Flakes	Post	90	210	5	13	5	3
13	10	Cap'n'Crunch	Quaker	120	220	0	12	12	2

of these cereals. *PHStat* provides a tool to generate a random set of values from a given population size *without replacement,* guaranteeing that each item in the sample will be unique (see *PHStat Note: Using the Random Sample Generator*). The Excel *Data Analysis* tool *Sampling* also allows you to select a simple random sample or a systematic sample from a list in the worksheet (see *Excel Note: Using the Sampling Tool*). However, this tool generates random samples *with replacement,* so you must be careful to check for duplicate observations in the sample.

Other methods of sampling include the following:

- **Systematic (Periodic) Sampling.** This is a sampling plan that selects items periodically from the population. For example, to sample 250 names from a list of 400,000, the first name could be selected at random from the first 1600, and then every 1,600th name could be selected. This approach can be used for telephone sampling when supported by an automatic dialer that is programmed to dial

PHSTAT NOTE
Using the Random Sample Generator

This tool can be used to generate a random list of integers between one and a specified population size or to randomly select values from a range of data on a worksheet without replacement. From the *PHStat* menu, select *Sampling,* then *Random Sample Generator.* Figure 4.2 shows the dialog box that appears. Enter the sample size desired in the *Sample Size* box. Click the first radio button if you want a list of random integers, and enter the population size in the box below this option. Click the second radio button to select a sample from data on a worksheet. The range of the data must be entered in the *Values Cell Range* box (checking *First cell contains label* if appropriate). This range must be a single column containing the values from which to draw the random sample. Figure 4.3 shows a sample of 10 cereals generated from the Excel file *Cereal Data* using this tool.

Figure 4.2 *PHStat Random Sample Generation* Dialog

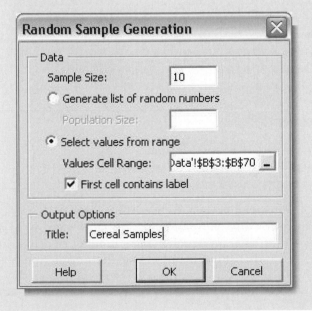

Figure 4.3 Random Sample of Ten Cereals

	A
1	**Cereal Samples**
2	Mueslix Crispy Blend
3	Life
4	Post Nat. Raisin Bran
5	Product 19
6	Crispix
7	Honey Comb
8	All-Bran w/Extra Fiber
9	Grape Nuts Flakes
10	Rice Chex
11	Corn Flakes

EXCEL NOTE
Using the Sampling Tool

Click on *Data Analysis* in the *Analysis* group of the *Data* tab and select *Sampling*. This brings up the dialog box shown in Figure 4.4. In the *Input Range* box, you specify the data range from which the sample will be taken—in this example, the list of product numbers. (This tool requires that the data sampled be numeric; the *PHStat Random Sample Generator* tool does not have this restriction.) The *Labels* box can be checked if the first row is a data set label. There are two options for sampling:

1. Sampling can be *periodic*, and you will be prompted for the *Period*, which is the interval between sample observations from the beginning of the data set. For instance, if a period of 5 is used, observations 5, 10, 15, etc. will be selected as samples.

2. Sampling can also be *random*, and you will be prompted for the *Number of Samples*. Excel will then randomly select this number of samples (with replacement) from the specified data set.

This tool was used to sample 10 products from the Excel file *Cereal Data*. Figure 4.5 shows the output. The tool only provides the list of product numbers in column K. We have extracted the product names by using the VLOOKUP function. For example, the formula in cell L2 is =VLOOKUP(K2,$A$4: $B$70,2). Note that product number 53 was selected twice because this tool samples with replacement.

Figure 4.4 Excel *Sampling* Tool Dialog

Figure 4.5 *Sampling* Tool Results

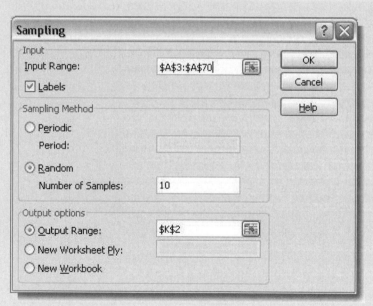

	K	L
1	**Sample**	**Product**
2	41	Lucky Charms
3	27	Fruit & Fibre
4	18	Count Chocula
5	53	Rice Chex
6	53	Rice Chex
7	10	Cap'n'Crunch
8	66	Wheaties
9	47	Product 19
10	7	Basic 4
11	6	Apple Jacks

numbers in a systematic manner. However, systematic sampling is not the same as simple random sampling because for any sample, every possible sample of a given size in the population does not have an equal chance of being selected. In some situations, this approach can induce significant bias if the population has some underlying pattern. For instance, sampling orders received every seven days may not yield a representative sample if customers tend to send orders on certain days every week.

- *Stratified Sampling.* This type of sampling applies to populations that are divided into natural subsets (strata) and allocates the appropriate proportion of samples to each stratum. For example, a large city may be divided into political districts called wards. Each ward has a different number of citizens. A stratified sample would choose a sample of individuals in each ward proportionate to its size. This approach ensures that each stratum is weighted by its size relative to the population and can provide better results than simple random sampling if the items in each stratum are not homogeneous. However, issues of cost or significance of certain strata might make a disproportionate sample more useful. For example, the ethnic or racial mix of each ward might be significantly different, making it difficult for a stratified sample to obtain the desired information.

- *Cluster Sampling.* This is based on dividing a population into subgroups (clusters), sampling a set of clusters, and (usually) conducting a complete census within the clusters sampled. For instance, a company might segment its customers into small geographical regions. A cluster sample would consist of a random sample of the geographical regions, and all customers within these regions would be surveyed (which might be easier because regional lists might be easier to produce and mail).

- *Sampling from a Continuous Process.* Selecting a sample from a continuous manufacturing process can be accomplished in two main ways. First, select a time at random; then select the next n items produced after that time. Second, select n times at random; then select the next item produced after these times. The first approach generally ensures that the observations will come from a homogeneous population; however, the second approach might include items from different populations if the characteristics of the process should change over time, so caution should be used.

Errors in Sampling

The purpose of sampling is to obtain statistics that estimate population parameters. Sample design can lead to two sources of errors. The first type of error, **nonsampling error**, occurs when the sample does not represent the target population adequately. This is generally a result of poor sample design, such as using a convenience sample when a simple random sample would have been more appropriate or choosing the wrong sampling frame. **Sampling (statistical) error** occurs because samples are only a subset of the total population. We observed such sampling error with the Monte Carlo simulation experiments in Chapter 3. Sampling error is inherent in any sampling process, and although it can be minimized, it cannot be totally avoided.

Sampling error depends on the size of the sample relative to the population. Thus, determining the number of samples to take is essentially a statistical issue that is based on the accuracy of the estimates needed to draw a useful conclusion. We discuss this later in this chapter. However, from a practical standpoint, one must also consider the cost of sampling and sometimes make a trade-off between cost and the information that is obtained.

ESTIMATION

Sample data provide the basis for many useful analyses to support decision making. **Estimation** involves assessing the value of an unknown population parameter— such as a population mean, population proportion, or population variance—using sample data. When we sample, the estimators we use—such as a sample mean, sample proportion, or sample variance—are random variables that are characterized by some sampling distribution. For example, in Chapter 3, we introduced the sampling distribution of the mean, and showed that it is approximately normally distributed with a mean equal to the mean of the population and a standard deviation equal to the standard deviation of the population divided by the square root of the sample size. By knowing the sampling distribution of $\bar{x}$, we can use probability theory to quantify the uncertainty associated with using the sample mean to estimate the population mean.

We use two types of estimates in statistics. **Point estimates** are single numbers used to estimate the value of a population parameter. However, because of sampling error, it is unlikely that a point estimate will equal the true value of the population parameter, and the point estimate alone does not provide any information of the magnitude of the sampling error. **Confidence interval estimates** provide a range of values between which the value of the population parameter is believed to be, and also provide an assessment of sampling error by specifying a probability that the interval correctly estimates the true (unknown) population parameter. Microsoft Excel and the *PHStat* add-in provide several options for supporting these analyses, as summarized in Table 4.1.

Point Estimates

The most common point estimates are the descriptive statistical measures we described in Chapter 2 and summarized in Table 4.2 along with their corresponding population parameters. They are used to estimate the population parameters, also listed in Table 4.2.

EXCEL FUNCTION	DESCRIPTION
CONFIDENCE(*alpha, standard_dev, size*)	Returns the confidence interval for a population mean
ANALYSIS TOOLPAK TOOLS	DESCRIPTION
Sampling	Creates a simple random sample with replacement or a systematic sample from a population
PHSTAT ADD-IN	DESCRIPTION
Random Sample Generator	Generates a random sample without replacement
Confidence Intervals	Computes confidence intervals for means with σ known or unknown, proportions, and population total
Sample Size	Determines sample sizes for means and proportions

Table 4.1
Sampling and Estimation Support in Excel

POINT ESTIMATE	POPULATION PARAMETER
Sample mean, $\bar{x}$	Population mean, μ
Sample variance, s^2	Population variance, σ^2
Sample standard deviation, s	Population standard deviation, σ
Sample proportion, $\hat{p}$	Population proportion, π

Table 4.2
Common Point Estimates

Figure 4.6 Point Estimates for Cereal Samples

	M	N	O	P	Q	R
1	**Cereal Samples**	**Calories**	**Sodium**	**Fiber**	**Carbs**	**Sugars**
2	Mueslix Crispy Blend	160	150	3	17	13
3	Life	100	150	2	12	6
4	Post Nat. Raisin Bran	120	200	6	11	14
5	Product 19	100	320	1	20	3
6	Crispix	110	220	1	21	3
7	Honey Comb	110	180	0	14	11
8	All-Bran w/Extra Fiber	50	140	14	8	0
9	Grape Nuts Flakes	100	140	3	15	5
10	Rice Chex	110	240	0	23	2
11	Corn Flakes	100	290	1	21	2
12	Sample Mean	106.00	203.00	3.10	16.20	5.90
13	Sample Standard Deviation	26.750	64.127	4.228	5.007	4.999
14						
15	**Population Parameters**					
16	Mean	105.52	167.31	2.19	14.77	6.96
17	Standard Deviation	18.631	79.978	2.487	3.831	4.376

Figure 4.6 shows 10 samples from the *Cereal Data* file that were selected using the *PHStat Random Sample Generator.* We calculated the sample mean and standard deviation for calories, sodium, fiber, carbohydrates, and sugars in the sample as well as the mean and standard deviation for the entire population in the data set. The sample statistics are point estimates. Notice that there are some considerable differences as compared to the population parameters because of sampling error. A point estimate alone does not provide any indication of the magnitude of the potential error in the estimate. A major metropolitan newspaper reported that college professors were the highest-paid workers in the region, with an average of $150,004, based on a Bureau of Labor Statistics survey. Actual averages for two local universities were less than $70,000. What happened? As reported in a follow-up story, the sample size was very small and included a large number of highly paid medical school faculty; as a result, the sampling error was huge. Interval estimates, which we will discuss soon, provide better information than point estimates alone.

Unbiased Estimators

It seems quite intuitive that the sample mean should provide a good point estimate for the population mean. However, it may not be clear why the formula for the sample variance that we introduced in Chapter 2 has a denominator of $n - 1$, particularly because it is different from the formula for the population variance [see formulas (2A.4) and (2A.5) in the appendix to Chapter 2]. Recall that the population variance is computed by:

$$\sigma^2 = \frac{\sum_{i=1}^{N}(x_i - \mu)^2}{N}$$

whereas the sample variance is computed by the formula:

$$s^2 = \frac{\sum_{i=1}^{n}(x_i - \bar{x})^2}{n - 1}$$

Why is this so? Statisticians develop many types of estimators, and from a theoretical as well as a practical perspective, it is important that they "truly estimate" the population parameters they are supposed to estimate. Suppose that we perform an experiment in which we repeatedly sampled from a population and computed a point estimate for a population parameter (similar to the Monte Carlo experiment we did in Chapter 3). Each individual point estimate will vary from the population parameter; however, we would hope that the long-term average (expected value) of all possible point estimates would equal the population parameter. If the expected value of an estimator equals the population parameter it is intended to estimate, the estimator is said to be **unbiased**. If this is not true, the estimator is called **biased** and will not provide correct results.

Fortunately, all the estimators in Table 4.2 are unbiased and, therefore, are meaningful for making decisions involving the population parameter. In particular, statisticians have shown that the denominator $n - 1$ used in computing s^2 is necessary to provide an unbiased estimator of σ^2. If we simply divided by the number of observations, the estimator would tend to underestimate the true variance.

Interval Estimates

An **interval estimate** provides a range within which we believe the true population parameter falls. A typical interval estimate is constructed by taking a point estimate and adding and subtracting a margin of error that we can calculate. For example, a Gallup poll might report that 56% of voters support a certain candidate with a margin of error of ±3%. We would conclude that the true percentage of voters that support the candidate is probably between 53% and 59%. Therefore, we would have a lot of confidence in predicting that the candidate would win a forthcoming election. If, however, the poll showed a 52% level of support with a margin of error of ±4%, we might not be as confident in predicting a win because the true percentage of supportive voters is probably somewhere between 48% and 56%.

The question you might be asking at this point is how to calculate the margin of error. In national surveys and political polls, such margins of error are usually stated, but they are never clearly explained! To understand them, we need to introduce the concept of confidence intervals.

CONFIDENCE INTERVALS: CONCEPTS AND APPLICATIONS

A **confidence interval** is an interval estimate that also specifies the likelihood that the interval contains the true population parameter. This probability is called the **level of confidence**, denoted by $1 - \alpha$, where α is a number between 0 and 1. The level of confidence is usually expressed as a percent; common values are 90%, 95%, or 99%. (Note that if the level of confidence is 90%, then $\alpha = 0.1$.) The margin of error depends on the level of confidence and the sample size. For example, suppose that the margin of error for some sample size and a level of confidence of 95% is calculated to be 2.0. One sample might yield a point estimate of 10. Then a 95% confidence interval would be [8, 12]. However, this interval may or may not include the true population mean. If we take a different sample, we will most likely have a different point estimate, say 10.4, which, given the same margin of error, would yield the interval estimate [8.4, 12.4]. Again, this may or may not include the true population mean. If we chose 100 different samples, leading to 100 different interval estimates, we would expect that 95% of them—the level of confidence—would contain the true population mean. We would say we are "95% confident" that the interval we obtain from sample data contains the true population mean. The higher the confidence level, the more assurance we have that the interval contains the true population parameter. As the confidence level increases, the confidence interval becomes larger to provide higher levels of assurance. You can view α as the risk of incorrectly concluding that the confidence interval contains the true mean.

When national surveys or political polls report an interval estimate, they are actually confidence intervals. However, the level of confidence is generally not stated because the average person would probably not understand the concept or terminology. While not stated, you can probably assume that the level of confidence is 95%, as this is the most common value used in practice.

Many different types of confidence intervals may be developed. The formulas used depend on the population parameter we are trying to estimate and possibly other characteristics or assumptions about the population. Table 4.3 provides a summary of the most common types of confidence intervals and *PHStat* tools available for computing them (no tools are available in Excel for computing confidence intervals). All tools can be found in the *Confidence Intervals* menu within *PHStat*. We will discuss other types of confidence intervals later in this chapter and in the appendix at the end of the chapter.

Confidence Interval for the Mean with Known Population Standard Deviation

The simplest type of confidence interval is for the mean of a population where the standard deviation is assumed to be known. You should realize, however, that in nearly all practical sampling applications, the population standard deviation will *not* be known. However, in some applications, such as measurements of parts from an automated machine, a process might have a very stable variance that has been established over a long history, and it can reasonably be assumed that the standard deviation is known.

To illustrate this type of confidence interval, we will use the cereal samples in Figure 4.6, as we have already calculated the population standard deviation of the cereal characteristics. For example, the point estimate for calories for the 10 samples of cereals was calculated as 106.0. The population standard deviation is known to be 18.631. In most practical applications, samples are drawn from very large populations. If the population is relatively small compared to the sample size, a modification must be made to the confidence interval. Specifically, when the sample size, n, is larger than 5% of the population size, N, a correction factor is needed in computing

Type of Confidence Interval	PHStat Tool
Mean, standard deviation known	Estimate for the mean, sigma known
Mean, standard deviation unknown	Estimate for the mean, sigma unknown
Proportion	Estimate for the proportion
Variance	Estimate for the population variance
Population total	Estimate for the population total

Table 4.3
Common
Confidence
Intervals

PHStat has two tools for finding confidence intervals for the mean: one assumes that the population standard deviation is known; the second assumes that it is unknown. Both input dialogs are similar. If the population standard deviation is assumed known, choose *Confidence Intervals* from the *PHStat* menu and select *Estimate for the Mean, Sigma Known*. The dialog box is shown in Figure 4.7. Enter the known population standard deviation. If the sample statistics (sample mean) has been calculated, you may enter it along with the sample size; otherwise you may check the radio button for *Sample Statistics Unknown* and enter the range of the data and the tool will perform the calculations. If the finite population correction is needed, check the box and enter the population size.

Figure 4.7 *PHStat* Dialog for Confidence Interval *Estimate for the Mean, Sigma Known*

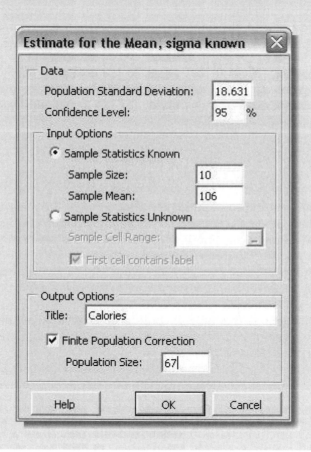

the margin of error. In this example, the population is only 67, so the 10 samples represent about 15% of the population. Therefore, in using the *PHStat* tool (see *PHStat Note: Confidence Intervals for the Mean*), we checked the box for the finite population correction.

Figure 4.8 shows the *PHStat* results. The 95% confidence interval (including the finite population correction) is [95.27, 116.73]. This means that we believe that the population mean of the number of calories per serving of breakfast cereals is

Figure 4.8 Results for a 95% Confidence
Interval for Calories

	A	B
1	Calories	
2		
3	Data	
4	Population Standard Deviation	18.631
5	Sample Mean	106
6	Sample Size	10
7	Confidence Level	95%
8		
9	Intermediate Calculations	
10	Standard Error of the Mean	5.891639509
11	Z Value	-1.95996398
12	Interval Half Width	11.54740125
13		
14	Confidence Interval	
15	Interval Lower Limit	94.45259875
16	Interval Upper Limit	117.5474012
17		
18		
19	Finite Populations	
20	Population Size	67
21	FPC Factor	0.929320377
22	Interval Half Width	10.73123528
23	Interval Lower Limit	95.26876472
24	Interval Upper Limit	116.7312353

somewhere between 95.27 and 116.73, with only a small likelihood (0.05) that the population mean is outside of this interval.

To fully understand these results, it is necessary to examine the formula used to compute the confidence interval. A $100(1 - \alpha)\%$ confidence interval for the population mean μ is given by:

$$\bar{x} \pm z_{\alpha/2}(\sigma/\sqrt{n}) \tag{4.1}$$

Note that this formula is simply the sample mean (point estimate) plus or minus a margin of error. The margin of error is a number $z_{\alpha/2}$ times the standard error of the sampling distribution of the mean, $\sigma/\sqrt{n}$, which was discussed in the previous chapter. The value $z_{\alpha/2}$ represents the value of a standard normal random variable that has a cumulative probability of $\alpha/2$ (the reasoning is explained in the appendix to this chapter). It may be found from the standard normal table (see Table A.1 in the appendix at the end of the book) or may be computed in Excel using the function NORMSINV($\alpha/2$). These values are shown in the *Intermediate Calculations* section of the *PHStat* results. The standard error is 5.8916 and $z_{\alpha/2}$ (z-value) is approximately -1.96. Note that the z-value is negative because $\alpha/2$ represents a small area in the left tail of the standard normal distribution; however, to calculate the margin of error, the positive value is used. Therefore, the margin of error, or interval half-width, is $(5.8916)(1.96) = 11.547$, resulting in the confidence interval 106.0 ± 11.547 or [94.45, 117.55], without consideration of the finite population correction. For finite populations, the finite population correction (FPC) factor is:

$$\sqrt{\frac{N - n}{N - 1}}$$

In this example, the FPC factor is 0.929 and is multiplied by the standard error in order to find the adjusted interval half-width and the confidence interval. That is, the adjusted standard error of the mean is:

$$\sigma_{\bar{x}} = \frac{\sigma}{\sqrt{n}} \sqrt{\frac{N - n}{N - 1}} \tag{4.2}$$

Note that if n is small relative to N, then the term $\sqrt{(N - n)/(N - 1)}$ is approximately 1 and the difference is minor.

Note that as the level of confidence, $1 - \alpha$, decreases, $z_{\alpha/2}$ decreases, and the confidence interval becomes smaller. For example, a 90% confidence interval will be smaller than a 95% confidence interval. Similarly, a 99% confidence interval will be larger than a 95% confidence interval. Essentially, you must trade off a higher level of accuracy with the risk that the confidence interval does not contain the true mean. Smaller risk will result in a larger confidence interval. However, you can also see that as the sample size increases, the standard error decreases, making the confidence interval smaller and providing a more accurate interval estimate for the same level of risk. So if you wish to reduce the risk, you should consider increasing the sample size.

Confidence Interval for the Mean with Unknown Population Standard Deviation

In most practical applications, the standard deviation of the population is unknown, and we need to calculate the confidence interval differently. For example, in the Excel file *Social Networking*, we have sample data about social networking habits of 50 students. Suppose that we want to estimate the mean number of hours online/week for the population of students from which this sample was taken (which we assume is large so that the finite population correction factor is not required). We could use the *PHStat* tool *Estimate for the Mean, Sigma Unknown* to find a confidence interval using these data. Figure 4.9 shows the results. The tool

	A	B
1	**Hours Online/Week**	
2		
3	**Data**	
4	Sample Standard Deviation	7.390120266
5	Sample Mean	12.28
6	Sample Size	50
7	Confidence Level	95%
8		
9	Intermediate Calculations	
10	Standard Error of the Mean	1.045120831
11	Degrees of Freedom	49
12	*t* Value	2.009575199
13	Interval Half Width	2.100248902
14		
15	Confidence Interval	
16	Interval Lower Limit	10.18
17	Interval Upper Limit	14.38

Figure 4.9 *PHStat* Results for Confidence Interval for Hours Online/Week

calculates the sample statistics and the confidence interval using the intermediate calculations. The confidence interval is the sample mean plus or minus the interval half-width, or 12.28 ± 2.10, or [10.18, 14.38].

You will notice that the intermediate calculations are somewhat different from the case in which the population standard deviation was known. Instead of using $z_{\alpha/2}$ based on the normal distribution, the tool uses a "*t*-value" with which to multiply the standard error to compute the interval half-width. The *t*-value comes from a new probability distribution called the **t-distribution**. The *t*-distribution is actually a family of probability distributions with a shape similar to the standard normal distribution. Different *t*-distributions are distinguished by an additional parameter, **degrees of freedom (df)**. The *t*-distribution has a larger variance than the standard normal, thus making confidence intervals wider than those obtained from the standard normal distribution, in essence correcting for the uncertainty about the true standard deviation. As the number of degrees of freedom increases, the *t*-distribution converges to the standard normal distribution (Figure 4.10). When sample sizes get to be as large as 120, the distributions are virtually identical; even for sample sizes as low as 30–35, it becomes difficult to distinguish between the two. Thus, for large sample sizes, many people use *z*-values to establish confidence intervals even when the standard deviation is unknown. We must point out, however, that for any sample size, the *true* sampling distribution of the mean is the *t*-distribution, so when in doubt, use the *t*.

The concept of "degrees of freedom" can be puzzling. It can best be explained by examining the formula for the sample variance:

$$s^2 = \frac{\sum_{i=1}^{n}(x_i - \overline{x})^2}{n - 1}$$

Note that to compute s^2 we need to first compute the sample mean, $\overline{x}$. If we know the value of the mean, then we need only know $n - 1$ distinct observations; the nth is completely determined. (For instance, if the mean of 3 values is 4, and you know that two of the values are 2 and 4, you can easily determine that the third number must be 6.) The number of sample values that are free to vary defines the number of degrees of

Figure 4.10 Comparison of the *t*-Distribution to the Standard Normal Distribution

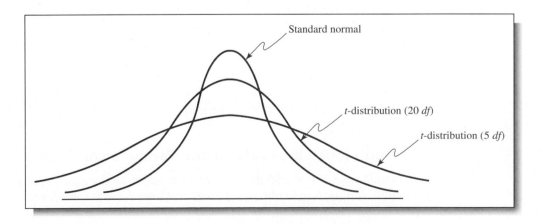

freedom; in general, *df* equals the number of sample values minus the number of esti-mated parameters. Because the sample variance uses one estimated parameter, the mean, the *t*-distribution used in confidence interval calculations has $n - 1$ degrees of freedom. Because the *t*-distribution explicitly accounts for the effect of the sample size in estimating the population variance, it is the proper one to use for any sample size. However, for large samples, the difference between *t*- and *z*-values is very small, as we noted earlier.

The formula for a $100(1 - \alpha)\%$ confidence interval for the mean μ when the population standard deviation is unknown is:

$$\bar{x} \pm t_{\alpha/2, n-1}(s/\sqrt{n}) \tag{4.3}$$

where $t_{\alpha/2, n-1}$ is the value from the *t*-distribution with $n - 1$ degrees of freedom, giving an upper-tail probability of $\alpha/2$. We may find *t*-values in Table A.2 in the appendix at the end of the book or by using the Excel function $\text{TINV}(\alpha, n - 1)$. However, take careful note that Table A.2 provides upper tail values only, whereas the TINV function accounts for both tails. Thus, $t_{\alpha/2, n-1}$ in Table A.2 is equivalent to $\text{TINV}(\alpha, n - 1)$. Thus, in our example, $t_{0.025, 49} = 2.0096$, and the standard error $s/\sqrt{n} = 1.045$, resulting in the confidence interval $12.28 \pm (2.0096)(1.045)$.

Confidence Interval for a Proportion

For categorical variables having only two possible outcomes, such as good or bad, male or female, and so on, we are usually interested in the *proportion* of observations in a sample that has a certain characteristic. An unbiased estimator of a population proportion π is the statistic $\hat{p} = x/n$ (the **sample proportion**), where x is the number in the sample having the desired characteristic and n is the sample size. For example, in the Excel file *Social Networking*, suppose we are interested in the proportion of stu-dents who use both Facebook and Myspace. We may easily confirm that 29 out of the 50 students used both social networking sites. Thus, $\hat{p} = 29/50 = 0.58$ is a point esti-mate for the proportion. Using the *PHStat* tool *Estimate for the Proportion* (see *PHStat Note: Confidence Interval for Proportions*), we find that a 95% confidence interval for the proportion of students who use both Facebook and MySpace is [0.443, 0.716]. Notice that this is a fairly large confidence interval, suggesting that we have quite a bit of uncertainty as to the true value of the population proportion. This is because of the relatively small sample size.

These calculations are based on the following: a $100(1 - \alpha)$ confidence interval for the proportion is:

$$\hat{p} \pm z_{a/2}\sqrt{\frac{\hat{p}(1 - \hat{p})}{n}} \tag{4.4}$$

Notice that as with the mean, the confidence interval is the point estimate plus or minus some margin of error. In this case, $\sqrt{\hat{p}(1 - \hat{p})}/n$ is the standard error for the sampling distribution of the proportion.

Confidence Intervals for the Variance and Standard Deviation

Understanding variability is critical to effective decisions. Thus, in many situations, one is interested in obtaining point and interval estimates for the variance or standard

PHSTAT NOTE
Confidence Interval for Proportions

From the *PHStat Confidence Intervals* menu, select *Estimate for the Proportion*. The dialog is shown in Figure 4.11. Enter the sample size and number of successes, that is, the number of observations having the characteristic of interest.

Figure 4.11 *PHStat* Dialog for Confidence Interval *Estimate for the Proportion* and Results

Estimate for the Proportion

Data
Sample Size: 50
Number of Successes: 29
Confidence Level: 95 %

Output Options
Title: Facebook/Myspace Users
☐ Finite Population Correction
Population Size:

Help   OK   Cancel

	A	B
1	Facebook/Myspace Users	
2		
3	Data	
4	Sample Size	50
5	Number of Successes	29
6	Confidence Level	95%
7		
8	Intermediate Calculations	
9	Sample Proportion	0.58
10	Z Value	-1.95996398
11	Standard Error of the Proportion	0.069799713
12	Interval Half Width	0.136804925
13		
14	Confidence Interval	
15	Interval Lower Limit	0.443195075
16	Interval Upper Limit	0.716804925

deviation. For example, the Excel file *Home Market Value* provides some data on a sample of houses in a residential neighborhood (see Figure 4.12). The standard deviation of market value provides information on the spread of home prices in this neighborhood. Using the sample standard deviation, $10,553, and a 2-standard deviation

Figure 4.12 Portion of Excel File *Home Market Value*

	A	B	C
1	Home Market Value		
2			
3	House Age	Square Feet	Market Value
4	33	1,812	$90,000.00
5	32	1,914	$104,400.00
6	32	1,842	$93,300.00
7	33	1,812	$91,000.00
8	32	1,836	$101,900.00
9	33	2,028	$108,500.00
10	32	1,732	$87,600.00
11	33	1,850	$96,000.00
12	32	1,791	$89,200.00

spread, we might predict that about 95% of home prices (assuming a normal distribution) would deviate about 2($10,553) = $21,106 from the average value. However, because of sampling error, this value could be quite different.

PHStat has a tool for computing confidence intervals for the variance and standard deviation. This is explained in PHStat Note: Confidence Intervals for the Population Variance. The calculations assume that the population from which the sample was drawn has an approximate normal distribution. If this assumption is not met, the confidence interval may not be accurate for the confidence level chosen. Using the PHStat tool for the Home Market Value data, we obtain the results in Figure 4.13. A 95% confidence interval for the standard deviation of market values is [$8683, $13457]. In other words, the standard deviation might be as low as about $8600 or over $13,000. Thus, deviation of home prices from the average might be as small as 2($8683) = $17,366 or as large as 2($13,457) = $26,914.

Calculation of the confidence intervals for variance and standard deviation is quite different from the other confidence intervals we have studied. Although we use the sample standard deviation s as a point estimate for σ, the sampling distribution of s is not normal, but is related to a special distribution called the **chi-square** (χ^2) **distribution**. The chi-square distribution is characterized by degrees of freedom, similar to the t-distribution. Table A.3 in the appendix in the back of this book provides critical values of the chi-square distribution for selected values of α.

Figure 4.13 PHStat Results for Confidence Interval for Population Variance of Home Market Values

	A	B	C	D	E
1	Home Market Value				
2					
3	Data				
4	Sample Size	42			
5	Sample Standard Deviation	10553.1			
6	Confidence Level	95%			
7					
8	Intermediate Calculations				
9	Degrees of Freedom	41			
10	Sum of Squares	4.57E+09			
11	Single Tail Area	0.025			
12	Lower Chi-Square Value	25.21452			
13	Upper Chi-Square Value	60.56057			
14					
15	Results				
16	Interval Lower Limit for Variance	7.5E+07			
17	Interval Upper Limit for Variance	1.8E+08			
18					
19	Interval Lower Limit for Standard Deviation	8683.13			
20	Interval Upper Limit for Standard Deviation	13456.9			
21					
22	*Assumption:*				
23	**Population from which sample was drawn has an approximate normal distribution.**				

The Excel function CHIDIST(x, *deg_freedom*) returns the probability above x for a given value of degrees of freedom. Also, the Excel function CHIINV(*probability, deg_freedom*) returns the value of x that has a right-tail area equal to *probability* for a specified degree of freedom.

However, unlike the normal or t-distributions, the chi-square distribution is not symmetric, which means that the confidence interval is not simply a point estimate plus or minus some number of standard errors. The point estimate is always closer to the left endpoint of the interval. A $100(1 - \alpha)\%$ confidence interval for the variance is:

$$\left[\frac{(n-1)s^2}{\chi^2_{n-1,\,\alpha/2}}, \frac{(n-1)s^2}{\chi^2_{n-1,\,1-\alpha/2}} \right] \tag{4.5}$$

We may compute $\chi^2_{41,0.025}$ using the Excel function CHIINV(0.025, 41) = 60.56057 and $\chi^2_{41,0.975}$ by CHIINV(0.975, 41) = 25.21452, as shown in the Intermediate Calculations section of Figure 4.13. The confidence interval limits for the standard deviation is simply the square root of the confidence interval limits of the variance.

PHSTAT NOTE
Confidence Intervals for the Population Variance

From the *PHStat* menu, select *Confidence Intervals* then *Estimate for the Population Variance*. The dialog box is simple, requiring only the sample size, sample standard deviation (which much be calculated beforehand), and confidence level. The tool provides confidence intervals for both the variance and standard deviation.

Confidence Interval for a Population Total

In some applications, we might be more interested in estimating the *total* of a population rather than the mean. For instance, an auditor might wish to estimate the total amount of receivables by sampling a small number of accounts. If a population of N items has a mean μ, then the population total is $N\mu$. We may estimate a population total from a random sample of size n from a population of size N by the point estimate $N\bar{x}$. For example, suppose that an auditor in a medical office wishes to estimate the total amount of unpaid reimbursement claims for 180 accounts over 60 days old. A sample of 20 from this population yielded a mean amount of unpaid claims of $185 and the sample standard deviation is $22. Using the *PHStat* tool *Estimate for the Population Proportion*, we obtain the results shown in Figure 4.14. A 95% confidence interval is [$31547.78, $35052.22].

These calculations are based on the following: a $100(1 - \alpha)\%$ confidence interval for the population total is:

$$N\bar{x} \pm t_{\alpha/2,\,n-1} N \frac{s}{\sqrt{n}} \sqrt{\frac{N-n}{N-1}} \tag{4.6}$$

If you examine this closely, it is almost identical to the formula used for the confidence interval for a mean with an unknown population standard deviation and a finite population correction factor, except that both the point estimate and the interval half-width are multiplied by the population size N to scale the result to a total, rather than an average.

	A	B
1	**Unpaid Reimbursement Claims**	
2		
3	**Data**	
4	**Population Size**	180
5	**Sample Mean**	185
6	**Sample Size**	20
7	**Sample Standard Deviation**	22
8	**Confidence Level**	95%
9		
10	**Intermediate Calculations**	
11	Population Total	33300.00
12	FPC Factor	0.945438918
13	Standard Error of the Total	837.1700134
14	Degrees of Freedom	19
15	*t* Value	2.09302405
16	Interval Half Width	1752.22
17		
18	**Confidence Interval**	
19	**Interval Lower Limit**	31547.78
20	**Interval Upper Limit**	35052.22

Figure 4.14 Confidence Interval for Population Total

USING CONFIDENCE INTERVALS FOR DECISION MAKING

Confidence intervals can be used in many ways to support business decisions. For example, in packaging some commodity product such as laundry soap, the manufacturer must ensure that the packages contain the stated amount to meet government regulations. However, variation may occur in the filling equipment. Suppose that the required weight for one product is 64 ounces. A sample of 30 boxes is measured and the sample mean is calculated to be 63.82 with a standard deviation of 1.05. Does this indicate that the equipment is underfilling the boxes? Not necessarily. A 95% confidence interval for the mean is [63.43, 64.21]. Although the true fill mean might be less than 64, the sample does not provide sufficient evidence to draw that conclusion because 64 is contained within the confidence interval. However, suppose that the sample standard deviation was only 0.46. The confidence interval for the mean would be [63.65, 63.99]. In this case, we would conclude that it is highly unlikely that the true filling weight is 64 ounces; the manufacturer should check and adjust the equipment to meet the standard.

As another example, suppose that an exit poll of 1,300 voters found that 692 voted for a particular candidate in a two-person race. This represents a proportion of 53.23% of the sample. Could we conclude that the candidate will likely win the election? A 95% confidence interval for the proportion is [0.505, 0.559]. This suggests that the population proportion of voters who favor this candidate will be larger than 50%, so it is safe to predict the winner. On the other hand, suppose that only 670 of the 1,300 voters voted for the candidate, a sample proportion of 0.515. The confidence interval for the population proportion is [0.488, 0.543]. Even though the sample proportion is larger than 50%, the confidence interval suggests that it is reasonably likely that the true population proportion will be less than 50%, so it would not be wise to predict the winner based on this information.

It is important not to confuse a confidence interval with a probability interval. A $100(1 - \alpha)\%$ **probability interval** for a random variable X is any interval $[A, B]$ such that $P(A < X < B) = 1 - \alpha$. A probability interval simply describes the probability that the random variable falls within the interval and is often used in describing risk. Probability intervals are often centered on the mean or median. For instance, in a normal distribution, the mean plus or minus 1 standard deviation describes an approximate 68% probability interval around the mean. As another example, the 5th and 95th percentiles in a data set constitute a 90% probability interval. A confidence interval provides an interval estimate of a population parameter, such as the mean or proportion. A confidence interval is a probability interval associated with the *sampling distribution* of a statistic, but it is not the same as a probability interval associated with the distribution of the random variable itself.

As a final note, confidence intervals are most appropriate for cross-sectional data. For time-series data, confidence intervals often make little sense because the mean and/or variance of such data typically change over time. However, for the case in which time-series data are *stationary*—that is, they exhibit a constant mean and constant variance—then confidence intervals can make sense. A simple way of determining whether time-series data are stationary is to plot them on a line chart. If the data do not show any trends or patterns and the variation remains relatively constant over time, then it is reasonable to assume the data are stationary. However, you should be cautious when attempting to develop confidence intervals for time-series data because high correlation between successive observations (called autocorrelation) can result in misleading confidence interval results.

CONFIDENCE INTERVALS AND SAMPLE SIZE

An important question in sampling is the size of the sample to take. Note that in all the formulas for confidence intervals, the sample size plays a critical role in determining the width of the confidence interval. As the sample size increases, the width of the confidence interval decreases, providing a more accurate estimate of the true population parameter. In many applications, we would like to control the margin of error in a confidence interval. For example, in reporting voter preferences, we might wish to ensure that the margin of error is $\pm$ 2%. Fortunately, it is relatively easy to determine the appropriate sample size needed to estimate the population parameter within a specified level of precision. *PHStat* provides tools for computing sample sizes for estimating means and proportions (see *PHStat Note: Determining Sample Size*).

PHSTAT NOTE
Determining Sample Size

From the *PHStat* menu, select *Sample Size* and either *Determination for the Mean* or *Determination for the Proportion*. The dialog boxes are shown in Figure 4.15. You need to enter either the population standard deviation (or at least an estimate if it is known) or estimate of the true proportion, sampling error desired, and confidence level. The sampling error is the desired half-width of the confidence interval. The output options also allow you to incorporate a finite population correction factor if appropriate.

Figure 4.15 *PHStat* Dialogs for *Sample Size Determination for the Mean* and *Sample Size Determination for the Proportion*

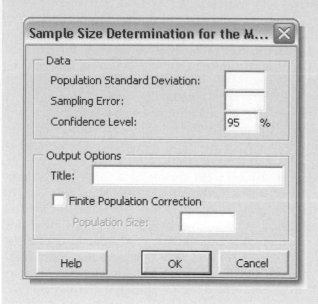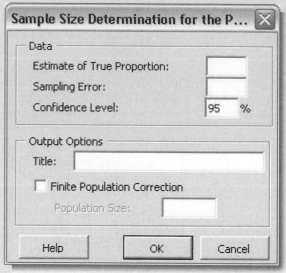

For example, in the first scenario of the soap product filling example described in the previous section, the confidence interval for the mean was [63.43, 64.21], having a sampling error (half-width of the confidence interval) of 0.39. Suppose the manufacturer would like the sampling error to be at most 0.15, resulting in a confidence interval width of 0.30. Using the *PHStat* tool for sample size determination for the mean with an estimated population standard deviation of 1.05 obtained from the original sample of 30, we find that 189 samples would be needed (see Figure 4.16). To verify this, Figure 4.17 shows that if a sample of 189 resulted in the same sample mean and

	A	B
1	**Sample Size Determination**	
2		
3	**Data**	
4	**Population Standard Deviation**	1.05
5	**Sampling Error**	0.15
6	**Confidence Level**	95%
7		
8	Intemediate Calculations	
9	Z Value	-1.95996398
10	Calculated Sample Size	188.2314822
11		
12	Result	
13	**Sample Size Needed**	189

Figure 4.16 *PHStat* Results for *Sample Size Determination* for the Mean

	A	B
1	Confidence Interval Estimate for the Mean	
2		
3	Data	
4	Sample Standard Deviation	1.05
5	Sample Mean	63.82
6	Sample Size	189
7	Confidence Level	95%
8		
9	Intermediate Calculations	
10	Standard Error of the Mean	0.076376262
11	Degrees of Freedom	188
12	t Value	1.972662649
13	Interval Half Width	0.150664599
14		
15	Confidence Interval	
16	Interval Lower Limit	63.67
17	Interval Upper Limit	63.97

Figure 4.17 Results for *Confidence Interval for the Mean* Using Sample Size = 189

standard deviation, the sampling error for the confidence interval [63.67, 63.97] is indeed ± 0.15.

Of course, we generally do not know the population standard deviation prior to finding the sample size. A commonsense approach would be to take an initial sample to estimate the population standard deviation using the sample standard deviation s and determine the required sample size, collecting additional data if needed. If the half-width of the resulting confidence interval is within the required margin of error, then we clearly have achieved our goal. If not, we can use the new sample standard deviation s to determine a new sample size and collect additional data as needed. Note that if s changes significantly, we still might not have achieved the desired precision and might have to repeat the process. Usually, however, this will be unnecessary.

For the voting example in the previous section, suppose that we wish to determine the number of voters to poll to ensure a sampling error of at most ±2%. The *PHStat* tool requires an estimate of the true proportion. In practice, this value will not be known. You could use the sample proportion from a preliminary sample as an estimate to plan the sample size, but this might require several iterations and additional samples to find the sample size that yields the required precision. When no information is available, the most conservative approach is to use 0.5 for the estimate of the true proportion. This will result in the sample size that will guarantee the required precision no matter what the true proportion is. However, if we do have a good estimate for the population proportion, then we should use it because it will result in smaller required sample sizes that will usually be less expensive to obtain. For example, using 0.5 as an estimate of the true proportion, to estimate the number of voters to poll to obtain a 95% confidence interval on the proportion of voters that choose a particular candidate with a precision of ±0.02 or less, we would need a sample of 2401 voters (see Figure 4.18).

	A	B
1	**Sample Size Determination**	
2		
3	**Data**	
4	**Estimate of True Proportion**	0.5
5	**Sampling Error**	0.02
6	**Confidence Level**	95%
7		
8	Intermediate Calculations	
9	Z Value	-1.95996398
10	Calculated Sample Size	2400.911763
11		
12	**Result**	
13	**Sample Size Needed**	2401

Figure 4.18 *PHStat Results for Sample Size Determination* for the Proportion

ADDITIONAL TYPES OF CONFIDENCE INTERVALS

Most confidence intervals have the same basic form: a point estimate of the population parameter of interest plus or minus some number of standard errors. Thus, establishing confidence intervals requires choosing the proper point estimate for a population parameter as well as an understanding of the sampling distribution of the parameter being estimated and, in particular, the standard error. In this section we summarize several additional types of confidence intervals. However, spreadsheet-based tools for computing them are not available in *PHStat* and you must resort to calculating them using the formulas, some of which are rather complex. The appendix to this chapter discusses the formulas and theory behind them. Optional skill-building exercises at the end of this chapter will ask you to create Excel templates for calculating these confidence intervals.

In many situations, we are interested in differences between two populations. For example, in the *Accounting Professionals* Excel file, we might be interested in the difference in mean years of service between females and males. Similarly, in the *Burglaries* Excel file, we might be interested in the difference between the mean number of burglaries per month before and after the citizen-police program was instituted. In both these examples, samples are drawn independently from the two populations.

To illustrate the application of this type of confidence interval, the means and standard deviations for both male and female employees in the *Accounting Professionals* Excel file were computed. The mean years of service for females is 10.07, and the mean years of service for males is 19.69. Thus, a point estimate for the difference in years of service is $10.07 - 19.69 = -9.62$, indicating that, on average, males have been working at the company over 9 years longer than females. Using the confidence interval formula described in the appendix, we find a 95% confidence interval for the mean difference in years of service between females and males for the population of accounting workers in the company is $[-15.118, -4.122]$. This clearly suggests that the male workers have more experience than females.

A second situation involves *paired samples*. For example, a deep-foundation engineering contractor has bid on a foundation system for a new world headquarters building for a Fortune 500 company. A part of the project consists of installing 311 auger cast piles. The contractor was given bid information for cost estimating purposes, which consisted of the estimated depth of each pile; however, actual drill footage of each pile could not be determined exactly until construction was performed. The Excel file *Pile Foundation* contains the estimates and actual pile lengths after the project was completed. We might be interested in the difference between the means of the actual and estimated pile lengths. From the sample data, the mean difference is found to be 6.38, indicating that, on average, the actual lengths were underestimated. Using the formula developed in the appendix, a 95% confidence interval for the mean difference is [5.234, 7.526]. This states that the true population difference does not appear to be zero, indicating a bias in estimating the pile depth.

A final type of confidence interval that has useful applications is the difference between proportions. For example in the *Accounting Professionals* data, we see that the proportion of females having a Certified Public Accountant (CPA) is 8/14 = 0.57, while the proportion of males having a CPA is 6/13 = 0.46. While this sample data suggests that a higher proportion of females have a CPA, a 95% confidence interval for the difference in proportions between females and males, using the formula provided in the appendix, is [−0.2650, 0.4850]. This suggests that we cannot conclusively state that the proportion of females having a CPA is higher than males, because a difference of zero falls within the confidence interval.

BASIC CONCEPTS REVIEW QUESTIONS

1. Explain the importance of sampling from a managerial perspective.
2. What is a sampling plan and what elements should be included in one?
3. How does a frame differ from a target population?
4. Describe the difference between subjective and probabilistic sampling methods. What are the advantages and disadvantages of each?
5. Explain how the following sampling approaches work:
   a. simple random sampling
   b. systematic sampling
   c. stratified sampling
   d. cluster sampling
   e. sampling from a continuous process
6. What is the difference between nonsampling error and sampling error? Why might each type of error occur?
7. Explain the difference between a point estimate and an interval estimate.

8. What do we mean by an unbiased estimator? Why is this important?
9. What is a confidence interval? How do you properly interpret the level of confidence, $1 - \alpha$?
10. How does the *t*-distribution differ from the standard normal distribution?
11. When is it important to apply the finite population correction factor to the standard error when developing confidence intervals?
12. Summarize the different types of confidence intervals that one may construct, and provide a practical application for each.
13. Discuss how confidence intervals can help in making decisions. Provide some examples different from those in the chapter.
14. Under what circumstances can confidence intervals be applied to time-series data?
15. Explain how a confidence interval changes with changes in the level of confidence and sample size.

SKILL-BUILDING EXERCISES

1. Use the *PHStat Random Sample Generator* to generate a list of 10 random integers from among 67 in the Excel file *Cereal Data*, and then use the VLOOKUP function to extract the names of the corresponding cereals.

2. In the Excel file *Cereal Data*, we have computed the population standard deviation of sugars as 4.376. Using a method to generate random samples, generate 20 random samples of 30 cereals and compute the

standard deviation of each sample using both the population formula and sample formula. Compare your results with the true population standard deviation and comment about the fact that the sample formula is an unbiased estimator.

3. Generate 50 random samples of size 10 from the Excel file *Cereal Data*, and compute a 90% confidence interval for the mean of Carbs for each sample, using the known population standard deviation of 3.831. Determine how many confidence intervals actually contain the true population mean 14.77.

4. Find a 99% confidence interval using *PHStat* for the number of hours checking email/week in the Excel file *Social Networking*, and verify the result using Equation (4.3).

5. Find a 99% confidence interval using *PHStat* for the proportion of Facebook users in the Excel file *Social Networking*, and verify the result using Equation (4.4).

6. Find a 95% confidence interval using *PHStat* for the variance of the number of hours checking email/week in the Excel file *Social Networking*, and verify the result using Equation (4.5).

7. Estimate the sample size needed to find a 95% confidence interval for the mean number of hours online/week in the Excel file *Social Networking* to ensure a sampling error of at most ±1.5.

8. Find the sample size needed to have a 95% confidence interval for the proportion of students using both social networking sites in the Excel file *Social Networking* to ensure a sampling error of 0.1. Use 0.58 as an estimate for the proportion.

Note: The following exercises require material found in the appendix to this chapter.

9. Develop an Excel template that will calculate a confidence interval for the difference in means for independent samples with unequal variances.

10. Develop an Excel template that will calculate a confidence interval for the difference in means for independent samples with equal variances.

11. Develop an Excel template that will calculate a confidence interval for the difference in means for paired samples.

12. Develop an Excel template that will calculate a confidence interval for the difference in proportions.

PROBLEMS AND APPLICATIONS

1. Your college or university wishes to obtain reliable information about student perceptions of administrative communication. Describe an appropriate sampling plan to implement for this situation.

2. Using the data in the Excel file *Accounting Professionals*, find and interpret 95% confidence intervals for the following:
   a. mean years of service
   b. proportion of employees who have a graduate degree

3. Using the data in the Excel file *MBA Student Survey*, find 95% confidence intervals for the mean number of nights out per week and mean number of study hours per week by gender. Based on the confidence intervals, would you conclude that there is a difference in social and study habits between males and females?

4. Using the data in the worksheet *Consumer Transportation Survey*, develop 90%, 95%, and 99% confidence intervals for the following:
   a. the mean hours per week that individuals spend in their vehicles
   b. the average number of miles driven per week
   c. the proportion of individuals who are satisfied with their vehicle
   d. the proportion of individuals who have at least one child
   Explain the differences as the level of confidence increases.

5. The Excel file *Salary Data* provides information on current salary, beginning salary, previous experience (in months) when hired, and total years of education for a sample of 100 employees in a firm. Develop 95% confidence intervals for each of the variables in the data set.

6. The Excel file *Infant Mortality* provides data on infant mortality rate (deaths per 1,000 births), female literacy (percent who read), and population density (people per square kilometer) for 85 countries. Develop a 90% confidence interval for infant mortality.

7. A manufacturer conducted a survey among 400 randomly selected target market households in the test market for its new disposable diapers. The objective of the survey was to determine the market share for its new brand. If the sample point estimate for market share is 20%, develop a 95% confidence interval. Can the company reasonably conclude that they have a 25% market share?

8. If, based on a sample size of 500, a political candidate finds that 254 people would vote for him in a two-person race, what is the 95% confidence interval for his expected proportion of the vote? Would he be confident of winning based on this poll?

9. If, based on a sample size of 100, a political candidate found that 62 people would vote for her in a two-person race, what is the 95% confidence interval for her expected proportion of the vote? Would she be confident of winning based on this poll?

10. The Excel file *Blood Pressure* shows diastolic blood pressure readings before and after a new medication. Find 95% confidence intervals for the variance for each of these groups. Based on these confidence intervals, would you conclude that the medication has kept the reading more stable?

11. Using data in the Excel file *Colleges and Universities*, find 95% confidence intervals for the standard deviation of the median SAT for each of the two groups, liberal arts colleges and research universities. Based on these confidence intervals, does there appear to be a difference in the variation of the median SAT scores between the two groups?

12. An auditor of a small business has sampled 50 of 700 accounts. The sample mean total receivables is $435, and the sample standard deviation is $86. Find a 95% confidence interval for the total amount of receivables.

13. The Excel file *New Account Processing* provides data for a sample of employees in a company. Assume that the company has 125 people in total assigned to new account processing. Find a 95% confidence interval for the total sales of the population of account representatives.

14. Trade associations such as the United Dairy Farmers Association frequently conduct surveys to identify characteristics of their membership. If this organization conducted a survey to estimate the annual per-capita consumption of milk and wanted to be 95% confident that the estimate was no more than ±0.5 gallons away from the actual average, what sample size is needed? Past data have indicated that the standard deviation is approximately four gallons.

15. If a manufacturer conducted a survey among randomly selected target market households and wanted to be 95% confident that the difference between the sample estimate and the actual market share for its new product was no more than ±4%, what sample size would be needed?

16. The Excel file *Baseball Attendance* shows the attendance in thousands at San Francisco Giants' baseball games for the 10 years before the Oakland A's moved to the Bay Area in 1968, as well as the combined attendance for both teams for the next 11 years.
    a. Do the data appear to be stationary?
    b. Develop 95% confidence intervals for the mean attendance of each of the two groups. Based on these confidence intervals, would you conclude that attendance has changed after the move?

17. The state of Ohio Department of Education has a mandated ninth-grade proficiency test that covers writing, reading, mathematics, citizenship (social studies), and science. The Excel file *Ohio Education Performance* provides data on success rates (defined as the percent of students passing) in school districts in the greater Cincinnati metropolitan area along with state averages. Find 50%

and 90% probability intervals centered on the median for each of the variables in the *Ohio Education Performance.xls* data.

Note: The following problems require material about Additional Confidence Intervals found in the appendix to this chapter.

18. A study of nonfatal occupational injuries in the United States found that about 31% of all injuries in the service sector involved the back. The National Institute for Occupational Safety and Health (NIOSH) recommended conducting a comprehensive ergonomics assessment of jobs and workstations. In response to this information, Mark Glassmeyer developed a unique ergonomic handcart to help field service engineers be more productive and also to reduce back injuries from lifting parts and equipment during service calls. Using a sample of 382 field service engineers who were provided with these carts, Mark collected the following data:

	YEAR 1 (WITHOUT CART)	YEAR 2 (WITH CART)
Average call time	8.05 hours	7.84 hours
Standard deviation call time	1.39 hours	1.34 hours
Proportion of back injuries	0.018	0.010

Find 95% confidence intervals for the difference in average call times and difference in proportion of back injuries. What conclusions would you reach?

19. A marketing study found that the mean spending in 15 categories of consumer items for 297 respondents in the 18–34 age group was $71.86 with a standard deviation of $70.90. For 736 respondents in the 35+ age group, the mean and standard deviation were $61.53 and $45.29, respectively. Develop 95% confidence intervals for the difference in mean spending amounts between each age group. What assumption did you make about equality of variances?

20. The Excel file *Mortgage Rates* contains time-series data on rates of three different mortgage instruments. Assuming that the data are stationary, construct a 95% confidence interval for the mean difference between the 30-year and 15-year fixed rate mortgage rates. Based on this confidence interval, would you conclude that there is a difference in the mean rates?

21. The Excel file *Student Grades* contains data on midterm and final exam grades in one section of a large statistics course. Construct a 95% confidence interval for the mean difference in grades between the midterm and final exams.

CASE

Analyzing a Customer Survey

A supplier of industrial equipment has conducted a survey of customers for one of its products in its principal marketing regions: North America, South America, Europe, and in its emerging market in China. The data, which tabulate the responses on a scale from 1–5 on dimensions of quality, ease of use, price, and service, are in the Excel file *Customer Survey*. Use point and interval estimates, as well as other data analysis tools such as charts, PivotTables, and descriptive statistics, to analyze these data and write a report to the marketing vice president. Specifically, you should address differences among regions and proportions of "top box" survey responses (which is defined as scale levels 4 and 5) for each of the product dimensions.

Appendix

Theory and Additional Topics

In this appendix, we will present the theory behind confidence intervals and also introduce several additional types of confidence intervals that do not have spreadsheet support tools.

Theory Underlying Confidence Intervals

Recall that the scale of the standard normal distribution is measured in units of standard deviations. We will define z_α to represent the value from the standard normal distribution that provides an upper tail probability of α. That is, the area to the right of z_α is equal to α. Some common values that we will use often include $z_{0.025} = 1.96$ and $z_{0.05} = 1.645$. You should check the standard normal table in Table A.1 in the appendix at the end of the book to verify where these numbers come from.

We stated that the sample mean, $\bar{x}$, is a point estimate for the population mean μ. We can use the central limit theorem (see Chapter 3) to quantify the sampling error in $\bar{x}$. Recall that the Central Limit theorem states that no matter what the underlying population, the distribution of sample means is approximately normal with mean μ and standard deviation (standard error) $\sigma_{\bar{x}} = \sigma/\sqrt{n}$. Therefore, based on our knowledge about the normal distribution, we can expect approximately 95% of sample means to fall within ± 2 standard errors of μ, and more than 99% of them to fall within ± 3 standard errors of μ. More specifically, $100(1 - \alpha)\%$ of sample means will fall within $\pm z_{\alpha/2}\,\sigma_{\bar{x}}$ of the population mean μ as illustrated in Figure 4A.1.

However, we do not know μ but estimate it by $\bar{x}$. Suppose that we construct an interval around $\bar{x}$ by adding and subtracting $z_{\alpha/2}\sigma_{\bar{x}}$. If $\bar{x}$ lies within $\pm z_{\alpha/2}\,\sigma_{\bar{x}}$ of the true mean μ as shown in Figure 4A.2, then you can see that this interval will contain the population mean μ. On the other hand, if $\bar{x}$ lies farther away than $z_{\alpha/2}\,\sigma_{\bar{x}}$ from the true mean μ (in one of the shaded regions in Figure 4A.3), then we see that the interval estimate does *not* contain the true population mean. Because $100(1 - \alpha)\%$ of sample means will fall within $z_{\alpha/2}\,\sigma_{\bar{x}}$ of the population mean μ, we can see that precisely $100(1 - \alpha)\%$ of the

Figure 4A.1 Sampling Distribution of the Mean

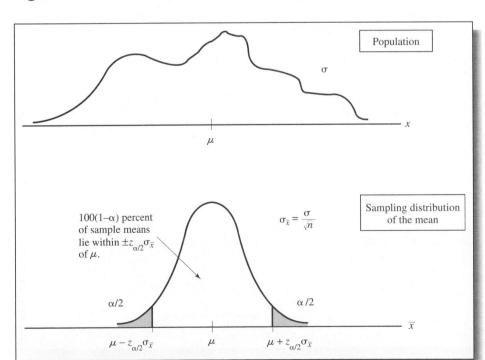

Figure 4A.2   An Interval Estimate that Contains the True Population Mean

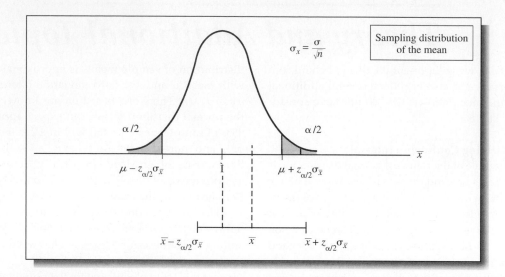

intervals we construct in this fashion will contain μ. Therefore, a $100(1 - \alpha)\%$ confidence interval for the population mean μ is given by:

$$\bar{x} \pm z_{\alpha/2}(\sigma/\sqrt{n}) \qquad (4A.1)$$

This confidence interval will contain the true population mean $100(1 - \alpha)\%$ of the time.

Sampling Distribution of the Proportion

For confidence intervals for the mean, we used the standard error based on the central limit theorem and the normal distribution to construct the confidence intervals.

To establish a confidence interval for a proportion, we need to know the sampling distribution of the proportion and its standard error. The **sampling distribution of the proportion** is analogous to the sampling distribution of the mean, and is the probability distribution of all possible values of $\hat{p}$. If we are sampling with replacement from a finite population, the sampling distribution of $\hat{p}$ follows the binomial distribution with mean $n\pi$ and variance $n\pi(1 - \pi)$. It follows that the sampling distribution of $\hat{p} = x/n$ has mean $n\pi/n = \pi$ and variance $n\pi(1 - \pi)/n^2 = \pi(1 - \pi)/n$. Thus, the standard error of the proportion is $\sqrt{\pi(1 - \pi)/n}$. When $n\pi$ and $n(1 - \pi)$

Figure 4A.3   An Interval Estimate that Does Not Contain the True Population Mean

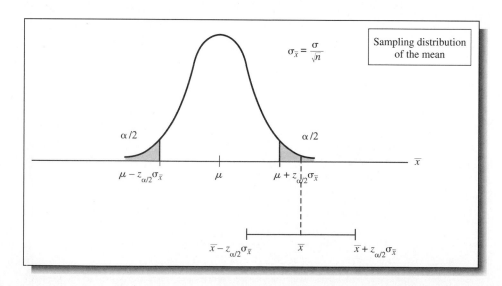

are at least 5, the sampling distribution of $\hat{p}$ approaches the normal distribution as a consequence of the Central Limit theorem. Therefore, under these conditions, we may use z-values to determine the range of sampling error for a specified confidence level. Because we generally use $\hat{p}$ as an estimate for π, the confidence interval becomes:

$$\hat{p} \pm z_{\alpha/2} \sqrt{\frac{\hat{p}(1 - \hat{p})}{n}} \qquad \text{(4A.2)}$$

Sample Size Determination

The formulas for determining sample sizes to achieve a given margin of error are based on the confidence interval half-widths. For example, consider the confidence interval for the mean with a known population standard deviation:

$$\bar{x} \pm z_{\alpha/2}(\sigma/\sqrt{n}) \qquad \text{(4A.3)}$$

Suppose we want the width of the confidence interval on either side of the mean to be at most E. In other words,

$$E \geq z_{\alpha/2}(\sigma/\sqrt{n})$$

Solving for n, we find:

$$n \geq (z_{\alpha/2})^2(\sigma^2)/E^2 \qquad \text{(4A.4)}$$

In a similar fashion, we can compute the sample size required to achieve a desired confidence interval half-width for a proportion by solving the following equation for n:

$$E \geq z_{\alpha/2} \sqrt{\frac{\pi(1 - \pi)}{n}}$$

This yields:

$$n \geq \frac{(z_{\alpha/2})^2\pi(1 - \pi)}{E^2} \qquad \text{(4A.5)}$$

In practice, the value of π will not be known. You could use the sample proportion from a preliminary sample as an estimate of π to plan the sample size, but this might require several iterations and additional samples to find the sample size that yields the required precision. When no information is available, the most conservative estimate is to set $\pi = 0.5$. This maximizes the quantity $\pi(1 - \pi)$ in the formula, resulting in the sample size that will guarantee the required precision no matter what the true proportion is.

Additional Confidence Intervals

In this section, we develop the formulas for additional types of confidence intervals for which no *PHStat* tool is available.

Difference in Means

An important type of confidence interval is one for estimating the difference in means of two populations. The method of constructing confidence intervals differs depending on whether the samples are independent or paired, and whether the variances of the populations can be assumed to be equal or not.

We will assume that we have random samples from two populations with the following:

	POPULATION 1	POPULATION 2
Mean	μ_1	μ_2
Standard deviation	σ_1	σ_2
Point estimate	$\bar{x}_1$	$\bar{x}_2$
Sample size	n_1	n_2

A point estimate for the difference in means, $\mu_1 - \mu_2$, is given by $\bar{x}_1 - \bar{x}_2$. We consider two different cases: when the variances of the two populations are unequal, and when they can be assumed to be equal.

Independent Samples with Unequal Variances

A confidence interval for independent samples with unequal variances is:

$$\bar{x}_1 - \bar{x}_2 \pm (t_{\alpha/2, \, df^*})\sqrt{\frac{s_1^2}{n_1} + \frac{s_2^2}{n_2}} \qquad \text{(4A.6)}$$

where the degrees of freedom for the t-distribution, df^*, is computed as

$$df^* = \frac{\left[\dfrac{s_1^2}{n_1} + \dfrac{s_2^2}{n_2}\right]^2}{\left[\dfrac{(s_1^2/n_1)^2}{n_1 - 1}\right] + \left[\dfrac{(s_2^2/n_2)^2}{n_2 - 1}\right]} \qquad \text{(4A.7)}$$

and fractional values are rounded down. This calculation may be eliminated by using a conservative estimate of the number of degrees of freedom as the minimum of n_1 and n_2, which results in a larger confidence interval.

To illustrate, we use we the *Accounting Professionals* data. Sorting the data by gender, the sample means and standard deviations were calculated. We found $s_1 = 4.39$ and $n_1 = 14$ (females), and $s_2 = 8.39$ and $n_2 = 13$ (males). Calculating df^*, we obtain $df^* = 17.81$, so use 17 as the degrees of freedom. A 95% confidence interval for the difference in years of service is:

$$10.07 - 19.69 \pm 2.1098 \sqrt{\frac{19.2721}{14} + \frac{70.3921}{13}}$$

$$= -9.62 \pm 5.498$$

$$\text{or} \, [-15.118, -4.122]$$

Independent Samples with Equal Variances

When we can assume that the variance of the two populations are equal, we can estimate a common ("pooled") standard deviation that is a weighted combination of the individual sample standard deviations, s_p:

$$s_p = \sqrt{\frac{(n_1 - 1)s_1^2 + (n_2 - 1)s_2^2}{n_1 + n_2 - 2}} \qquad (4A.8)$$

Then the sampling distribution of $\bar{x}_1 - \bar{x}_2$ has a t-distribution with $n_1 + n_2 - 2$ degrees of freedom and standard error:

$$s_p\sqrt{\frac{1}{n_1} + \frac{1}{n_2}} \qquad (4A.9)$$

Therefore, a $100(1 - \alpha)\%$ confidence interval is:

$$\bar{x}_1 - \bar{x}_2 \pm (t_{\alpha/2, \, n_1+n_2-2})s_p\sqrt{\frac{1}{n_1} + \frac{1}{n_2}} \qquad (4A.10)$$

If we assume equal population variances for the *Accounting Professionals* data, the pooled standard deviation is:

$$s_p = \sqrt{\frac{(14 - 1)(4.39)^2 + (13 - 1)(8.39)^2}{14 + 13 - 2}} = 6.62$$

Then, a 95% confidence interval for the difference in mean years of service between females and males is:

$$10.07 - 19.69 \pm (2.0595)6.62\sqrt{\frac{1}{14} + \frac{1}{13}}$$

$$= -9.62 \pm 5.25 \text{ or } [-14.87, -4.37]$$

Note that there is little difference in the confidence interval from the unequal variance case for this example. In general, assume equal population variances unless you have evidence that the variances are significantly different.

Paired Samples

For paired samples, we first compute the difference between each pair of observations, D_i, for $i = 1, \ldots, n$. If we average these differences, we obtain $\bar{D}$, a point estimate for the mean difference between the populations. The standard deviation of the differences is similar to calculating an ordinary standard deviation:

$$s_D = \sqrt{\frac{\sum_{i=1}^{n}(D_i - \bar{D})^2}{n - 1}} \qquad (4A.11)$$

A $100(1 - \alpha)\%$ confidence interval is:

$$\bar{D} \pm (t_{n-1, \, \alpha/2}) s_D / \sqrt{n} \qquad (4A.12)$$

For the *Pile Foundation* data described in the main text, we compute the difference for each pile by subtracting the estimated value from the actual value, as shown in Figure 4A.4, then calculated $\bar{D} = 6.38$ and $s_D = 10.31$. Note that because the sample size exceeds the largest degrees of freedom listed in the table, we must use the critical value of t with an infinite number of degrees of freedom in Table A.2 in the appendix at the end of the book. For $\alpha/2 = 0.025$, this value is 1.96, which is the same as the z-value. Thus, a 95% confidence interval is:

$$6.38 \pm 1.96(10.31/\sqrt{311}) = 6.38 \pm 1.146$$

$$\text{or } [5.234, 7.526]$$

Figure 4A.4   Difference Calculations for Portion of *Pile Foundation Data*

	A	B	C	D
1	Pile Foundation Data			
2				
3	Pile	Estimated	Actual	
4	Number	Pile Length (ft.)	Pile Length (ft.)	Actual - Estimated
5	1	10.58	18.58	8.00
6	2	10.58	18.58	8.00
7	3	10.58	18.58	8.00
8	4	10.58	18.58	8.00
9	5	10.58	28.58	18.00
10	6	10.58	26.58	16.00
11	7	10.58	17.58	7.00
12	8	10.58	27.58	17.00
13	9	10.58	27.58	17.00
14	10	10.58	37.58	27.00

Differences between Proportions

Let $\hat{p}_1$ and $\hat{p}_2$ be sample proportions from two populations using sample sizes n_1 and n_2, respectively. For reasonably large sample sizes, that is, when $n_i \hat{p}_i$ and $n_i(1 - \hat{p}_i)$ are greater than 5 for $i = 1, 2$, the distribution of the statistic $\hat{p}_1 - \hat{p}_2$ is approximately normal. A confidence interval for differences between proportions of two populations is computed as follows:

$$\hat{p}_1 - \hat{p}_2 \pm z_{\alpha/2}\sqrt{\frac{\hat{p}_1(1 - \hat{p}_1)}{n_1} + \frac{\hat{p}_2(1 - \hat{p}_2)}{n_2}} \qquad (4A.13)$$

For example, in the *Accounting Professionals* data, the proportion of females having a CPA is $8/14 = 0.57$, while the proportion of males having a CPA is $6/13 = 0.46$. A 95% confidence interval for the difference in proportions between females and males is:

$$0.57 - 0.46 \pm 1.96\sqrt{\frac{0.57(1 - 0.57)}{14} + \frac{0.46(1 - 0.46)}{13}}$$

$$= 0.11 \pm 0.3750 \text{ or } [-0.2650, 0.4850]$$

Table 4A.1 provides a complete summary of all confidence interval formulas we have discussed.

Table 4A.1
Summary of Confidence Interval Formulas

Type of Confidence Interval	Formula
Mean, standard deviation known	$\bar{x} \pm z_{\alpha/2}(\sigma/\sqrt{n})$
Mean, standard deviation unknown	$\bar{x} \pm t_{\alpha/2, n-1}(s/\sqrt{n})$
Proportion	$\hat{p} \pm z_{\alpha/2}\sqrt{\dfrac{\hat{p}(1 - \hat{p})}{n}}$
Population total	$N\bar{x} \pm t_{\alpha/2, n-1} N\dfrac{s}{\sqrt{n}} \sqrt{\dfrac{N - n}{N - 1}}$
Difference between means, independent samples, equal variances	$\bar{x}_1 - \bar{x}_2 \pm (t_{\alpha/2, n_1+n_2-2})s_p\sqrt{\dfrac{1}{n_1} + \dfrac{1}{n_2}}$ $s_p = \sqrt{\dfrac{(n_1 - 1)s_1^2 + (n_2 - 1)s_2^2}{n_1 + n_2 - 2}}$
Difference between means, independent samples, unequal variances	$\bar{x}_1 - \bar{x}_2 \pm (t_{\alpha/2, df^*})\sqrt{\dfrac{s_1^2}{n_1} + \dfrac{s_2^2}{n_2}}$ $df^* = \dfrac{\left[\dfrac{s_1^2}{n_1} + \dfrac{s_2^2}{n_2}\right]^2}{\left[\dfrac{(s_1^2/n_1)^2}{n_1 - 1}\right] + \left[\dfrac{(s_2^2/n_2)^2}{n_2 - 1}\right]}$
Difference between means, paired samples	$\bar{D} \pm (t_{n-1, \alpha/2})s_D/\sqrt{n}$
Differences between proportions	$\hat{p}_1 - \hat{p}_2 \pm z_{\alpha/2}\sqrt{\dfrac{\hat{p}_1(1 - \hat{p}_1)}{n_1} + \dfrac{\hat{p}_2(1 - \hat{p}_2)}{n_2}}$
Variance	$\left[\dfrac{(n - 1)s^2}{\chi_{n-1, \alpha/2}^2}, \dfrac{(n - 1)s^2}{\chi_{n-1, 1-\alpha/2}^2}\right]$

Chapter 5

Hypothesis Testing and Statistical Inference

INTRODUCTION

*S*tatistical inference is the process of drawing conclusions about populations from sample data. For example, in the worksheet *Burglaries*, we find that the average number of monthly burglaries before the citizen-police program is 64.317, while the average after the program began was 60.647. Although the average number of monthly burglaries appears to have fallen, we cannot tell whether the difference is significant or simply due to sampling error. *Hypothesis testing* is a tool that allows you to draw valid statistical conclusions about the value of population parameters or differences between them. In this chapter, we focus on hypothesis testing and other important tools for statistical inference:

➤ *Principles of hypothesis testing, including proper formulation of statistical hypotheses, decision rules for drawing conclusions, and spreadsheet tools*

➤ *Applications of one-sample and two-sample hypothesis tests for drawing conclusions about means, variances, and proportions*

➤ *An introduction to analysis of variance (ANOVA) for testing for differences in means among multiple populations*

➤ *The chi-square test for independence*

BASIC CONCEPTS OF HYPOTHESIS TESTING

Hypothesis testing involves drawing inferences about two contrasting propositions (hypotheses) relating to the value of a population parameter, such as a mean, proportion, or variance, one of which is assumed to be true in the absence of contradictory data. In conducting a hypothesis test, we seek evidence to determine if the assumed hypothesis can be rejected; if not, we can only *assume* it to be true. The evidence we seek is based on sample data. For instance, a producer of computer-aided design software for the aerospace industry receives numerous calls for technical support (see the Excel file *Customer Support Survey*). The company has a goal of responding to customers in less than 30 minutes on average. Without data-based statistical evidence, the company would have no basis to determine if it is meeting its goal. Similarly, in the Excel file *Burglaries*, local government officials might wish to determine whether a federally funded citizen-police program had a significant effect in reducing the rate of burglaries. To verify the impact of the program to the funding agency, statistical evidence is needed to show positive results. Without such evidence, we could only conclude that the rate was at least the same as before.

However, what does "statistical evidence" mean? In looking at descriptive statistics of the burglary data, we find that the average number of burglaries before the program was 64.32 per month, whereas the average after the program began was 60.65 per month. Can we draw the conclusion that the program was beneficial based simply on these averages? Absolutely not! We need to consider the variability of sampling in our decision, recognizing that each sample mean is only one from an extremely large number of possibilities that can be drawn from the sampling distribution of the mean.

Conducting a hypothesis test involves several steps:

1. Formulating the hypotheses to test
2. Selecting a *level of significance*, which defines the risk of drawing an incorrect conclusion about the assumed hypothesis that is actually true

3. Determining a decision rule on which to base a conclusion
4. Collecting data and calculating a test statistic
5. Applying the decision rule to the test statistic and drawing a conclusion

Hypothesis Formulation

Hypothesis testing begins by defining two alternative, mutually exclusive propositions. The first is called the **null hypothesis,** denoted by H_0, which represents a theory or statement about the status quo that is accepted as correct. The second is called the **alternative hypothesis**, denoted by H_1, which must be true if we conclude that the null hypothesis is false. In the *Customer Support Survey* example, the null and alternative hypotheses would be:

$$H_0: \textit{mean response time} \geq 30 \textit{ minutes}$$

$$H_1: \textit{mean response time} < 30 \textit{ minutes}$$

If we can find evidence that the mean response time is less than 30 minutes, then we would reject the null hypothesis and conclude that H_1 is true. If we cannot find such evidence, we would simply have to assume that the mean response time is 30 minutes or more because of the lack of contradictory data; however, we will not have proven this hypothesis in a statistical sense.

This example hypothesis test involves a single population parameter—the mean response time—and is called a *one-sample hypothesis test* because we will base our conclusion on one sample drawn from the population. We could also formulate hypotheses about the parameters of two populations, called *two-sample tests*, which involve drawing two samples, one from each of the two populations. For instance, in the burglaries example, we might define the null hypothesis to be:

$$H_0: \textit{Mean number of burglaries after program}$$
$$- \textit{mean number of burglaries before program} \geq 0$$

$$H_1: \textit{Mean number of burglaries after program}$$
$$- \textit{mean number of burglaries before program} < 0$$

If we find statistical evidence to conclude that H_1 is true, then we can state—although not conclusively—that the mean number of burglaries after the program commenced has been reduced.

Table 5.1 summarizes the types of one-sample and two-sample hypothesis tests we may conduct, and the proper way to formulate the hypotheses. One-sample tests always compare a population parameter—typically a mean, proportion, or variance—to some constant. For one-sample tests, note that the statements of the null hypotheses are expressed as either $\geq$, $\leq$, or $=$. It is not correct to formulate a null hypothesis using $>$, $<$, or $\neq$. For two-sample tests, the proper statement of the null hypothesis is always a difference between the population parameters, again expressing the null hypotheses as either $\geq$, $\leq$, or $=$. In most applications we compare the difference in means to zero; however, any constant may be used on the right side of the hypothesis.

How do we determine the proper form of the null and alternative hypotheses? Hypothesis testing always *assumes* that H_0 is true, and uses sample data to determine whether H_1 is more likely to be true. Statistically, we cannot "prove" that H_0 is true; we can only *fail to reject* it. Thus, if we cannot reject the null hypothesis, we have only shown that there is insufficient evidence to conclude that the alternative hypothesis is true. However, rejecting the null hypothesis provides strong evidence (in a statistical sense) that the null hypothesis is not true and that the alternative hypothesis is true. The legal

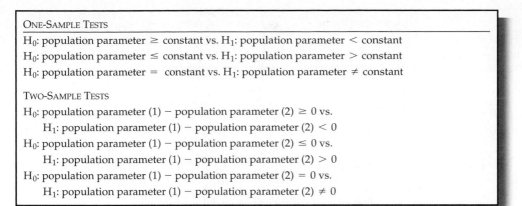

Table 5.1
Types of Hypothesis Tests

ONE-SAMPLE TESTS

H_0: population parameter $\geq$ constant vs. H_1: population parameter $<$ constant
H_0: population parameter $\leq$ constant vs. H_1: population parameter $>$ constant
H_0: population parameter $=$ constant vs. H_1: population parameter $\neq$ constant

TWO-SAMPLE TESTS

H_0: population parameter (1) $-$ population parameter (2) $\geq$ 0 vs.
   H_1: population parameter (1) $-$ population parameter (2) $<$ 0
H_0: population parameter (1) $-$ population parameter (2) $\leq$ 0 vs.
   H_1: population parameter (1) $-$ population parameter (2) $>$ 0
H_0: population parameter (1) $-$ population parameter (2) $=$ 0 vs.
   H_1: population parameter (1) $-$ population parameter (2) $\neq$ 0

analogy of "burden of proof" provides a way to understand this. In our legal system, an individual is assumed innocent (H_0) unless evidence demonstrates guilt (H_1).

Burden of proof revolves around the alternative hypothesis. For example, if we wish to show statistically that the citizen-police program has had an effect in reducing the rate of burglaries, it would be *incorrect* to state the hypotheses as:

H_0: *Mean number of burglaries after program*
   — mean number of burglaries before program $\leq$ 0

H_1: *Mean number of burglaries after program*
   — mean number of burglaries before program $>$ 0

If we fail to reject H_0, we have no conclusive evidence to support it; we have only failed to find evidence to support H_1. Therefore, we cannot conclude that the mean after the program started is necessarily any smaller than before. However, using the original forms of our hypotheses, if we have evidence that shows that the mean number of burglaries after the program began minus the mean number of burglaries before the program is less than zero, then we can conclude that the program was beneficial. A useful way of thinking about this is whatever claim is made (i.e., what you would like to prove to be true) should define the *alternative* hypothesis. Thus, in the *Customer Support Survey* example, the claim that the firm is meeting its goal of a mean response time of less than 30 minutes defines H_1.

Significance Level

Hypothesis testing can result in four different outcomes:

1. The null hypothesis is actually true, and the test correctly fails to reject it.
2. The null hypothesis is actually false, and the hypothesis test correctly reaches this conclusion.
3. The null hypothesis is actually true, but the hypothesis test incorrectly rejects it (called **Type I error**).
4. The null hypothesis is actually false, but the hypothesis test incorrectly fails to reject it (called **Type II error**).

The probability of making a Type I error is generally denoted by α and is called the **level of significance** of the test. This probability is essentially the risk that you can afford to take in making the incorrect conclusion that the alternative hypothesis is true when in fact the null hypothesis is true. The **confidence coefficient** is $1 - \alpha$, which is the probability of *correctly failing to reject* the null hypothesis. For a confidence coefficient of 0.95, we mean that we expect 95 out of 100 cases to support the

	Test Rejects H_0	Test Fails to Reject H_0	Table 5.2
Alternative hypothesis (H_1) is true	Correct	Type II error (β)	Error Types in
Null hypothesis (H_0) is true	Type I error (α)	Correct	Hypothesis Testing

null hypothesis rather than the alternate hypothesis. Commonly used levels for α are 0.10, 0.05, and 0.01, resulting in confidence levels of 0.90, 0.95, and 0.99, respectively.

The probability of a Type II error is denoted by β. Unlike α, this cannot be specified in advance but depends on the true value of the (unknown) population parameter. To see this, consider the hypotheses in the customer survey example:

$$H_0: \textit{mean response time} \geq 30 \textit{ minutes}$$

$$H_1: \textit{mean response time} < 30 \textit{ minutes}$$

If the true mean response is, say, 20 minutes, we would expect to have a much lower probability of incorrectly concluding that the null hypothesis is true than when the true mean response is 28 minutes, for example. In the first case, the sample mean would very likely be much less than 30, leading us to reject H_0. In the second case, however, even though the true mean response is less than 30, we would have a much higher probability of failing to reject H_0 because a higher likelihood exists that the sample mean would be greater than 30 due to sampling error. Thus, the farther away the true mean response time is from the hypothesized value, the smaller is β. Generally, as α decreases, β increases, so the decision maker must consider the trade-offs of these risks.

The value $1 - \beta$ is called the **power of the test** and represents the probability of *correctly rejecting* the null hypothesis when it is indeed false. We would like the power of the test to be high to allow us to make a valid conclusion. If the power of the test is deemed to be too small, it can be increased by taking larger samples. Larger samples enable us to detect small differences between the sample statistics and population parameters with more accuracy. However, a larger sample size incurs higher costs, giving more meaning to the adage, "There is no such thing as a free lunch." Table 5.2 summarizes this discussion.

Decision Rules

The decision to reject or fail to reject a null hypothesis is based on computing a test statistic from sample data that is a function of the mean, variance, or proportion and comparing it to a critical value from the hypothesized sampling distribution of the test statistic. The sampling distribution is usually the normal distribution, t-distribution, or some other well-known distribution. The sampling distribution is divided into two parts, a *rejection region* and a *nonrejection region*. If the null hypothesis is true, it is unlikely that the test statistic will fall into the rejection region. Thus, if the test statistic falls into the rejection region, we reject the null hypothesis; otherwise, we fail to reject it. The probability of it falling into the rejection region if H_0 is true is the probability of a Type I error, α.

The rejection region generally occurs in the tails of the sampling distribution of the test statistic. For tests in which we reject the null hypothesis if the test statistic is either significantly high or low—for example, in testing whether the mean amount of soap in a 32-ounce box meets the target as reflected in the hypotheses:

$$H_0: \textit{mean weight} = 32 \textit{ ounces}$$

$$H_1: \textit{mean weight} \neq 32 \textit{ ounces}$$

the rejection region will occur in *both* the upper and lower tail of the distribution [see Figure 5.1(a)]. This is called a **two-tailed test of hypothesis**. Because the probability that the test statistic falls into the rejection region, given that H_0 is true, the combined area of both tails must be α. Usually, each tail has an area of $\alpha/2$.

The other type of hypothesis, which specifies a direction of relationship, such as:

$$H_0: \textit{mean response time} \geq 30\textit{ minutes}$$

$$H_1: \textit{mean response time} < 30\textit{ minutes}$$

is a **one-tailed test of hypothesis**. In this case, the rejection region occurs in one tail of the distribution [see Figure 5.1(b)]. Determining the correct tail of the distribution to use as the rejection region for a one-tailed test is easy. If H_1 is stated as "<," the rejection region is in the lower tail; if H_1 is stated as ">," the rejection region is in the upper tail (just think of the inequality as an arrow pointing to the proper tail direction!).

The rejection region is defined by a *critical value* of the test statistic, which is the value that divides the rejection region from the rest of the distribution. Two-tailed tests have both upper and lower critical values, while one-tailed tests have either a lower or upper critical value. For standard normal and *t*-distributions, which have a mean of zero, lower-tail critical values are negative; upper-tail critical values are positive.

Figure 5.1   Illustration of Rejection Regions in Hypothesis Testing

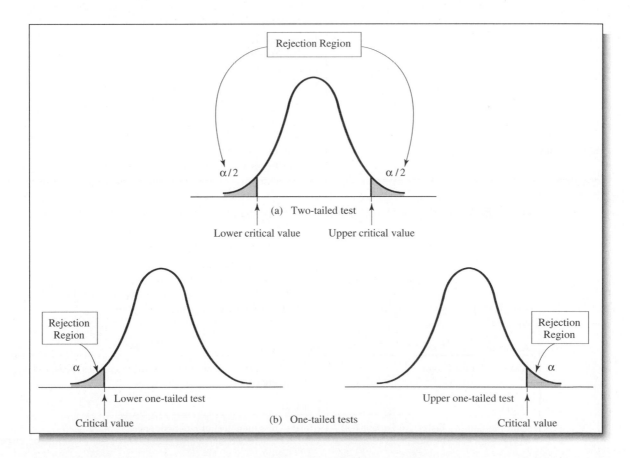

Decision rules are based on a comparison of the test statistic to the critical value(s) for the proper sampling distribution. For example, for an upper one-tailed test based on the normal distribution, if the test statistic is greater than the critical value, the decision would be to reject the null hypothesis. Similarly, for a lower one-tailed test based on the normal distribution, if the test statistic is less than the critical value, we would reject the null hypothesis. For a two-tailed test, if the test statistic is either greater than the upper critical value or less than the lower critical value, the decision would be to reject the null hypothesis.

Spreadsheet Support for Hypothesis Testing

Both Excel and *PHStat* have numerous tools for conducting hypothesis tests. In some cases, only one procedure is available; in others, both Excel and *PHStat* provide tools. Table 5.3 summarizes some of the more common types of hypothesis tests and Excel and *PHStat* tools for conducting these tests. Each test can be applied to different forms of the hypotheses (that is, when the null hypothesis is stated as "≥," "≤," or "="). The challenge is to identify the proper test statistic and decision

Table 5.3 Common Types of Hypothesis Tests

TYPE OF TEST	EXCEL/*PHSTAT* PROCEDURE
One-sample test for mean, σ known	*PHStat*: One-Sample Test—*z*-test for the Mean, Sigma Known
One-sample test for mean, σ unknown	*PHStat*: One-Sample Test—*t*-test for the Mean, Sigma Unknown
One-sample test for proportion	*PHStat*: One-Sample Test—*z*-test for the Proportion
Two-sample test for means, σ^2 known	Excel *z*-test: Two-Sample for Means
	PHStat: Two-Sample Tests—*z*-test for Differences in Two Means
Two-sample test for means, σ^2 unknown, unequal	Excel *t*-test: Two-Sample Assuming Unequal Variances
Two-sample test for means, σ^2 unknown, assumed equal	Excel *t*-test: Two-Sample Assuming Equal Variances
	PHStat: Two Sample Tests—*t*-test for Differences in Two Means
Paired two-sample test for means	Excel *t*-test: Paired Two-Sample for Means
Two-sample test for proportions	*PHStat*: Two-Sample Tests—*z*-test for Differences in Two Proportions
Equality of variances	Excel *F*-test Two-Sample for Variances
	PHStat: Two-Sample Tests—*F*-Test for Differences in Two Variances

Table 5.4 Additional Excel Support for Hypothesis Testing and Statistical Inference

EXCEL FUNCTION	DESCRIPTION
CHITEST(*actual_range, expected_range*)	Returns the test for independence, the value of the chi-square distribution, and the appropriate degrees of freedom
TTEST(*array1, array2, tails, type*)	Returns the probability associated with a *t*-test
ZTEST(*array, x, sigma*)	Returns the two-tailed *p*-value of a *z*-test
ANALYSIS TOOLPAK TOOLS	**DESCRIPTION**
ANOVA: Single Factor	Tests hypothesis that means of two or more samples measured on one factor are equal
***PHSTAT* ADD-IN**	**DESCRIPTION**
Multiple-Sample Tests/Chi-Square Test	Performs chi-square test of independence

rule and to understand the information provided in the output. Table 5.4 provides a summary of additional spreadsheet support for hypothesis testing. We will illustrate these tests through examples in the next several sections. Many other types of hypothesis tests exist, some of which are described in the appendix to this chapter; however, spreadsheet-based procedures are not available.

ONE-SAMPLE HYPOTHESIS TESTS

In this section, we will discuss several hypothesis tests for means, proportions, and variances involving a single sample.

One-Sample Tests for Means

We will first consider one-sample tests for means. The appropriate sampling distribution and test statistic depends on whether the population standard deviation is known or unknown. If the population standard deviation is known, then the sampling distribution is normal; if not, we use a t-distribution, which was introduced in the previous chapter when discussing confidence intervals for the mean with an unknown population standard deviation. In most practical applications, the population standard deviation will not be known but is estimated from the sample, so we will only illustrate this case.

For the *Customer Support Survey* data, we will test the hypotheses:

$$H_0: \text{mean response time} \geq 30 \text{ minutes}$$

$$H_1: \text{mean response time} < 30 \text{ minutes}$$

with a level of significance of 0.05. This is a lower-tailed, one-sample test for the mean with an unknown standard deviation.

We will use the *PHStat* procedure for a one-sample test for the mean with an unknown population standard deviation (see *PHStat Note: One-Sample Test for the Mean, Sigma Unknown*). From the data, we computed the sample mean to be 21.91 and the sample standard deviation as 19.49 for the 44 observations. Figure 5.3 shows the output provided by *PHStat*. The *Data* portion of the output simply summarizes the hypothesis we are testing, level of significance specified, and sample statistics. The *t-Test Statistic* is calculated using the formula:

$$t = \frac{\bar{x} - \mu_0}{s/\sqrt{n}} \tag{5.1}$$

where μ_0 is the hypothesized value and $s/\sqrt{n}$ is the standard error of the sampling distribution of the mean. Applied to this example, we have:

$$t = \frac{\bar{x} - \mu_0}{s/\sqrt{n}} = \frac{21.91 - 30}{19.49/\sqrt{44}} = \frac{-8.09}{2.938} = -2.75$$

Observe that the numerator is the distance between the sample mean (21.91) and the hypothesized value (30). By dividing by the standard error, the value of t represents the number of standard errors the sample mean is from the hypothesized value. In this case, the sample mean is 2.75 standard errors below the hypothesized value of 30.

This notion provides the fundamental basis for the hypothesis test—if the sample mean is "too far" away from the hypothesized value, then the null hypothesis should be rejected. The decision is based on the level of significance, α. For a one-tailed test,

PHSTAT NOTE
One-Sample Test for the Mean, Sigma Unknown

From the *PHStat One-Sample Tests* menu, select *One-Sample Tests* then *t-Test for the Mean, Sigma Unknown.* The resulting dialog box, shown in Figure 5.2, first asks you to input the value of the null hypothesis and significance level. You may either specify the sample statistics or let the tool compute them from the data. The sample statistics for the *Customer Support Survey* data have been entered in the dialog box shown in Figure 5.2. Under *Test Options* you may choose among a two-tailed test, upper one-tailed test, or lower one-tailed test.

Figure 5.2 *PHStat* Dialog for *t-Test for the Mean, sigma unknown*

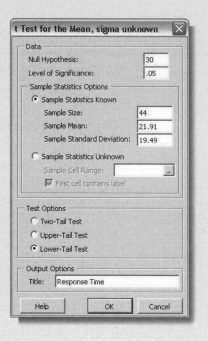

the critical value is the number of standard errors away from the hypothesized value for which the probability of exceeding it is at most α. If $\alpha = 0.05$, for example, then we are saying that there is only a 5% chance that a sample mean will be that far away from the hypothesized value purely because of sampling error, and that the small likelihood of this occurring suggests that the true population mean is different from what was hypothesized.

The *t-Test Statistic* has a *t*-distribution with $n - 1$ degrees of freedom. If the level of significance is 0.05, then the critical value for a lower-tail test is the value of the *t*-distribution with $n - 1$ degrees of freedom that provides a lower tail area of 0.05; that is, $t_{\alpha/2, n-1}$. We may find *t*-values in Table A.2 in the appendix at the end of the book or by using the Excel function $\text{TINV}(\alpha, n - 1)$. However, as noted in the previous chapter, Table A.2 provides upper-tail values only, whereas the TINV function accounts for both tails. Thus, $t_{\alpha/2, n-1}$ in Table A.2 is equivalent to $\text{TINV}(\alpha, n - 1)$. Hence, to find the critical value, we find $t_{0.05, 43} = \text{TINV}(0.10, 43) = 1.68$. Because this is a lower-tail test, we use the negative of this number because the *t*-distribution is symmetric with a mean of 0.

By comparing the *t-Test Statistic* with the *Lower Critical Value*, we see that the test statistic falls below the critical value in the rejection region, and we reject H_0 and conclude that the mean response time is less than 30 minutes. Figure 5.4 illustrates the conclusion we reached.

Figure 5.3 Results for One-Sample Test for the Mean

	A	B	C	D	E
1	Response Time				
2					
3	Data				
4	Null Hypothesis μ=	30			
5	Level of Significance	0.05			
6	Sample Size	44			
7	Sample Mean	21.91			
8	Sample Standard Deviation	19.49			
9					
10	Intermediate Calculations				
11	Standard Error of the Mean	2.938228053			
12	Degrees of Freedom	43			
13	*t* Test Statistic	-2.753360139			
14					
15	Lower-Tail Test			Calculations Area	
16	Lower Critical Value	-1.681070704		For one-tailed tests:	
17	*p*-Value	0.004303659		TDIST value	0.004304
18	Reject the null hypothesis			1-TDIST value	0.995696

Using *p*-Values

In the *PHStat* output in Figure 5.3, we see something called a *p-value*. An alternative approach to comparing a test statistic to a critical value in hypothesis testing is to find the probability of obtaining a test statistic value equal to or more extreme than that obtained from the sample data when the null hypothesis is true. This probability is commonly called a ***p*-value**, or **observed significance level**. For example, the *t-Test Statistic* for the hypothesis test in the response time example is −2.75. If the true mean is really 30, then what is the probability of obtaining a test statistic of −2.75 or less? Equivalently, what is the probability that a sample mean from a population with a mean of 30 will be at least 2.75 standard errors away from 30? We can calculate this using the Excel function TDIST(*x, degrees_freedom, tails*), with *x* = 2.75, *degrees_freedom* = 43, and *tails* = 1. This turns out to be 0.0043. In other words, there is less than a 0.5% chance that the test statistic would be −2.75 or smaller if the null hypothesis were true. With such a low probability, we would not attribute this to sampling error alone, but would conclude that the true mean is probably less than 30 and reject the null hypothesis. In general, we compare the *p*-value to the chosen level of

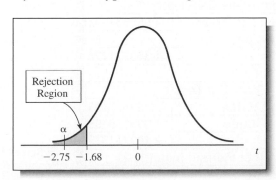

Figure 5.4 *t*-Test for Mean Response Time

significance; whenever $p < \alpha$, we reject the null hypothesis. *P*-values make it easy to draw conclusions about hypothesis tests.

Next, we illustrate a two-tailed test using the data for 50 respondents in the Excel file *Consumer Transportation Survey*. Suppose that the sponsor of the survey wanted to target individuals who were approximately 40 years old. Thus, we wish to test whether the average age of respondents is equal to 40. The hypothesis to test is:

$$H_0: \textit{mean age of respondents} = 40$$

$$H_1: \textit{mean age of respondents} \neq 40$$

The sample mean is computed to be 37.9, and the sample standard deviation is 11. Figure 5.5 shows the results using the *PHStat t-Test for the Mean, Sigma Unknown*.

The test statistic is the same as in the previous example:

$$t = \frac{\bar{x} - \mu_0}{s/\sqrt{n}} = (37.9 - 40)/(11/\sqrt{50}) = -1.35$$

In this case, the sample mean is 1.35 standard errors below the hypothesized mean of 40. However, because this is a two-tailed test, the rejection region and decision rule are different. For a level of significance α, we reject H_0 if the *t-Test Statistic* falls either below the negative critical value, $-t_{n-1,\alpha/2}$, or above the positive critical value, $t_{n-1,\alpha/2}$. Using the Excel function TINV(.05,49) we obtain 2.00958 for $t_{49,.025}$; thus, the critical values are $\pm$ 2.00958. Because the *t-Test Statistic* falls between these values, we cannot reject the null hypothesis that the average age is 40 (see Figure 5.6).

The *p*-value for this test is 0.1832, which can also be computed by TDIST(1.35, 49, 2). This represents the probability of obtaining a result equal to or more extreme *in either tail* of the distribution than the computed test statistic. In other words, there is a 0.1832 probability that the *t*-test statistic will be either greater than 1.35 or less than -1.35 when the null hypothesis is true. Since this is *larger* than the chosen significance level of 0.05, we would fail to reject H_0.

	A	B
1	**Respondent Age**	
2		
3	**Data**	
4	**Null Hypothesis** μ=	40
5	**Level of Significance**	0.05
6	**Sample Size**	50
7	**Sample Mean**	37.9
8	**Sample Standard Deviation**	11
9		
10	Intermediate Calculations	
11	Standard Error of the Mean	1.555634919
12	Degrees of Freedom	49
13	*t* Test Statistic	-1.349931128
14		
15	Two-Tail Test	
16	**Lower Critical Value**	-2.009575199
17	**Upper Critical Value**	2.009575199
18	*p*-Value	0.183241372
19	Do not reject the null hypothesis	

Figure 5.5 *PHStat* Results for Two-Tailed *t*-Test

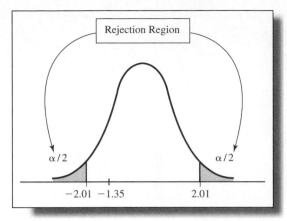

Figure 5.6 Illustration of Two-Tailed *t*-Test

One-Sample Tests for Proportions

Many important business measures, such as market share or the fraction of deliveries received on time, are expressed as proportions. For example, in generating the *Customer Support Survey* data, one question asks the customer to rate the overall quality of the company's software product using a scale of

> 0—Very poor
> 1—Poor
> 2—Good
> 3—Very good
> 4—Excellent

These data are in column G in the Excel file *Customer Support Survey*. The firm tracks customer satisfaction of quality by measuring the proportion of responses in the top two categories. Over the past year, this proportion has averaged about 75%. For these data, 30 of the 44 responses, or 68.2%, are in the top two categories. Is there sufficient evidence to conclude that this satisfaction measure has dropped below 75%? Answering this question involves testing the hypotheses about the population proportion π:

$$H_0: \pi \geq 0.75$$
$$H_1: \pi < 0.75$$

PHStat has a tool for performing this test (see *PHStat Note: One-Sample Test for Proportions*), but Excel does not. Applying this tool to the *Customer Support Survey*

PHSTAT NOTE
One-Sample Test for Proportions

One-sample tests involving a proportion can be found by selecting the menu item *One-Sample Tests* and choosing *z-Test for the Proportion*. Figure 5.7 shows the dialog box. The tool requires you to enter the value for the null hypothesis, significance level, number of successes, sample size, and the type of test. Note that you cannot enter the proportion alone because the test depends on the sample size. Thus, if you only know the sample proportion and the sample size, you must convert it into the number of successes by multiplying the sample proportion by the sample size.

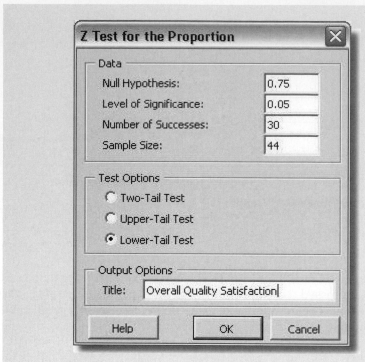

Figure 5.7 *PHStat* Dialog for One-Sample Test for Proportions

data we obtain the results shown in Figure 5.8. The *Z-Test Statistic* for a one-sample test for proportions is:

$$z = \frac{\hat{p} - \pi_0}{\sqrt{\pi_0(1 - \pi_0)/n}} \tag{5.2}$$

	A	B
1	**Overall Quality Satisfaction**	
2		
3	**Data**	
4	**Null Hypothesis** $\pi =$	**0.75**
5	**Level of Significance**	**0.05**
6	**Number of Successes**	**30**
7	**Sample Size**	**44**
8		
9	Intermediate Calculations	
10	Sample Proportion	0.681818182
11	Standard Error	0.065279121
12	Z Test Statistic	-1.044465936
13		
14	**Lower-Tail Test**	
15	**Lower Critical Value**	-1.644853627
16	*p*-Value	0.148134936
17	**Do not reject the null hypothesis**	

Figure 5.8 *PHStat* Results for *z*-Test for Proportions

where π_0 is the hypothesized value. The denominator represents the standard error for the sampling distribution of the proportion and is shown in the Intermediate Calculations. For this example,

$$z = \frac{0.682 - 0.75}{\sqrt{0.75(1 - 0.75)/44}} = -1.04$$

Similar to the test statistic for means, the *Z-Test Statistic* shows the number of standard errors that the sample proportion is from the hypothesized value. In this case, the sample proportion of 0.68 is 1.04 standard errors below the hypothesized value of 0.75. Because this is a lower one-tailed test, we reject H_0 if the *Z-Test Statistic* is less than the lower critical value. The sampling distribution of z is a standard normal; therefore, for a level of significance of 0.05, the critical value of z is found by the Excel function NORMSINV(0.05) $= -1.645$. Because the *Z-Test Statistic* is not less than the *Lower Critical Value*, we cannot reject the null hypothesis that the proportion is at least 0.75 and attribute the low proportion of responses in the top two boxes to sampling error. Note that the p-value is greater than the significance level, leading to the same conclusion of not rejecting the null hypothesis.

Type II Errors and the Power of a Test

The probability of a Type I error, α, can be specified by the experimenter. However, the probability of a Type II error, β, (the probability of failing to reject H_0 when it indeed is false) and the power of the test $(1 - \beta)$ are the result of the hypothesis test itself. Understanding the power of a test is important to interpret and properly apply the results of hypothesis testing. The power of the test depends on the true value of the population mean, the level of confidence used, and the sample size. This is illustrated in Figure 5.9. Suppose that we are testing the null hypothesis $H_0: \mu \geq \mu_0$ against the alternative $H_1: \mu < \mu_0$, and suppose that the true mean is actually μ_1. We specify the probability of a Type I error, α, in the lower tail of the hypothesized distribution; this defines the rejection region. However, the sampling distributions overlap, and it is possible that the test statistic will fall into the acceptance region, even though the sample comes from a distribution with mean μ_1 leading us to not reject H_0. The area that overlaps the hypothesized distribution in the acceptance region is β, the probability of a Type II error.

From the figure, it would seem intuitive that β would be smaller the farther away the true mean is from μ_0. In other words, if the true mean is close to μ_0, it would be difficult to distinguish much of a difference and the probability of concluding that H_0

Figure 5.9 Finding the Probability of a Type II Error

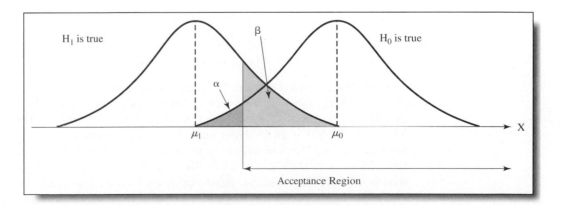

Figure 5.10 How β Depends on H_1

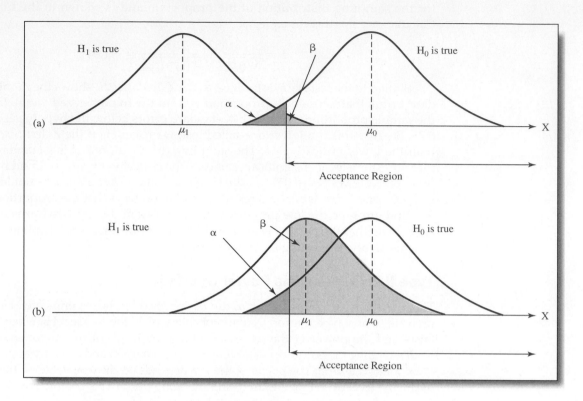

is true would be high. However, if the true mean is far away, the chances that sampling error alone would lead us to conclude that H_0 was true would probably be quite small. This is illustrated in Figure 5.10.

In a similar fashion, β also depends on the sample size. We know that the standard error of the mean decreases as the sample size increases. This makes the sampling distribution narrower, and reduces the overlap shown in Figure 5.10, thus reducing the value of β (see Figure 5.11).

Clearly, we would like the power of the test to be high so we can more easily discriminate between the two hypotheses and avoid Type II errors. However, because the true mean is unknown, we cannot determine β exactly. All we can do is to calculate it for different values of the true mean and assess the potential for committing a Type II error. This is relatively easy to do.

We will illustrate this using the *Customer Support Survey* data in the context of Figure 5.9. Recall that the hypothesis test is:

$$H_0: \textit{mean response time} \geq 30 \textit{ minutes}$$

$$H_1: \textit{mean response time} < 30 \textit{ minutes}$$

and that the standard error of the mean was calculated to be 2.932. Because the sample size is large, we will use z-values instead of t. If $\alpha = 0.05$, then the left-hand limit of the acceptance region is 1.96 standard errors to the left of 30, or $30 - 1.96(2.932) = 24.25$. Now suppose that the true mean is $\mu_1 = 27$. The area to the right of 24.25 for a distribution with a mean of 27 is β. The z-value is calculated as $z = (24.25 - 27)/2.932 = -0.94$. Because $\beta = 1 - \text{NORMSINV}(-0.94)$, we find that is $1 - 0.1736 = 0.8264$. Now suppose that the true mean is only 15. Then the z-value corresponding to 24.25 for this distribution is $(24.25 - 15)/2.932 = 3.14$. The

Figure 5.11 How β Depends on Sample Size

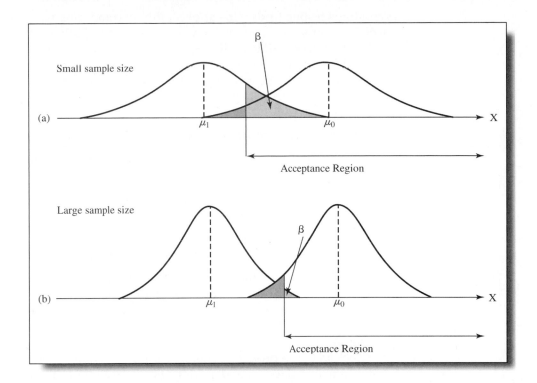

area to the right of this value is $1 - \text{NORMSINV}(3.14) = 0.00084$. If we perform these calculations for a range of values for μ_1 (using the NORMDIST function), we could draw a **power curve** that shows $(1 - \beta)$ as a function of μ_1, shown in Figure 5.12. The relative steepness of the curve will help assess the risk of a Type II error.

Figure 5.12 Power Curve for *Customer Support Survey* Hypothesis

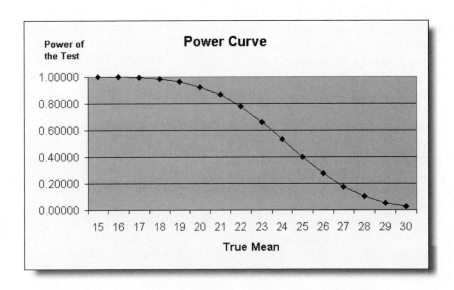

TWO-SAMPLE HYPOTHESIS TESTS

Hypothesis testing finds wide applicability in comparing two populations for differences in means, proportions, or variances. The hypothesis-testing procedures are similar to those previously discussed in the sense of formulating one- or two-tailed tests and identifying the critical test statistic and rejection region. The formulas for the test statistics are more complicated than for one-sample tests, and are discussed in the appendix to this chapter.

Two-Sample Tests for Means

In the Excel file *Burglaries* that we described earlier, we might wish to test the hypotheses

$$H_0: mean\ burglaries/month\ after\ program\ (\mu_1)$$
$$- mean\ burglaries/month\ before\ program\ (\mu_2) \geq 0$$

$$H_1: mean\ burglaries/month\ after\ program\ (\mu_1)$$
$$- mean\ burglaries/month\ before\ program\ (\mu_2) < 0$$

Rejecting the null hypothesis suggests that the citizen-police program was effective in reducing the number of burglaries. If we cannot reject the null hypothesis, then even though the mean number of monthly burglaries is smaller since the program began, the difference would most likely be due to sampling error and is not statistically significant.

Selection of the proper test statistic for a two-sample test for means depends on whether the population standard deviations are known, and if not, whether they are assumed to be equal. *PHStat* and Excel have procedures for conducting these tests (see *Excel/PHStat Note: Using Two-Sample t-Test Tools*). Please read this carefully because care must be taken in interpreting the Excel output.

EXCEL/*PHSTAT* NOTE
Using Two-Sample *t*-Test Tools

To test differences between the means of two populations, three types of hypothesis tests are available.

1. *Population variance is known.* Both Excel and *PHStat* have tools for a two-sample test for means with known population variances. The dialogs are shown in Figure 5.13. In Excel, choose *z-Test: Two Sample for Means* from the *Data Analysis* menu. The dialog prompts you for the range of each variable, hypothesized mean difference, known variances for each variable, whether the ranges have labels, and the level of significance (α). If you leave the box *Hypothesized Mean Difference* blank or enter zero, the test is for equality of means. However, the tool allows you to specify a value d to test the hypothesis $H_0: \mu_1 - \mu_2 = d$. In *PHStat*, from the *Two-Sample Test* menu, choose *Z Test for Differences in Two*

Means. However, *PHStat* requires you to enter a value for the sample mean instead of calculating it from the data range. You must also enter the hypothesized difference and level of significance. For *PHStat*, you may specify the type of test (two-tail, upper-tail, or lower-tail).

2. *Population variance is unknown and unequal.* Only Excel has a tool for a two-sample test for means with unequal variances. From the *Data Analysis* menu, choose *t-test: Two-Sample Assuming Unequal Variances*. Data input is straightforward: The dialog box prompts you to enter the ranges for each sample, whether the ranges have labels, the hypothesized mean difference, and the level of significance (α), which defaults to 0.05. As in the z-test, the tool also allows you to specify a value d to test the hypothesis $H_0: \mu_1 - \mu_2 = d$.

3. *Population variance unknown but assumed equal.* In Excel, choose *t-test: Two-Sample Assuming Equal Variance;* in *PHStat*, choose *t-test for Differences in Two Means.* These tools conduct the test for equal variances only. The dialogs are similar to those in Figure 5.13. *PHStat* also provides an option to compute a confidence interval.

Although the *PHStat* output is straightforward, you must be *very careful* in interpreting the information from Excel. If *t Stat* is negative, $P(T \le t)$ provides the correct *p*-value for a lower-tail test; however, for an upper-tail test, you must subtract this number from 1.0 to get the correct *p*-value. If *t Stat* is nonnegative, then $P(T \le t)$ provides the correct *p*-value for an upper tail test; consequently, for a lower tail test, you must subtract this number from 1.0 to get the correct *p*-value. Also, for a lower tail test, you must change the sign on *t* Critical one-tail.

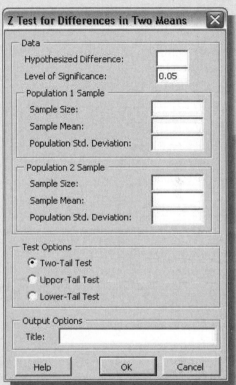

Figure 5.13 Excel and *PHStat* Dialogs for Two-Sample Test for Means, Sigma Known

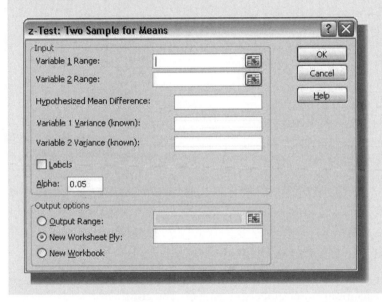

For the burglary data, the population variances are unknown, but the sample variances are close enough that we will assume equality for selecting the proper tool. Figure 5.14 shows the output from both the *PHStat* and Excel tests for the difference in means with unknown but equal variances. From the *PHStat* output, we see that the *t*-Test Statistic is −0.768. When compared to the lower critical value of −1.67252, we cannot reject the null hypothesis and conclude that there is insufficient evidence to claim that the mean number of burglaries has decreased after the citizen-police program was initiated. The *p*-value of 0.226 confirms this, as it is much larger than the level of significance.

The Excel results are more difficult to interpret. As stated in the Excel/*PHStat* note about these tools, the *p*-values may be misleading. Thus, for a lower-tail test, we must change the sign on the critical value (*t Critical one-tail*) to −1.67252. For a two-tailed test, the critical value is specified for the upper tail only, and you must also realize that the lower-tail critical value is the negative; however, the *p*-value provided is correct.

Figure 5.14 Excel and *PHStat* Results for Differences in Means

	A	B	C	D	E	F
1	t Test for Differences in Two Means			t-Test: Two-Sample Assuming Equal Variances		
2	(assumes equal population variances)					
3	Data				Variable 1	Variable 2
4	Hypothesized Difference	0		Mean	60.64705882	64.31707
5	Level of Significance	0.05		Variance	253.8676471	282.472
6	Population 1 Sample			Observations	17	41
7	Sample Size	17		Pooled Variance	274.2992929	
8	Sample Mean	60.65		Hypothesized Mean Differ	0	
9	Sample Standard Deviation	16.81		df	56	
10	Population 2 Sample			t Stat	-0.768170828	
11	Sample Size	41		P(T<=t) one-tail	0.222806507	
12	Sample Mean	64.32		t Critical one-tail	1.672522304	
13	Sample Standard Deviation	16.81		P(T<=t) two-tail	0.445613014	
14				t Critical two-tail	2.003240704	
15	Intermediate Calculations					
16	Population 1 Sample Degrees of Freedom	16				
17	Population 2 Sample Degrees of Freedom	40				
18	Total Degrees of Freedom	56				
19	Pooled Variance	282.5761				
20	Difference in Sample Means	-3.67				
21	t Test Statistic	-0.75683				
22						
23	Lower-Tail Test					
24	Lower Critical Value	-1.67252				
25	p-Value	0.226161				
26	Do not reject the null hypothesis					

Two-Sample Test for Means with Paired Samples

The Excel file *Pile Foundation* contains the estimates used in a bid and actual auger cast pile lengths for a foundation engineering project. The contractor's past experience suggested that the bid information was generally accurate, so the average difference between the actual pile lengths and estimated lengths should be close to zero. After this project was completed, the contractor found that the average difference was 6.38. Did he receive poor bid information, or was this difference simply the result of sampling error?

In this situation, the data from the two samples are naturally paired. When this is the case, a paired *t*-test is more accurate than assuming that the data come from independent populations, as was the case for the two-sample tests for means that we discussed earlier. The hypothesis to test is:

$$H_0: \text{average difference} = 0$$
$$H_1: \text{average difference} \neq 0$$

Excel has a *Data Analysis* tool, *t-Test: Paired Two Sample for Means*. In the dialog, you need only enter the variable ranges and hypothesized mean difference. Figure 5.15 shows the results.

Figure 5.15 Results for Paired Two-Sample *t*-Test for Means

	A	B	C
1	t-Test: Paired Two Sample for Means		
2			
3		*Variable 1*	*Variable 2*
4	Mean	34.55623794	28.17755627
5	Variance	267.0113061	255.8100385
6	Observations	311	311
7	Pearson Correlation	0.79692836	
8	Hypothesized Mean Difference	0	
9	df	310	
10	t Stat	10.91225025	
11	P(T<=t) one-tail	5.59435E-24	
12	t Critical one-tail	1.649783823	
13	P(T<=t) two-tail	1.11887E-23	
14	t Critical two-tail	1.967645863	

We compute the difference for each pile, D_i, by subtracting the estimated value from the actual value and finding the average difference, $\overline{D}$, and standard deviation of the differences, s_D. The test statistic is:

$$t = \frac{\overline{D} - \mu_D}{s_D/\sqrt{n}} \tag{5.3}$$

with $n - 1$ degrees of freedom. Here, μ_D is the hypothesized difference between the means, or 0. For the data, we found the average difference to be 6.3 and $s_D = 10.31$. Therefore, $t = 6.38/(10.31/\sqrt{311}) = 10.91$. This is a two-tailed test, so in Figure 5.15 we interpret the results using only the two-tail information. As before, Excel only provides the upper-tail critical value. Because t is much larger than the upper critical value of 1.96 (assuming a 0.05 significance level), we must reject the null hypothesis and conclude that the mean of the estimates is significantly different from the mean of the actual pile lengths. Note that the *p*-value is essentially zero, verifying this conclusion.

Two-Sample Tests for Proportions

Similar to means, we may conduct hypothesis tests for differences in proportions. For example, in the Excel file *Accounting Professionals*, we might wish to test whether the proportion of females who have a graduate degree is the same as the proportion of males. We test the hypotheses:

$$H_0: \pi_1 - \pi_2 = 0$$
$$H_1: \pi_1 - \pi_2 \neq 0$$

where π_1 is the population proportion of females who have a graduate degree and π_2 is the population proportion of males who have a graduate degree. *PHStat* has a tool for conducting this test, *z-Test for Differences in Two Proportions*, found in the

Figure 5.16 *PHStat* Results for Differences in Proportions

	A	B	C	D	E
1	Graduate Degree Differences by Gender				
2					
3	Data			Confidence Interval Estimate	
4	Hypothesized Difference	0		of the Difference Between Two Proportions	
5	Level of Significance	0.05			
6	Group 1			Data	
7	Number of Successes	46		Confidence Level	95%
8	Sample Size	141			
9	Group 2			Intermediate Calculations	
10	Number of Successes	88		Z Value	-1.959963985
11	Sample Size	256		Std. Error of the Diff. between two Proportions	0.049397531
12				Interval Half Width	0.096817381
13	Intermediate Calculations				
14	Group 1 Proportion	0.326241135		Confidence Interval	
15	Group 2 Proportion	0.34375		Interval Lower Limit	-0.114326246
16	Difference in Two Proportions	-0.017508865		Interval Upper Limit	0.079308516
17	Average Proportion	0.337531486			
18	Z Test Statistic	-0.353063254			
19					
20	Two-Tail Test				
21	Lower Critical Value	-1.959963985			
22	Upper Critical Value	1.959963985			
23	p-Value	0.72404102			
24	Do not reject the null hypothesis				

Two-Sample Tests menu. The dialog prompts you to enter the hypothesized difference, and the number of successes and sample size for each sample. From data, we have 46 out of 141 females having a graduate degree, and 88 out of 256 males. Figure 5.16 shows the results, including the optional calculation of a confidence interval.

The formula used for calculating the *Z-Test Statistic* is explained in the appendix to this chapter. Note that the test statistic value falls between the lower and upper critical values, indicating that we should not reject the null hypothesis that the proportion of females who have a graduate degree is the same as the proportion of males. Again, the *p*-value exceeds the level of significance, leading to the same conclusion. The confidence interval shows that the estimated difference in the population proportions can be either positive or negative.

Hypothesis Tests and Confidence Intervals

You may have been thinking that a relationship exists between hypothesis tests and confidence intervals because they rely on essentially the same information. For instance, in the previous example for differences in proportions, we observed that the confidence interval did not conclusively show that the difference in proportions between females and males is either positive or negative, and we did not reject the null hypothesis. In contrast for the paired test for means with the pile foundation data, we rejected the null hypothesis that the population means are the same, and found in the previous chapter that a 95% confidence interval for the mean difference was (5.234, 7.526). This confidence interval suggests a positive difference between the actual and estimated pile lengths. Thus, in testing the difference between two population parameters, if the confidence interval for the difference contains 0, then we would not reject the corresponding null hypothesis. However, if both confidence limits are either positive or negative, then we would reject the null hypothesis.

For one-tail hypothesis tests, we need to examine on which side the confidence interval falls. For example, we tested the hypotheses:

H_0: *mean response time* $\geq$ 30 *minutes*

H_1: *mean response time* $<$ 30 *minutes*

for the *Customer Support Survey* data using a level of significance of 0.05. We rejected the null hypothesis and concluded that the mean is less than 30 minutes. If we construct a 95% confidence interval for the mean response time, we obtain (15.98, 27.84). Note that this confidence interval lies entirely below 30; therefore, we can reach the same conclusion.

Test for Equality of Variances

As we have seen, Excel supports two different *t*-tests for differences in means, one assuming equal variances and the other assuming unequal variances. We can test for equality of variances between two samples using a new type of test, the *F*-test. To use this test, we must assume that both samples are drawn from normal populations. To illustrate the *F*-test, we will use the data in the Excel file *Burglaries* and test whether the variance in the number of monthly burglaries is the same before and after the citizen-police program began.

$$H_0: \sigma^2_{before} = \sigma^2_{after}$$
$$H_1: \sigma^2_{before} \neq \sigma^2_{after}$$

The *F*-test can be applied using either the *PHStat* tool found in the menu by choosing *Two-Sample Tests . . . F-test for Differences in Two Variances,* or the Excel *Data Analysis* tool *F-test for Equality of Variances* (see *Excel/PHStat Note: Testing for Equality of Variances*).

Figure 5.17 shows the output from both tests. Both tests compute an *F*-test statistic. The *F*-test statistic is the ratio of the variances of the two samples:

$$F = \frac{s_1^2}{s_2^2} \tag{5.4}$$

For this example, the test statistic is $F = s_1^2/s_2^2 = 282.47/253.87 = 1.113$. This statistic is compared to critical values from the *F*-distribution for a given confidence level.

EXCEL/*PHSTAT* NOTE
Testing for Equality of Variances

The *PHStat* tool *F-test for Differences in Two Variances* requires that you enter the level of significance and the sample size and sample standard deviation for each sample, which you must compute from the data. As with all *PHStat* tools, you can specify whether the test is two-tail, upper-tail, or lower-tail.

To use the Excel tool, select *F-test for Equality of Variances* from the *Data Analysis* menu. Specify the *Variable 1 Range* and the *Variable 2 Range* for both data sets and a value for the significance level. Note, however, that this tool provides results for only a one-tailed test. Thus, for a two-tailed test of equality of variances, *you must use $\alpha/2$ for the significance level* in the Excel dialog box. If the variance of variable 1 is greater than the variance of variable 2, the output will specify the upper tail; otherwise, you obtain the lower-tail information. For ease of interpretation, we suggest that you ensure that variable 1 has the larger variance. See the example for how to properly interpret the Excel results.

Figure 5.17 *PHStat* and Excel Results for Equality of Variances Test

	A	B	C	D	E	F
1	**PHStat Results for Equality of Variances**			**Excel Results for Equality of Variances**		
2						
3	**Data**			F-Test Two-Sample for Variances		
4	**Level of Significance**	0.05				
5	**Population 1 Sample**				*Variable 1*	*Variable 2*
6	**Sample Size**	41		Mean	64.31707317	60.64705882
7	**Sample Standard Deviation**	16.81		Variance	282.4719512	253.8676471
8	**Population 2 Sample**			Observations	41	17
9	**Sample Size**	17		df	40	16
10	**Sample Standard Deviation**	15.93		F	1.112674082	
11				P(F<=f) one-tail	0.424390536	
12	**Intermediate Calculations**			F Critical one-tail	2.508529216	
13	***F* Test Statistic**	1.113535				
14	Population 1 Sample Degrees of Freedom	40				
15	Population 2 Sample Degrees of Freedom	16				
16						
17	**Two-Tail Test**					
18	**Lower Critical Value**	0.464213				
19	**Upper Critical Value**	2.508529				
20	***p*-Value**	0.847369				
21	**Do not reject the null hypothesis**					

Like the t-distribution, the F-distribution is characterized by degrees of freedom. However, the F-statistic has *two* values of degrees of freedom—one for the sample variance in the numerator and the other for the sample variance in the denominator. In both cases, the number of degrees of freedom is equal to the respective sample size minus 1. The *PHStat* output provides both lower and upper critical values from the F-distribution. We see that the F-Test Statistic falls between these critical values; therefore, we do not reject the null hypothesis. The p-value of 0.847 confirms this, as it is larger than the level of significance.

Proper interpretation of the Excel results depends on how we take the ratio of sample variances. As suggested in the note, we recommend ensuring that variable 1 has the larger variance. Note that when we do this, $F > 1$. If the variances differ significantly, we would expect F to be much larger than 1; the closer F is to 1, the more likely it is that the variances are the same. Therefore, we need only compare F to the upper-tail critical value. However, because Excel provides results for a one-tail test only, we must use $\alpha/2$ as the input value in the Excel dialog.

In comparing the results, note that the computed F-test statistic is the same (to within some differences because of rounding the calculations of the sample standard deviations in *PHStat*) and that the upper critical values are both the same. Therefore, the same conclusions are reached by comparing the test statistic to the upper-tail critical value. However, note that the p-values differ, because in *PHStat* the level of significance is entered as α, whereas in Excel, the level of significance was entered as $\alpha/2$ for a one-tailed test. Thus, the p-value for Excel is half that of *PHStat*. These are both correct, as long as you realize that you compare the p-values to the proper levels of significance; that is, in *PHStat*, $0.847369 > 0.05$, while in Excel, $0.42439 > 0.025$. The same conclusions are reached in both cases.

ANOVA: TESTING DIFFERENCES OF SEVERAL MEANS

To this point, we have discussed hypothesis tests that compare a population parameter to a constant value or that compare the means of two different populations. Often, we would like to compare the means of several different groups to determine if all are equal, or if any are significantly different from the rest. For example, in the Excel file *Social Networking*, we might be interested in whether any significant differences exist in the hours spent online each week by freshmen, sophomores, juniors, seniors, and graduate students. In statistical terminology, the college year is called a **factor**, and we have five categorical levels of this factor. Thus, it would appear that we will have to perform ten different pairwise tests to establish whether any significant differences exist among them. As the number of factor levels increases, you can easily see that the number of pairwise tests grows large very quickly. Fortunately, other statistical tools exist that eliminate the need for such a tedious approach. **Analysis of variance (ANOVA)** provides a tool for doing this. The null hypothesis is that the population means of all groups are equal; the alternative hypothesis is that at least one mean differs from the rest:

$$H_0: \mu_1 = \mu_2 = \ldots = \mu_m$$

H₁: *at least one mean is different from the others*

Let us apply ANOVA to test the hypothesis that the mean hours online per week for all student groups in the Excel file *Social Networking* are equal against the alternative that at least one mean is different. The table below shows the summary statistics:

	FRESHMAN	SOPOHOMORE	JUNIOR	SENIOR	GRAD STUDENT
	10	20	30	7	35
	15	5	20	30	5
	10	7	10	10	10
	8	10	10	15	4
	15	6	7	8	4
	10	8	5	12	8
	25	10	2	7	4
	20	10	7	15	8
	15	10		25	
	12	10		10	
	20	10			
		20			
		20			
AVERAGE	14.545	11.231	11.375	13.900	9.750
COUNT	11	13	8	10	8

Excel provides a *Data Analysis* tool, *ANOVA: Single Factor* (see *Excel Note: Single-Factor Analysis of Variance*) to conduct analysis of variance. The results for the *Social Networking* data are given in Figure 5.18. The output report begins with a summary report of basic statistics for each group. The ANOVA section reports the details of the hypothesis test. ANOVA derives its name from the fact that we are analyzing variances in the data. The theory is described more fully in the appendix to this chapter, but essentially ANOVA computes a measure of the variance between the means of

Figure 5.18 ANOVA Results for *Social Networking* Data

	A	B	C	D	E	F	G
1	Anova: Single Factor						
2							
3	SUMMARY						
4	*Groups*	*Count*	*Sum*	*Average*	*Variance*		
5	Freshman	11	160	14.54545	28.07273		
6	Sopohomore	13	146	11.23077	27.85897		
7	Junior	8	91	11.375	84.55357		
8	Senior	10	139	13.9	60.98889		
9	Grad Student	8	78	9.75	109.3571		
10							
11							
12	ANOVA						
13	*Source of Variation*	*SS*	*df*	*MS*	*F*	*P-value*	*F crit*
14	Between Groups	154.77	4	38.69251	0.690579	0.602324	2.578739
15	Within Groups	2521.31	45	56.02911			
16							
17	Total	2676.08	49				

EXCEL NOTE
Single-Factor Analysis of Variance

To use ANOVA to test for difference in sample means, click on *Tools, Data Analysis*, and select *ANOVA: Single Factor*. This displays the dialog box shown in Figure 5.19. You need only specify the input range of the data (in contiguous columns) and whether it is stored in rows or columns (i.e., whether each factor level or group is a row or column in the range). You must also specify the level of significance (alpha level) and the output options.

Figure 5.19 *ANOVA: Single-Factor* Dialog

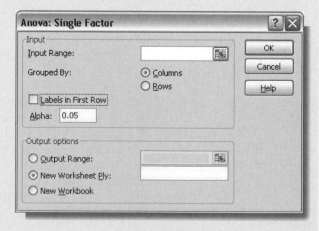

each group, and a measure of the variance within the groups. These measures are in the MS column (MS stands for "mean square"). If the null hypothesis is true, the between group variance should be small, as each of the group means should be close to one another. If the means among groups are significantly different, then MS Between Groups will be significantly larger than MS Within Groups. The ratio of the variances, MS Between Groups/MS Within Groups, is a test statistic from an F-distribution (similar to the test for equality of variances). Thus, when MS Between Groups is large relative to MS Within Groups, F will be large. If the F-statistic is large enough based on the level of significance chosen and exceeds a critical value, we would reject the null hypothesis. In this example, $F = 38.69251/56.02911 = 0.690579$ and the critical value (F crit) from the F-distribution is 2.578739. Here $F < F$ crit; therefore, we cannot reject the null hypothesis and conclude that there are no significant differences in the means of the groups; that is, the mean number of hours spent online for social networking does not differ among students in different years. Alternatively, we see that the p-value is larger than the chosen level of significance, 0.05, leading to the same conclusion.

Assumptions of ANOVA

ANOVA requires assumptions that the m groups or factor levels being studied represent populations whose outcome measures

1. are randomly and independently obtained;
2. are normally distributed; and
3. have equal variances.

If these assumptions are violated, then the level of significance and the power of the test can be affected. Usually, the first assumption is easily validated when random samples are chosen for the data. ANOVA is fairly robust to departures from normality, so in most cases this isn't a serious issue. The third assumption is required in order to pool the variances within groups. If sample sizes are equal, this assumption does not have serious effects on the statistical conclusions; however, with unequal sample sizes, it can. When you suspect this, you can use the *Levene test* to investigate the hypothesis:

$$H_0: \sigma_1^2 = \sigma_2^2 = \ldots \sigma_m^2$$
$$H_1: \text{Not all } \sigma_j^2 \text{ are equal}$$

The Levene test can be found in the *PHStat Multiple Sample Test* menu.

When the assumptions underlying ANOVA are violated, you may use a **nonparametric test** that does not require these assumptions; for example, the Kruskal–Wallis rank test for determining whether multiple populations have equal medians. This test is available in *PHStat* also, and we refer you to more comprehensive texts on statistics for further information and examples.

Tukey–Kramer Multiple Comparison Procedure

Figure 5.20 shows an ANOVA for the customer ratings of four product dimensions in the Excel file *Customer Survey*. The F-test statistic is much larger than the critical value (and the p-value is essentially zero), leading us to reject the null hypothesis. Although ANOVA can identify a difference among the means of multiple populations, it cannot determine which of the means are significantly different from the rest. To do this, we may use the **Tukey–Kramer multiple comparison procedure**. This method compares the absolute value of the difference in means for all pairs of groups and compares these to a critical range. Values exceeding the critical range identify

Figure 5.20 Analysis of Variance for *Customer Survey* Data

	A	B	C	D	E	F	G
1	Anova: Single Factor						
2							
3	SUMMARY						
4	*Groups*	*Count*	*Sum*	*Average*	*Variance*		
5	Quality	200	879	4.395	0.581884		
6	Ease of Use	200	833	4.165	0.610829		
7	Price	200	734	3.67	1.136784		
8	Service	200	828	4.14	0.794372		
9							
10							
11	ANOVA						
12	*Source of Variation*	*SS*	*df*	*MS*	*F*	*P-value*	*F crit*
13	Between Groups	55.505	3	18.50167	23.6907	1.08E-14	2.616089
14	Within Groups	621.65	796	0.780967			
15							
16	Total	677.155	799				

those populations whose means differ. *PHStat* provides this procedure (see *PHStat Note: Using the Tukey–Kramer Multiple Comparison Procedure*). Figure 5.21 shows the results of the procedure for the customer survey data. We see that in comparing one group to another, all means are significantly different except for Group 2 and Group 4 (Ease of Use and Service). Note that although the means of Quality, Ease of Use, and Service all appear to be close to one another, the large sample size provides the ability to differentiate between these seemingly similar customer ratings.

Analysis of variance may be extended to more than one factor. For example, suppose that a company wants to investigate whether changes in temperature and pressure settings affect output quality or yield in some production process. Temperature

Figure 5.21 Results from Tukey–Kramer Procedure

	A	B	C	D	E	F	G	H	I	J	K
1	Customer Survey										
2											
3		Sample	Sample			Absolute	Std. Error	Critical			
4	Group	Mean	Size		Comparison	Difference	of Difference	Range	Results		
5	1	4.395	200		Group 1 to Group 2	0.23	0.06248869	0.2268	Means are different		
6	2	4.165	200		Group 1 to Group 3	0.725	0.06248869	0.2268	Means are different		
7	3	3.67	200		Group 1 to Group 4	0.255	0.06248869	0.2268	Means are different		
8	4	4.14	200		Group 2 to Group 3	0.495	0.06248869	0.2268	Means are different		
9					Group 2 to Group 4	0.025	0.06248869	0.2268	Means are not different		
10	Other Data				Group 3 to Group 4	0.47	0.06248869	0.2268	Means are different		
11	Level of significance	0.05									
12	Numerator d.f.	4									
13	Denominator d.f.	796									
14	MSW	0.780967									
15	Q Statistic	3.63									

The Tukey–Kramer procedure is selected from the *Multiple Sample Tests* menu. You need only enter the cell range of the data. However, *PHStat* prompts you to enter a value for "Q Statistic" manually in the worksheet before displaying the conclusions of the test. This value is based on the numerator *df* and denominator *df* shown in the output and not the degrees of freedom for the ANOVA test, which are different. The Q-statistic may be found in Table A.5 in the appendix at the back of this book.

and pressure represent two factors, and multiple samples might be selected for each of three different combinations of temperature levels and pressure settings. This would be an example of a two-factor ANOVA and would allow the investigator to test hypotheses about whether differences exist among the levels of each factor individually and also whether any interactions exist between the factors; that is, whether the effect of one factor depends on the level of the other. Further discussion is beyond the scope of this book, but additional information may be found in more comprehensive statistics texts and books devoted exclusively to the subject.

CHI-SQUARE TEST FOR INDEPENDENCE

A common problem in business is to determine whether two categorical variables are independent. We discussed the notion of statistical independence in Chapter 3 and why it is important in marketing applications. For example, a consumer study might collect data on preferences for three different energy drinks of both male and female high school students. The objective of the study might be to determine if energy drink preferences are independent of gender. Independence would mean that the proportion of individuals who prefer one drink over another would be essentially the same no matter if the individual is male or female. On the other hand, if males have different preferences than females, the variables would be dependent. Knowing this can help marketing personnel better target advertising campaigns to different demographic groups.

In Chapter 3, we described how to determine if two random variables are statistically independent by examining the joint and marginal probabilities. However, with empirical data, sampling error can make it difficult to properly assess independence of categorical variables. We would never expect the joint probabilities to be exactly the same as the product of the marginal probabilities because of sampling error even if the two variables are statistically independent. However, we can draw a conclusion using a hypothesis test called the *chi-square test for independence.* The chi-square test for independence tests the following hypotheses:

H_0: *the two categorical variables are independent*

H_1: *the two categorical variables are dependent*

For example, the data below show the sample data used in Chapter 3 for brand preferences of energy drinks.

CROSS-TABULATION	BRAND 1	BRAND 2	BRAND 3	TOTAL
MALE	25	17	21	63
FEMALE	9	6	22	37
TOTAL	34	23	43	100

The chi-square test for independence tests whether the proportion of males who prefer a particular brand is no different from the proportion of females. For instance, of the 63 male students, 25 (40%) prefer Brand 1. If gender and brand preference are indeed independent, we would expect that about the same proportion of the sample of female students would also prefer Brand 1. In actuality, only 9 of 37 (24%) prefer Brand 1. However, we do not know whether this is simply due to sampling error or represents a significant difference.

PHStat provides a tool for conducting this test (see *PHStat Note: Chi-Square Test for Independence*). Figure 5.22 shows the completed worksheet for the test. The test calculates the expected frequencies that should be observed if the null hypothesis is true. For example, if the variables are independent, we would expect to find about 21 or 22 males and about 12 or 13 females who prefer Brand 1. To compute the expected frequency for a particular cell in the table, simply multiply the row total by the column

Figure 5.22 Chi-Square Test for Independence Results

	A	B	C	D	E
1	**Brand Preferences**				
2					
3		**Observed Frequencies**			
4			**Column variable**		
5	**Row variable**	**Brand 1**	**Brand 2**	**Brand 3**	**Total**
6	**Male**	25	17	21	63
7	**Female**	9	6	22	37
8	**Total**	34	23	43	100
9					
10		**Expected Frequencies**			
11			Column variable		
12	Row variable	Brand 1	Brand 2	Brand 3	Total
13	Male	21.42	14.49	27.09	63
14	Female	12.58	8.51	15.91	37
15	Total	34	23	43	100
16					
17	**Data**				
18	**Level of Significance**	0.05			
19	Number of Rows	2			
20	Number of Columns	3			
21	Degrees of Freedom	2			
22					
23	**Results**				
24	Critical Value	5.99146			
25	Chi-Square Test Statistic	6.49243			
26	*p*-Value	0.03892			
27	**Reject the null hypothesis**				

total and divide by the grand total. Thus, the expected frequency for Male and Brand 1 is $(63)(34)/100 = 21.42$. One caution: All expected frequencies should be at least 1.0. If not, then you should combine rows or columns to meet this requirement. Drawing from the discussion in Chapter 3, you can verify that the product of the marginal probabilities equals the joint probabilities in the Expected Frequencies table.

The procedure uses the observed and expected frequencies to compute a test statistic, called a **chi-square statistic**, which is the sum of the squares of the differences between observed frequency, f_o, and expected frequency, f_e, divided by the expected frequency in each cell:

$$\chi^2 = \sum \frac{(f_0 - f_e)^2}{f_e} \tag{5.5}$$

The closer the observed frequencies are to the expected frequencies, the smaller will be the value of the chi-square statistic. We compare this statistic for a specified level of significance α to the critical value from a chi-square distribution with $(r - 1)(c - 1)$ degrees of freedom, where r and c are the number of rows and columns in the contingency table, respectively. If the test statistic exceeds the critical value for a specified level of significance, we reject H_0. In this case, the Chi-Square Test Statistic is 6.49243 and the critical value if 5.99146. Because the test statistic exceeds the critical value, we reject the null hypothesis that the two categorical variables are independent. The small p-value of 0.03892 also confirms this conclusion. The Excel function CHITEST(*actual_range, expected_range*) computes the p-value for the Chi-Square test, but you must calculate the expected frequencies.

PHSTAT NOTE
Chi-Square Test for Independence

From the *PHStat Multiple Sample Tests* menu, select *Chi-Square Test*. In the dialog box, enter the level of significance, number of rows, and number of columns in the contingency table. *PHStat* will create a worksheet in which you will need to enter the data for the observed frequencies. You may also customize the names of the rows and columns for your specific application. After you complete the table, the calculations are performed automatically. *PHStat* includes an optional check box for the *Marascuilo procedure*. This makes comparisons among all pairs of groups to determine if the proportions are significantly different, similar to the way the Tukey–Kramer procedure does for ANOVA. This is useful if there are more than two levels of a category so you can identify which pairs are significantly different from each other.

BASIC CONCEPTS REVIEW QUESTIONS

1. Explain the notion of hypothesis testing. What is the general process that one should follow in conducting a hypothesis test?
2. Explain the difference between the null and alternative hypothesis. Which one can be proven in a statistical sense?
3. What are the differences between one- and two-sample hypothesis tests? Describe the correct mathematical form for specifying the null and alternative hypotheses for both types of tests.
4. Explain Type I and Type II errors. Which one is controllable by the experimenter?
5. Explain how the rejection region is determined for a hypothesis test. How does it differ between one- and two-tailed tests?
6. How can you determine when to use a lower one-tailed test of hypothesis versus an upper one-tailed test?
7. In one-sample tests for means, why is it important to know whether the standard deviation is known or not?

8. What is a *p*-value? How is it used to reach a conclusion about a hypothesis test?

9. How are Type II errors influenced by the true (unknown) population mean and sample size used in a hypothesis test?

10. Explain the peculiar nuances associated with the Excel tools for two-sample *t*-tests. What issues must you consider to use the tests and interpret the results properly?

11. What is the difference between paired and independent samples?

12. Explain the relationship between confidence intervals and hypothesis testing.

13. Explain how to interpret the Excel results for a hypothesis test for equality of variances.

14. What is analysis of variance? What hypothesis does it test? Provide some practical examples.

15. What are the key assumptions of ANOVA? What should be done if these are seriously violated?

16. Explain the purpose of the chi-square test for independence. Provide some practical examples where this test might be used in business.

SKILL-BUILDING EXERCISES

1. For the *Customer Support Survey* data, test the null hypothesis that the mean response time is less than or equal to 30 minutes against the alternative hypothesis that it is greater than 30 minutes. Compare your results to Figure 5.3. What is the relationship between the *p*-values for the two tests?

2. For the *Customer Support Survey* data, use the concepts in Figure 5.9 to calculate the probability of a Type II error if the hypothesized value is 30 and the true mean is either 20 or 25. Compare your answers to the power curve in Figure 5.12.

3. For the burglary data used in the example for testing the difference in two means, how much difference would the results have been if you assumed that the variances were unequal? Using *PHStat* tools, devise

and run an experiment to vary the assumed standard deviations to see how the results of the two tools differ.

4. Use the example in the chapter for testing for equality of variances, but run the test by switching the variables so that the value of *F* is less than 1. Using both the *PHStat* and Excel tools, explain how to interpret the Excel results properly in this situation.

5. Using the information in the appendix to this chapter, verify the calculations for the sums of squares and mean squares for the ANOVA example.

6. Using the example for the chi-square test for independence, verify the calculation of the chi-square test statistic using formula (5.5).

PROBLEMS AND APPLICATIONS

For all hypothesis tests, assume that the level of significance is 0.05 unless otherwise stated.

1. Call centers typically have high turnover. The director of human resources for a large bank has compiled data on about 70 former employees at one of the bank's call centers in the Excel file *Call Center Data*. In writing an article about call center working conditions, a reporter has claimed that the average tenure is no more than two years. Formulate and test a hypothesis to determine if this claim has a statistical basis from this sample.

2. The state of Ohio Department of Education has a mandated ninth-grade proficiency test that covers writing, reading, mathematics, citizenship (social studies), and science. The Excel file *Ohio Education Performance* provides data on success rates (defined as the percent of students passing) in school districts in the greater Cincinnati metropolitan area along with state averages. Test the hypothesis that the average score in the Cincinnati area is equal to the state average in each test and for the composite score.

3. Formulate and test a hypothesis to determine if statistical evidence suggests that the graduation rate for either top liberal arts colleges or research universities in the sample *Colleges and Universities* exceeds 90%. Do the data support a conclusion that the graduation rate exceeds 85%? Would your conclusions change if the level of significance were 0.01 instead of 0.05?

4. The Excel file *Sales Data* provides data on a sample of customers. An industry trade publication stated that the average profit per customer for this industry was $4,500. Using a test of hypothesis, can this company claim that they are meeting the industry standard? Can the company claim that they are exceeding the standard?

5. The Excel file *Room Inspection* provides data for 100 room inspections at 25 hotels in a major chain. Management would like the proportion of nonconforming rooms to be less than 2%. Test an appropriate hypothesis to determine if management can make this claim.

6. Using the data in the Excel file *Consumer Transportation Survey*, test the following null hypotheses:
   a. Individuals spend at least 10 hours per week in their vehicles.
   b. Individuals drive an average of 450 miles per week.
   c. The average age of SUV drivers is no greater than 35.
   d. At least 75% of individuals are satisfied with their vehicles.

7. A marketing study found that the mean spending in 15 categories of consumer items for 297 respondents in the 18 to 34 age group was $75.86 with a standard deviation of $50.90. For 736 respondents in the 35+ age group, the mean and standard deviation were $68.53 and $45.29, respectively. Test the hypothesis that there is no difference in the mean spending between these two populations.

8. The Excel file *Baseball Attendance* shows the attendance in thousands at San Francisco Giants' baseball games for the 10 years before the Oakland A's moved to the Bay Area in 1968, as well as the combined attendance for both teams for the next 11 years. Is there a statistical reason to believe that overall attendance has changed after the A's move?

9. Determine if there is evidence to conclude that the mean number of vacations taken by married individuals is less than the number taken by single/divorced individuals using the data in the Excel file *Vacation Survey*.

10. The Excel file *Accounting Professionals* provides the results of a survey of 27 employees in a tax division of a *Fortune* 100 company.
    a. Test the null hypothesis that the average number of years of service is the same for males and females.
    b. Test the null hypothesis that the average years of undergraduate study is the same for males and females.

11. Determine if there is evidence to conclude that the proportion of males who plan to attend graduate school is larger than the proportion of females who plan to attend graduate school using the data in the Excel file *Graduate School Survey*. Use a level of significance of 0.10.

12. A study of nonfatal occupational injuries in the United States found that about 31% of all injuries in the service sector involved the back. The National Institute for Occupational Safety and Health (NIOSH) recommended conducting a comprehensive ergonomics assessment of jobs and workstations. In response to this information, Mark Glassmeyer developed a unique ergonomic handcart to help field service engineers be more productive and also to reduce back injuries from lifting parts and equipment during service calls. Using a sample of 382 field service engineers who were provided with these carts, Mark collected the following data:

	Year 1 (without cart)	Year 2 (with cart)
Average call time	8.05 hours	7.84 hours
Standard deviation call time	1.39 hours	1.34 hours
Proportion of back injuries	0.018	0.010

a. Determine if there is statistical evidence that average call time has decreased as a result of using the cart. What other factors might account for any changes?

b. Determine if there is statistical evidence that the proportion of back injuries has decreased as a result of using the cart.

13. The director of human resources for a large bank has compiled data on about 70 former employees at one of the bank's call centers (see the Excel file *Call Center Data*). For each of the following, assume equal variances of the two populations.

a. Test the null hypothesis that the average length of service for males is the same as for females.

b. Test the null hypothesis that the average length of service for individuals without prior call center experience is the same as those with experience.

c. Test the null hypothesis that the average length of service for individuals with a college degree is the same as for individuals without a college degree.

d. Now conduct tests of hypotheses for equality of variances. Were your assumptions of equal variances valid? If not, repeat the test(s) for means using the unequal variance test.

14. A producer of computer-aided design software for the aerospace industry receives numerous calls for technical support. Tracking software is used to monitor response and resolution times. In addition, the company surveys customers who request support using the following scale: 0—Did not exceed expectations; 1—Marginally met expectations; 2—Met expectations; 3—Exceeded expectations; 4—Greatly exceeded expectations. The questions are as follows:

Q1: Did the support representative explain the process for resolving your problem?

Q2: Did the support representative keep you informed about the status of progress in resolving your problem?

Q3: Was the support representative courteous and professional?

Q4: Was your problem resolved?

Q5: Was your problem resolved in an acceptable amount of time?

Q6: Overall, how did you find the service provided by our technical support department?

A final question asks the customer to rate the overall quality of the product using a scale of 0—Very poor; 1—Poor; 2—Good; 3—Very good; 4—Excellent. A sample of survey responses and associated resolution and response data are provided in the Excel file *Customer Support Survey*.

a. The company has set a service standard of one day for the mean resolution time. Does evidence exist that the response time is less than one day? How do the outliers in the data affect your result? What should you do about them?

b. Test the hypothesis that the average service index is equal to the average engineer index.

15. Jim Aspenwall, a NASCAR enthusiast, has compiled some key statistics for NASCAR Winston Cup racing tracks across the United States. These tracks range in shape, length, and amount of banking on the turns and straightaways (see the Excel file *NASCAR Track Data*). Test the hypothesis that there is no difference in qualifying record speed between oval and other shapes of tracks.

16. Using the data in the Excel file *Ohio Education Performance,* test the hypotheses that the mean difference in writing and reading scores is zero, and that the mean difference in math and science scores is zero. Use the paired sample procedure.

17. The Excel file *Blood Pressure* shows the monthly blood pressure (diastolic) readings for a patient before and after medication.

a. Test the hypothesis that the variances of monthly blood pressure readings before and after medication are the same.

b. Conduct an appropriate hypothesis test to determine if there is evidence that the medication has lowered the patient's blood pressure.

18. The Excel file *Unions and Labor Law Data* reports the percent of public- and private-sector employees in unions in 1982 for each state, along with indicators whether the states had a bargaining law that covered public employees or right-to-work laws.

a. Test the hypothesis that the percent of employees in unions for both the public sector and private sector is the same for states having bargaining laws as for those who do not.

b. Test the hypothesis that the percent of employees in unions for both the public sector and private sector is the same for states having right-to-work laws as for those who do not.

19. Using the data in the Excel file *Student Grades*, which represent exam scores in one section of a large statistics course, test the hypothesis that the variance in grades is the same for both tests.

20. An engineer measured the surface finish of 35 parts produced on a lathe, noting the revolutions per minute of the spindle and the type of tool used (see the Excel file *Surface Finish*). Use ANOVA to test the hypothesis that the mean surface finish is the same for each tool. If the null hypothesis is rejected, apply the Tukey–Kramer multiple comparison procedure to identify significant differences.

21. Using the data in the Excel file *Freshman College Data*, use ANOVA to determine whether the mean retention rate is the same for all colleges over the four-year period. Second, use ANOVA to determine if the mean ACT and SAT scores are the same each year over all colleges. If the null hypothesis is rejected, apply the Tukey–Kramer multiple comparison procedure to identify significant differences.

22. A mental health agency measured the self-esteem score for randomly selected individuals with disabilities who were involved in some work activity within the past year. The Excel file *Self Esteem* provides the data, including the individuals' marital status, length of work, type of support received (direct support includes job-related services such as job coaching and counseling), education, and age.

a. Apply ANOVA to determine if self-esteem is the same for all marital status levels. If the null hypothesis is rejected, apply the Tukey–Kramer multiple comparison procedure to identify significant differences.

b. Use the chi-square test to determine if marital status is independent of support level.

23. For the data in the Excel file *Accounting Professionals*, perform a chi-square test of independence to determine if age group is independent of having a graduate degree.

24. For the data in the Excel file *Graduate School Survey*, perform a chi-square test for independence to determine if plans to attend graduate school are independent of gender.

25. For the data in the Excel file *New Account Processing*, perform chi-square tests for independence to determine if certification is independent of gender, and if certification is independent of having prior industry background.

CASE

HATCO, Inc.

The Excel file *HATCO*[1] consists of data related to predicting the level of business obtained from a survey of purchasing managers of customers of an industrial supplier, HATCO. The variables are as follows:

- *Delivery Speed*—amount of time it takes to deliver the product once an order is confirmed
- *Price Level*—perceived level of price charged by product suppliers
- *Price Flexibility*—perceived willingness of HATCO representatives to negotiate price on all types of purchases
- *Manufacturing Image*—overall image of the manufacturer or supplier
- *Overall Service*—overall level of service necessary for maintaining a satisfactory relationship between supplier and purchaser
- *Sales Force Image*—overall image of the manufacturer's sales force
- *Product Quality*—perceived level of quality of a particular product
- *Size of Firm*—size relative to others in this market (0 = small; 1 = large)

- *Usage Level*—percentage of total product purchased from HATCO

Responses to the first seven variables were obtained using a graphic rating scale, where a 10-centimeter line was drawn between endpoints labeled "poor" and "excellent." Respondents indicated their perceptions using a mark on the line, which was measured from the left endpoint. The result was a scale from 0 to 10 rounded to one decimal place. Management considers a score of at least 7 to be its benchmark; anything less will trigger some type of intervention.

You have been asked to analyze these data. Your analysis should include a descriptive summary of the data using appropriate charts and visual displays, recommendations on whether any interventions are necessary based on the benchmark targets, whether any differences exist between firm size and customer ratings, and between firm size and usage level. Use appropriate hypothesis tests to support your conclusions and summarize your findings in a formal report to the Vice President of Purchasing.

Appendix

Hypothesis-Testing Theory and Computation

In this appendix, we provide some additional details for two-sample hypothesis tests and the theory behind analysis of variance.

Two-Sample Tests for Differences in Means

The test statistics for the various two sample tests for means are summarized in the following text:

- **Population Variance Known** When the population variance is known, the sampling distribution of the difference in means is normal. Thus, the test statistic is a z-value, and the critical value is found in the normal distribution table.

$$z = \frac{\bar{x}_1 - \bar{x}_2}{\sqrt{\sigma_1^2/n_1 + \sigma_2^2/n_2}} \qquad (5A.1)$$

- **Population Variance Unknown, Assumed Equal** When the population variance is unknown, the sampling distribution of the difference in means has a t-distribution with $n_1 + n_2 - 2$ degrees of freedom. If we assume that the two population variances are equal, the test statistic is:

$$t = \frac{\bar{x}_1 - \bar{x}_2}{\sqrt{\frac{(n_1 - 1)s_1^2 + (n_2 - 1)s_2^2}{n_1 + n_2 - 2} \left(\frac{n_1 + n_2}{n_1 n_2} \right)}} \qquad (5A.2)$$

- **Population Variance Unknown, Unequal** When the population variance is unknown but unequal, the sampling distribution of the difference in means

---

[1]Adopted from Hair, Anderson, Tatham, and Black in *Multivariate Analysis*, 5th ed., Prentice-Hall, 1998.

also has a t-distribution, but the degrees of freedom is more complicated to compute. The test statistic is:

$$t = (\bar{x}_1 - \bar{x}_2) \Big/ \sqrt{\frac{s_1^2}{n_1} + \frac{s_2^2}{n_2}} \qquad (5A.3)$$

and the degrees of freedom for the t-distribution, df^*, is computed as:

$$df^* = \frac{\left[\dfrac{s_1^2}{n_1} + \dfrac{s_2^2}{n_2}\right]^2}{\left[\dfrac{(s_1^2/n_1)^2}{n_1 - 1}\right] + \left[\dfrac{(s_2^2/n_2)^2}{n_2 - 1}\right]} \qquad (5A.4)$$

Two-Sample Test for Differences in Proportions

For a two-sample test for the difference in proportions, the test statistic is:

$$z = \frac{\hat{p}_1 - \hat{p}_2}{\sqrt{\bar{p}(1 - \bar{p})\left(\dfrac{1}{n_1} + \dfrac{1}{n_2}\right)}} \qquad (5A.5)$$

where $\bar{p}$ = number of successes in both samples/$(n_1 + n_2)$. This statistic has an approximate standard normal distribution; therefore, the critical value is chosen from a standard normal distribution.

Test for Equality of Variances

In this chapter, we explained that the ratio of variances of two samples is an F-statistic with $df_1 = n_1 - 1$ (numerator) and $df_2 = n_2 - 1$ (denominator) degrees of freedom. If F exceeds the critical value $F_{\alpha/2, df_1, df_2}$ of the F-distribution, then we reject H_0. As noted, this is really a two-tailed test, but as long as F is the ratio of the larger to the smaller variance, we can base our conclusion on the upper-tail value only. *PHStat* provides both the lower and upper critical values. Table A.4 in the appendix at the end of the book provides only upper-tail critical values, and the distribution is *not* symmetric. To find the lower-tail critical value, reverse the degrees of freedom, find the upper-tail value, and take the reciprocal. Thus, for the example in the chapter lower-tail critical value is approximately $1/F_{0.025, 16, 40} = 1/2.18 = 0.46$ (we used the numerator $df = 15$ to give an approximation for the critical value because 16 is not included in Table A.4). This is close to the value shown in the *PHStat* results.

Suppose we took the ratio of the smaller variance to the larger one; that is, $F = 253.87/282.47 = 0.899$. In this case the closer that F is to zero, the greater the likelihood that the population variances differ, so we need only

compare F to the lower-tail critical value and reject if F is less than this value. Thus, in this case, with a numerator $df = 16$ and the denominator $df = 40$, the lower-tail critical value is 0.464. Since $F = 0.899 > 0.464$, we cannot reject the null hypothesis and reach the same conclusion as before.

Theory of Analysis of Variance

We define n_j as the number of observations in sample j. ANOVA examines the variation among and within the m groups or factor levels. Specifically, the total variation in the data is expressed as the variation between groups plus the variation within groups:

$$SST = SSB + SSW \qquad (5A.6)$$

where

SST = total variation in the data
SSB = variation between groups
SSW = variation within groups

We compute these terms using the following formulas.

$$SST = \sum_{j=1}^{m} \sum_{i=1}^{n_j} (X_{ij} - \bar{\bar{X}})^2 \qquad (5A.7)$$

$$SSB = \sum_{j=1}^{m} n_j (\bar{X}_j - \bar{\bar{X}})^2 \qquad (5A.8)$$

$$SSW = \sum_{j=1}^{m} \sum_{i=1}^{n_j} (X_{ij} - \bar{X}_j)^2 \qquad (5A.9)$$

where

n = total number of observations
$\bar{\bar{X}}$ = overall or grand mean
X_{ij} = ith observation in group j
$\bar{X}_j$ = sample mean of group j

From these formulas, you can see that each term is a "sum of squares" of elements of the data; hence, the notation "SST," which can be thought of as the "Sum of Squares Total," SSB is the "Sum of Squares Between" groups, and SSW is the "Sum of Squares Within" groups. Observe that if the means of each group are indeed equal (H_0 is true), then the sample means of each group will be essentially the same as the overall mean, and SSB would be very small, and most of the total variation in the data is due to sampling variation within groups. The sum of squares is computed in the ANOVA section of the Excel output (Figure 5.18). By dividing the sums of squares by the degrees of freedom, we compute the mean squares (MS).

Chapter

Regression Analysis

INTRODUCTION

*I*n Chapter 2, we discussed statistical relationships and introduced correlation as a measure of the strength of a linear relationship between two numerical variables. Decision makers are often interested in predicting the value of a dependent variable from the value of one or more independent, or explanatory, variables. For example, many colleges try to predict the success of their students as measured by their college GPA (the dependent variable) from various independent variables such as SAT scores, high school GPA, type of school (public or private), and other variables. Such a model might be used for admission decisions and might have the form:

$$\text{College GPA} = a + b \times \text{SAT} + c \times \text{GPA} + d \times \text{SchoolType}$$

Using historical data on the high school and college performance of students, we could derive values of the coefficients a, b, c, and d that would best explain the College GPA as a function of the other variables. Another example is predicting the demand for a product (the dependent variable) as a function of price and advertising (the independent variables). The model might look like:

$$\text{Demand} = a + b \times \text{Price} + c \times \sqrt{\text{Advertising}}$$

This function might be incorporated into a decision model for maximizing profit.

Regression analysis is a tool for building statistical models that characterize relationships among a dependent variable and one or more independent variables, all of which are numerical. Two broad categories of regression models are used often in business settings: (1) regression models of cross-sectional data, such as those just described, and (2) regression models of time-series data, in which the independent variables are time or some function of time and the focus is on predicting the future. Time-series regression is an important tool in *forecasting,* which is the subject of Chapter 7.

A regression model that involves a single independent variable is called *simple regression.* In the Excel file *Home Market Value* (see Figure 6.1), data obtained from a county auditor provides information about the age, square footage, and current market value of houses in a particular subdivision. We might wish to develop a simple regression model for predicting the market value of homes as a function of the size of the home. Market Value would be the dependent variable, and Square Feet would be the independent variable. A regression model that involves two or more independent variables is called *multiple regression.* In the Excel file *Colleges and Universities* (see Figure 6.2), we might wish to predict the graduation rate as a function of the other variables—median SAT, acceptance rate, expenditures/student, and percent in the top 10% of their high school class. In a multiple regression model, the graduation rate would be the dependent variable, and the remaining variables would be the independent variables.

In this chapter, we describe how to develop and analyze both simple and multiple regression models. Our principal focus is to gain a basic understanding of the assumptions of regression models, statistical issues associated with interpreting regression results, and practical

	A	B	C
1	Home Market Value		
2			
3	**House Age**	**Square Feet**	**Market Value**
4	33	1,812	$90,000.00
5	32	1,914	$104,400.00
6	32	1,842	$93,300.00
7	33	1,812	$91,000.00
8	32	1,836	$101,900.00
9	33	2,028	$108,500.00
10	32	1,732	$87,600.00
11	33	1,850	$96,000.00
12	32	1,791	$89,200.00

Figure 6.1 Portion of *Home Market Value*

Figure 6.2 Portion of *Colleges and Universities*

	A	B	C	D	E	F	G
1	Colleges and Universities						
2							
3	**School**	**Type**	**Median SAT**	**Acceptance Rate**	**Expenditures/Student**	**Top 10% HS**	**Graduation %**
4	Amherst	Lib Arts	1315	22%	$ 26,636	85	93
5	Barnard	Lib Arts	1220	53%	$ 17,653	69	80
6	Bates	Lib Arts	1240	36%	$ 17,554	58	88
7	Berkeley	University	1176	37%	$ 23,665	95	68
8	Bowdoin	Lib Arts	1300	24%	$ 25,703	78	90
9	Brown	University	1281	24%	$ 24,201	80	90

issues in using regression as a tool for making and evaluating decisions. We will investigate the following things:

➤ Simple linear regression, focusing on the development of the model, understanding the significance of regression and the use of hypothesis tests, analyzing residuals, and interpreting key statistics such as the coefficient of determination and standard error of the estimate, and understanding statistical assumptions

➤ Multiple linear regression, including understanding how to select variables in models and evaluate goodness of fit, use stepwise and best-subsets procedures, understand multicollinearity, and build useful models

➤ Regression models with categorical variables and curvilinear models

SIMPLE LINEAR REGRESSION

The simplest type of regression model involves one independent variable, X, and one dependent variable, Y. The relationship between two variables can assume many forms, as illustrated in Figure 6.3. The relationship may be linear or nonlinear, or

Figure 6.3 Examples of Variable Relationships

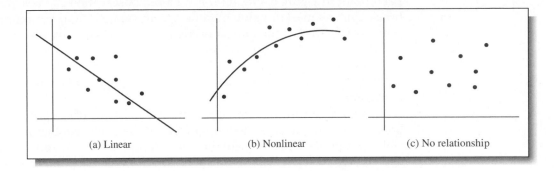

(a) Linear   (b) Nonlinear   (c) No relationship

there may be no relationship at all. We will focus our discussion on linear regression models; therefore, the first thing to do is to verify that the relationship is linear as in Figure 6.3(a). If the relationship is clearly nonlinear as in Figure 6.3(b), then alternate approaches must be used, and if no relationship is evident as in Figure 6.3(c), then it is pointless to even consider developing a regression model.

The type of relationship can usually be visualized in a scatter chart (see Chapter 2), and we always recommend that you create one first to gain some understanding of the nature of any potential relationship. For example, Figure 6.4 shows a scatter chart of the market value in relation to the size of the home in the Excel file *Home Market Value*. The independent variable, X, is the number of square feet, and the dependent variable, Y, is the market value. In general, we see that higher market values are associated with larger home sizes. The relationship clearly is not perfect; but our goal is to build a good model to estimate the market value as a function of the number of square feet by finding the best-fitting straight line that represents the data.

Figure 6.4 Scatter Chart of Market Value versus Home Size

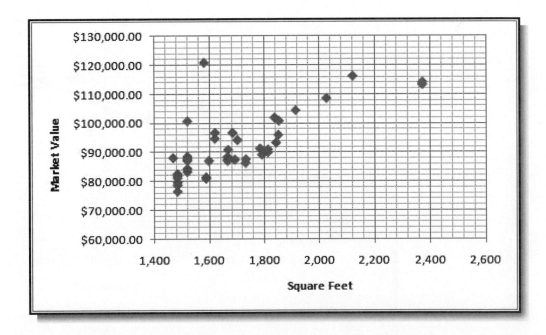

Least-Squares Regression

Each point in Figure 6.4 represents a paired observation of the square footage of the house and its market value obtained from sample data. We represent the ith data point as (X_i, Y_i). A simple linear regression model is the equation of a straight line $\hat{Y} = b_0 + b_1 X$, where b_0 is the intercept and b_1 is the slope, which in some way best fits the data. Thus, when the value of the independent variable is X_i, then $\hat{Y}_i = b_0 + b_1 X_i$ is the estimated value of the dependent variable that lies on the regression line.

If we draw a straight line through the data, some of the points will fall above the line, some below it, and a few might fall on the line itself. Figure 6.5 shows two possible straight lines to represent the relationship between X and Y. Clearly, you would choose A as the better fitting line over B because all the points are "closer" to the line and the line appears to be in the "middle" of the data. The only difference between the lines is the value of the slope and intercept; thus, we seek to determine the values of the slope and intercept that provide the best-fitting line.

One way to quantify the relationship between each point and the line is to measure the vertical distance between them, $Y_i - \hat{Y}_i$ (see Figure 6.6). We can think of these differences as the observed errors (often called **residuals**), e_i, associated with estimating the value of the dependent variable using the regression line. The best-fitting line should minimize some measure of these errors. Because some errors will be negative and others positive, we might take their absolute value, or simply square them. Mathematically, it is easier to work with the squares of the errors.

Adding the squares of the errors, we obtain the following function:

$$\sum_{i=1}^{n} e_i^2 = \sum_{i=1}^{n} (Y_i - (b_0 + b_1 X_i))^2 \tag{6.1}$$

Figure 6.5 Two Possible Regression Lines

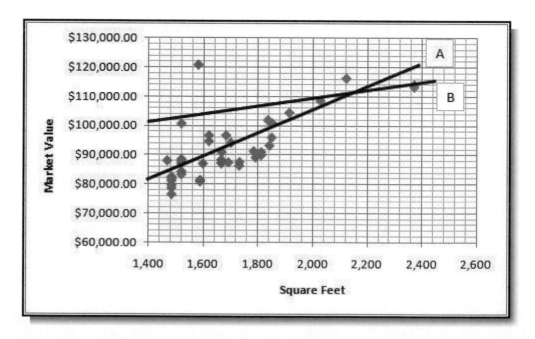

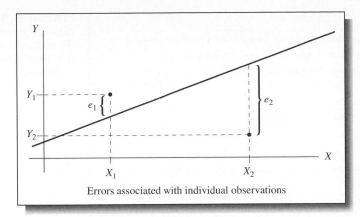

Errors associated with individual observations

Figure 6.6 Measuring the Errors in a Regression Model

If we can find the best values of the slope and intercept that minimizes the sum of squares of the observed errors e_i, we will have found the best-fitting regression line. This is called **least-squares regression**. Note that X_i and Y_i are the known observations (our sample data) and that b_0 and b_1 are unknowns. Using calculus, we can show that the solution that minimizes the sum of squares of the observed errors is:

$$b_1 = \frac{\sum_{i=1}^{n} X_i Y_i - n\overline{X}\,\overline{Y}}{\sum_{i=1}^{n} X_i^2 - n\overline{X}^2} \tag{6.2}$$

$$b_0 = \overline{Y} - b_i \overline{X} \tag{6.3}$$

It is important to understand that the least-squares regression line is derived from sample data. If we select another sample, the regression line will be slightly different. Thus, for any given value of X, we have a distribution of values of Y—a population having the conditional probability distribution $f(Y|X)$, and whose mean is $E(Y|X) = \beta_0 + \beta_1 X$. The equation $Y = \beta_0 + \beta_1 X$ repesents the true (unknown) regression line that runs through the means of these populations (see Figure 6.7). Because individual observations drawn from these populations will not all lie on the line itself, we include an error term in the model: $Y = \beta_0 + \beta_1 X + \varepsilon$. The least-squares regression line estimates this true regression line using the sample data. Thus, b_0 and b_1 are estimates of β_0 and β_1.

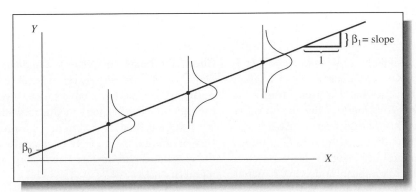

Figure 6.7 True Regression Line

Table 6.1
Spreadsheet Functions and Tools for Regression Analysis

EXCEL FUNCTION	DESCRIPTION
INTERCEPT(*known_y's, known_x's*)	Calculates the intercept for a least-squares regression line
SLOPE(*known_y's, known_x's*)	Calculates the slope of a linear regression line
TREND(*known_y's, known_x's, new_x's*)	Computes the value on a linear regression line for specified values of the independent variable
ANALYSIS TOOLPAK TOOLS	DESCRIPTION
Regression	Performs linear regression using least squares
PHSTAT ADD-IN	DESCRIPTION
Simple Linear Regression	Generates a simple linear regression analysis
Multiple Regression	Generates a multiple linear regression analysis
Best Subsets	Generates a best-subsets regression analysis
Stepwise Regression	Generates a stepwise regression analysis

Although the calculations for the least-squares coefficients appear to be somewhat complicated, they can easily be performed on an Excel spreadsheet. Fortunately, Excel has built-in capabilities to do this. Table 6.1 summarizes useful spreadsheet functions and tools that we will use in this chapter. For example, you may use the functions INTERCEPT and SLOPE to find the least-squares coefficients b_0 and b_1. Using these functions for the *Home Market Value* data, $b_0 = 32673$ and the slope, $b_1 = 35.036$. The slope tells us that for every additional square foot, the market value increases by \$35.036. Such a model can be used to estimate the market value for data not in the sample. Thus, for a house with 1750 square feet, the estimated market value is 35.036(1750) + 32673 = \$93,986. This can also be found by using the Excel function TREND(*known_y's, known_x's, new_x's*). Thus, using the *Home Market Value* data, we would estimate $\hat{Y}(1750)$ using the function = TREND(C4:C45, B4:B45, 1750).

We may also determine and plot the least-squares regression line from the data on a scatter chart using the *Trendline* option (see *Excel Note: Using the Trendline Option*). For the *Home Market Value* data, the results are shown in Figure 6.9 The regression model is $\hat{Y} = 35.036X + 32673$. (Excel expresses the model in the form $\hat{Y} = b_1 X + b_0$).

One point we should make is that it is dangerous to extrapolate a regression model outside the ranges covered by the observations. For instance, if you wanted to predict the market value of a house that has 3,000 square feet, the results may or may not be accurate, because the regression model estimates did not use any observations larger than 2,400 square feet. We cannot be sure that a linear extrapolation will hold and should not use the model to make such predictions.

EXCEL NOTE
Using the Trendline Option

First, click the chart to which you wish to add a trendline to display the *Chart Tools* menu. The *Trendline* option is selected from the *Analysis* group under the *Layout* tab in the *Chart Tools* menu. Click the *Trendline* button and then *More Trendline Options. . . .* This brings up the *Format Trendline* dialog shown in Figure 6.8. Make sure that the radio button for Linear is selected. We will discuss the other nonlinear models in Chapter 9.

Check the boxes for *Display Equation on chart* and *Display R-squared value on chart.* Excel will display the results on the chart you have selected; you may move the equation and *R*-squared value for better readability by dragging them to a different location. A simpler way of doing this is to right-click on the data series in the chart and choose *Add trendline* from the pop-up menu (try it!).

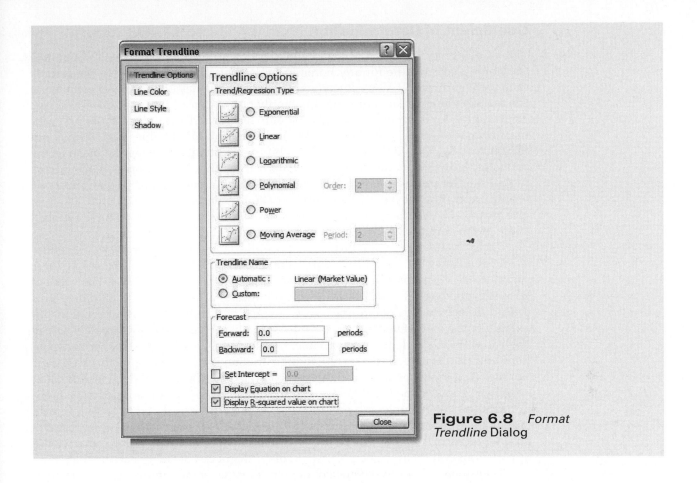

Figure 6.8 *Format Trendline* Dialog

Figure 6.9 Least-Squares Regression Line for *Home Market Value* Data

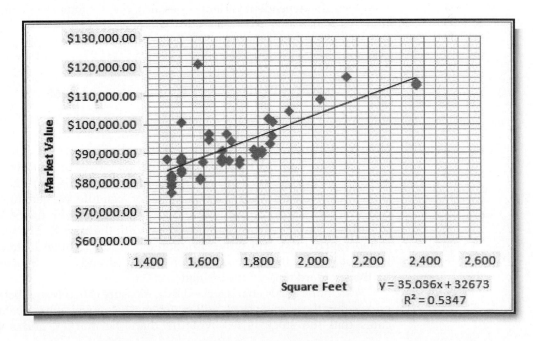

Coefficient of Determination

The value of regression can be explained as follows. Suppose we wanted to estimate the home market value for any home in the neighborhood from which the sample data were gathered. If all we knew were the market values, then the best estimate of the market value for any home would simply be the sample mean, which is $92,069. Thus, no matter if the house has 1,500 square feet, or 2,200 square feet, the best estimate of market value would be $92,069. Because the market values vary from about $75,000 to over $120,000, there is quite a bit of uncertainty in using the mean as the estimate. However, from the scatter chart, we see that larger homes tend to have higher market values. Therefore, if we know that a home has 2,200 square feet, we would expect the market value estimate to be higher, probably over $100,000, than one that has only 1,500 square feet. By knowing the home size, we have reduced some of the variation associated with simply using the mean value.

The R^2 value provides information about the strength of the regression relationship. Specifically, R^2, called the **coefficient of determination**, gives the *proportion of variation that is explained by the independent variable of the regression model*. The value of R^2 will be between 0 and 1. A value of 1.0 indicates a perfect fit and all data points would lie on the regression line, while a value of 0 indicates that no relationship exists. For the market value data, $R^2 = 0.5347$. This means that approximately 53% of the variation in the dependent variable, Market Value, is explained by the independent variable, Square Feet. The remaining variation is due to other factors that were not included in the model. Although we would like high values of R^2, it is difficult to specify a "good" value that signifies a strong relationship because this depends on the application. For example, in scientific applications such as calibrating physical measurement equipment, R^2 values close to 1 would be expected; in marketing research studies, an R^2 of 0.6 or more is considered very good; however, in many social science applications, values in the neighborhood of 0.3 might be considered acceptable.

The square root of the coefficient of determination is the **sample correlation coefficient**, *r* [see formula (2A.16) in Chapter 2]. Values of *r* range from −1 to 1, where the sign is determined by the sign of the slope of the regression line. A correlation coefficient of $r = 1$ indicates perfect positive correlation; that is, as the independent variable increases, the dependent variable does also; $r = -1$ indicates perfect negative correlation—as X increases, Y decreases. As with R^2, a value of $r = 0$ indicates no correlation. Because R^2 measures the actual *proportion* of the variation explained by regression, it is generally easier to interpret than *r*.

Application of Regression to Investment Risk

Investing in the stock market is highly attractive to everyone. However, stock investments do carry an element of risk. Risk associated with an individual stock can be measured in two ways. The first is **systematic risk**, which is the variation in stock price explained by the market—as the market moves up or down, the stock tends to move in the same direction. The Standard & Poor's (S&P's) 500 index is most commonly used as a measure of the market. For example, we generally see that stocks of consumer products companies are highly correlated with the S&P index, while utility stocks generally show less correlation with the market. The second type of risk is called **specific risk** and is the variation that is due to other factors, such as the earnings potential of the firm, acquisition strategies, and so on. Specific risk is measured by the standard error of the estimate.

Systematic risk is characterized by a measure called *beta*. A beta value equal to 1.0 means that the specific stock will match market movements, a beta less than 1.0 indicates that the stock is less volatile than the market, and a beta greater than

Figure 6.10 Illustration of Beta Risk

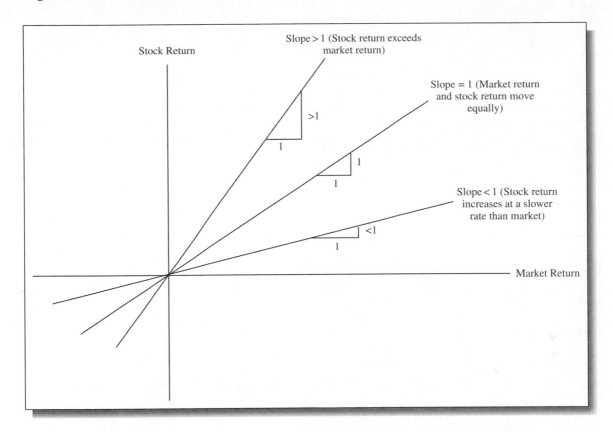

1.0 indicates that the stock has greater variance than the market. Thus, stocks with large betas are riskier than those with lower beta values. Beta values can be calculated by developing a regression model of a particular stock's returns (the dependent variable) against the average market returns (the independent variable). The slope of the regression line is the beta risk. This can be explained through the chart in Figure 6.10. If we plot the daily market returns against the returns of the individual stock and find the regression line, we would observe that if the slope equals 1, the stock changes at the same rate as the market. However, if the stock price changes are less than the market changes, the slope of the regression line would be less than 1, while the slope would be greater than 1 when the stock price changes exceed that of the market. A negative slope would indicate a stock that moves in the opposite direction to the market (e.g., if the market goes up, the stock price goes down).

For example, suppose that we collected data on a specific stock and the S&P 500 index over an extended period of time and computed the percentage change from one day to the next for both variables. If there appears to be a linear relationship between the change in the S&P 500 index and change in the stock price, we could fit the model:

$$\text{Daily Change in Stock Price} = \beta_0 + \beta_1 \text{S\&P Change}$$

Suppose that the least-squares fit resulted in the model:

$$\text{Daily Change in Stock Price} = 0.002 + 1.205 \text{ S\&P Change}$$

The slope of the regression line, 1.205, is the beta risk of the stock. This indicates that the stock is more risky than the average S&P 500 stock.

EXCEL/*PHSTAT* NOTE
Using Regression Tools

Both Excel and *PHStat* provide tools for regression analysis. The Excel *Regression* tool can be used for single or multiple linear regressions. From the *Data Analysis* menu in the *Analysis* group under the *Data* tab, select the *Regression* tool. The dialog box shown in Figure 6.11 is displayed. In the box for the *Input Y Range*, specify the range of the dependent variable values. In the box for the *Input X Range*, specify the range for the independent variable values. Check *Labels* if your data range contains a descriptive label (we highly recommend using this). You have the option of forcing the intercept to zero by checking *Constant is Zero*; however, you will usually not check this box because adding an intercept term allows a better fit to the data. You also can set a *Confidence Level* (the default of 95% is commonly used) to provide confidence intervals for the intercept and slope parameters. In the *Residuals* section, you have the option of including a residuals output table by checking *Residuals, Standardized Residuals, Residual Plots,* and *Line Fit Plots.* The *Residual Plots* generates a chart for each independent variable versus the residual, and the *Line Fit Plot* generates a scatter chart with the values predicted by the regression model included (a scatter chart with an added trendline is visually superior). Finally, you may also choose to have Excel construct a normal probability plot for the dependent variable.

PHStat provides two separate tools, one for *Simple Linear Regression* and one for *Multiple Regression*, both found in the *Regression* menu of *PHStat*. In the *Simple Linear Regression* tool (see Figure 6.12), the data input is identical to the Excel tool. However, the output options are somewhat different. Excel automatically provides the Regression Statistics Table and Analysis of Variance (ANOVA) and Coefficients Table, which are options in the *PHStat* tool. Note that while Excel provides Standardized Residuals and Line Fit Plots, *PHStat* does not. Other output options in *PHStat* include a scatter diagram, Durbin–Watson statistic (also discussed later), and confidence and prediction intervals. Results are generated on separate worksheets in the active workbook. The dialog box for the *Multiple Regression* tool is nearly identical, but has some additional check boxes for information related exclusively to multiple regression, and we will explain them later in this chapter.

Figure 6.12 *PHStat Simple Linear Regression* Tool Dialog

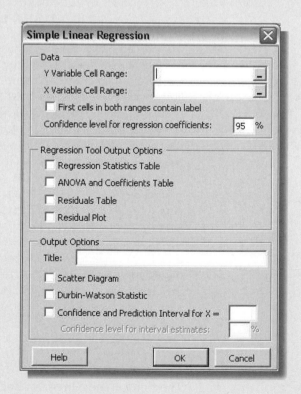

Figure 6.11 Excel *Regression* Tool Dialog

Interpreting Regression Analysis Output

Regression analysis software tools available in Excel and *PHStat* provide a variety of information concerning regression analysis (see *Excel/PHStat Note: Using Regression Tools*). In this section, we will describe how to interpret the results from these tools using the *Home Market Value* example.

Figure 6.13 shows the basic regression analysis output provided by both the Excel (by default) and *PHStat* (by checking the *Regression Statistics Table* and *ANOVA and Coefficients Table* boxes) tools. The least-squares estimates of the slope and intercept are found in the Coefficients column in the bottom section of the output. We see that the intercept is 32,673, and the slope (coefficient of the independent variable, Square Feet) is 35.036.

Regression Statistics

Both Excel and *PHStat* tools provide basic regression statistics, including the coefficient of determination (*R-Square*), correlation coefficient (called *Multiple R*), which we have previously discussed. The standard deviation of the observations around the regression line is called the **standard error of the estimate**, S_{YX}. If the data are clustered close to the regression line, then the standard error will be small; the more scattered the data are, the larger the standard error. The standard error of the estimate is $7287.72. The computed standard deviation of the market values is $10,553. You can see that the variation around the regression line is less than the variation around the mean because the independent variable explains some of the variation. The **adjusted R^2** is a statistic that modifies the value of R^2 by incorporating the sample size and the number of explanatory variables in the model. Although it does not give the actual percent of variation explained by the model as R^2 does, it is useful when comparing this model with other models that include additional explanatory variables. We will discuss it more fully in the context of multiple linear regression later in this chapter.

Figure 6.13 Basic Regression Analysis Output

	A	B	C	D	E	F	G
1	Regression Analysis						
2							
3	*Regression Statistics*						
4	Multiple R	0.731255223					
5	R Square	0.534734202					
6	Adjusted R Square	0.523102557					
7	Standard Error	7287.722712					
8	Observations	42					
9							
10	ANOVA						
11		*df*	*SS*	*MS*	*F*	*Significance F*	
12	Regression	1	2441633669	2441633669	45.97236277	3.79802E-08	
13	Residual	40	2124436093	53110902.32			
14	Total	41	4566069762				
15							
16		*Coefficients*	*Standard Error*	*t Stat*	*P-value*	*Lower 95%*	*Upper 95%*
17	Intercept	32673.2199	8831.950745	3.699434116	0.000649604	14823.18178	50523.25802
18	Square Feet	35.03637258	5.16738385	6.780292234	3.79802E-08	24.59270036	45.48004481

Hypothesis Testing

In Chapter 5, we introduced analysis of variance (ANOVA), which conducts an *F*-test to determine whether variation due to a particular factor, such as the differences in sample means, is significantly larger than that due to error. ANOVA is commonly applied to regression to test for *significance of regression*. For a simple linear regression model, **significance of regression** is simply a hypothesis test of whether the regression coefficient β_1 (slope of the independent variable) is zero:

$$H_0: \beta_1 = 0$$
$$H_1: \beta_1 \neq 0$$

If we reject the null hypothesis, then we may conclude that the slope of the independent variable is not zero, and therefore is statistically significant in the sense that it explains some of the variation of the dependent variable around the mean. The value of *Significance F* is the *p*-value for the *F*-test; if this is less than the level of significance, we would reject the null hypothesis. In this example the *p*-value is essentially zero (3.798×10^{-8}) and would lead us to conclude that home size is a significant variable in explaining the variation in market value.

In the remaining columns of the rows associated with the intercept and slope are the standard errors of the estimated regression coefficients, a *t*-statistic for testing the null hypothesis that the coefficient is zero against the alternative hypothesis that the coefficient is not zero, the *p*-value for the test, and confidence interval limits for the coefficients. The standard errors provide a measure of the uncertainty of the slope and intercept and are used to compute the lower and upper confidence interval limits in the last two columns.

The *t*-statistics are used to test the hypothesis that β_0 or β_1 equal zero. The value of *t Stat* is the value of the coefficient divided by the standard error. Usually, it makes little sense to test or interpret the hypothesis that $\beta_0 = 0$ unless the intercept has a significant physical meaning in the context of the application. Testing the null hypothesis $H_0: \beta_1 = 0$ is the same as the significance of regression test that we described earlier. Note that the *p*-value associated with the test for the independent variable coefficient, Square Feet, is equal to the *Significance F* value. This will always be true for a regression model with one independent variable because it is the only explanatory variable. However, as we shall see, this will not be the case for multiple regression models.

Finally, the confidence intervals provide information about the potential values of the true regression coefficients, accounting for sampling error. For example, a 95% confidence interval for the slope is [24.59, 45.48], suggesting a bit of uncertainty about the influence of house size on market value.

Residual Analysis

Recall that residuals are the observed errors, which are the differences between the actual values and the estimated values of the dependent variable using the regression line. Figure 6.14 shows a portion of the residual table from the Excel tool. The *PHStat* residual output is identical, except that it does not provide the last column, *Standard Residuals*. The residual output includes, for each observation, the predicted value using the estimated regression equation and the residual. For example, the first home has a market value of $90,000 and the regression model predicts $96,159.13. Thus, the residual is –$6,159.13. Both Excel and *PHStat* provide options for displaying a plot of the residuals. Figure 6.15 shows the *PHStat* residual plot. While essentially identical to the Excel residual plot, it cannot be customized as can the Excel chart.

Figure 6.14 Portion of Residual Output

	A	B	C	D
22	RESIDUAL OUTPUT			
23				
24	*Observation*	*Predicted Market Value*	*Residuals*	*Standard Residuals*
25	1	96159.12702	-6159.127018	-0.855636403
26	2	99732.83702	4667.162978	0.64837022
27	3	97210.2182	-3910.218196	-0.543214164
28	4	96159.12702	-5159.127018	-0.716714702
29	5	96999.99996	4900.00004	0.680716341
30	6	103726.9835	4773.016503	0.663075572
31	7	93356.21721	-5756.217212	-0.799663487
32	8	97490.50918	-1490.509177	-0.20706407
33	9	95423.36319	-6223.363194	-0.864560202
34	10	91043.81662	-2643.816621	-0.367283503

Figure 6.15 *PHStat* Residual Plot

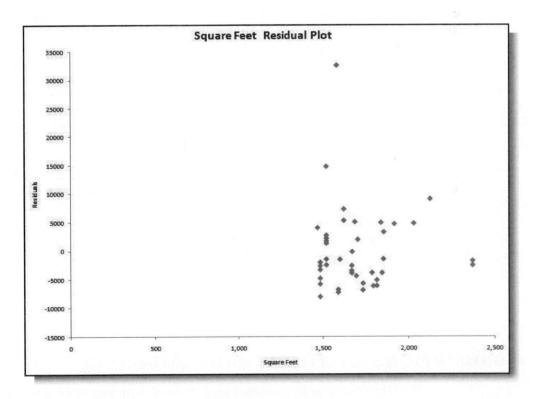

Standard residuals are residuals divided by the standard error S_{YX} and describe how far each residual is from its mean in units of standard deviations (similar to a z-value for a standard normal distribution). For example, the first observation is about 0.85 standard errors below the regression line. Standard residuals are useful in checking assumptions underlying regression analysis, which we will address shortly, and to detect outliers that may bias the results. Some consider a standardized residual outside of ± 2 standard errors as an outlier. However, this can be misleading, and commercial software has more sophisticated techniques for identifying outliers. A more conservative rule of thumb would be to consider outliers outside of a ± 3 standard error range.

For example, if you look back at Figure 6.4, you will notice that one point (with a market value of around $120,000) appears to be quite different from the rest of the data. In fact, the standard residual for this observation is 4.53 (you can see that it appears to be an outlier in the residual plot in Figure 6.15). You might question whether this observation belongs in the data, because the house has large value despite a relatively small size. A reason might be an outdoor pool or an unusually large plot of land. Because this value will influence the regression results and may not be representative of the other homes in the neighborhood, you might consider dropping this observation and recomputing the results (try this and examine the differences in the results).

Confidence and Prediction Intervals

We can develop a point estimate of the market value for a specific value of square feet using the regression model. For instance, we saw that a house with 1750 square feet would have an estimated market value of 35.036(1750) + 32673 = $93,986. As we know, such a point estimate does not tell us anything about the uncertainty of this estimate. An interval estimate would provide an assessment of the potential variability of the estimate.

We may develop two types of interval estimates: a confidence interval for the average value of the dependent variable, and a prediction interval for the value of an individual response. A confidence interval provides an interval estimate that quantifies the uncertainty about a population parameter, for example, the *mean value* of the population of homes having 1,750 square feet. A **prediction interval** is a probability interval that quantifies the uncertainty for a *single future observation*, for example, the market value of an individual home with 1,750 square feet. *PHStat* provides these interval estimates by checking the appropriate box and entering the value of the independent variable. Figure 6.16 shows the results. A 95% confidence interval for the average market value is [$91,643, $96,330], while a 95% prediction interval for the market value of an individual home is [$79,073, $108,901]. Notice that the prediction interval is wider than a confidence interval because the variance of the mean value is smaller than the variance of individual values (recall our discussion of the difference between the standard error of the mean for the sampling distribution compared to the standard deviation of individual observations in Chapter 3).

ASSUMPTIONS OF REGRESSION ANALYSIS

The statistical hypothesis tests associated with regression analysis are predicated on some key assumptions about the data. Clearly, the first assumption is linearity. This is usually checked by examining a scatter diagram of the data or examining the residual plot. If the model is appropriate, then the residuals should appear to be randomly scattered about zero, with no apparent pattern. If the residuals exhibit some well-defined pattern, such as a linear trend, a U-shape, and so on, then there is good evidence that some other functional form might better fit the data. The scatter diagram of

	A	B
1	**Confidence Interval Estimate**	
2		
3	**Data**	
4	**X Value**	**1750**
5	**Confidence Level**	**95%**
6		
7	Intermediate Calculations	
8	Sample Size	42
9	Degrees of Freedom	40
10	t Value	2.021075
11	XBar, Sample Mean of X	1695.262
12	Sum of Squared Differences from XBar	1989034
13	Standard Error of the Estimate	7287.723
14	h Statistic	0.025316
15	Predicted Y (YHat)	93986.87
16		
17	**For Average Y**	
18	Interval Half Width	2343.533
19	**Confidence Interval Lower Limit**	**91643.3**
20	**Confidence Interval Upper Limit**	**96330.4**
21		
22	**For Individual Response Y**	
23	Interval Half Width	14914.31
24	**Prediction Interval Lower Limit**	**79072.6**
25	**Prediction Interval Upper Limit**	**108901**

Figure 6.16 *PHStat* Interval Estimates

the market value data appears to be linear; looking at the residual plot in Figure 6.15 also confirms no pattern in the residuals.

The next key assumption is that the errors for each individual value of X are normally distributed, with a mean of zero and a constant variance. This can be verified by examining a histogram of the standard residuals and inspecting for a bell-shaped distribution or using more formal goodness-of-fit tests. Figure 6.17 shows a histogram of the standard residuals for the market value data. The distribution appears to be somewhat positively skewed. Another way of verifying normality is to examine a normal probability plot of the residuals, one of the options in the Excel tool. A normal probability plot transforms the cumulative probability scale (vertical axis) so that the graph of the cumulative normal distribution will be a straight line. The closer the points are to a straight line, the better the fit to a normal distribution. Figure 6.18 shows the normal probability plot generated by the Excel regression tool. The curvature suggests a skewed distribution and a lack of normality. However, the departure from normality does not appear to be that great, and regression analysis is fairly robust against departures from normality.

The third assumption is **homoscedasticity**, which means that the variation about the regression line is constant for all values of the independent variable. This can also be evaluated by examining the residual plot and looking for large differences in the variances at different values of the independent variable. In Figure 6.15,

Figure 6.17 Histogram of Standard Residuals

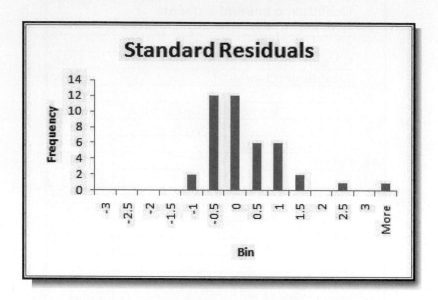

Figure 6.18 Normal Probability Plot

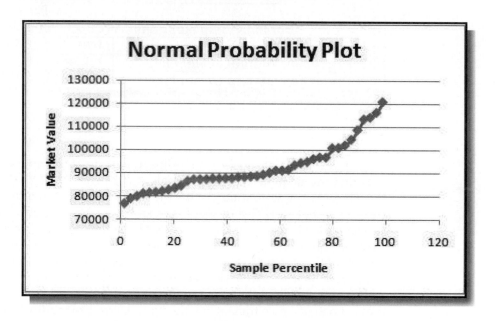

we see no serious differences in the spread of the data for different values of X, particularly if the outlier is eliminated. Caution should be exercised when looking at residual plots. In many applications, the model is derived from limited data, and multiple observations for different values of X are not available, making it difficult to draw definitive conclusions about homoscedasticity. If this assumption is seriously violated, then other techniques should be used instead of least squares for estimating the regression model.

Finally, residuals should be independent for each value of the independent variable. For cross-sectional data, this assumption is usually not a problem. However, when time is the independent variable, this is an important assumption. If successive observations appear to be correlated—for example, by becoming larger over time or exhibiting a cyclical type of pattern—then this assumption is violated. Correlation among successive observations over time is called **autocorrelation** and can be identified by residual plots having clusters of residuals with the same sign. Autocorrelation can be evaluated more formally using a statistical test based on the **Durbin–Watson statistic**. The Durbin–Watson statistic is:

$$D = \frac{\sum_{i=2}^{n}(e_i - e_{i-1})^2}{\sum_{i=1}^{n}e_i^2} \tag{6.4}$$

This is a ratio of the squared differences in successive residuals to the sum of the squares of all residuals. D will range from 0 to 4. When successive residuals are positively autocorrelated, D will approach 0. Critical values of the statistic have been tabulated based on the sample size and number of independent variables that allow you to conclude that there is either evidence of autocorrelation, no evidence of autocorrelation, or the test is inconclusive. *PHStat* computes the Durbin–Watson statistic, and for most practical purposes, values below 1 suggest autocorrelation; values above 1.5 and below 2.5 suggest no autocorrelation; and values above 2.5 suggest negative autocorrelation. This can become an issue when using regression in forecasting, as we discuss in the next chapter.

When assumptions of regression are violated, then statistical inferences drawn from the hypothesis tests may not be valid. Thus, before drawing inferences about regression models and performing hypothesis tests, these assumptions should be checked. However, other than linearity, these assumptions are not needed solely for model fitting and estimation purposes.

MULTIPLE LINEAR REGRESSION

A linear regression model with more than one independent variable is called a **multiple linear regression** model. Simple linear regression is just a special case of multiple linear regression. Multiple regression has been effectively used in many business applications. For example, Kimes and Fitzsimmons[1] developed a model for La Quinta Motor Inns to evaluate proposed sites for new motels. This model had 35 variables that included 6 variables about competition, 18 variables about demand, 3 demographic variables, 4 market-related variables, and 4 physical variables. The characteristics of each proposed site could be entered into a spreadsheet containing the regression model and evaluated immediately.

Consider the data in the Excel file *Colleges and Universities* (see Figure 6.2). We might believe that the graduation rate is related to the other variables. For example, it is logical to propose that schools with students who have higher SAT scores, a lower acceptance rate, a larger budget, and a higher percentage of students in the top 10% of their high school classes will tend to retain and graduate more students. To determine if this is the case, we could develop a multiple regression model.

A multiple linear regression model has the form:

$$Y = \beta_0 + \beta_1 X_1 + \beta_2 X_2 + \ldots + \beta_k X_k + \varepsilon \tag{6.5}$$

---

[1]S.E. Kimes and J.A. Fitzsimmons, "Selecting Profitable Hotel Sites at La Quinta Motor Inns," *Interfaces* 19, no. 6 (1990): 83–94.

where

> Y is the dependent variable
> $X_1 \ldots X_k$ are the independent (explanatory) variables
> β_0 is the intercept term
> $\beta_1 \ldots \beta_k$ are the regression coefficients for the independent variables
> ε is the error term

Similar to simple linear regression, we estimate the regression coefficients—called **partial regression coefficients**—b_0, b_1, b_2, $\ldots$, b_k, then use the model:

$$\hat{Y} = b_0 + b_1 X_1 + b_2 X_2 + \ldots + b_k X_k \tag{6.6}$$

to predict the value of the dependent variable. The partial regression coefficients represent the expected change in the dependent variable when the associated independent variable is increased by one unit *while the values of all other independent variables are held constant.*

For the college and university data, the proposed model would be:

$$\text{Graduation\%} = b_0 + b_1\, \text{SAT} + b_2\, \text{ACCEPTANCE} + b_3\, \text{EXPENDITURES} + b_4\, \text{TOP10\% HS}$$

Thus, b_2 would represent an estimate of the change in the graduation rate for a unit increase in the acceptance rate while holding all other variables constant.

As with simple linear regression, multiple linear regression uses least squares to estimate the intercept and slope coefficients that minimize the sum of squared error terms over all observations. The principal assumptions discussed for simple linear regression also hold here. The Excel *Regression* and *PHStat Multiple Regression* tools can easily perform multiple linear regression; you need only specify the full range for the independent variable data. One caution when using the tools: *The independent variables in the spreadsheet must be in contiguous columns.*

The multiple regression results for the college and university data are shown in Figure 6.19. From the *Coefficients* section, we see that the model is:

$$\text{Graduation\%} = 17.92 + 0.072\, \text{SAT} - 24.859\, \text{ACCEPTANCE} - 0.000136\, \text{EXPENDITURES} - 0.163\, \text{TOP10\% HS}$$

The signs of some coefficients make sense; higher SAT scores and lower acceptance rates suggest higher graduation rates. However, we had expected that larger student expenditures and a higher percentage of top high school students would also positively influence the graduation rate. Perhaps some of the best students are more demanding and transfer schools if their needs are not being met, some entrepreneurial students might pursue other interests before graduation, or it might simply be the result of sampling error. As with simple linear regression, the model should be used only for values of the independent variables within the range of the data.

Interpreting Results from Multiple Linear Regression

The results from the *Regression* tool are in the same format as we saw for simple linear regression. *Multiple R*, the **multiple correlation coefficient**, and *R Square*, the **coefficient of multiple determination**, indicate the strength of association between the dependent and independent variables. The value of R^2 (0.53) indicates that 53% of the variation in the dependent variable is explained by these independent variables. This suggests that other factors not included in the model, perhaps campus living conditions, social opportunities, and so on, might also influence the graduation rate.

Figure 6.19 Multiple Regression Results

	A	B	C	D	E	F	G
1	Regression Analysis						
2							
3	*Regression Statistics*						
4	Multiple R	0.731044486					
5	R Square	0.534426041					
6	Adjusted R Square	0.492101135					
7	Standard Error	5.30833812					
8	Observations	49					
9							
10	ANOVA						
11		*df*	*SS*	*MS*	*F*	*Significance F*	
12	Regression	4	1423.209266	355.8023166	12.62675098	6.33158E-07	
13	Residual	44	1239.851958	28.1784536			
14	Total	48	2663.061224				
15							
16		*Coefficients*	*Standard Error*	*t Stat*	*P-value*	*Lower 95%*	*Upper 95%*
17	Intercept	17.92095587	24.55722367	0.729763108	0.469402466	-31.57087575	67.4127875
18	Median SAT	0.072006285	0.017983915	4.003927007	0.000236106	0.035762085	0.108250484
19	Acceptance Rate	-24.8592318	8.315184822	-2.989618672	0.004559569	-41.61738544	-8.10107817
20	Expenditures/Student	-0.00013565	6.59314E-05	-2.057438385	0.045600176	-0.000268526	-2.77379E-06
21	Top 10% HS	-0.162764489	0.079344518	-2.051364015	0.046213846	-0.322672855	-0.002856122

The ANOVA section in Figure 6.19 tests for significance of the *entire model.* That is, it computes an *F*-statistic for testing the hypotheses:

$$H_0: \beta_1 = \beta_2 = \ldots = \beta_k = 0$$
$$H_1: at\ least\ one\ \beta_j\ is\ not\ 0$$

The null hypothesis states that no linear relationship exists between the dependent and *any* of the independent variables, while the alternative hypothesis states that the dependent variable has a linear relationship with *at least* one independent variable. If the null hypothesis is rejected, we cannot conclude that a relationship exists with every independent variable individually. At a 5% significance level, we reject the null hypothesis for this example because *Significance F* is essentially zero.

The last section in Figure 6.19 provides information to test hypotheses about each of the individual regression coefficients, which define the marginal contributions of the independent variables in the model. For example, to test the hypothesis that the population slope β_1 (associated with SAT score) is zero, we examine the *p*-value and compare it to the level of significance, assumed to be 0.05. Because $p < 0.05$, we reject the null hypothesis that this partial regression coefficient is zero and conclude that SAT score is significant in predicting graduation rate. Similarly, the *p*-values for all other coefficients are less than 0.05, indicating that each of them is significant. This will not always be the case, and we will learn how to deal with large *p*-values later.

For multiple regression models, a residual plot is generated for each independent variable. This allows you to assess the linearity and homoscedasticity assumptions of regression. Figure 6.20 shows one of the residual plots from the Excel output. The assumptions appear to be met, and the other residual plots (not shown) also validate these assumptions. The normal probability plot (also not shown) does not suggest any serious departures from normality.

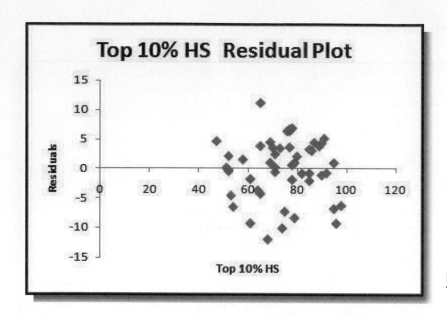

Figure 6.20 Residual Plot of Top 10% HS Variable

Correlation and Multicollinearity

As discussed in Chapter 2, correlation, a numerical value between −1 and +1, measures the linear relationship between pairs of variables. The higher the absolute value of the correlation, the greater the strength of the relationship. The sign simply indicates whether variables tend to increase together (positive) or not

EXCEL NOTE
Using the Correlation Tool

From the *Tools* menu, select *Data Analysis* and the *Correlation* tool. The dialog box shown in Figure 6.21 is displayed. In the box for the *Input Range*, specify the range of the data for which you want correlations. As with other tools, check *Labels in First Row* if your data range contains a descriptive label.

Figure 6.21 *Correlation* Tool Dialog

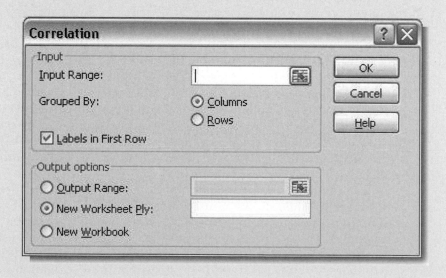

Figure 6.22   Correlation Matrix for *Colleges and Universities* Data

	A	B	C	D	E	F
1		*Median SAT*	*Acceptance Rate*	*Expenditures/Student*	*Top 10% HS*	*Graduation %*
2	Median SAT	1				
3	Acceptance Rate	-0.601901959	1			
4	Expenditures/Student	0.572741729	-0.284254415	1		
5	Top 10% HS	0.503467995	-0.609720972	0.505782049	1	
6	Graduation %	0.564146827	-0.55037751	0.042503514	0.138612667	1

(negative). Therefore, examining correlations between the dependent and independent variables can be useful in selecting variables to include in a multiple regression model because a strong correlation indicates a strong linear relationship. The Excel *Data Analysis Correlation* tool computes the correlation between all pairs of variables (see *Excel Note: Using the Correlation Tool*). However, strong correlations *among the independent variables* can be problematic.

Figure 6.22 shows the correlation matrix for the variables in the *Colleges and Universities* data. You can see that SAT and Acceptance Rate have moderate correlations with the dependent variable, Graduation%, but the correlation between Expenditures/Student and Top 10% HS with Graduation% are relatively low. The strongest correlation, however, is between two independent variables: Top 10% HS and Acceptance Rate. This can potentially signify a phenomenon called **multicollinearity**, a condition occurring when two or more independent variables in the same regression model contain high levels of the same information and, consequently, are strongly correlated with one another and can predict each other better than the dependent variable. When significant multicollinearity is present, the coefficients of the independent variables in the regression model can be unstable, and even the signs of these coefficients may change when different variables are included, making it difficult to interpret regression coefficients. Also, *p*-values can be inflated, resulting in the conclusion not to reject the null hypothesis for significance of regression when it should be rejected.

Multicollinearity is best measured using the **variance inflation factor (VIF)** for each independent variable, rather than simply examining the correlation matrix. The variance inflation factor can be computed in the *PHStat Multiple Regression* tool by checking the appropriate box in the dialog. Figure 6.23 shows the results for the *Colleges and Universities* data. If the independent variables are not correlated, then $VIF_j = 1$. Conservative guidelines suggest that a maximum VIF of 5 or more suggests too much multicollinearity. For the example, we see that all VIF values are below the suggested guideline. Thus, the correlation matrix alone might have misled us into believing that significant multicollinearity existed. Thus, VIFs should be checked along with the assumptions of regression analysis that we discussed. If some VIFs are too large, then you should consider dropping the variables from the model.

BUILDING GOOD REGRESSION MODELS

In the colleges and universities regression example, all of the independent variables were found to be significant by evaluating the *p*-values of the regression analysis. This will not always be the case, and leads to the question of how to build good regression models that include the "best" set of variables.

Figure 6.23 Variance Inflation Factors

	A	B	C	D	E
1	Regression Analysis			Regression Analysis	
2	Median SAT and all other X			Expenditures/Student and all other X	
3	*Regression Statistics*			*Regression Statistics*	
4	Multiple R	0.733444235		Multiple R	0.659705397
5	R Square	0.537940446		R Square	0.435211211
6	Adjusted R Square	0.507136476		Adjusted R Square	0.397558625
7	Standard Error	44.0015595		Standard Error	12002.17094
8	Observations	49		Observations	49
9	VIF	2.164223188		VIF	1.770573388
10					
11	Regression Analysis			Regression Analysis	
12	Top 10% HS and all other X			Acceptance Rate and all other X	
13	*Regression Statistics*			*Regression Statistics*	
14	Multiple R	0.701553357		Multiple R	0.724669621
15	R Square	0.492177112		R Square	0.525146059
16	Adjusted R Square	0.458322253		Adjusted R Square	0.49348913
17	Standard Error	9.973219924		Standard Error	0.095165693
18	Observations	49		Observations	49
19	VIF	1.969190488		VIF	2.10591071

Figure 6.24 shows a portion of the Excel file *Banking Data*, which provides data acquired from banking and census records for different zip codes in the bank's current market. Such information can be useful in targeting advertising for new customers or for choosing locations for branch offices. The data show the median age of the population, median years of education, median income, median home value, median household wealth, and average bank balance.

Figure 6.25 shows the results for regression analysis to predict the average bank balance as a function of the other variables. While the independent variables

Figure 6.24 Portion of *Banking Data*

	A	B	C	D	E	F
1	Banking Data					
2						
3	Median	Median Years	Median	Median	Median Household	Average Bank
4	Age	Education	Income	Home Value	Wealth	Balance
5	35.9	14.8	$91,033	$183,104	$220,741	$38,517
6	37.7	13.8	$86,748	$163,843	$223,152	$40,618
7	36.8	13.8	$72,245	$142,732	$176,926	$35,206
8	35.3	13.2	$70,639	$145,024	$166,260	$33,434
9	35.3	13.2	$64,879	$135,951	$148,868	$28,162
10	34.8	13.7	$75,591	$155,334	$188,310	$36,708
11	39.3	14.4	$80,615	$181,265	$201,743	$38,766

Figure 6.25   Regression Analysis Results for *Banking Data*

	A	B	C	D	E	F	G
1	SUMMARY OUTPUT						
2							
3	*Regression Statistics*						
4	Multiple R	0.97309221					
5	R Square	0.946908448					
6	Adjusted R Square	0.944143263					
7	Standard Error	2055.64333					
8	Observations	102					
9							
10	ANOVA						
11		*df*	*SS*	*MS*	*F*	*Significance F*	
12	Regression	5	7235179873	1447035975	342.4394584	1.5184E-59	
13	Residual	96	405664271.9	4225669.499			
14	Total	101	7640844145				
15							
16		*Coefficients*	*Standard Error*	*t Stat*	*P-value*	*Lower 95%*	*Upper 95%*
17	Intercept	-10710.64278	4260.976308	-2.513659314	0.013613178	-19168.6137	-2252.671867
18	Age	318.6649626	60.98611242	5.225205378	1.01152E-06	197.6084892	439.721436
19	Education	621.8603472	318.9595184	1.949652891	0.054135369	-11.26927724	1254.989972
20	Income	0.146323453	0.040781001	3.588029937	0.000526666	0.065373808	0.227273099
21	Home Value	0.009183067	0.011038075	0.831944635	0.407504891	-0.012727338	0.031093473
22	Wealth	0.074331533	0.011189265	6.643111131	1.84838E-09	0.052121018	0.096542049

explain over 94% of the variation in the average bank balance, you can see that at a 0.05 significance level, both Education and Home Value do not appear to be significant. A good regression model should include only significant independent variables. However, it is not always clear exactly what will happen when we add or remove variables from a model; variables that are (or are not) significant in one model may (or may not) be significant in another. Therefore, you should *not* consider dropping all insignificant variables at one time, but rather take a more structured approach.

Adding an independent variable to a regression model will *always* result in R^2 equal to or greater than the R^2 of the original model. Conversely, dropping an independent variable will always cause R^2 to decrease. This is true even when the new independent variable has little true relationship with the dependent variable. Thus, trying to maximize R^2 is not a useful criterion. A better way of evaluating the relative fit of different models is to use adjusted R^2. Adjusted R^2 reflects both the number of independent variables and the sample size, and may either increase or decrease when an independent variable is added or dropped, thus providing an indication of the value of adding or removing independent variables in the model. An increase in adjusted R^2 indicates that the model has improved.

This suggests a systematic approach to building good regression models:

1. Construct a model with all available independent variables. Check for significance of the independent variables by examining the *p*-values.
2. Identify the independent variable having the largest *p*-value that exceeds the chosen level of significance.

3. Remove the variable from the model and evaluate adjusted R^2.

4. Continue until all variables are significant.

In essence, this approach seeks to find a significant model that has the highest adjusted R^2. Another criterion to determine if a variable should be removed is the t-statistic. If $t < 1$, then the standard error will decrease and adjusted R^2 will increase if the variable is removed. If $t > 1$, then the opposite will occur. In the banking regression results, we see that the t-statistic for Home Value is less than 1; therefore, we expect the adjusted R^2 to increase if we remove this variable.

Figure 6.26 shows the regression results after removing Home Value. Note that the adjusted R^2 has increased slightly, while the R^2 value decreased slightly because we removed a variable from the model. Also notice that although the p-value for Education was larger than 0.05 in the first regression analysis, this variable is now significant after Home Value was removed.

This approach may involve considerable experimentation to identify the best set of variables that result in the largest adjusted R^2 using the p-values or t-statistics. For large numbers of independent variables, the number of potential models can be overwhelming. For example, there are $2^{10} = 1,024$ possible models that can be developed from a set of 10 independent variables. This can make it difficult to effectively screen out insignificant variables. Fortunately, automated methods exist that facilitate this process.

Stepwise Regression

Stepwise regression is a search procedure that attempts to find the best regression model without examining all possible regression models (see *PHStat Note: Stepwise Regression*). In stepwise regression, variables are either added to or deleted from the

Figure 6.26 Regression Results without Home Value

	A	B	C	D	E	F	G
1	SUMMARY OUTPUT						
2							
3	*Regression Statistics*						
4	Multiple R	0.97289551					
5	R Square	0.946525674					
6	Adjusted R Square	0.944320547					
7	Standard Error	2052.378536					
8	Observations	102					
9							
10	ANOVA						
11		*df*	*SS*	*MS*	*F*	*Significance F*	
12	Regression	4	7232255152	1808063788	429.2386497	9.68905E-61	
13	Residual	97	408588992.5	4212257.655			
14	Total	101	7640844145				
15							
16		*Coefficients*	*Standard Error*	*t Stat*	*P-value*	*Lower 95%*	*Upper 95%*
17	Intercept	-12432.45673	3718.674319	-3.343249681	0.001177705	-19812.99569	-5051.917773
18	Age	325.0652837	60.40284468	5.381622098	5.1267E-07	205.1823604	444.9482071
19	Education	773.3800418	261.4330936	2.958233142	0.003886994	254.5077323	1292.252351
20	Income	0.159747379	0.037393587	4.272052794	4.52422E-05	0.085531461	0.233963297
21	Wealth	0.072988791	0.011054665	6.602532898	2.16051E-09	0.051048341	0.094929241

From the *PHStat* menu, select *Regression* then *Stepwise Regression*. The dialog box that appears (Figure 6.27) prompts you to enter the range for the dependent variable and the independent variables, as well as the confidence level for the regression. One of two criteria may be chosen to guide the stepwise selection: *p*-values or *t*-values. Choosing *p*-values, for example, would result in the procedure selecting the variable with the smallest *p*-value below a threshold to include in the model or a variable with a *p*-value greater than a threshold to be removed. Other options include *General Stepwise, Forward Selection,* and *Backward Elimination. General Stepwise* considers including or deleting variables that meet the criteria, as appropriate. *Forward Selection* begins with a model having no independent variables and successively adds one at a time until no additional variable makes a significant contribution. *Backward Elimination* begins with all independent variables in the model and deletes one at a time until the best model is identified. The procedure produces two worksheets: one that contains the multiple regression model that includes all independent variables and another with a table of stepwise results.

Figure 6.27   *Stepwise Regression* Dialog

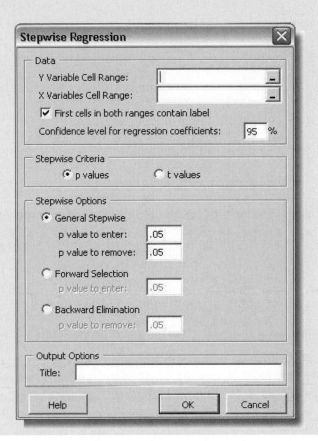

current regression model at each step of the process based on the *p*-values or *t*-statistics. The process continues until no addition or removal of variables can improve the model based on the value of the adjusted R^2. Essentially, the stepwise procedure automates the trial-and-error process we described. However, it is a myopic search process that will not guarantee finding the model having the highest adjusted R^2. Each option might find a different model.

Figure 6.28 shows the results of applying the *PHStat Stepwise Regression* tool to the banking data. Note that using *p*-value thresholds of 0.05 results in the inclusion of Income, Wealth, Age, and Education in the final model, as we had identified earlier. Each of these variables is significant at the 0.05 level. However, note that this procedure does not provide the R^2 or adjusted R^2 values, so you would need to run the multiple regression analysis procedure to find these values.

Best-Subsets Regression

Best-subsets regression evaluates either all possible regression models for a set of independent variables or the best subsets of models for a fixed number of independent variables. *PHStat* includes a useful tool for performing best-subsets regression (see *PHStat Note: Best-Subsets Regression*). Figure 6.29 shows the *PHStat* output for the banking

Figure 6.28 Stepwise Regression Results

	A B	C	D	E	F	G	H
1	Banking Data General Stepwise						
2	Table of Results for General Stepwise						
3							
4	Income entered.						
5							
6		*df*	*SS*	*MS*	*F*	*Significance F*	
7	Regression	1	6920338342	6920338342	960.4833602	4.43329E-53	
8	Residual	100	720505802.5	7205058.025			
9	Total	101	7640844145				
10							
11		*Coefficients*	*Standard Error*	*t Stat*	*P-value*	*Lower 95%*	*Upper 95%*
12	Intercept	4020.251548	723.8864893	5.553704355	2.31287E-07	2584.081409	5456.421688
13	Income	0.427519104	0.013794647	30.99166598	4.43329E-53	0.400150917	0.454887291
14							
15							
16	Wealth entered.						
17							
18		*df*	*SS*	*MS*	*F*	*Significance F*	
19	Regression	2	7088382281	3544191140	635.1115723	3.37651E-57	
20	Residual	99	552461863.7	5580422.866			
21	Total	101	7640844145				
22							
23		*Coefficients*	*Standard Error*	*t Stat*	*P-value*	*Lower 95%*	*Upper 95%*
24	Intercept	6279.906804	758.5622772	8.278696414	6.04778E-13	4774.754713	7785.058894
25	Income	0.231792518	0.037676923	6.152108514	1.62772E-08	0.157033331	0.306551705
26	Wealth	0.066901189	0.012191467	5.487542179	3.13968E-07	0.042710674	0.091091703
27							

	A B	C	D	E	F	G	H
29	Age entered.						
30							
31		*df*	*SS*	*MS*	*F*	*Significance F*	
32	Regression	3	7195393082	2398464361	527.6662846	2.54304E-60	
33	Residual	98	445451062.9	4545419.01			
34	Total	101	7640844145				
35							
36		*Coefficients*	*Standard Error*	*t Stat*	*P-value*	*Lower 95%*	*Upper 95%*
37	Intercept	-3115.402529	2053.813954	-1.516886436	0.132514586	-7191.129374	960.3243157
38	Income	0.211937005	0.034249258	6.188075842	1.41707E-08	0.143970469	0.279903541
39	Wealth	0.063805426	0.011021439	5.78921007	8.5209E-08	0.041933739	0.085677112
40	Age	301.8731327	62.21535975	4.852067623	4.60805E-06	178.4087793	425.3374862
41							
42							
43	Education entered.						
44							
45		*df*	*SS*	*MS*	*F*	*Significance F*	
46	Regression	4	7232255152	1808063788	429.2386497	9.68905E-61	
47	Residual	97	408588992.5	4212257.655			
48	Total	101	7640844145				
49							
50		*Coefficients*	*Standard Error*	*t Stat*	*P-value*	*Lower 95%*	*Upper 95%*
51	Intercept	-12432.45673	3718.674319	-3.343249681	0.001177705	-19812.99569	-5051.917773
52	Income	0.159747379	0.037393587	4.272052794	4.52422E-05	0.085531461	0.233963297
53	Wealth	0.072988791	0.011054665	6.602532898	2.16051E-09	0.051048341	0.094929241
54	Age	325.0652837	60.40284468	5.381622098	5.1267E-07	205.1823604	444.9482071
55	Education	773.3800418	261.4330936	2.958233142	0.003886994	254.5077323	1292.252351
56							
57							
58	No other variables could be entered into the model. Stepwise ends.						

data example (X1 = Age, X2 = Education, X3 = Income, X4 = Home Value, X5 = Wealth). Best Subsets screens models using a statistic called *Cp*, which is called the Bonferroni criterion. *Cp* estimates the bias introduced in the estimates of the responses by having an *underspecified model* (a model with important predictors missing). If *Cp* is much greater than $k + 1$ (the number of independent variables plus 1) there is substantial bias. For the largest model with all possible predictors, $Cp = k + 1$. The full model always has $Cp = k + 1$, so this usually doesn't help us much. If all models except the full model have large *Cp*s, it suggests that important predictor variables are missing. If more than one model has a *Cp* near $k + 1$, in general, choose the simpler model or use adjusted R^2 to select the better model. The one with the lowest *Cp* approximately equal to $k + 1$ is the most "adequate" model in terms of not having important predictors missing. This is highlighted in Figure 6.29 and corresponds to the model we have previously identified.

The Art of Model Building in Regression

The procedures we have described identify the best regression model from a purely "technical" perspective, focusing on the adjusted R^2 and significance of the independent variables. Other issues, such as multicollinearity, should be considered. For

PHSTAT NOTE
Best-Subsets Regression

From the *PHStat* menu, select *Regression*, then *Best Subsets*. The dialog box that appears prompts you to enter the range for the dependent variable and the independent variables, as well as the confidence level for the regression. The tool creates a *Best Subsets* worksheet that contains a summary of the models analyzed. With a large number of variables, the analysis can take a significant amount of time and memory and may cause a fatal error in Excel, depending on the processor capability and amount of memory available, so this tool should be used cautiously. The tool also provides worksheets with ANOVA output for each of the combinations for further analysis.

Figure 6.29 Best-Subsets Results for *Banking Data*

	A	B	C	D	E	F
10	**Model**	**Cp**	**k+1**	**R Square**	**Adj. R Square**	**Std. Error**
11	X1	1132.021	2	0.319753	0.312950268	7209.482
12	X2	1153.467	2	0.307893	0.300971473	7272.06
13	X3	72.5069	2	0.905703	0.90476041	2684.224
14	X4	648.154	2	0.587349	0.583222723	5615.158
15	X5	82.72217	2	0.900054	0.899054494	2763.462
16	X1X2	744.972	3	0.534911	0.525515617	5991.298
17	X1X3	45.46647	3	0.921764	0.920183276	2457.293
18	X1X4	496.0324	3	0.672584	0.665969731	5026.931
19	X1X5	50.6053	3	0.918922	0.917283904	2501.526
20	X2X3	74.36336	3	0.905783	0.903879382	2696.611
21	X2X4	648.0156	3	0.588532	0.58021935	5635.354
22	X2X5	56.89208	3	0.915445	0.913736844	2554.599
23	X3X4	74.06647	3	0.905947	0.904046887	2694.26
24	X3X5	34.73949	3	0.927696	0.926235544	2362.292
25	X4X5	46.7294	3	0.921065	0.919470723	2468.237
26	X1X2X3	46.14721	4	0.922493	0.920120756	2458.255
27	X1X2X4	497.0943	4	0.673103	0.663095931	5048.509
28	X1X2X5	20.88464	4	0.936465	0.93451958	2225.695
29	X1X3X4	47.11486	4	0.921958	0.919569227	2466.727
30	X1X3X5	11.4155	4	0.941701	0.939916675	2131.999
31	X1X4X5	20.80774	4	0.936507	0.934563412	2224.95
32	X2X3X4	76.06601	4	0.905947	0.903068036	2707.968
33	X2X3X5	31.56207	4	0.93056	0.928433803	2326.826
34	X2X4X5	46.41904	4	0.922343	0.919965823	2460.638
35	X3X4X5	30.37914	4	0.931214	0.929108031	2315.84
36	X1X2X3X4	48.13093	5	0.922502	0.919306637	2470.75
37	X1X2X3X5	4.692132	5	0.946526	0.944320547	2052.379
38	X1X2X4X5	16.87396	5	0.939789	0.937305731	2177.83
39	X1X3X4X5	7.801146	5	0.944806	0.942530244	2085.113
40	X2X3X4X5	31.30277	5	0.931809	0.928997006	2317.652
41	X1X2X3X4X5	6	6	0.946908	0.944143263	2055.643

instance, Figure 6.30 shows the correlation matrix for all the data in the banking example. Note that large correlations exist between Education and Home Value and also between Wealth and Income. In fact, the variance inflation factors are Age: 1.34, Education: 2.45, Income: 14.9, Home Value: 4.38, and Wealth: 10.71, indicating significant multicollinearity. However, after removing Home Value from the model, multicollinearity still exits, as the variance inflation factors are Age: 1.32, Education: 1.66, Income: 12.57, and Wealth: 10.49. Thus, the model suggested by stepwise and best-subsets regression may not be adequate. The VIF values suggest that either Income or Wealth may not be appropriate variables to keep in the model.

If we remove Wealth from the model, the adjusted R^2 drops to 0.9201 and all VIFs are less than 2, but we discover that Education is no longer significant. Dropping Education and leaving only Age and Income in the model results in an adjusted R^2 of 0.9202. However, if we remove Income from the model instead of Wealth, the adjusted R^2 drops to only 0.9345, all VIFs are less than 2, and all remaining variables (Age,

Figure 6.30 Correlation Matrix for *Banking Data*

	A	B	C	D	E	F	G
1		*Age*	*Education*	*Income*	*Home Value*	*Wealth*	*Balance*
2	Age	1					
3	Education	0.173407147	1				
4	Income	0.4771474	0.57539402	1			
5	Home Value	0.386493114	0.753521067	0.795355158	1		
6	Wealth	0.468091791	0.469413035	0.946665447	0.698477789	1	
7	Balance	0.565466834	0.55488066	0.951684494	0.766387128	0.948711734	1

Education, and Wealth) are significant (see Figure 6.31). The R^2 value for the model with these three variables is 0.936.

So what should a model builder do? The independent variables selected should make some sense in attempting to explain the dependent variable (i.e., you should have some reason to believe that changes in the independent variable will cause changes in the dependent variable even though causation cannot be proven statistically). Logic should guide your model development. In many applications, economic or physical theory might suggest that certain variables should belong in a model. Remember that additional variables do contribute to a higher R^2 and, therefore, help to explain a larger proportion of the variation. Even though a variable with a large *p*-value is significant, it could simply be the result of sampling error and a modeler might wish to keep it.

Good modelers also try to have as simple a model as possible—an age-old principle known as **parsimony**—with the fewest number of explanatory variables that

Figure 6.31 Regression Results for Age, Education, and Wealth as Independent Variables

	A	B	C	D	E	F	G
1	Banking Data						
2	X1 X2 X5						
3	*Regression Statistics*						
4	Multiple R	0.967710981					
5	R Square	0.936464543					
6	Adjusted R Square	0.93451958					
7	Standard Error	2225.695322					
8	Observations	102					
9							
10	ANOVA						
11		*df*	*SS*	*MS*	*F*	*Significance F*	
12	Regression	3	7155379617	2385126539	481.4819367	1.71667E-58	
13	Residual	98	485464527.3	4953719.667			
14	Total	101	7640844145				
15							
16		*Coefficients*	*Standard Error*	*t Stat*	*P-value*	*Lower 95%*	*Upper 95%*
17	Intercept	-17732.45142	3801.662822	-4.664393517	9.79978E-06	-25276.72737	-10188.17547
18	Age	367.8214086	64.59823831	5.693985134	1.2977E-07	239.6283103	496.0145069
19	Education	1300.308712	249.9731413	5.201793703	1.08292E-06	804.2451615	1796.372263
20	Wealth	0.116467903	0.004679827	24.88722652	3.75813E-44	0.10718094	0.125754866

will provide an adequate interpretation of the dependent variable. In the physical and management sciences, some of the most powerful theories are the simplest. Thus, a model for the banking data that only includes Age, Education, and Wealth is simpler than one with four variables; because of the multicollinearity issue, there would be little gain to include Income in the model. Whether the model explains 93% or 94% of the variation in bank deposits would probably make little difference. Therefore, building good regression models relies as much on experience and judgment as it does on technical analysis.

REGRESSION WITH CATEGORICAL INDEPENDENT VARIABLES

Some data of interest in a regression study may be ordinal or nominal. For instance, the Excel file *Employee Salaries* shown in Figure 6.32 provides salary and age data for 35 employees, along with an indicator of whether or not the employees have an MBA (Yes or No). The MBA indicator variable, however, is categorical. Since regression analysis requires numerical data, we could include categorical variables by *coding* the variables. For example, we might code "No" as 0 and "Yes" as 1. Such variables are often called **dummy variables**.

If we are interested in predicting salary as a function of the other variables, we would propose the model:

$$Y = \beta_0 + \beta_1 X_1 + \beta_2 X_2 + \varepsilon$$

where

Y = salary
X_1 = age
X_2 = MBA indicator (0 or 1)

After coding the MBA indicator column in the data file, we begin by running a regression on the entire data set, yielding the output shown in Figure 6.33. Note that the model explains about 95% of the variation, and the *p*-values of both variables are significant. The model is:

$$\text{Salary} = 893.59 + 1044.15 \times \text{Age} + 14767.23 \times \text{MBA}$$

	A	B	C	D
1	Salary Data			
2				
3	Employee	Salary	Age	MBA
4	1	$ 28,260	25	No
5	2	$ 43,392	28	Yes
6	3	$ 56,322	37	Yes
7	4	$ 26,086	23	No
8	5	$ 36,807	32	No
9	6	$ 57,119	57	No

Figure 6.32 Portion of *Employee Salaries*

Figure 6.33 Initial Regression Model for *Employee Salaries*

	A	B	C	D	E	F	G
1	SUMMARY OUTPUT						
2							
3	*Regression Statistics*						
4	Multiple R	0.976118476					
5	R Square	0.952807278					
6	Adjusted R Square	0.949857733					
7	Standard Error	2941.914352					
8	Observations	35					
9							
10	ANOVA						
11		*df*	*SS*	*MS*	*F*	*Significance F*	
12	Regression	2	5591651177	2795825589	323.0353318	6.05341E-22	
13	Residual	32	276955521.7	8654860.054			
14	Total	34	5868606699				
15							
16		*Coefficients*	*Standard Error*	*t Stat*	*P-value*	*Lower 95%*	*Upper 95%*
17	Intercept	893.5875971	1824.575283	0.489751015	0.627650922	-2822.950618	4610.125812
18	Age	1044.146043	42.14128238	24.77727265	1.8878E-22	958.3070603	1129.985026
19	MBA	14767.23159	1351.801764	10.92411031	2.49752E-12	12013.70151	17520.76166

Thus, a 30-year-old with an MBA would have an estimated salary of:

$$\text{Salary} = 893.59 + 1044.15 \times 30 + 14767.23 \times 1 = \$46,985.32$$

This model suggests that having an MBA increases the salary of this group of employees by almost \$15,000. Note that by substituting either 0 or 1 for MBA, we obtain two models:

$$\text{No MBA: Salary} = 893.59 + 1044.15 \times \text{Age}$$
$$\text{MBA: Salary} = 15,660.82 + 1044.15 \times \text{Age}$$

The only difference between them is the intercept. The models suggest that the rate of salary increase for age is the same for both groups. Of course, this may not be true. Individuals with MBAs might earn relatively higher salaries as they get older. In other words, the slope of Age may depend on the value of MBA. Such a dependence is called an **interaction**.

We can test for interactions by defining a new variable, $X_3 = X_1 \times X_2$ and testing whether this variable is significant. If so, then the original model should not be used. With the interaction term, the new model is:

$$Y = \beta_0 + \beta_1 X_1 + \beta_2 X_2 + \beta_3 X_3 + \varepsilon$$

In the worksheet, we need to create a new column (called Interaction) by multiplying MBA by Age for each observation (see Figure 6.34). The regression results are shown in Figure 6.35.

From Figure 6.35, we see that the adjusted R^2 increases; however, the p-value for the MBA indicator variable is 0.33, indicating that this variable is not significant. Therefore, we drop this variable and run a regression using only Age and the interaction term. The results are shown in Figure 6.36. Adjusted R^2 increased slightly, and both Age and the interaction term are significant. The final model is:

$$\text{Salary} = 3323.11 + 984.25 \times \text{Age} + 425.58 \times \text{MBA} \times \text{Age}$$

Figure 6.34 Portion of *Employee Salaries* Modified for Interaction Term

	A	B	C	D	E
1	Salary Data				
2					
3	Employee	Salary	Age	MBA	Interaction
4	1	$ 28,260	25	0	0
5	2	$ 43,392	28	1	28
6	3	$ 56,322	37	1	37
7	4	$ 26,086	23	0	0
8	5	$ 36,807	32	0	0

The models for employees with and without an MBA are:

$$\text{No MBA: Salary} = 3323.11 + 984.25 \times \text{Age} + 425.58 \times 0 \times \text{Age}$$
$$= 3323.11 + 984.25 \times \text{Age}$$

$$\text{MBA: Salary} = 3323.11 + 984.25 \times \text{Age} + 425.58 \times 1 \times \text{Age}$$
$$= 3323.11 + 1409.83 \times \text{Age}$$

Here we see that salary not only depends on whether an employee holds an MBA, but also on age.

Figure 6.35 Regression Results with Interaction Term

	A	B	C	D	E	F	G
1	SUMMARY OUTPUT						
2							
3	*Regression Statistics*						
4	Multiple R	0.989321416					
5	R Square	0.978756863					
6	Adjusted R Square	0.976701076					
7	Standard Error	2005.37675					
8	Observations	35					
9							
10	ANOVA						
11		*df*	*SS*	*MS*	*F*	*Significance F*	
12	Regression	3	5743939086	1914646362	476.098288	5.31397E-26	
13	Residual	31	124667613.2	4021535.91			
14	Total	34	5868606699				
15							
16		*Coefficients*	*Standard Error*	*t Stat*	*P-value*	*Lower 95%*	*Upper 95%*
17	Intercept	3902.509386	1336.39766	2.920170772	0.006467654	1176.908399	6628.110372
18	Age	971.3090382	31.06887722	31.26308786	5.23658E-25	907.9436456	1034.674431
19	MBA	-2971.080074	3026.24236	-0.98177202	0.333812767	-9143.142034	3200.981887
20	Interaction	501.8483604	81.55221742	6.153705887	7.9295E-07	335.5215171	668.1752038

Figure 6.36 Final Regression Model with Interaction Term

	A	B	C	D	E	F	G
1	SUMMARY OUTPUT						
2							
3	*Regression Statistics*						
4	Multiple R	0.98898754					
5	R Square	0.978096355					
6	Adjusted R Square	0.976727377					
7	Standard Error	2004.24453					
8	Observations	35					
9							
10	ANOVA						
11		*df*	*SS*	*MS*	*F*	*Significance F*	
12	Regression	2	5740062823	2870031411	714.4720368	2.80713E-27	
13	Residual	32	128543876.4	4016996.136			
14	Total	34	5868606699				
15							
16		*Coefficients*	*Standard Error*	*t Stat*	*P-value*	*Lower 95%*	*Upper 95%*
17	Intercept	3323.109564	1198.353141	2.773063675	0.009184278	882.1441051	5764.075022
18	Age	984.2455409	28.12039088	35.00113299	4.40388E-27	926.9661794	1041.524902
19	Interaction	425.5845915	24.81794165	17.14826304	1.08793E-17	375.0320988	476.1370841

Categorical Variables with More Than Two Levels

When a categorical variable has only two levels, as in the previous example, we coded the levels as 0 and 1 and added a new variable to the model. However, when a categorical variable has $k > 2$ levels, we need to add $k - 1$ additional variables to the model. To illustrate this, the Excel file *Surface Finish* provides measurements of the surface finish of 35 parts produced on a lathe, along with the revolutions per minute (RPM) of the spindle and one of four types of cutting tools used (see Figure 6.37). The engineer who collected the data is interested in predicting the surface finish as a function of RPM and type of tool.

Intuition might suggest defining a dummy variable for each tool type; however, doing so will cause numerical instability in the data and cause the regression tool to crash. Instead, we will need $k - 1 = 3$ dummy variables corresponding to three of the levels of the categorical variable. The level left out will correspond to a reference, or baseline value. Therefore, because we have $k = 4$ levels of tool type, we will define a regression model of the form:

$$Y = \beta_0 + \beta_1 X_1 + \beta_2 X_2 + \beta_3 X_3 + \beta_4 X_4 + \varepsilon$$

where

Y = surface finish
X_1 = RPM
X_2 = 1 if tool type is B and 0 if not
X_3 = 1 if tool type is C and 0 if not
X_4 = 1 if tool type is D and 0 if not

Note that when $X_2 = X_3 = X_4 = 0$, then by default, the tool type is A. Substituting these values for each tool type into the model, we obtain:

Tool Type A: $Y = \beta_0 + \beta_1 X_1 + \varepsilon$
Tool Type B: $Y = \beta_0 + \beta_1 X_1 + \beta_2 + \varepsilon$

Figure 6.37 Portion of *Surface Finish*

	A	B	C	D
1	**Surface Finish Data**			
2				
3	**Part**	**Surface Finish**	**RPM**	**Cutting Tool**
4	1	45.44	225	A
5	2	42.03	200	A
6	3	50.10	250	A
7	4	48.75	245	A
8	5	47.92	235	A
9	6	47.79	237	A
10	7	52.26	265	A
11	8	50.52	259	A
12	9	45.58	221	A
13	10	44.78	218	A
14	11	33.50	224	B
15	12	31.23	212	B
16	13	37.52	248	B
17	14	37.13	260	B
18	15	34.70	243	B

Tool Type C: $Y = \beta_0 + \beta_1 X_1 + \beta_3 + \varepsilon$
Tool Type D: $Y = \beta_0 + \beta_1 X_1 + \beta_4 + \varepsilon$

For a fixed value of RPM (X_1), the slopes corresponding to the dummy variables represent the difference between the surface finish using that tool type and the baseline using tool type A.

To incorporate these dummy variables into the regression model, we add three columns to the data as shown in Figure 6.38. Using these data, we obtain the regression results shown in Figure 6.39. The resulting model is:

$$\text{Surface Finish} = 24.49 + 0.098 \times \text{RPM} - 13.31 \times \text{Type B} - 20.49 \times \text{Type C} - 26.04 \times \text{Type D}$$

Almost 99% of the variation in surface finish is explained by the model, and all variables are significant. The models for each individual tool are:

$$
\begin{aligned}
\text{Tool A: Surface Finish} &= 24.49 + 0.098 \times \text{RPM} - 13.31 \times 0 - 20.49 \times 0 \\
&\quad - 26.04 \times 0 \\
&= 24.49 + 0.098 \times \text{RPM}
\end{aligned}
$$

$$
\begin{aligned}
\text{Tool B: Surface Finish} &= 24.49 + 0.098 \times \text{RPM} - 13.31 - 20.49 \times 0 \\
&\quad - 26.04 \times 0 \\
&= 11.18 + 0.098 \times \text{RPM}
\end{aligned}
$$

$$
\begin{aligned}
\text{Tool C: Surface Finish} &= 24.49 + 0.098 \times \text{RPM} - 13.31 \times 0 - 20.49 \times 1 \\
&\quad - 26.04 \times 0 \\
&= 4.00 + 0.098 \times \text{RPM}
\end{aligned}
$$

$$
\begin{aligned}
\text{Tool D: Surface Finish} &= 24.49 + 0.098 \times \text{RPM} - 13.31 - 20.49 \times 0 \\
&\quad - 26.04 \times 1 \\
&= -1.55 + 0.098 \times \text{RPM}
\end{aligned}
$$

	A	B	C	D	E	F
1	**Surface Finish Data**					
2						
3	**Part**	**Surface Finish**	**RPM**	**Type B**	**Type C**	**Type D**
4	1	45.44	225	0	0	0
5	2	42.03	200	0	0	0
6	3	50.10	250	0	0	0
7	4	48.75	245	0	0	0
8	5	47.92	235	0	0	0
9	6	47.79	237	0	0	0
10	7	52.26	265	0	0	0
11	8	50.52	259	0	0	0
12	9	45.58	221	0	0	0
13	10	44.78	218	0	0	0
14	11	33.50	224	1	0	0
15	12	31.23	212	1	0	0
16	13	37.52	248	1	0	0
17	14	37.13	260	1	0	0
18	15	34.70	243	1	0	0
19	16	33.92	238	1	0	0
20	17	32.13	224	1	0	0
21	18	35.47	251	1	0	0
22	19	33.49	232	1	0	0
23	20	32.29	216	1	0	0
24	21	27.44	225	0	1	0
25	22	24.03	200	0	1	0
26	23	27.33	250	0	1	0
27	24	27.20	245	0	1	0
28	25	27.10	235	0	1	0
29	26	27.30	237	0	1	0
30	27	28.30	265	0	1	0
31	28	28.40	259	0	1	0
32	29	26.80	221	0	1	0
33	30	26.40	218	0	1	0
34	31	21.40	224	0	0	1
35	32	20.50	212	0	0	1
36	33	21.90	248	0	0	1
37	34	22.13	260	0	0	1
38	35	22.40	243	0	0	1

Figure 6.38 Data Matrix for *Surface Finish* with Dummy Variables

Note that the only differences among these models are the intercepts; the slopes associated with RPM are the same. This suggests that we might wish to test for interactions between the type of cutting tool and RPM; we leave this to you as an exercise.

Figure 6.39 *Surface Finish* Regression Model Results

	A	B	C	D	E	F	G
1	SUMMARY OUTPUT						
2							
3	*Regression Statistics*						
4	Multiple R	0.994447053					
5	R Square	0.988924942					
6	Adjusted R Square	0.987448267					
7	Standard Error	1.089163115					
8	Observations	35					
9							
10	ANOVA						
11		*df*	*SS*	*MS*	*F*	*Significance F*	
12	Regression	4	3177.784271	794.4460678	669.6973322	7.32449E-29	
13	Residual	30	35.58828875	1.186276292			
14	Total	34	3213.37256				
15							
16		*Coefficients*	*Standard Error*	*t Stat*	*P-value*	*Lower 95%*	*Upper 95%*
17	Intercept	24.49437244	2.473298088	9.903526211	5.73134E-11	19.4432239	29.54552099
18	RPM	0.097760627	0.010399996	9.400064035	1.89415E-10	0.076521002	0.119000251
19	Type B	-13.31056756	0.487142953	-27.32374035	9.37003E-23	-14.30544619	-12.31568893
20	Type C	-20.487	0.487088553	-42.06011387	3.12134E-28	-21.48176753	-19.49223247
21	Type D	-26.03674519	0.596886375	-43.62094073	1.06415E-28	-27.25574979	-24.81774059

REGRESSION MODELS WITH NONLINEAR TERMS

Linear regression models are not appropriate for every situation. A scatter chart of the data might show a nonlinear relationship, or the residuals for a linear fit might result in a nonlinear pattern. In such cases, we might propose a nonlinear model to explain the relationship. For instance, a second-order polynomial model would be:

$$Y = \beta_0 + \beta_1 X + \beta_2 X^2 + \varepsilon$$

Sometimes this is called a **curvilinear regression model**. In this model, β_1 represents the linear effect of X on Y, and β_2 represents the curvilinear effect. However, although this model appears to be quite different from ordinary linear regression models, it is still *linear in the parameters* (the betas, which are the unknowns that we are trying to estimate). In other words, all terms are a product of a beta coefficient and some function of the data. In such cases we can still apply least squares to estimate the regression coefficients.

To illustrate this, the Excel file *Beverage Sales* provides data on the sales of cold beverages at a small restaurant with a large outdoor patio during the summer months (see Figure 6.40). The owner has observed that sales tend to increase on hotter days. Figure 6.41 shows linear regression results for these data. The U-shape of the residual

	A	B
1	**Beverage Sales**	
2		
3	**Temperature**	**Sales**
4	85	$ 1,810
5	90	$ 4,825
6	79	$ 438
7	82	$ 775
8	84	$ 1,213
9	96	$ 8,692
10	88	$ 2,356
11	76	$ 266
12	93	$ 4,930
13	97	$ 9,138
14	89	$ 2,714
15	83	$ 1,082
16	85	$ 1,290
17	90	$ 3,970
18	82	$ 894
19	91	$ 2,906
20	90	$ 4,615
21	84	$ 1,168
22	79	$ 462
23	81	$ 1,018
24	95	$ 5,950

Figure 6.40 Excel File *Beverage Sales*

Figure 6.41   Linear Regression Results for *Beverage Sales*

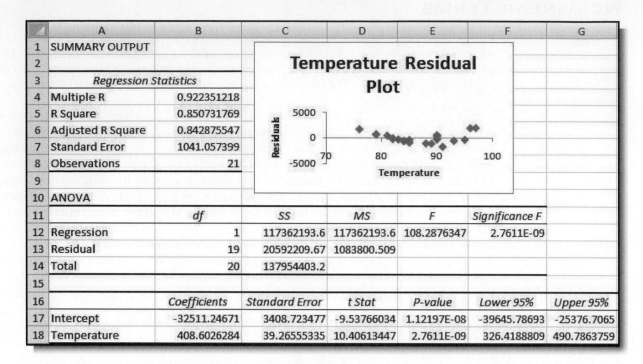

plot suggests that a linear relationship is not appropriate. To apply a curvilinear regression model, add a column to the data matrix by squaring the temperatures. Now, both temperature and temperature squared are the independent variables. Figure 6.42 shows the results for the curvilinear regression model. The model is:

$$\text{Sales} = 142850 - 3643.17 \times \text{Temperature} + 23.2 \times \text{Temperature}^2$$

Figure 6.42   Curvilinear Regression Results for *Beverage Sales*

	A	B	C	D	E	F	G
1	SUMMARY OUTPUT						
2							
3	*Regression Statistics*						
4	Multiple R	0.973326989					
5	R Square	0.947365428					
6	Adjusted R Square	0.941517142					
7	Standard Error	635.1365123					
8	Observations	21					
9							
10	ANOVA						
11		*df*	*SS*	*MS*	*F*	*Significance F*	
12	Regression	2	130693232.2	65346616.12	161.9902753	3.10056E-12	
13	Residual	18	7261171.007	403398.3893			
14	Total	20	137954403.2				
15							
16		*Coefficients*	*Standard Error*	*t Stat*	*P-value*	*Lower 95%*	*Upper 95%*
17	Intercept	142850.3406	30575.70155	4.672021683	0.000189738	78613.17542	207087.5058
18	Temperature	-3643.171723	705.2304165	-5.165931075	6.492E-05	-5124.805846	-2161.5376
19	Temp^2	23.30035581	4.053196314	5.748637374	1.89343E-05	14.78490636	31.81580527

Note that the adjusted R^2 has increased significantly from the linear model and that the residual plots now show more random patterns.

Curvilinear regression models are also often used in forecasting when the independent variable is time. This and other applications of regression in forecasting will be discussed in the next chapter.

BASIC CONCEPTS REVIEW QUESTIONS

1. Provide some practical examples from your work experience where regression models might be used.
2. Explain the difference between simple and multiple linear regression.
3. What is least-squares regression? Describe how least-squares regression defines the best-fitting regression line.
4. Explain what the coefficient of determination, R^2, measures. How is it related to the sample correlation coefficient?
5. How can regression analysis be applied to understand risk associated with stocks and mutual funds?
6. What is the standard error of the estimate? What information does it provide?
7. Explain the difference between a confidence interval and a prediction interval.
8. What do we mean by "significance of regression"?
9. How are standardized residuals calculated, and how can they help identify outliers?
10. Explain the assumptions of linear regression. How can you determine if each of these assumptions holds?
11. What are partial regression coefficients and what information do they provide?
12. Explain how adjusted R^2 is used in evaluating the fit of multiple regression models.
13. What is the difference in the hypotheses tested in single and multiple regression using the *Significance F* statistic?
14. What is multicollinearity, and how is it measured? What problems can it cause in regression results?
15. Describe the systematic process for building a good regression model using *p*-values to guide independent variable selection.
16. Describe the differences and advantages/disadvantages of using stepwise and best-subsets approaches in building regression models.
17. How should you balance technical criteria and practical judgment in your decisions?
18. Explain how to include categorical variables in regression models.
19. What is interaction, and why is it important to test for it?
20. What is a curvilinear regression model? How do you know when to use one?

SKILL-BUILDING EXERCISES

1. For the *Home Market Value* data, use formulas (6.2) and (6.3) on a spreadsheet to calculate the least-squares regression coefficients.
2. For the *Colleges and Universities* data, plot scatter charts for each variable individually against the graduation rate and add linear trendlines, finding both the equation and R^2 values.
3. Use the regression tools to find simple linear regression models for each independent variable for the *Colleges and Universities* data.
4. Use the INTERCEPT and SLOPE functions in Excel to develop regression models for each of the independent variables in the *Colleges and Universities* example.
5. Find the standard residuals for the *Colleges and Universities* regression model and construct a histogram for them to evaluate normality.
6. Apply stepwise and best-subsets regression to the *Colleges and Universities* data. Do you obtain the same results as the original model?
7. Run the example for *Employee Salaries* to verify the results with categorical variables.
8. For the *Surface Finish* example in the chapter, run models to examine interactions between the type of cutting tool and RPM. (*Hint:* Add interaction terms between each dummy variable and RPM.) Are any interactions significant?
9. Run the linear regression for *Beverage Sales* using the Excel tool. Using the *Trendline* option, fit a second-order polynomial to the residual plot for temperature. Also, construct a scatter chart for the original data and add a second-order polynomial trendline.

1. Using the data in the Excel in the file *Home Market Value*, construct a scatter chart for market value versus age and add a linear trendline. Compare the coefficient of determination with the model developed in this chapter. What do you conclude?

2. Using the data in the Excel file *Demographics*, determine if a linear relationship exists between unemployment rates and cost of living indexes by constructing a scatter chart and adding a trendline. What is the regression model and R^2?

3. Using the data in the Excel file *Student Grades*, construct a scatter chart for midterm versus final exam grades and add a linear trendline. What is the regression model and R^2? If a student scores 85 on the midterm, what would you estimate her grade on the final exam to be?

4. (From Horngren, Foster, and Datar, *Cost Accounting: A Managerial Emphasis*, 9th ed., Prentice-Hall, 1997, p. 371.) The managing director of a consulting group has the following monthly data on total overhead costs and professional labor hours to bill to clients:

TOTAL	BILLABLE
$340,000	3,000
$400,000	4,000
$435,000	5,000
$477,000	6,000
$529,000	7,000
$587,000	8,000

Develop a regression model to identify the fixed overhead costs to the consulting group.
   a. What is the constant component of the consultant group's overhead?
   b. If a special job requiring 1,000 billable hours that would contribute a margin of $38,000 before overhead was available, would the job be attractive?

5. The Excel file *National Football League* provides various data on professional football for the 2007 season.
   a. Construct a scatter diagram for Points/Game and Yards/Game in the Excel file. Does there appear to be a linear relationship?
   b. Develop a regression model for predicting Points/Game as a function of Yards/Game. Explain the statistical significance of the model and draw conclusions about the validity of the regression analysis assumptions.
   c. Find a 95% confidence interval for the mean number of Points/Game for teams with 300 Yards/Game.
   d. Find a 95% prediction interval for a team having 300 Yards/Game.

6. The Excel file *Hypothetical Stock Data* shows daily closing prices for a particular stock and the S&P 500 index.
   a. Apply regression analysis to find the beta associated with the stock.
   b. Test the hypothesis that beta is equal to one.
   c. Find a 95% confidence interval for beta.

7. Find real data on daily changes in the S&P 500 (or the Dow Jones Industrial Average) and a stock of your interest. Use regression analysis to estimate the beta risk of the stock.

8. Data collected in 1960 from the National Cancer Institute provides the per capita numbers of cigarettes sold along with death rates for various forms of cancer (see the Excel file *Smoking and Cancer*). Use simple linear regression to determine if a significant relationship exists between the number of cigarettes sold and each form of cancer. Examine the residuals for assumptions and outliers.

9. A deep-foundation engineering contractor has bid on a foundation system for a new world headquarters building for a *Fortune* 500 company. A part of the project consists of installing 311 auger cast piles. The contractor was given bid information for cost-estimating purposes, which consisted of the estimated depth of each pile; however, actual drill footage of each pile could not be determined exactly until construction was performed. The Excel file *Pile Foundation* contains the estimates and actual pile lengths after the project was completed. Develop a linear regression model to estimate the actual pile length as a function of the estimated pile lengths. What do you conclude?

10. Using the data in the Excel file *Home Market Value*, develop a multiple linear regression model for estimating the market value as a function of both the age and size of the house. Find a 95% confidence interval for the mean market value for houses that are 32 years old and have 1,900 square feet, and a 95% prediction interval for a house that is 28 years old with 1,500 square feet.

11. The Excel file *Infant Mortality* provides data on infant mortality rate (deaths per 1,000 births), female literacy (percent who read), and population density (people per square kilometer) for 85 countries. Develop simple and multiple regression models for the relationship between mortality, population density, and literacy. Explain all statistical output.

12. The Excel file *Concert Sales* provides data on sales dollars and the number of radio and TV and newspaper ads promoting the concerts for a group of cities. Develop simple linear regression models for predicting sales as a function of the number of each type of ad. Compare these results to a multiple linear regression model using both independent variables. Examine the residuals of the best model for regression assumptions and possible outliers.

13. The Excel file *Cereal Data* provides a variety of nutritional information about 67 cereals and their shelf location in a supermarket. Use regression analysis to determine if a relationship exists between calories and the other variables. Investigate the model assumptions and clearly explain your conclusions.

14. The Excel file *Salary Data* provides information on current salary, beginning salary, previous experience (in months) when hired, and total years of education for a sample of 100 employees in a firm.
   a. Develop a multiple regression model for predicting current salary as a function of the other variables.
   b. Find the best model for predicting current salary.

15. The Excel file *Major League Baseball* provides data on the 2007 season.
   a. Construct and examine the correlation matrix. Is multicollinearity a potential problem? Find the variance inflation factors to check your intuition.
   b. Suggest an appropriate set of independent variables that predict the number of wins by examining the correlation matrix and variance inflation factors.
   c. Find the best multiple regression model for predicting the number of wins. How good is your model? Does it use the same variables you thought were appropriate in part (b)?

16. Apply stepwise regression to find a good model for predicting the number of points scored per game by football teams using the data in the Excel file *National Football League*. Use a regression analysis tool for the stepwise model to examine the residuals and regression assumptions.

17. The Excel file *Credit Approval Decisions* provides information on credit history for a sample of banking customers. Apply the best-subsets regression tool to identify the best model for predicting the credit score as a function of the other numerical variables. For the model you select, conduct further analysis to check for significance of the independent variables and for multicollinearity.

18. Using the data in the Excel file *Freshman College Data*, use best-subsets regression to identify the best model for predicting the first year retention rate. For the model you select, conduct further analysis to check for significance of the independent variables and for multicollinearity.

19. The State of Ohio Department of Education has a mandated ninth-grade proficiency test that covers writing, reading, mathematics, citizenship (social studies), and science. The Excel file *Ohio Education Performance* provides data on success rates (defined as the percent of students passing) in school districts in the greater Cincinnati metropolitan area along with state averages.

   a. Develop a multiple regression model to predict math success as a function of success in all other subjects using the systematic approach described in this chapter. Is multicollinearity a problem?

   b. Suggest the best regression model to predict math success as a function of success in the other subjects by examining the correlation matrix; then run the regression tool for this set of variables.

   c. Develop the best regression model to predict math success as a function of success in the other subjects using best-subsets regression.

   d. Develop the best regression model to predict math success as a function of success in the other subjects using stepwise regression.

   e. Compare the results for parts (a) through (d). Are the models the same? Why or why not?

20. A mental health agency measured the self-esteem score for randomly selected individuals with disabilities who were involved in some work activity within the past year. The Excel file *Self Esteem* provides the data, including the individuals' marital status, length of work, type of support received (direct support includes job-related services such as job coaching and counseling), education, and age.

   a. Use simple linear regression to determine if there is a relationship between self-esteem and length of work.

   b. Use multiple linear regression for predicting self-esteem as a function of the other variables.

   c. Investigate possible interaction effects, and determine the best model.

21. A national homebuilder builds single-family homes and condominium-style townhouses. The Excel file *House Sales* provides information on the selling price, lot cost, type of home, and region of the country (M = Midwest, S = South) for closings during one month.

   a. Develop a multiple regression model for sales price as a function of lot cost, region of country, and type of home.

   b. Determine if any interactions exist between lot cost, region, and type of home.

22. (From Horngren, Foster, and Datar, *Cost Accounting: A Managerial Emphasis*, 9th ed., Prentice-Hall, 1997, p. 349.) Cost functions are often nonlinear with volume because production facilities are often able to produce larger quantities at lower rates than smaller quantities. Using the following data, apply simple linear regression, and examine the residual plot. What do you conclude? Construct a scatter chart and use the Excel *Trendline* feature to identify the best type of trendline that maximizes R^2.

Units Produced	Costs
500	$12,500
1,000	$25,000
1,500	$32,500
2,000	$40,000
2,500	$45,000
3,000	$50,000

23. (From Horngren, Foster, and Datar, *Cost Accounting: A Managerial Emphasis,* 9th ed., Prentice-Hall, 1997, p. 349.) The Helicopter Division of Aerospatiale is studying assembly costs at its Marseilles plant. Past data indicates the following labor hours per helicopter:

Helicopter Number	Labor Hours
1	2,000
2	1,400
3	1,238
4	1,142
5	1,075
6	1,029
7	985
8	957

Use regression to compare the results of simple linear regression with a second-order polynomial regression model.

CASE

Hatco

The Excel file *HATCO* (adapted from Hair, Anderson, Tatham, and Black in *Multivariate Analysis,* 5th ed., Prentice-Hall, 1998) consists of data related to predicting the level of business (Usage Level) obtained from a survey of purchasing managers of customers of an industrial supplier, HATCO. The independent variables are the following:

- *Delivery Speed*—amount of time it takes to deliver the product once an order is confirmed
- *Price Level*—perceived level of price charged by product suppliers
- *Price Flexibility*—perceived willingness of HATCO representatives to negotiate price on all types of purchases
- *Manufacturing Image*—overall image of the manufacturer or supplier
- *Overall Service*—overall level of service necessary for maintaining a satisfactory relationship between supplier and purchaser

- *Sales Force Image*—overall image of the manufacturer's sales force
- *Product Quality*—perceived level of quality of a particular product
- *Size of Firm*—size relative to others in this market (0 = small; 1 = large)

Responses to the first seven variables were obtained using a graphic rating scale, where a 10-centimeter line was drawn between endpoints labeled "poor" and "excellent." Respondents indicated their perceptions using a mark on the line, which was measured from the left endpoint. The result was a scale from 0 to 10 rounded to one decimal place.

Using the tools in this chapter, conduct a complete analysis to predict Usage Level. Be sure to investigate the impact of the categorical variable Size of Firm (coded as 0 for small firms, and 1 for large firms) and possible interactions. Also stratify the data by firm size to account for any differences between small and large firms. Write up your results in a formal report to HATCO management.

Appendix

Regression Theory and Computation

In this appendix, we present the basic underlying theory behind regression analysis, and also show how some of the key statistics are computed.

Regression as Analysis of Variance

The objective of regression analysis is to explain the variation of the dependent variable around its mean value as the independent variable changes. Figure 6A.1 helps to understand this. In Figure 6A.1(a), the scatter plot does not show any linear relationship. If we tried to fit a regression line, β_1 would be zero and the model would reduce to $Y = \beta_0 + \varepsilon$. The intercept β_0 would simply be the sample mean of the dependent variable observation, $\overline{Y}$. Thus,

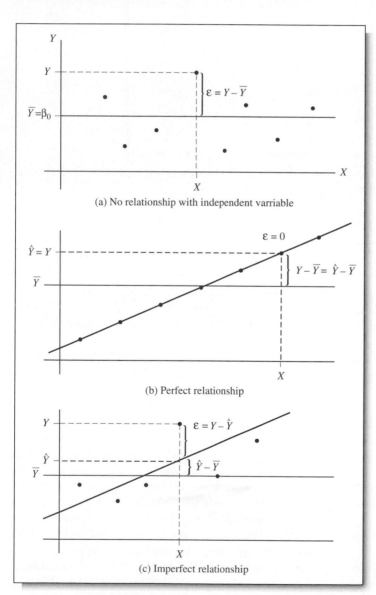

(a) No relationship with independent varriable

(b) Perfect relationship

(c) Imperfect relationship

Figure 6A.1 Illustrations of Variation in Regression

an estimate of the population mean of Y for any value of X would be $\overline{Y}$. We could measure the error by subtracting the estimate of the mean from each observation $(Y - \overline{Y})$. In Figure 6A.1(b), we have the opposite extreme in which all points lie on the regression line; that is, all of the variation in the data around the mean is explained by the independent variable. Note that if we estimate Y using $\overline{Y}$ alone, we would have a large amount of error. However, by knowing the value of X, we can predict the value of Y accurately using the regression line $\hat{Y} = b_0 + b_1 X$. In this case the value of each observation is equal to its predicted value and the error—the difference between $\hat{Y}$ and Y—is zero. Because $(Y - \overline{Y}) = (\hat{Y} - \overline{Y})$, all the variation from the mean is explained by the regression line. Finally, Figure 6A.1(c) shows the typical case in which some of the variation around the mean, $(Y - \overline{Y})$, is explained by the independent variable $(\hat{Y} - \overline{Y})$, while some is also due to error, $(Y - \hat{Y})$ because of the fact that the points do not line on the regression line.

We have defined three measures of variation:

1. variation between the observations and the mean: $(Y - \overline{Y})$
2. variation between the predicted values using the regression line and the mean, which is explained by the regression line: $(\hat{Y} - \overline{Y})$
3. variation between the individual observations and the predicted values, which is the remaining unexplained variation: $(Y - \hat{Y})$

Because some terms will be positive and others will be negative, we need to square them to obtain a useful measure of the total variation in all the data; otherwise, they will sum to zero.

The sums of the squares of the deviations of individual observations from the mean, summed over all observations, $\Sigma(Y - \overline{Y})^2$, is called the *total sum of squares*, or SST. Mathematically, this can be shown to equal $\Sigma(\hat{Y} - \overline{Y})^2 + \Sigma(Y - \hat{Y})^2$, which is simply the sums of the squares of the variation *explained by regression*, called SSR, and the sum of squares of the errors, or *unexplained variation*, SSE. In other words,

$$SST = SSR + SSE \qquad (6A.1)$$

These sums of squares are shown in the ANOVA section of the regression analysis output.

Looking back at Figure 6A.1, notice that in (a), SST = SSE because the independent variable did not explain any variation in the dependent variable. In (b), SST = SSR; all the variation is explained by the regression line. In (c), both SSR and SSE are positive, indicating that some variation is explained by regression while some error still exists. As points become clustered closer around a straight line, SSR increases while SSE decreases.

In a simple linear regression model with one explanatory variable, the total sum of squares, $SST = \Sigma(Y - \overline{Y})^2$, has $n - 1$ degrees of freedom because we estimate the mean. SSR has 1 degree of freedom because there is one independent variable. This leaves $n - 2$ degrees of freedom for SSE. In Chapter 5, we noted that sums of squares divided by their appropriate degrees of freedom provide estimates of variances, which we called mean squares. Thus, $MSR = SSR/1$ represents the variance between observations explained by regression, while $MSE = SSE/(n - 2)$ represents the remaining variance due to error. These are shown in the MS column of the ANOVA section of the regression output. By dividing MSR by MSE, we obtain an F-statistic. If this number is higher than the critical value from the F-distribution for a chosen level of significance, then we would reject the null hypothesis. Logically, if the null hypothesis is true, then SST = SSE, and SSR (and MSR) would be ideally zero. Therefore, the smaller the F-ratio, the greater is the likelihood that H_0 is true. Likewise, the larger the F-ratio, the greater the likelihood is that $\beta_1 \neq 0$ and that the independent variable helps to explain variation in the data about the mean.

Standard Error of the Estimate

The standard error of the estimate, S_{YX}, is the standard deviation of the errors about the regression line. This is computed as:

$$S_{YX} = \sqrt{\frac{SSE}{n - 2}} \qquad (6A.2)$$

Adjusted R-Square

Adjusted R^2 is computed as:

$$R^2_{adj} = 1 - \left[(1 - R^2)\frac{n - 1}{n - 2} \right] \qquad (6A.3)$$

Confidence Intervals

We can develop a $100(1 - \alpha)\%$ confidence interval for the *mean value* of Y using the following formula:

$$\hat{Y} \pm t_{\alpha/2, n-2} S_{YX} \sqrt{h_i} \qquad (6A.4)$$

where $\hat{Y} = b_0 + b_1 X_i$ is the predicted mean of Y for a given value, X_i, of the independent variable, and:

$$h_i = \frac{1}{n} + \frac{(X_i - \overline{X})^2}{\displaystyle\sum_{j=1}^{n}(X_j - \overline{X})^2} \qquad (6A.5)$$

This is a bit different from the types of confidence intervals that we computed in Chapter 4. It turns out that the true standard error and the width of the confidence interval actually depend on the value of X_i as expressed by the h_i statistic. The further X_i deviates from the mean of X, the larger the value of h_i, and hence, the larger the confidence interval. Therefore, we actually have a collection of confidence intervals for each value of X. Such a collection of confidence intervals is called a **confidence band** around the regression line.

Prediction Intervals

A $100(1 - \alpha)\%$ prediction interval for $\hat{Y}$ for a given value, X_i, of the independent variable is:

$$\hat{Y} \pm t_{\alpha/2, n-2} S_{YX} \sqrt{1 + h_i} \qquad (6A.6)$$

In many cases, the value of h_i is quite small so that the term $\sqrt{1 + h_i}$ is close to one. Therefore, we can develop an *approximate* $100(1 - \alpha)\%$ prediction interval simply by

$$\hat{Y} \pm t_{\alpha/2, n-2} S_{YX}$$

Chapter 7

Forecasting

INTRODUCTION

One of the major problems that managers face is forecasting future events in order to make good decisions. For example, forecasts of interest rates, energy prices, and other economic indicators are needed for financial planning; sales forecasts are needed to plan production and workforce capacity; and forecasts of trends in demographics, consumer behavior, and technological innovation are needed for long-term strategic planning. The government also invests significant resources on predicting short-run U.S. business performance using the Index of Leading Indicators. This index focuses on the performance of individual businesses, which often is highly correlated with the performance of the overall economy, and is used to forecast economic trends for the nation as a whole. In this chapter, we introduce some common methods and approaches to forecasting, including both qualitative and quantitative techniques.

Managers may choose from a wide range of forecasting techniques. Selecting the appropriate method depends on the characteristics of the forecasting problem, such as the time horizon of the variable being forecast, as well as available information on which the forecast will be based. Three major categories of forecasting approaches are *qualitative and judgmental techniques*, *statistical time-series models*, and *explanatory/causal methods*.

Qualitative and judgmental techniques rely on experience and intuition; they are necessary when historical data are not available or when the decision maker needs to forecast far into the future. For example, a forecast of when the next generation of a microprocessor will be available and what capabilities it might have will depend greatly on the opinions and expertise of individuals who understand the technology.

Statistical time-series models find greater applicability for short-range forecasting problems. A **time series** is a stream of historical data, such as weekly sales. Time-series models assume that whatever forces have influenced sales in the recent past will continue into the near future; thus, forecasts are developed by extrapolating these data into the future.

Explanatory/causal models seek to identify factors that explain statistically the patterns observed in the variable being forecast, usually with regression analysis. While time-series models use only time as the independent variable, explanatory/causal models generally include other factors. For example, forecasting the price of oil might incorporate independent variables such as the demand for oil (measured in barrels), the proportion of oil stock generated by OPEC countries, and tax rates. Although we can never prove that changes in these variables actually cause changes in the price of oil, we often have evidence that a strong influence exists.

Surveys of forecasting practices have shown that both judgmental and quantitative methods are used for forecasting sales of product lines or product families, as well as for broad company and industry forecasts. Simple time-series models are used for short- and medium-range forecasts, whereas regression analysis is the most popular method for long-range forecasting. However, many companies rely on judgmental methods far more than quantitative methods, and almost half judgmentally adjust quantitative forecasts.

In this chapter, we focus on these three approaches to forecasting. Specifically, we will discuss the following:

➤ *Historical analogy and the Delphi method as approaches to judgmental forecasting*

➤ *Moving average and exponential smoothing models for time-series forecasting, with a discussion of evaluating the quality of forecasts*

➤ *A brief discussion of advanced time-series models and the use of Crystal Ball (CB) Predictor for optimizing forecasts*

➤ *The use of regression models for explanatory/causal forecasting*

➤ *Some insights into practical issues associated with forecasting*

QUALITATIVE AND JUDGMENTAL METHODS

Qualitative, or judgmental, forecasting methods are valuable in situations for which no historical data are available or for those that specifically require human expertise and knowledge. One example might be identifying future opportunities and threats as part of a SWOT (Strengths, Weaknesses, Opportunities, and Threats) analysis within a strategic planning exercise. Another use of judgmental methods is to incorporate nonquantitative information, such as the impact of government regulations or competitor behavior, in a quantitative forecast. Judgmental techniques range from such simple methods as a manager's opinion or a group-based jury of executive opinion to more structured approaches such as historical analogy and the Delphi method.

Historical Analogy

One judgmental approach is **historical analogy**, in which a forecast is obtained through a comparative analysis with a previous situation. For example, if a new product is being introduced, the response of similar previous products to marketing campaigns can be used as a basis to predict how the new marketing campaign might fare. Of course, temporal changes or other unique factors might not be fully considered in such an approach. However, a great deal of insight can often be gained through an analysis of past experiences. For example, in early 1998, the price of oil was about $22 a barrel. However, in mid-1998, the price of a barrel of oil dropped to around $11. The reasons for this price drop included an oversupply of oil from new production in the Caspian Sea region, high production in non-OPEC regions, and lower-than-normal demand. In similar circumstances in the past, OPEC would meet and take action to raise the price of oil. Thus, from historical analogy, we might forecast a rise in the price of oil. OPEC members did in fact meet in mid-1998 and agreed to cut their production, but nobody believed that they would actually cooperate effectively, and the price continued to drop for a time. Subsequently, in 2000, the price of oil rose dramatically, falling again in late 2001. Analogies often provide good forecasts, but you need to be careful to recognize new or different circumstances. Another analogy is international conflict relative to the price of oil. Should war break out, the price would be expected to rise, analogous to what it has done in the past.

The Delphi Method

A popular judgmental forecasting approach, called the **Delphi method**, uses a panel of experts, whose identities are typically kept confidential from one another, to respond to a sequence of questionnaires. After each round of responses, individual opinions, edited to ensure anonymity, are shared, allowing each to see what the other experts think. Seeing other experts' opinions helps to reinforce those in agreement and to influence those who did not agree to possibly consider other factors. In the next round, the experts revise their estimates, and the process is repeated, usually for no more than two or three rounds. The Delphi method promotes unbiased exchanges of ideas and discussion and usually results in some convergence of opinion. It is one of the better approaches to forecasting long-range trends and impacts.

Indicators and Indexes for Forecasting

Indicators and indexes generally play an important role in developing judgmental forecasts. Indicators are measures that are believed to influence the behavior of a variable we wish to forecast. By monitoring changes in indicators, we expect to gain insight about the future behavior of the variable to help forecast the future. For example, one variable that is important to the nation's economy is the Gross Domestic Product (GDP), which is a measure of the value of all goods and services produced in the United States. Despite its shortcomings (for instance, unpaid work such as housekeeping and child care is not measured; production of poor-quality output inflates the measure, as does work expended on corrective action), it is a practical and useful measure of economic performance. Like most time series, the GDP rises and falls in a cyclical fashion. Predicting future trends in the GDP is often done by analyzing *leading indicators*—series that tend to rise and fall some predictable length of time prior to the peaks and valleys of the GDP. One example of a leading indicator is the formation of business enterprises; as the rate of new businesses grows, one would expect the GDP to increase in the future. Other examples of leading indicators are the percent change in the money supply (M1) and net change in business loans. Other indicators, called *lagging indicators*, tend to have peaks and valleys that follow those of the GDP. Some lagging indicators are the Consumer Price Index, prime rate, business investment expenditures, or inventories on hand. The GDP can be used to predict future trends in these indicators.

Indicators are often combined quantitatively into an index. The direction of movement of all the selected indicators are weighted and combined, providing an index of overall expectation. For example, financial analysts use the Dow Jones Industrial Average as an index of general stock market performance. Indexes do not provide a complete forecast, but rather a better picture of direction of change, and thus play an important role in judgmental forecasting.

The Department of Commerce began an Index of Leading Indicators to help predict future economic performance. Components of the index include the following:

- average weekly hours, manufacturing
- average weekly initial claims, unemployment insurance
- new orders, consumer goods and materials
- vendor performance—slower deliveries
- new orders, nondefense capital goods
- building permits, private housing
- stock prices, 500 common stocks (Standard & Poor)
- money supply

- interest rate spread
- index of consumer expectations (University of Michigan)

Business Conditions Digest included more than 100 time series in seven economic areas. This publication was discontinued in March 1990, but information related to the Index of Leading Indicators was continued in *Survey of Current Business.* In December 1995, the U.S. Department of Commerce sold this data source to The Conference Board, which now markets the information under the title *Business Cycle Indicators*; information can be obtained at its Web site (www.conference-board.org). The site includes excellent current information about the calculation of the index, as well as its current components.

STATISTICAL FORECASTING MODELS

Many forecasts are based on analysis of historical time-series data and are predicated on the assumption that the future is an extrapolation of the past. We will assume that a time series consists of T periods of data, $A_t, t = 1, 2, \ldots, T$. A naive approach is to eyeball a **trend**—a gradual shift in the value of the time series—by visually examining a plot of the data. For instance, Figure 7.1 shows a chart of total energy production from the data in the Excel file *Energy Production & Consumption.* We see that energy production was rising quite rapidly during the 1960s; however, the slope appears to have decreased after 1970. It appears that production is increasing by about 500,000 each year and that this can provide a reasonable forecast provided that the trend continues.

Figure 7.1 Total Energy Production Time Series

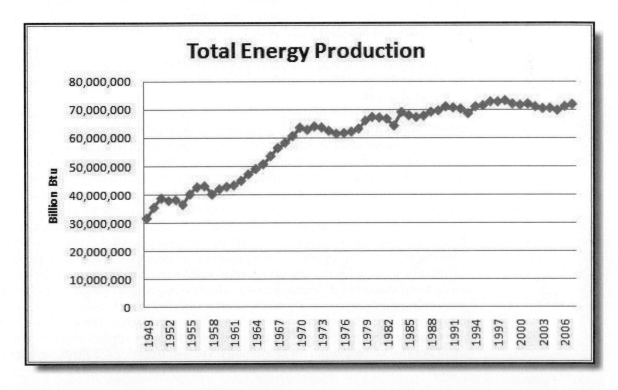

Figure 7.2 Federal Funds Rate Time Series

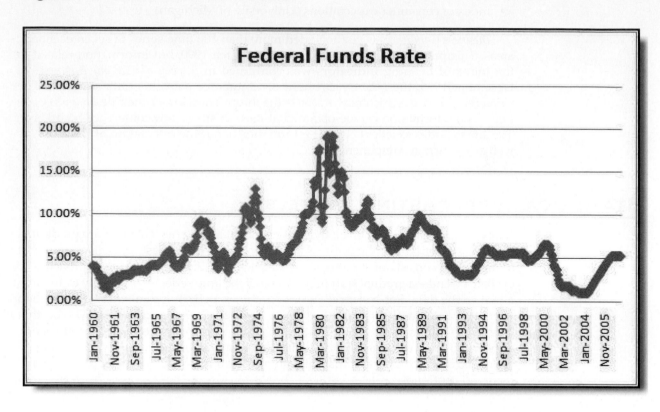

Time series may also exhibit short-term *seasonal effects* (over a year, month, week, or even a day) as well as longer-term *cyclical effects* or nonlinear trends. At a neighborhood grocery store, for instance, short-term seasonal patterns may occur over a week, with the heaviest volume of customers on weekends, and even during the course of a day. Cycles relate to much longer-term behavior, such as periods of inflation and recession or bull and bear stock market behavior. Figure 7.2 shows a chart of the data in the Excel file *Federal Funds Rate.* We see some evidence of long-term cycles in the time series.

Of course, unscientific approaches such as the "eyeball method" may be a bit unsettling to a manager making important decisions. Subtle effects and interactions of seasonal and cyclical factors may not be evident from simple visual extrapolation of data. Statistical methods, which involve more formal analyses of time series, are invaluable in developing good forecasts. A variety of statistically based forecasting methods for time series are commonly used. Among the most popular are *moving average methods*, *exponential smoothing*, and *regression analysis*. These can be implemented very easily on a spreadsheet using basic functions available in Microsoft Excel and its *Data Analysis* tools; these are summarized in Table 7.1. Moving average and exponential smoothing models work best for stationary time series. For time series that involve trends and/or seasonal factors, other techniques have been developed. These include double moving average and exponential smoothing models, seasonal additive and multiplicative models, and Holt–Winters additive and multiplicative models. We will review each of these types of models. This book provides an Excel add-in, *CB Predictor,* that applies these methods and incorporates some intelligent technology. We will describe *CB Predictor* later in this chapter.

Table 7.1
Excel Support for
Forecasting

EXCEL FUNCTIONS	DESCRIPTION
TREND (*known_y's, known_x's, new_x's, constant*)	Returns values along a linear trend line
LINEST(*known_y's, known_x's, new_x's, constant, stats*)	Returns an array that describes a straight line that best fits the data
FORECAST(*x, known_y's, known_x's*)	Calculates a future value along a linear trend

ANALYSIS TOOLPAK	DESCRIPTION
Moving average	Projects forecast values based on the average value of the variable over a specific number of preceding periods
Exponential smoothing	Predicts a value based on the forecast for the prior period, adjusted for the error in that prior forecast
Regression	Used to develop a model relating time-series data to a set of variables assumed to influence the data

FORECASTING MODELS FOR STATIONARY TIME SERIES

Two simple approaches that are useful over short time periods when trend, seasonal, or cyclical effects are not significant are moving average and exponential smoothing models.

Moving Average Models

The **simple moving average** method is based on the idea of averaging random fluctuations in the time series to identify the underlying direction in which the time series is changing. Because the moving average method assumes that future observations will be similar to the recent past, it is most useful as a short-range forecasting method. Although this method is very simple, it has proven to be quite useful in stable environments, such as inventory management, in which it is necessary to develop forecasts for a large number of items.

Specifically, the simple moving average forecast for the next period is computed as the average of the most recent k observations. The value of k is somewhat arbitrary, although its choice affects the accuracy of the forecast. The larger the value of k, the more the current forecast is dependent on older data; the smaller the value of k, the quicker the forecast responds to changes in the time series. (In the next section, we discuss how to select k by examining errors associated with different values.)

For instance, suppose that we want to forecast monthly burglaries from the Excel file *Burglaries* since the citizen-police program began. Figure 7.3 shows a chart of these data. The time series appears to be relatively stable, without trend, seasonal, or cyclical effects; thus, a moving average model would be appropriate. Setting $k = 3$, the three-period moving average forecast for month 59 is:

$$\text{Month 59 forecast} = \frac{82 + 71 + 50}{3} = 67.67$$

Moving average forecasts can be generated easily on a spreadsheet. Figure 7.4 shows the computations for a three-period moving average forecast of burglaries. Figure 7.5 shows a chart that contrasts the data with the forecasted values. Moving

Figure 7.3 Monthly Burglaries Chart

average forecasts can also be obtained from Excel's *Data Analysis* options (see *Excel Note: Forecasting with Moving Averages*).

In the simple moving average approach, the data are weighted equally. This may not be desirable because we might wish to put more weight on recent observations

Figure 7.4 Excel Implementation of Moving Average Forecast

	C	D	E	F
3	**After Citizen-Police Program**		**Moving Average**	
4	**Month**	**Monthly burglaries**	**Forecast**	
5	42	88		
6	43	44		
7	44	60		
8	45	56	64.00 ←	Forecast for month 45 =AVERAGE(D5:D7)
9	46	70	53.33	
10	47	91	62.00	
11	48	54	72.33	
12	49	60	71.67	
13	50	48	68.33	
14	51	35	54.00	
15	52	49	47.67	
16	53	44	44.00	
17	54	61	42.67	
18	55	68	51.33	
19	56	82	57.67	
20	57	71	70.33	
21	58	50	73.67	
22	59		67.67 ←	Forecast for month 59 =AVERAGE(D19:D21)

Figure 7.5 Chart of Burglaries and Moving Average Forecast

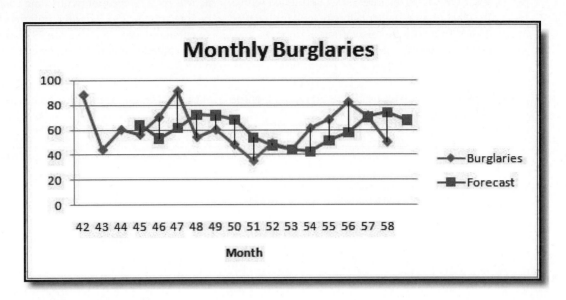

EXCEL NOTE
Forecasting with Moving Averages

From the *Analysis* group, select *Data Analysis* then *Moving Average*. Excel displays the dialog box shown in Figure 7.6. You need to enter the *Input Range* of the data, the *Interval* (the value of k), and the first cell of the *Output Range*. To align the actual data with the forecasted values in the worksheet, select the first cell of the *Output Range* to be one row below the first value. You may also obtain a chart of the data and the moving averages, as well as a column of standard errors, by checking the appropriate boxes. However, we do not recommend using the chart or error options because the forecasts generated by this tool are not properly

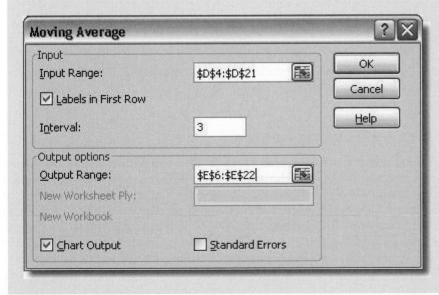

Figure 7.6 Excel *Moving Average* Tool Dialog

aligned with the data (the forecast value aligned with a particular data point represents the forecast for the *next* month) and, thus, can be misleading. Rather, we recommend that you generate your own chart as we did in Figure 7.5. Figure 7.7 shows the results produced by the *Moving Average* tool (with some customization of the forecast chart to show the months on the *x*-axis). Note that the forecast for month 59 is aligned with the actual value for month 58 on the chart. Compare this to Figure 7.5 and you can see the difference.

Figure 7.7 Results of Excel *Moving Average* Tool (note misalignment of forecasts with actual in the chart)

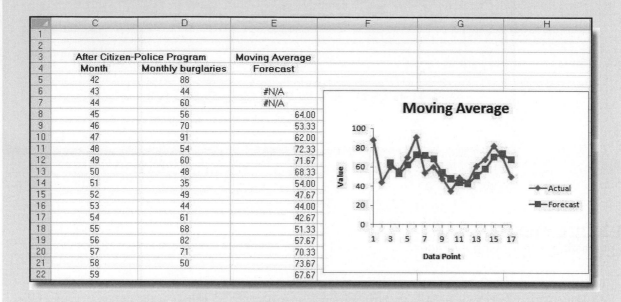

than on older observations, particularly if the time series is changing rapidly. Such models are called **weighted moving averages**. For example, you might assign a 60% weight to the most recent observation, 30% to the second most recent observation, and the remaining 10% of the weight to the third most recent observation. In this case, the three-period weighted moving average forecast for month 59 would be:

$$\text{Month 59 Forecast} = \frac{0.1 \times 82 + 0.3 \times 71 + 0.6 \times 50}{0.1 + 0.3 + 0.6} = \frac{59.5}{1} = 59.5$$

Different weights can easily be incorporated into Excel formulas. This leads us to the questions of how to measure forecast accuracy and also how to select the best parameters for a forecasting model.

Error Metrics and Forecast Accuracy

The quality of a forecast depends on how accurate it is in predicting future values of a time series. The error in a forecast is the difference between the forecast and the actual value of the time series (once it is known!). In Figure 7.5, the forecast error is simply the vertical distance between the forecast and the data for the same time period. In the simple moving average model, different values for k will produce different forecasts. How do we know, for example, if a two- or three-period moving average forecast or a three-period weighted moving average model (or others) would be the best predictor for burglaries? We might first generate different forecasts using each of these models, as shown in Figure 7.8, and compute the errors associated with each model.

Figure 7.8 Alternative Moving Average Forecasting Models

	A	B	C	D	E	F	G	H
1	After Citizen-Police Program						3 Period	
2	Month	Monthly burglaries	k = 2	Error	k = 3	Error	Weighted	Error
3	42	88						
4	43	44						
5	44	60	66.00	-6.00				
6	45	56	52.00	4.00	64.00	-8.00	58.00	-2.00
7	46	70	58.00	12.00	53.33	16.67	56.00	14.00
8	47	91	63.00	28.00	62.00	29.00	64.80	26.20
9	48	54	80.50	-26.50	72.33	-18.33	81.20	-27.20
10	49	60	72.50	-12.50	71.67	-11.67	66.70	-6.70
11	50	48	57.00	-9.00	68.33	-20.33	61.30	-13.30
12	51	35	54.00	-19.00	54.00	-19.00	52.20	-17.20
13	52	49	41.50	7.50	47.67	1.33	41.40	7.60
14	53	44	42.00	2.00	44.00	0.00	44.70	-0.70
15	54	61	46.50	14.50	42.67	18.33	44.60	16.40
16	55	68	52.50	15.50	51.33	16.67	54.70	13.30
17	56	82	64.50	17.50	57.67	24.33	63.50	18.50
18	57	71	75.00	-4.00	70.33	0.67	75.70	-4.70
19	58	50	76.50	-26.50	73.67	-23.67	74.00	-24.00
20	59		60.50		67.67		59.50	

To analyze the accuracy of these models, we can define *error metrics*, which compare quantitatively the forecast with the actual observations. Three metrics that are commonly used are the *mean absolute deviation, mean square error,* and *mean absolute percentage error.* The **mean absolute deviation (MAD)** is the absolute difference between the actual value and the forecast, averaged over a range of forecasted values:

$$\text{MAD} = \frac{\sum_{t=1}^{n} |A_t - F_t|}{n} \qquad (7.1)$$

where A_t is the actual value of the time series at time t, F_t is the forecast value for time t, and n is the number of forecast values (*not* the number of data points since we do not have a forecast value associated with the first k data points). MAD provides a robust measure of error and is less affected by extreme observations.

Mean square error (MSE) is probably the most commonly used error metric. It penalizes larger errors because squaring larger numbers has a greater impact than squaring smaller numbers. The formula for MSE is:

$$\text{MSE} = \frac{\sum_{t=1}^{n} (A_t - F_t)^2}{n} \qquad (7.2)$$

Again, n represents the number of forecast values used in computing the average. Sometimes the square root of MSE, called the **root mean square error (RMSE)**, is used.

	$k = 2$	$k = 3$	3-PERIOD WEIGHTED
MAD	13.63	14.86	13.70
MSE	254.38	299.84	256.31
MAPE	23.63%	26.53%	24.46%

Table 7.2 Error Metrics for Moving Average Models of Burglary Data

A third commonly used metric is **mean absolute percentage error (MAPE)**. MAPE is the average of absolute errors divided by actual observation values.

$$\text{MAPE} = \frac{\sum\limits_{t=1}^{n} \left| \dfrac{A_t - F_t}{A_t} \right|}{n} \times 100 \tag{7.3}$$

The values of MAD and MSE depend on the measurement scale of the time-series data. For example, forecasting profit in the range of millions of dollars would result in very large MAD and MSE values, even for very accurate forecasting models. On the other hand, market share is measured in proportions; therefore, even bad forecasting models will have small values of MAD and MSE. Thus, these measures have no meaning except in comparison with other models used to forecast the same data. Generally, MAD is less affected by extreme observations and is preferable to MSE if such extreme observations are considered rare events with no special meaning. MAPE is different in that the measurement scale is eliminated by dividing the absolute error by the time-series data value. This allows a better relative comparison. Although these comments provide some guidelines, there is no universal agreement on which measure is best.

These measures can be used to compare the moving average forecasts in Figure 7.8. The results, shown in Table 7.2, verify that the two-period moving average model provides the best forecast among these alternatives.

Exponential Smoothing Models

A versatile, yet highly effective approach for short-range forecasting is **simple exponential smoothing**. The basic simple exponential smoothing model is:

$$\begin{aligned} F_{t+1} &= (1 - \alpha)F_t + \alpha A_t \\ &= F_t + \alpha(A_t - F_t) \end{aligned} \tag{7.4}$$

where F_{t+1} is the forecast for time period $t + 1$, F_t is the forecast for period t, A_t is the observed value in period t, and α is a constant between 0 and 1, called the **smoothing constant**. To begin, the forecast for period 2 is set equal to the actual observation for period 1.

Using the two forms of the forecast equation just given, we can interpret the simple exponential smoothing model in two ways. In the first model, the forecast for the next period, F_{t+1}, is a weighted average of the forecast made for period t, F_t, and the actual observation in period t, A_t. The second form of the model, obtained by simply rearranging terms, states that the forecast for the next period, F_{t+1}, equals the forecast for the last period, F_t, plus a fraction α of the forecast error made in period t, $A_t - F_t$. Thus, to make a forecast once we have selected the smoothing constant, we need only know the previous forecast and the actual value. By repeated substitution for F_t in the equation, it is easy to demonstrate that F_{t+1} is a decreasingly weighted average of all past time-series data. Thus, the forecast actually reflects *all* the data, provided that α is strictly between 0 and 1.

For the burglary data, the forecast for month 43 is 88, the actual observation for month 42. Suppose we choose $\alpha = 0.7$; then the forecast for month 44 would be:

$$\text{Month 44 Forecast} = (1 - 0.7)(88) + (0.7)(44) = 57.2$$

The actual observation for month 44 is 60; thus, the forecast for month 45 would be:

$$\text{Month 45 Forecast} = (1 - 0.7)(57.2) + (0.7)(60) = 59.16$$

Since the simple exponential smoothing model requires only the previous forecast and the current time-series value, it is very easy to calculate; thus, it is highly suitable for environments such as inventory systems where many forecasts must be made. The smoothing constant α is usually chosen by experimentation in the same manner as choosing the number of periods to use in the moving average model. Different values of α affect how quickly the model responds to changes in the time series. For instance, a value of $\alpha = 0$ would simply repeat last period's forecast, while $\alpha = 1$ would forecast last period's actual demand. The closer α is to 1, the quicker the model responds to changes in the time series because it puts more weight on the actual current observation than on the forecast. Likewise, the closer α is to 0, the more weight is put on the prior forecast, so the model would respond to changes more slowly.

An Excel spreadsheet for evaluating exponential smoothing models for the burglary data using values of α between 0.1 and 0.9 is shown in Figure 7.9. A smoothing constant of $\alpha = 0.6$ provides the lowest error for all three metrics. Excel has a *Data Analysis* tool for exponential smoothing (see *Excel Note: Forecasting with Exponential Smoothing*).

Figure 7.9 Exponential Smoothing Forecasts for *Burglaries*

	C	D	E	F	G	H	I	J	K	L	M
3	After Citizen-Police Program		Smoothing Constant								
4	Month	Monthly burglaries	0.1	0.2	0.3	0.4	0.5	0.6	0.7	0.8	0.9
5	42	88	88.00	88.00	88.00	88.00	88.00	88.00	88.00	88.00	88.00
6	43	44	88.00	88.00	88.00	88.00	88.00	88.00	88.00	88.00	88.00
7	44	60	83.60	79.20	74.80	70.40	66.00	61.60	57.20	52.80	48.40
8	45	56	81.24	75.36	70.36	66.24	63.00	60.64	59.16	58.56	58.84
9	46	70	78.72	71.49	66.05	62.14	59.50	57.86	56.95	56.51	56.28
10	47	91	77.84	71.19	67.24	65.29	64.75	65.14	66.08	67.30	68.63
11	48	54	79.16	75.15	74.37	75.57	77.88	80.66	83.53	86.26	88.76
12	49	60	76.64	70.92	68.26	66.94	65.94	64.66	62.86	60.45	57.48
13	50	48	74.98	68.74	65.78	64.17	62.97	61.87	60.86	60.09	59.75
14	51	35	72.28	64.59	60.45	57.70	55.48	53.55	51.86	50.42	49.17
15	52	49	68.55	58.67	52.81	48.62	45.24	42.42	40.06	38.08	36.42
16	53	44	66.60	56.74	51.67	48.77	47.12	46.37	46.32	46.82	47.74
17	54	61	64.34	54.19	49.37	46.86	45.56	44.95	44.70	44.56	44.37
18	55	68	64.00	55.55	52.86	52.52	53.28	54.58	56.11	57.71	59.34
19	56	82	64.40	58.04	57.40	58.71	60.64	62.63	64.43	65.94	67.13
20	57	71	66.16	62.83	64.78	68.03	71.32	74.25	76.73	78.79	80.51
21	58	50	66.65	64.47	66.65	69.22	71.16	72.30	72.72	72.56	71.95
22	59		64.98	61.57	61.65	61.53	60.58	58.92	56.82	54.51	52.20
23		MAD	19.33	17.16	16.15	15.36	14.93	14.71	14.72	14.88	15.36
24		MSE	496.07	390.84	359.18	346.56	340.77	338.41	339.03	343.32	352.38
25		MAPE	38.28%	32.71%	30.12%	28.36%	27.54%	27.09%	27.09%	27.38%	28.23%

EXCEL NOTE
Forecasting with Exponential Smoothing

From the *Analysis* group, select *Data Analysis* then *Exponential Smoothing*. In the dialog (Figure 7.10), as in the *Moving Average* dialog, you must enter the *Input Range* of the time-series data, the *Damping Factor* $(1 - \alpha)$—*not* the smoothing constant as we have defined it (!)—and the first cell of the *Output Range*, which should be adjacent to the first data point. You also have options for labels, to chart output, and to obtain standard errors. As opposed to the *Moving Average* tool, the chart generated by this tool does correctly align forecasts with the actual data, as shown in Figure 7.11. You can see that the exponential smoothing model follows the pattern of the data quite closely, although it tends to lag with an increasing trend in the data.

Figure 7.10 *Exponential Smoothing* Tool Dialog

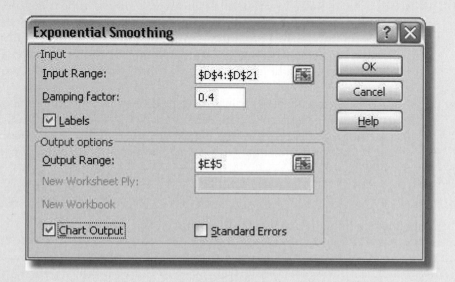

Figure 7.11 Exponential Smoothing Forecasts for $\alpha = 0.6$

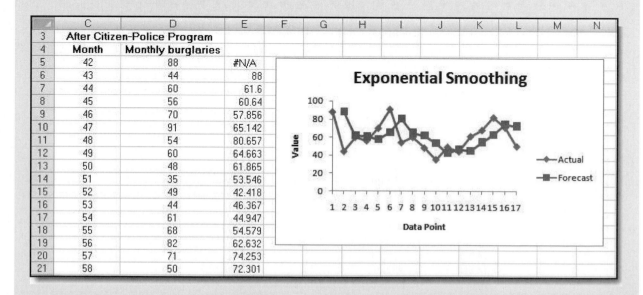

	C	D	E
3	After Citizen-Police Program		
4	Month	Monthly burglaries	
5	42	88	#N/A
6	43	44	88
7	44	60	61.6
8	45	56	60.64
9	46	70	57.856
10	47	91	65.142
11	48	54	80.657
12	49	60	64.663
13	50	48	61.865
14	51	35	53.546
15	52	49	42.418
16	53	44	46.367
17	54	61	44.947
18	55	68	54.579
19	56	82	62.632
20	57	71	74.253
21	58	50	72.301

FORECASTING MODELS FOR TIME SERIES WITH TREND AND SEASONALITY

When time series exhibit trend and/or seasonality, different techniques provide better forecasts than the basic moving average and exponential smoothing models we have described. The computational theory behind these models are presented in the appendix to this chapter as they are quite a bit more complicated than the simple moving average and exponential smoothing models. However, a basic understanding of these techniques is useful in order to apply *CB Predictor* software for forecasting, which we introduce in the next section.

Models for Linear Trends

For time series with a linear trend but no significant seasonal components, **double moving average** and **double exponential smoothing** models are more appropriate. Both methods are based on the linear trend equation:

$$F_{t+k} = a_t + b_t k \tag{7.5}$$

This may look familiar from simple linear regression. That is, the forecast for k periods into the future from period t is a function of a base value a_t, also known as the *level*, and a *trend*, or slope, b_t. Double moving average and double exponential smoothing differ in how the data are used to arrive at appropriate values for a_t and b_t.

Models for Seasonality

Seasonal factors (with no trend) can be incorporated into a forecast by adjusting the level, a_t, in one of two ways. The **seasonal additive** model is:

$$F_{t+k} = a_t + S_{t-s+k} \tag{7.6}$$

and the **seasonal multiplicative** model is:

$$F_{t+k} = a_t S_{t-s+k} \tag{7.7}$$

In both models, S_{t-s+k} is the seasonal factor for period $t - s + k$ and s is the number of periods in a season. A "season" can be a year, quarter, month, or even a week, depending on the application. In any case, the forecast for period $t + k$ is adjusted up or down from a level (a_t) by the seasonal factor. The multiplicative model is more appropriate when the seasonal factors are increasing or decreasing over time. This is evident when the amplitude of the time series changes over time.

Models for Trend and Seasonality

Many time series exhibit both trend and seasonality. Such might be the case for growing sales of a seasonal product. The methods we describe are based on the work of two researchers, C.C. Holt, who developed the basic approach, and P.R. Winters, who extended Holt's work. Hence, these approaches are commonly referred to as **Holt–Winters models**. These models combine elements of both the trend and seasonal models described above. The Holt-Winters additive model is based on the equation:

$$F_{t+1} = a_t + b_t + S_{t-s+1} \tag{7.8}$$

	NO SEASONALITY	SEASONALITY
No Trend	Single moving average or single exponential smoothing	Seasonal additive or seasonal multiplicative model
Trend	Double moving average or double exponential smoothing	Holt–Winters additive or Holt–Winters multiplicative model

Table 7.3
Forecasting Model Choice

and the Holt-Winters multiplicative model is:

$$F_{t+1} = (a_t + b_t)S_{t-s+1} \qquad (7.9)$$

The additive model applies to time series with relatively stable seasonality, while the multiplicative model applies to time series whose amplitude increases or decreases over time.

Table 7.3 summarizes the choice of models based on characteristics of the time series.

CHOOSING AND OPTIMIZING FORECASTING MODELS USING *CB PREDICTOR*

CB Predictor is an Excel add-in for forecasting that is part of the *Crystal Ball* suite of applications. We introduced *Crystal Ball* for distribution fitting in Chapter 3. *CB Predictor* can be used as a stand-alone program for forecasting, and can also be integrated with Monte Carlo simulation, which we discuss in Chapter 10. *CB Predictor* includes all the time-series forecasting approaches we have discussed. See *Excel Note: Using CB Predictor* for basic information on using the add-in.

We will illustrate the use of *CB Predictor* first for the data in the worksheet *Burglaries* after the citizen-police program commenced. Only the single moving

EXCEL NOTE
Using *CB Predictor*

After *Crystal Ball* has been installed, *CB Predictor* may be accessed in Excel from the *Crystal Ball* tab. Click on the *Tools* menu and then *CB Predictor*. *CB Predictor* guides you through four dialog boxes, the first of which is shown in Figure 7.12. These can be selected by clicking the *Next* button or by clicking on the tabs. *Input Data* allows you to specify the data range on which to base your forecast; *Data Attributes* allows you to specify the type of data and whether or not seasonality is present (see Figure 7.13); *Method Gallery* allows you to select one or more of eight time-series methods—single moving average, double moving average, single exponential smoothing, double exponential smoothing, seasonal additive, seasonal multiplicative, Holt–Winters additive, or Holt–Winters multiplicative (see Figure 7.14). The charts shown in the *Method Gallery* suggest the method that is best suited for the data similar to Table 7.3. However, *CB Predictor* can run each method you select and will recommend the one that best forecasts your data. Not only does it select the best type of model, it also optimizes the forecasting parameters to minimize forecasting errors. The *Advanced* button allows you to change the error metric on which the models are ranked. The final dialog, *Results*, allows you to specify a variety of reporting options (see Figure 7.15). The *Preferences* button allows you to customize these results.

Figure 7.12 *CB Predictor Input Data* Dialog

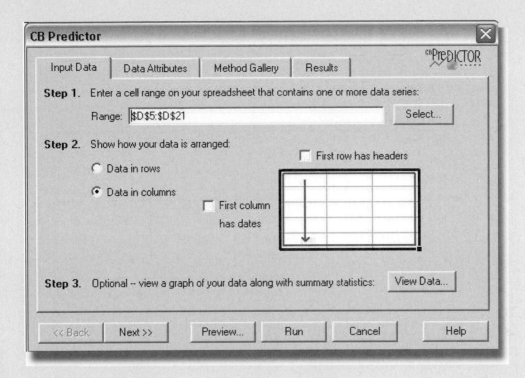

Figure 7.13 *CB Predictor Data Attributes* Dialog

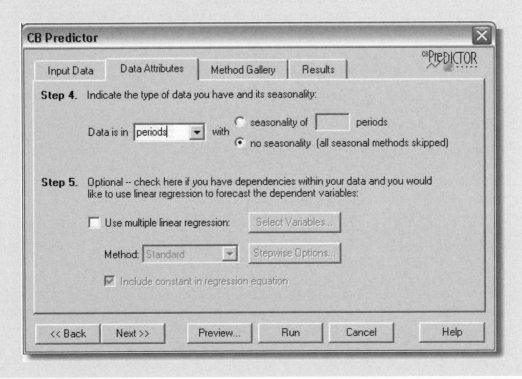

Figure 7.14 *CB Predictor Method Gallery* Dialog

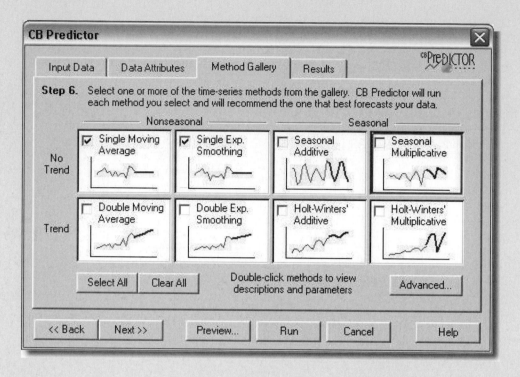

Figure 7.15 *CB Predictor Results* Dialog

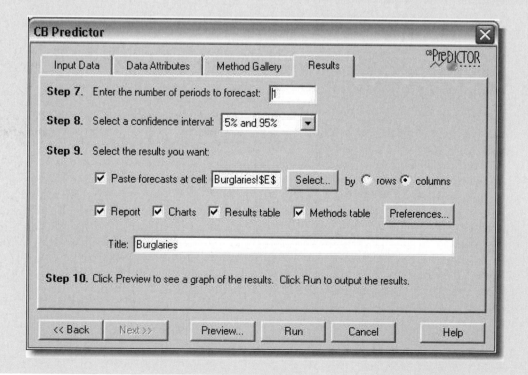

Figure 7.16 *CB Predictor* Output—Methods Table

	A	B	C	D	E	F	G	H	I
1	**Methods Table for Burglaries - Academic Edition (not for commercial use)**								
2	Created: 9/3/2008 at 12:56:30 PM								
3									
19	Series	ColumnD							
20									
21		Table Items							
22	Methods	Rank	RMSE	MAD	MAPE	Durbin-Watson	Theil's U	Periods	Alpha
23	Single Exponential Smoothing	2	18.392	14.612	26.917	1.448	0.93		0.631
24	**Single Moving Average**	**1**	**15.949**	**13.633**	**23.626**	**1.457**	**0.959**	**2**	

average and single exponential methods were chosen in the *Method Gallery* for this example. *CB Predictor* creates a worksheet for each of the results checked in the *Results* dialog. Figure 7.16 shows the Methods Table, which summarizes the forecasting methods used and ranks them according to the lowest RMSE error

Figure 7.17 *CB Predictor* Output—Results Table

	A	B	C	D	E	F
1	**Results Table for Burglaries - Academic Edition (not for commercial use)**					
2	Created: 9/3/2008 at 12:56:30 PM					
3						
23	Series	ColumnD				
24						
25		Data				
26	Date	Historical Data	Lower: 5%	Fit & Forecast	Upper: 95%	Residuals
27	Period 1	88				88
28	Period 2	44				44
29	Period 3	60		66		-6
30	Period 4	56		52		4
31	Period 5	70		58		12
32	Period 6	91		63		28
33	Period 7	54		80.5		-26.5
34	Period 8	60		72.5		-12.5
35	Period 9	48		57		-9
36	Period 10	35		54		-19
37	Period 11	49		41.5		7.5
38	Period 12	44		42		2
39	Period 13	61		46.5		14.5
40	Period 14	68		52.5		15.5
41	Period 15	82		64.5		17.5
42	Period 16	71		75		-4
43	Period 17	50		76.5		-26.5
44	Period 18		34.26323839	60.5	86.73676161	

criterion. In this example, *CB Predictor* found the best fit to be a 2-period moving average. This method was also the best for the MAD and MAPE error metrics. The Durbin–Watson statistic checks for autocorrelation (see the discussion of autocorrelation in regression in Chapter 6), with values of 2 indicating no autocorrelation. Theil's U statistic is a relative error measure that compares the results with a naive forecast. A value less than 1 means that the forecasting technique is better than guessing, a value equal to 1 means that the technique is about as good as guessing, and a value greater than 1 means that the forecasting technique is worse than guessing. Note that *CB Predictor* identifies the best number of periods for the moving average or the best smoothing constants as appropriate. For instance, in Figure 7.16, we see that the best-fitting single exponential smoothing model has alpha = 0.631.

The Results Table (Figure 7.17) provides the historical data, fitted forecasts, and residuals. For future forecasts, it also provides a confidence interval based on Step 8 in the *Results* dialog. Thus, the forecast for month 59 is 60.5, with a 95% confidence interval between 34.26 and 86.74. *CB Predictor* also creates a chart showing the data and fitted forecasts, and a summary report of all results.

As a second example, the data in the Excel file *Gas & Electric* provides two years of data for natural gas and electric usage for a residential property (see Figure 7.18). In the *Data Attributes* tab of *CB Predictor*, we select a seasonality of 12 months. Although the data are clearly seasonal, we will select all the time-series methods in the *Method Gallery* tab. Figure 7.19 shows the results. In this example the Seasonal Multiplicative method was ranked first, although you will notice that the top four methods provide essentially the same quality of results. Figure 7.20 shows the forecasts generated for the next 12 months.

Figure 7.18   *Gas & Electric* Data

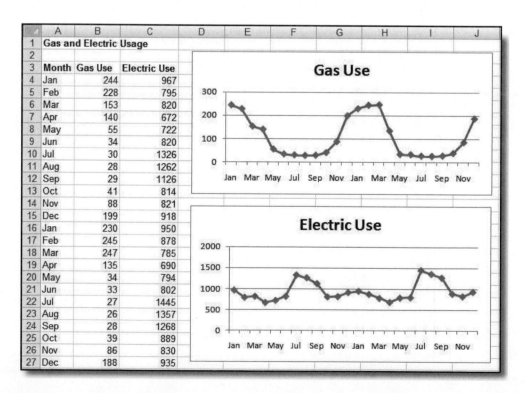

Figure 7.19 Methods Table for Gas Use

	A	B	C	D	E	F	G	H	I	J	K
1	**Methods Table for Gas Use - Academic Edition (not for commercial use)**										
2	Created: 9/3/2008 at 2:07:00 PM										
3											
68	Series	Gas Use									
69											
70		Table Items									
71	Methods	Rank		RMSE	MAD	MAPE	Durbin-Watson	Theil's U	Periods	Alpha	Beta	Gamma
72	Double Exponential Smoothing	5	50.055	39.205	56.673	1.866	1.368		0.999	0.807	
73	Double Moving Average	8	57.467	44.655	76.398	1.237	1.795	2			
74	Holt-Winters' Additive	4	28.779	14.472	13.07	1.756	0.241		0.001	0.001	0.618
75	Holt-Winters' Multiplicative	2	28.77	14.427	12.955	1.757	0.24		0.001	0.001	0.618
76	Seasonal Additive	3	28.779	14.471	13.069	1.756	0.241		0.001		0.618
77	**Seasonal Multiplicative**	**1**	**28.77**	**14.427**	**12.955**	**1.757**	**0.24**		**0.001**		**0.618**
78	Single Exponential Smoothing	7	53.237	35.596	44.305	0.87	1		0.999		
79	Single Moving Average	6	53.211	35.565	44.252	0.871	1	1			

	A	B
28	Jan	**235.32048**
29	Feb	**238.4683**
30	Mar	**210.96728**
31	Apr	**136.83973**
32	May	**42.01053**
33	Jun	**33.373238**
34	Jul	**28.140303**
35	Aug	**26.759744**
36	Sep	**28.378119**
37	Oct	**39.759713**
38	Nov	**86.755971**
39	Dec	**192.19026**

Figure 7.20 Gas Use Forecasts

REGRESSION MODELS FOR FORECASTING

We introduced regression in the previous chapter as a means of developing relationships between dependent and independent variables. Simple linear regression can be applied to forecasting using time as the independent variable. For example, Figure 7.21 shows a portion of the Excel file *Coal Production*, which provides data on total tons produced from 1960 through 2007. A linear trendline shows an R^2 value of 0.969 (the fitted model assumes that the years are numbered 1 through 48, not as actual dates). The actual values of the coefficients in the model:

$$\text{Tons} = 416,896,322.7 + 16,685,398.57 \times \text{Year}$$

Thus, a forecast for 2008 would be:

$$\text{Tons} = 416,896,322.7 + 16,685,398.57 \times (49) = 1,234,480,853$$

CB Predictor can also use linear regression for forecasting, and provides additional information. To apply it, first add a column to the spreadsheet to number the years beginning with 1 (corresponding to 1960). In Step 1 of the *Input Data* tab, select

Figure 7.21 Portion of *Coal Production*

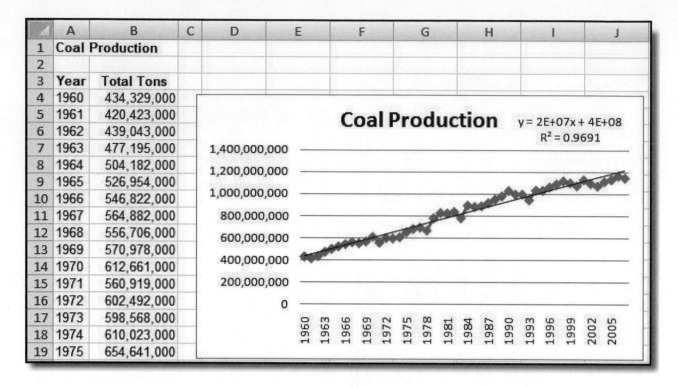

the ranges of both this new Year column and Total Tons. In the *Data Attributes* tab, check the box for multiple linear regression in Step 5, and click the *Select Variables* button; this will allow you to specify which are the independent and dependent variables. Figure 7.22 shows a portion of the output showing forecasts for the next 5 years and 95% confidence intervals. However, note that the Durbin–Watson statistic (see Chapter 6) suggests that the data are autocorrelated, indicating that other approaches, called *autoregressive models,* are more appropriate.

Autoregressive Forecasting Models

An autoregressive forecasting model incorporates correlations between consecutive values in a time series. A first-order autocorrelation refers to the correlation among data values one period apart, a second-order autocorrelation refers to the correlation among data values two periods apart, and so on. Autoregressive models improve forecasting when autocorrelation is present in data. A first-order autoregressive model is:

$$Y_i = a_0 + a_1 Y_{i-1} + \delta_i \tag{7.10}$$

where Y_i is the value of the time series in period i and δ_i is a nonautocorrelated random error term having 0 mean and constant variance. A second-order autoregressive model is:

$$Y_i = a_0 + a_1 Y_{i-1} + a_2 Y_{i-2} + \delta_i \tag{7.11}$$

Additional terms may be added for higher-order models.

To build an autoregressive model using multiple linear regression, we simply add additional columns to the data matrix for the dependent variable that lag the original data by some number of periods. Thus, for a second-order autoregressive model, we add columns that lag the dependent variable by one and two periods. For

Figure 7.22 Portion of *CB Predictor* Output for Regression Forecasting

Series: Total Tons   **Range: C4:C51**

Method: Multiple Linear Regression

Statistics:
    R-squared: 0.969
    Adjusted R-squared: 0.9685
    SSE: 8.16E+16
    F Statistic: 1444.3
    F Probability: 2.15E-36
    Durbin-Watson: 0.768
    No. of Values: 48
    Independent variables: 1 included out of 1 selected

Series Statistics:
    Mean: 825,688,588
    Std. Dev.: 237,286,293
    Minimum: 420,423,000
    Maximum: 1,162,749,659
    Ljung-Box: 293.9327

Forecast:

Date	Lower: 5%	Forecast	Upper: 95%
Period 49	1,165,162,688	1,234,480,853	1,303,799,017
Period 50	1,181,848,086	1,251,166,251	1,320,484,416
Period 51	1,198,533,485	1,267,851,650	1,337,169,814
Period 52	1,215,218,884	1,284,537,048	1,353,855,213
Period 53	1,231,904,282	1,301,222,447	1,370,540,611

the coal production data, a portion of this data matrix is shown in Figure 7.23. Using these additional columns as independent variables, we run the multiple regression tool, obtaining the results shown in Figure 7.24.

Note that the *p*-value for the second-order term exceeds 0.05 (although not by much), indicating that this variable is not significant. Dropping it and re-running the regression using only the first-order term results in the model shown in Figure 7.25. However, the adjusted R^2 is less than that of the second-order model, indicating a poorer fit. Thus, we use the second-order model:

$$\text{Tons} = 136{,}892{,}640 + 0.608 \times (\text{Year} - 1) + 0.259 \times (\text{Year} - 2)$$

A forecast for year 49 (2008) would be:

$$\begin{aligned}
\text{Tons} &= 136{,}892{,}640 + 0.608 \times 1{,}162{,}749{,}659 + 0.259 \times 1{,}131{,}498{,}099 \\
&= 1{,}136{,}902{,}440
\end{aligned}$$

A forecast for year 50 (2009) would be:

$$\begin{aligned}
\text{Tons} &= 136{,}892{,}640 + 0.608 \times 1{,}136{,}902{,}440 + 0.259 \times 1{,}162{,}749{,}659 \\
&= 1{,}129{,}281{,}485
\end{aligned}$$

Figure 7.23 Portion of Data Matrix for Autoregressive Forecasting of *Coal Production* Data

	A	B	C	D
3	Year	Total Tons	Year - 1	Year - 2
4	1960	434,329,000	0	0
5	1961	420,423,000	434,329,000	0
6	1962	439,043,000	420,423,000	434,329,000
7	1963	477,195,000	439,043,000	420,423,000
8	1964	504,182,000	477,195,000	439,043,000
9	1965	526,954,000	504,182,000	477,195,000
10	1966	546,822,000	526,954,000	504,182,000

Incorporating Seasonality in Regression Models

Quite often time-series data exhibit seasonality, especially on an annual basis, as we saw in the *Gas & Electric* data. Multiple linear regression models with categorical variables can be used for time series with seasonality. To do this, we use dummy categorical variables for the seasonal components. With monthly data, as we have for

Figure 7.24 Results for Second-Order Autoregressive Model

	A	B	C	D	E	F	G
1	SUMMARY OUTPUT						
2							
3	*Regression Statistics*						
4	Multiple R	0.967152069					
5	R Square	0.935383124					
6	Adjusted R Square	0.932511263					
7	Standard Error	61643644.02					
8	Observations	48					
9							
10	ANOVA						
11		*df*	*SS*	*MS*	*F*	*Significance F*	
12	Regression	2	2.47533E+18	1.23766E+18	325.7062473	1.70917E-27	
13	Residual	45	1.70997E+17	3.79994E+15			
14	Total	47	2.64632E+18				
15							
16		*Coefficients*	*Standard Error*	*t Stat*	*P-value*	*Lower 95%*	*Upper 95%*
17	Intercept	136892640.5	29908384.85	4.577065635	3.7061E-05	76654062.1	197131218.9
18	Year - 1	0.608116691	0.138335098	4.395968195	6.6595E-05	0.329495505	0.886737876
19	Year - 2	0.258737894	0.128915436	2.007035796	0.050775425	-0.000911118	0.518386906

Figure 7.25 First-Order Autoregressive Forecasting Model

	A	B	C	D	E	F	G
1	SUMMARY OUTPUT						
2							
3	*Regression Statistics*						
4	Multiple R	0.964157103					
5	R Square	0.929598919					
6	Adjusted R Square	0.92806846					
7	Standard Error	63640316.39					
8	Observations	48					
9							
10	ANOVA						
11		*df*	*SS*	*MS*	*F*	*Significance F*	
12	Regression	1	2.46002E+18	2.46002E+18	607.3990547	3.78177E-28	
13	Residual	46	1.86304E+17	4.05009E+15			
14	Total	47	2.64632E+18				
15							
16		*Coefficients*	*Standard Error*	*t Stat*	*P-value*	*Lower 95%*	*Upper 95%*
17	Intercept	122487305.3	29974836.48	4.086337731	0.000173734	62151089.81	182823520.8
18	Year - 1	0.877003564	0.035584781	24.64546722	3.78177E-28	0.805375115	0.948632013

natural gas usage, we have a seasonal categorical variable with $k = 12$ levels. As discussed in Chapter 6, we construct the regression model using $k - 1$ dummy variables. We will use January as the reference month; therefore, this variable does not appear in the model:

$$\text{Gas usage} = \beta_0 + \beta_1 \text{Time} + \beta_2 \text{February} + \beta_3 \text{March} + \beta_4 \text{April} + \beta_5 \text{May}$$
$$+ \beta_6 \text{June} + \beta_7 \text{July} + \beta_8 \text{August} + \beta_9 \text{September} + \beta_{10} \text{October}$$
$$+ \beta_{11} \text{November} + \beta_{12} \text{December}$$

This coding scheme results in the data matrix shown in Figure 7.26. This model picks up trends from the regression coefficient for time, and seasonality from the dummy variables for each month. The forecast for the next January will be $\beta_0 + \beta_1(25)$. The variable coefficients (betas) for each of the other 11 months will show the adjustment relative to January. For example, forecast for next February would be $\beta_0 + \beta_1(25) + \beta_2(1)$, and so on.

Figure 7.27 shows the results of using the *Regression* tool in Excel after eliminating insignificant variables (Time and Feb). Because the data shows no clear linear trend, the variable Time could not explain any significant variation in the data. The dummy variable for February was probably insignificant because the historical gas usage for both January and February were very close to each other. The R^2 for this model is 0.971, which is very good. The final regression model is:

$$\text{Gas Usage} = 236.75 - 36.75 \text{ March} - 99.25 \text{ April} - 192.25 \text{ May} - 203.25 \text{ June}$$
$$- 208.25 \text{ July} - 209.75 \text{ August} - 208.25 \text{ September} - 196.75 \text{ October}$$
$$- 149.75 \text{ November} - 43.25 \text{ December}$$

Figure 7.26 Data Matrix for Seasonal Regression Model

	A	B	C	D	E	F	G	H	I	J	K	L	M	N
1	**Gas and Electric Usage**													
2														
3	**Month**	**Gas Use**	**Time**	**Feb**	**Mar**	**Apr**	**May**	**Jun**	**Jul**	**Aug**	**Sep**	**Oct**	**Nov**	**Dec**
4	Jan	244	1	0	0	0	0	0	0	0	0	0	0	0
5	Feb	228	2	1	0	0	0	0	0	0	0	0	0	0
6	Mar	153	3	0	1	0	0	0	0	0	0	0	0	0
7	Apr	140	4	0	0	1	0	0	0	0	0	0	0	0
8	May	55	5	0	0	0	1	0	0	0	0	0	0	0
9	Jun	34	6	0	0	0	0	1	0	0	0	0	0	0
10	Jul	30	7	0	0	0	0	0	1	0	0	0	0	0
11	Aug	28	8	0	0	0	0	0	0	1	0	0	0	0
12	Sep	29	9	0	0	0	0	0	0	0	1	0	0	0
13	Oct	41	10	0	0	0	0	0	0	0	0	1	0	0
14	Nov	88	11	0	0	0	0	0	0	0	0	0	1	0
15	Dec	199	12	0	0	0	0	0	0	0	0	0	0	1
16	Jan	230	13	0	0	0	0	0	0	0	0	0	0	0
17	Feb	245	14	1	0	0	0	0	0	0	0	0	0	0
18	Mar	247	15	0	1	0	0	0	0	0	0	0	0	0
19	Apr	135	16	0	0	1	0	0	0	0	0	0	0	0
20	May	34	17	0	0	0	1	0	0	0	0	0	0	0
21	Jun	33	18	0	0	0	0	1	0	0	0	0	0	0
22	Jul	27	19	0	0	0	0	0	1	0	0	0	0	0
23	Aug	26	20	0	0	0	0	0	0	1	0	0	0	0
24	Sep	28	21	0	0	0	0	0	0	0	1	0	0	0
25	Oct	39	22	0	0	0	0	0	0	0	0	1	0	0
26	Nov	86	23	0	0	0	0	0	0	0	0	0	1	0
27	Dec	188	24	0	0	0	0	0	0	0	0	0	0	1

Regression Forecasting with Causal Variables

In many forecasting applications, other independent variables such as economic indexes or demographic factors may influence the time series, and can be incorporated into a regression model. For example, a manufacturer of hospital equipment might include such variables as hospital capital spending and changes in the proportion of people over the age of 65 in building models to forecast future sales.

To illustrate the use of multiple linear regression for forecasting with causal variables, suppose that we wish to forecast gasoline sales. Figure 7.28 shows the sales over 10 weeks during June through August along with the average price per gallon and a chart of the gasoline sales time series with a fitted trendline (Excel file *Gasoline Sales*). During the summer months, it is not unusual to see an increase in sales as more people go on vacations. The chart shows a linear trend, although R^2 is not very high.

The trendline is:

$$Sales = 4790.1 + 812.99 \text{ Week}$$

Figure 7.27 Final Regression Model for Forecasting Gas Use

	A	B	C	D	E	F	G
1	SUMMARY OUTPUT						
2							
3	*Regression Statistics*						
4	Multiple R	0.985480895					
5	R Square	0.971172595					
6	Adjusted R Square	0.948997667					
7	Standard Error	19.54432831					
8	Observations	24					
9							
10	ANOVA						
11		*df*	*SS*	*MS*	*F*	*Significance F*	
12	Regression	10	167292.2083	16729.22083	43.79597661	2.33344E-08	
13	Residual	13	4965.75	381.9807692			
14	Total	23	172257.9583				
15							
16		*Coefficients*	*Standard Error*	*t Stat*	*P-value*	*Lower 95%*	*Upper 95%*
17	Intercept	236.75	9.772164157	24.22697738	3.33921E-12	215.6385229	257.8614771
18	Mar	-36.75	16.92588482	-2.171230656	0.04901621	-73.31615098	-0.183849024
19	Apr	-99.25	16.92588482	-5.863799799	5.55744E-05	-135.816151	-62.68384902
20	May	-192.25	16.92588482	-11.35834268	4.02824E-08	-228.816151	-155.683849
21	Jun	-203.25	16.92588482	-12.00823485	2.07264E-08	-239.816151	-166.683849
22	Jul	-208.25	16.92588482	-12.30364038	1.54767E-08	-244.816151	-171.683849
23	Aug	-209.75	16.92588482	-12.39226204	1.41949E-08	-246.316151	-173.183849
24	Sep	-208.25	16.92588482	-12.30364038	1.54767E-08	-244.816151	-171.683849
25	Oct	-196.75	16.92588482	-11.62420766	3.05791E-08	-233.316151	-160.183849
26	Nov	-149.75	16.92588482	-8.847395666	7.30451E-07	-186.316151	-113.183849
27	Dec	-43.25	16.92588482	-2.555257847	0.023953114	-79.81615098	-6.683849024

Figure 7.28 *Gasoline Sales* Data and Trendline

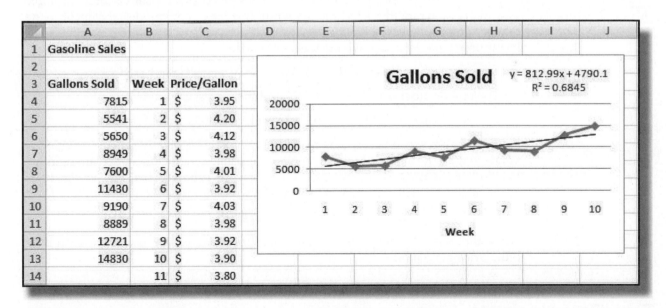

Using this model, we would predict sales for week 11 as:

$$\text{Sales} = 4790.1 + 812.99(11) = 13{,}733 \text{ gallons}$$

However, we also see that the average price per gallon changes each week, and this may influence consumer sales. Therefore, the sales trend might not simply be a factor

Figure 7.29 Regression Results for Gas Sales

	A	B	C	D	E	F	G
1	SUMMARY OUTPUT						
2							
3	*Regression Statistics*						
4	Multiple R	0.930528528					
5	R Square	0.865883342					
6	Adjusted R Square	0.827564297					
7	Standard Error	1235.400329					
8	Observations	10					
9							
10	ANOVA						
11		*df*	*SS*	*MS*	*F*	*Significance F*	
12	Regression	2	68974748.7	34487374.35	22.59668368	0.000883465	
13	Residual	7	10683497.8	1526213.972			
14	Total	9	79658246.5				
15							
16		*Coefficients*	*Standard Error*	*t Stat*	*P-value*	*Lower 95%*	*Upper 95%*
17	Intercept	72333.08447	21969.92267	3.292368642	0.013259225	20382.47253	124283.6964
18	Week	508.6681395	168.1770861	3.024598364	0.019260863	110.9925233	906.3437558
19	Price/Gallon	-16463.19901	5351.082403	-3.076611005	0.017900405	-29116.49823	-3809.899789

of steadily increasing demand, but might also be influenced by the average price per gallon. The average price per gallon can be considered as a **causal variable**. Multiple linear regression provides a technique for building forecasting models that incorporate not only time, but other potential causal variables also. Thus, to forecast gasoline sales, we propose a model using two independent variables (Week and Price/Gallon).

$$\text{Sales} = \beta_0 + \beta_1 \text{Week} + \beta_2 \text{Price/Gallon}$$

The results are shown in Figure 7.29 and the regression model is:

$$\text{Sales} = 72333.08 + 508.67\,\text{Week} - 16463.2\,\text{Price/Gallon}$$

This makes sense because as price changes, sales typically reflect the change. Notice that the R^2 value is higher when both variables are included, explaining more than 86% of the variation in the data. If the company estimates that the average price for the next week will drop to $3.80, the model would forecast the sales for week 11 as:

$$\text{Sales} = 72333.08 + 508.67(11) - 16463.2(3.80) = 15{,}368 \text{ gallons}$$

Notice that this is higher than the pure time-series forecast because of the sensitivity to the price per gallon.

THE PRACTICE OF FORECASTING

In practice, managers use a variety of judgmental and quantitative forecasting techniques. Statistical methods alone cannot account for such factors as sales promotions, unusual environmental disturbances, new product introductions, large one-time orders, and so on. Many managers begin with a statistical forecast and adjust it to account for intangible factors. Others may develop independent judgmental and statistical forecasts then combine them, either objectively by averaging or in a subjective manner. It is impossible to provide universal guidance as to which approaches are best, for they depend on a variety of factors, including the presence or absence of trends and seasonality, the number of data points available, length of the forecast time horizon, and the experience and knowledge of the forecaster. Often, quantitative approaches will miss significant changes in the data, such as reversal of trends, while qualitative forecasts may catch them, particularly when using indicators as discussed earlier in this chapter.

Here we briefly highlight three practical examples of forecasting and encourage you to read the full articles cited for better insight into the practice of forecasting.

- Allied-Signal's Albuquerque Microelectronics Operation (AMO) produced radiation-hardened microchips for the U.S. Department of Energy (DOE). In 1989 a decision was made to close a plant, but operations at AMO had to be phased out over several years because of long-term contractual obligations. AMO experienced fairly erratic yields in the production of some of its complex microchips, and accurate forecasts of yields were critical. Overestimating yields could lead to an inability to meet contractual obligations in a timely manner, requiring the plant to remain open longer. Underestimates would cause AMO to produce more chips than actually needed. AMO's yield forecasts had previously been made by simply averaging all historical data. More sophisticated forecasting techniques were implemented, resulting in improved forecasts of wafer fabrication. Using more accurate yield forecasts and optimization models, AMO was able to close the plant sooner, resulting in significant cost savings.[1]

- More than 70% of the total sales volume at L.L. Bean is generated through orders to its call center. Calls to the L.L. Bean call center are classified into two types: telemarketing (TM), which involves placing an order, and telephone inquiry (TI), which involves customer inquiries such as order status or order problems. Accurately forecasting TM and TI calls helps the company better plan the number of agents to have on hand at any point in time. Analytical forecasting models for both types of calls take into account historical trends, seasonal factors, and external explanatory variables such as holidays and catalog mailings. The estimated benefit from better precision from the two forecasting models is approximately $300,000 per year.[2]

- DIRECTV was founded in 1991 to provide subscription satellite television. Prior to launching this product, it was vital to forecast how many homes in the United States would subscribe to satellite television, and when. A forecast was developed using the Bass diffusion model, which describes the adoption pattern of new products and technologies. The model is based on the proposition that the conditional probability of adoption by potential consumers of a new product at a given time will be a linear-increasing function of the number of previous adopters. The model was supported by forecasting analogies with cable TV. The 1992 forecast proved to be quite good in comparison with actual data over the five-year period from 1994 through 1999.[3]

---

[1] D.W. Clements and R.A. Reid, "Analytical MS/OR Tools Applied to a Plant Closure," *Interfaces* 24, no. 2 (March–April, 1994): 1–12.
[2] B.H. Andrews and S.M. Cunningham, "L.L. Bean Improves Call-Center Forecasting," *Interfaces* 25, no. 6 (November–December, 1995): 1–13.
[3] Frank M. Bass, Kent Gordon, and Teresa L. Ferguson, "DIRECTV: Forecasting Diffusion of a New Technology Prior to Product Launch," *Interfaces* 31, no. 3 (May–June 2001): Part 2 of 2, S82–S93.

BASIC CONCEPTS REVIEW QUESTIONS

1. Explain the differences between qualitative and judgmental, statistical time-series, and explanatory/causal forecasting models.
2. Describe some common forecasting approaches for judgmental forecasting.
3. How are indicators and indexes used in judgmental forecasting?
4. What are the primary components of time series?
5. Summarize statistical methods used in forecasting and the types of time series to which they are most appropriate.
6. Explain how a simple moving average is calculated.
7. List and define the three principal ways of measuring forecast accuracy. What are the key differences among them?

8. Explain the differences between moving average and exponential smoothing models.
9. What types of forecasting models are best for time series with trends and/or seasonality?
10. What are the advantages of using *CB Predictor* for forecasting?
11. What are autoregressive models, and when should they be used?
12. How are dummy variables used in regression forecasting models with seasonality?
13. What is a causal variable in forecasting? Provide an example from your experience of some applications where causal variables might be used in a forecast.
14. Summarize some of the practical issues in using forecasting tools and approaches.

SKILL-BUILDING EXERCISES

1. Find a 4-period moving average forecast for the monthly burglaries data, compute MAD, MSE, and MAPE error metrics, and determine if this model is better than the 2-period moving average discussed in the chapter (Table 7.2).
2. Try to identify the best set of weights for a 3-period moving average model for the burglary data that minimizes the MAD error metric.
3. Find the best value of the smoothing constant between 0.5 and 0.7 (in increments of 0.05) for exponential smoothing for the burglary data.

4. Use *CB Predictor* to find the best forecasting model for Electric Use in the *Gas & Electric* Excel file.
5. Set up and fit a third-order autoregressive model for the coal production example. Compare the results to the example in the chapter. What do you find?
6. Find the best multiple regression model for Electric Use in the *Gas & Electric* Excel file using the approach for incorporating seasonality.

PROBLEMS AND APPLICATIONS

1. The Excel file *Closing Stock Prices* provides data for four stocks over a six-month period.
   a. Develop spreadsheet models for forecasting each of the stock prices using single moving average and single exponential smoothing.
   b. Using MAD, MSE, and MAPE as guidance, find the best number of moving average periods and best smoothing constant for exponential smoothing. (You might consider using data tables to facilitate your search.)
   c. Compare your results to the best moving average and exponential smoothing models found by *CB Predictor*.
2. For the data in the Excel file *Baseball Attendance* do the following:
   a. Develop spreadsheet models for forecasting attendance using single moving average and single exponential smoothing.
   b. Using MAD, MSE, and MAPE as guidance, find the best number of moving average periods and best smoothing constant for exponential smoothing.
   c. Compare your results to the best moving average and exponential smoothing models found by *CB Predictor*.

3. For the data in the Excel file *Ohio Prison Population* do the following:
   a. Develop spreadsheet models for forecasting both male and female populations using single moving average and single exponential smoothing.
   b. Using MAD, MSE, and MAPE as guidance, find the best number of moving average periods and best smoothing constant for exponential smoothing.
   c. Compare your results to the best moving average and exponential smoothing models found by *CB Predictor*.
4. For the data in the Excel file *Gasoline Prices* do the following:
   a. Develop spreadsheet models for forecasting prices using single moving average and single exponential smoothing.
   b. Using MAD, MSE, and MAPE as guidance, find the best number of moving average periods and best smoothing constant for exponential smoothing.
   c. Compare your results to the best moving average and exponential smoothing models found by *CB Predictor*.
5. Construct a line chart for the data in the Excel file *Arizona Population*.
   a. Suggest the best-fitting functional form for forecasting these data.
   b. Use *CB Predictor* to find the best forecasting model.
6. Construct a line chart for each of the variables in the data file *Death Cause Statistics,* and suggest the best forecasting technique. Then apply *CB Predictor* to find the best forecasting models for these variables.
7. The Excel file *Olympic Track and Field Data* provides the gold medal–winning distances for the high jump, discus, and long jump for the modern Olympic Games. Develop forecasting models for each of the events. What does the model predict for the next Olympics and what are the confidence intervals?
8. Use *CB Predictor* to find the best forecasting model for the data in the following Excel files:
   a. *New Car Sales*
   b. *Housing Starts*
   c. *Coal Consumption*
   d. *DJIA December Close*
   e. *Federal Funds Rates*
   f. *Mortgage Rates*
   g. *Prime Rate*
   h. *Treasury Yield Rates*
9. Consider the data in the Excel file *Consumer Price Index*.
   a. Use simple linear regression in *CB Predictor* to forecast the data. What would be the forecasts for the next six months?
   b. Are the data autocorrelated? Construct first- and second-order autoregressive models and compare the results to part (a).
10. Consider the data in the Excel file *Nuclear Power*.
    a. Use simple linear regression in *CB Predictor* to forecast the data. What would be the forecasts for the next three years?
    b. Are the data autocorrelated? Construct first- and second-order autoregressive models and compare the results to part (a).
11. Develop a multiple regression model with categorical variables that incorporate seasonality for forecasting the temperature in Washington, D.C., using the data for years 1999 and 2000 in the Excel file *Washington DC Weather*. Use the model to generate forecasts for the next 9 months and compare the forecasts to the actual observations in the data for the year 2001.
12. Develop a multiple regression model with categorical variables that incorporate seasonality for forecasting sales using the last three years of data in the Excel file *New Car Sales*.

13. Develop a multiple regression model with categorical variables that incorporate seasonality for forecasting housing starts beginning in June 2006 using the data in the Excel file *Housing Starts*.
14. Data in the Excel File *Microprocessor Data* shows the demand for one type of chip used in industrial equipment from a small manufacturer.
    a. Construct a chart of the data. What appears to happen when a new chip is introduced?
    b. Develop a causal regression model to forecast demand that includes both time and the introduction of a new chip as explanatory variables.
    c. What would the forecast be for the next month if a new chip is introduced? What would it be if a new chip is not introduced?

CASE

Energy Forecasting

The Excel file *Energy Production & Consumption* provides data on energy production, consumption, imports, and exports. You have been hired as an analyst for a government agency and have been asked to forecast these variables over the next 10 years. Apply forecasting tools and appropriate visual aids, and write a formal report to the agency director that explains these data and the future forecasts.

Appendix

Advanced Forecasting Models—Theory and Computation

In this appendix, we present computational formulas for advanced models for time-series forecasting. The calculations are somewhat complex, but can be implemented on spreadsheets with a bit of effort.

Double Moving Average

Double moving average involves taking averages of averages. Let M_t be the simple moving average for the last k periods (including period t):

$$M_t = [A_{t-k+1} + A_{t-k+2} + \dots + A_t]/k \quad (7A.1)$$

The double moving average, D_t, for the last k periods (including period t) is the average of the simple moving averages:

$$D_t = [M_{t-k+1} + M_{t-k+2} + \dots + M_t]/k \quad (7A.2)$$

Using these values, the double moving average method estimates the values of a_t and b_t in the linear trend model $F_{t+k} = a_t + b_t k$ as:

$$a_t = 2M_t - D_t$$
$$b_t = (2/(k - 1))[M_t - D_t] \quad (7A.3)$$

These equations are derived essentially by minimizing the sum of squared errors using the last k periods of data. Once these parameters are determined, forecasts beyond the end of the observed data (time period T) are calculated using the linear trend model with values of a_T and b_T. That is, for k periods beyond period T, the forecast is $F_{T+k} = a_T + b_T k$. For instance, the forecast for the next period would be $F_{T+1} = a_T + b_T(1)$.

Double Exponential Smoothing

Like double moving average, double exponential smoothing is also based on the linear trend equation, $F_{t+k} = a_t + b_t k$, but the estimates of a_t and b_t are obtained from the following equations:

$$a_t = \alpha F_t + (1 - \alpha)(a_{t-1} + b_{t-1})$$
$$b_t = \beta(a_t - a_{t-1}) + (1 - \beta)b_{t-1} \tag{7A.4}$$

In essence, we are smoothing both parameters of the linear trend model. From the first equation, the estimate of the level in period t is a weighted average of the observed value at time t and the predicted value at time t, $a_{t-1} + b_{t-1}$ based on single exponential smoothing. For large values of α, more weight is placed on the observed value. Lower values of α put more weight on the smoothed predicted value. Similarly, from the second equation, the estimate of the trend in period t is a weighted average of the differences in the estimated levels in periods t and $t - 1$ and the estimate of the level in period $t - 1$. Larger values of β place more weight on the differences in the levels, while lower values of β put more emphasis on the previous estimate of the trend.

To initialize the double exponential smoothing process, we need values for a_1 and b_1. One approach is to let $a_1 = A_1$ and $b_1 = A_2 - A_1$; that is, estimate the initial level with the first observation and the initial trend with the difference in the first two observations. As with single exponential smoothing we are free to choose the values of α and β. MAD, MSE, or MAPE may be used to find good values for these smoothing parameters. We leave it to you as an exercise to implement this model on a spreadsheet for the total energy consumption data.

Additive Seasonality

The seasonal additive model is:

$$F_{t+k} = a_t + S_{t-s+k} \tag{7A.5}$$

The level and seasonal factors are estimated in the additive model using the following equations:

$$a_t = \alpha(A_t - S_{t-s}) + (1 - \alpha)a_{t-1}$$
$$S_t = \gamma(A_t - a_t) + (1 - \gamma)S_{t-s} \tag{7A.6}$$

where α and γ are smoothing constants. The first equation estimates the level for period t as a weighted average of the deseasonalized data for period t, $(A_t - S_{t-s})$, and the previous period's level. The seasonal factors are updated as well using the second equation. The seasonal factor is a weighted average of the estimated seasonal component for period t, $(A_t - a_t)$, and the seasonal factor for the last period of that season type. Then the forecast for the next

period is $F_{t+1} = a_t + S_{t-s+1}$. For k periods out from the final observed period T, the forecast is:

$$F_{T+k} = a_T + S_{T-s+k} \tag{7A.7}$$

To initialize the model, we need to estimate the level and seasonal factors for the first s periods (e.g., for an annual season with quarterly data this would be the first 4 periods; for monthly data, it would be the first 12 periods, etc.). We will use the following approach:

$$a_s = \sum_{t=1}^{s} A_t/s$$
$$a_t = a_s \quad \text{for} \quad t = 1, 2, \ldots s \tag{7A.8}$$

and

$$S_t = A_t - a_t \quad \text{for} \quad t = 1, 2, \ldots s \tag{7A.9}$$

That is, we initialize the level for the first s periods to the average of the observed values over these periods and the seasonal factors to the difference between the observed data and the estimated levels. Once these have been initialized, the smoothing equations can be implemented for updating.

Multiplicative Seasonality

The seasonal multiplicative model is:

$$F_{t+k} = a_t S_{t-s+k} \tag{7A.10}$$

This model has the same basic smoothing structure as the additive seasonal model but is more appropriate for seasonal time series that increase in amplitude over time. The smoothing equations are:

$$a_t = \alpha(A_t/S_{t-s}) + (1 - \alpha)a_{t-1}$$
$$S_t = \gamma(A_t/a_t) + (1 - \gamma)S_{t-s} \tag{7A.11}$$

where α and γ are again the smoothing constants. Here, A_t/S_{t-s} is the deseasonalized estimate for period t. Large values of α put more emphasis on this term in estimating the level for period t. The term A_t/a_t is an estimate of the seasonal factor for period t. Large values of γ put more emphasis on this in the estimate of the seasonal factor.

The forecast for the period $t + 1$ is $F_{t+1} = a_t S_{t-s+1}$. For k periods out from the final observed period T, the forecast is:

$$F_{t+k} = a_T S_{T-s+k} \tag{7A.12}$$

As in the additive model, we need initial values for the level and seasonal factors. We do this as follows:

$$a_s = \sum_{t=1}^{s} A_t/s$$
$$a_t = a_s \quad \text{for} \quad t = 1, 2, \ldots s \tag{7A.13}$$

and

$$S_t = A_t/a_t \quad \text{for} \quad t = 1, 2, \ldots s \tag{7A.14}$$

Once these have been initialized, the smoothing equations can be implemented for updating.

Holt–Winters Additive Model

The Holt–Winters additive model is based on the equation:

$$F_{t+1} = a_t + b_t + S_{t-s+1} \tag{7A.15}$$

This model is similar to the additive model incorporating seasonality that we described in the previous section, but it also includes a trend component. The smoothing equations are:

$$a_t = \alpha(A_t - S_{t-s}) + (1 - \alpha)(a_{t-1} + b_{t-1})$$
$$b_t = \beta(a_t - a_{t-1}) + (1 - \beta)b_{t-1} \tag{7A.16}$$
$$S_t = \gamma(A_t - a_t) + (1 - \gamma)S_{t-s}$$

Here, α, β, and γ are the smoothing parameters for level, trend, and seasonal components, respectively. The forecast for period $t + 1$ is:

$$F_{t+1} = a_t + b_t + S_{t-s+1} \tag{7A.17}$$

The forecast for k periods beyond the last period of observed data (period T) is:

$$F_{T+k} = a_T + b_T k + S_{T-s+k} \tag{7A.18}$$

The initial values of level and trend are estimated in the same fashion as in the additive model for seasonality. The initial values for the trend are $b_t = b_s$, for $t = 1, 2, \ldots s$, where:

$$b_s = [(A_{s+1} - A_1)/s + (A_{s+2} - A_s)/s + \cdots$$
$$(A_{s+s} - A_s)/s]/s \tag{7A.19}$$

Note that each term inside the brackets is an estimate of the trend over one season. We average these over the first $2s$ periods.

Holt–Winters Multiplicative Model

The Holt-Winters multiplicative model is:

$$F_{t+1} = (a_t + b_t)S_{t-s+1} \tag{7A.20}$$

This model parallels the additive model:

$$a_t = \alpha(A_t/S_{t-s}) + (1 - \alpha)(a_{t-1} + b_{t-1})$$
$$b_t = \beta(a_t - a_{t-1}) + (1 - \beta)b_{t-1} \tag{7A.21}$$
$$S_t = \gamma(A_t/a_t) + (1 - \gamma)S_{t-s}$$

The forecast for period $t + 1$ is:

$$F_{t+1} = (a_t + b_t)S_{t-s+1} \tag{7A.22}$$

The forecast for k periods beyond the last period of observed data (period T) is:

$$F_{T+k} = (a_T + b_T k)S_{T-s+k} \tag{7A.23}$$

The initial values of level and trend are estimated in the same fashion as in the multiplicative model for seasonality, and the trend component as in the Holt–Winters additive model.

Chapter 8

Statistical Quality Control

INTRODUCTION

An important application of statistics and data analysis in both manufacturing and service operations is in the area of *quality control*.
Quality control methods help employees monitor production operations
to ensure that output conforms to specifications. This is important in
manufactured goods since product performance depends on achieving
design tolerances. It is also vital to service operations to ensure that customers receive error-free, consistent service.

Why is quality control necessary? The principal reason is that no two
outputs from any production process are exactly alike. If you measure any
quality characteristic—such as the diameters of machined parts, the
amount of soft drink in a bottle, or the number of errors in processing
orders at a distribution center—you will discover some variation. Variation
is the result of many small differences in those factors that comprise a
process: people, machines, materials, methods, and measurement systems. Taken together, they are called **common causes of variation**.

269

Other causes of variation occur sporadically and can be identified and either eliminated or at least explained. For example, when a tool wears down, it can be replaced; when a machine falls out of adjustment, it can be reset; when a bad lot of material is discovered, it can be returned to the supplier. Such examples are called **special causes of variation**. Special causes of variation cause the distribution of process output to change over time. Using statistical tools, we can identify when they occur and take appropriate action, thus preventing unnecessary quality problems. Equally important is knowing when to leave the process alone and not react to common causes over which we have no control.

In this chapter, we introduce basic ideas of *statistical process control* and *process capability analysis*—two important tools in helping to achieve quality. The key concepts that we will describe are as follows:

➤ *The statistical basis and the application of control charts for monitoring both attributes and variables*
➤ *Rules for determining when data signal the need for action*
➤ *The use of statistics for measuring the capability of a process to meet specifications*

The applications of statistics to quality control are far more extensive than we can present; much additional information may be found in many other sources.

THE ROLE OF STATISTICS AND DATA ANALYSIS IN QUALITY CONTROL

We can learn a lot about the common causes of variation in a process and their effect on quality by studying process output. For example, suppose that a company such as General Electric Aircraft Engines produces a critical machined part. Key questions might be: What is the average dimension? How much variability occurs in the output of the process? What does the distribution of part dimensions look like? What proportion of output, if any, does not conform to design specifications? These are fundamental questions that can be addressed with statistics.

The role of statistics is to provide tools to analyze data collected from a process and enable employees to make informed decisions when the process needs short-term corrective action or long-term improvements. Statistical methods have been used for quality control since the 1920s, when they were pioneered at the Western Electric Company. They became a mainstay of Japanese manufacturing in the early 1950s; however, they did not become widely used in the United States until the quality management movement of the 1980s, led by pioneers such as W. Edwards Deming and Joseph M. Juran, both of whom were instrumental in the adoption of these methods in Japan. Since then, statistical quality control has been shown to be a proven means of improving customer satisfaction and reducing costs in many industries.

To illustrate the applications of statistics to quality control, we will use an example from a Midwest pharmaceutical company that manufactures individual syringes with a self-contained, single dose of an injectable drug.[1] In the manufacturing

---

[1]Adapted from LeRoy A. Franklin and Samar N. Mukherjee, "An SPC Case Study on Stabilizing Syringe Lengths," *Quality Engineering* 12, no. 1 (1999–2000): 65–71.

process, sterile liquid drug is poured into glass syringes and sealed with a rubber stopper. The remaining stage involves insertion of the cartridge into plastic syringes and the electrical "tacking" of the containment cap at a precisely determined length of the syringe. A cap that is tacked at a shorter-than-desired length (less than 4.920 inches) leads to pressure on the cartridge stopper and, hence, partial or complete activation of the syringe. Such syringes must then be scrapped. If the cap is tacked at a longer-than-desired length (4.980 inches or longer), the tacking is incomplete or inadequate, which can lead to cap loss and potentially a cartridge loss in shipment and handling. Such syringes can be reworked manually to attach the cap at a lower position. However, this process requires a 100% inspection of the tacked syringes and results in increased cost for the items. This final production step seemed to be producing more and more scrap and reworked syringes over successive weeks. At this point, statistical consultants became involved in an attempt to solve this problem and recommended statistical process control for the purpose of improving the tacking operation.

STATISTICAL PROCESS CONTROL

In quality control, measures that come from counting are called *attributes.* Examples of attributes are the number of defective pieces in a shipment of components, the number of errors on an invoice, and the percent of customers rating service a 6 or 7 on a seven-point satisfaction scale. Measures that are based on a continuous measurement scale are called *variables.* Examples of variables data are the inside diameter of a drilled hole, the weight of a carton, and the time between order and delivery. This distinction is important because different statistical process control tools must be used for each type of data.

When a process operates under ideal conditions, variation in the distribution of output is due to common causes. When only common causes are present, the process is said to be **in control**. A controlled process is stable and predictable to the extent that we can predict the likelihood that future output will fall within some range once we know the probability distribution of outcomes. Special causes, however, cause the distribution to change. The change may be a shift in the mean, an increase or decrease in the variance, or a change in shape. Clearly, if we cannot predict how the distribution may change, then we cannot compute the probability that future output will fall within some range. When special causes are present, the process is said to be **out of control** and needs to be corrected to bring it back to a stable state. **Statistical process control (SPC)** provides a means of identifying special causes as well as telling us when the process is in control and should be left alone. Control charts were first used by Dr. Walter Shewhart at Bell Laboratories in the 1920s (they are sometimes called *Shewhart charts*). Shewhart was the first to make a distinction between common causes of variation and special causes of variation.

SPC consists of the following:

1. Selecting a sample of observations from a production or service process
2. Measuring one or more quality characteristics
3. Recording the data
4. Making a few calculations
5. Plotting key statistics on a *control chart*
6. Examining the chart to determine if any unusual patterns, called **out-of-control conditions**, can be identified
7. Determining the cause of out-of-control conditions and taking corrective action

When data are collected, it is important to clearly record the data, the time the data were collected, the measuring instruments that were used, who collected the data, and any other important information such as lot numbers, machine numbers, and the like. By having a record of such information, we can trace the source of quality problems more easily.

Control Charts

A **run chart** is a line chart in which the independent variable is time and the dependent variable is the value of some sample statistic, such as the mean, range, or proportion. A **control chart** is a run chart that has two additional horizontal lines, called **control limits**, as illustrated in Figure 8.1. Control limits are chosen statistically so that there is a high probability (usually greater than 0.99) that sample statistics will fall randomly within the limits *if the process is in control.*

To understand the statistical basis for control charts, let us assume that we are dealing with a variables measurement that is normally distributed with a mean μ and standard deviation σ. If the process is stable, or in control, then each individual measurement will stem from this distribution. In high-volume production processes, it is generally difficult, if not impossible, to measure each individual output, so we take samples at periodic intervals. For samples of a fixed size, n, we know from Chapter 3 that the sampling distribution will be normal with mean μ and standard deviation (standard error) $\sigma_{\bar{x}} = \sigma/\sqrt{n}$. We would expect that about 99.7% of sample means will lie within ± 3 standard errors of the mean, or between $\mu - 3\sigma_{\bar{x}}$ and $\mu + 3\sigma_{\bar{x}}$, provided the process remains in control. These values become the theoretical control limits for a control chart to monitor the centering of a process using the sample mean. Of course, we do not know the true population parameters, so we estimate them by the sample mean $\bar{x}$ and sample standard deviation, s. Thus, the actual control limits would be:

$$\text{Lower control limit: } \bar{x} - 3\,s_{\bar{x}}$$
$$\text{Upper control limit: } \bar{x} + 3\,s_{\bar{x}}$$

Figure 8.1 Structure of a Control Chart

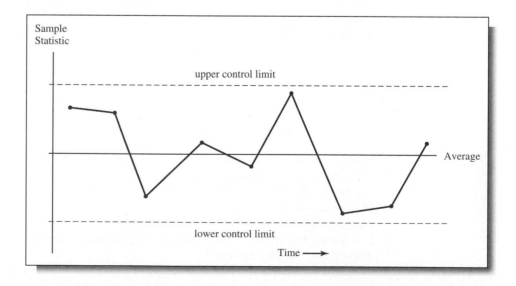

In general, control limits are established as ± 3 standard errors ($s_{\bar{x}}$) from the mean of the sampling distribution of the statistic we plot on the chart. Statisticians have devised various formulas for computing these limits in a practical manner that is easy for shop floor workers to understand and use. However, the theory is based on understanding the sampling distribution of the statistic we measure.

There are many different types of control charts. We will introduce three basic types of control charts in this chapter: $\bar{x}$- and R-charts for variables data and p-charts for attributes data. Discussions of other types of charts may be found in the references.

$\bar{x}$- and R-Charts

The $\bar{x}$-chart monitors the centering of process output for variables data over time by plotting the mean of each sample. In manufacturing, for example, the permissible variation in a dimension is usually stated by a **nominal specification** (target value) and some **tolerance**. For example, the specifications on the syringe length are 4.950 ± 0.030 inches. The nominal is 4.950 and the tolerance is ± 0.030. Therefore, the lengths should be between 4.920 to 4.980 inches. The $\bar{x}$-chart is used to monitor the centering of a process. The **R-chart**, or range chart, monitors the variability in the data as measured by the range of each sample. Thus, the R-chart monitors the uniformity or consistency of the process. The smaller the value of R, the more uniform is the process. Any increase in the average range is undesirable; this would mean that the variation is getting larger. However, decreases in variability signify improvement. We could use the standard deviation of each sample instead of the range to provide a more accurate characterization of variability; however, for small samples (around eight or less), little differences will be apparent, and if the calculations are done manually by a worker on the shop floor, R-charts are much easier to apply.

The basic procedure for constructing and using any control chart is first to gather at least 25 to 30 samples of data with a fixed sample size n from a production process, measure the quality characteristic of interest, and record the data. We will illustrate the construction of a control chart using the data in the Excel file *Syringe Samples* (see Figure 8.2), which shows 47 samples that were taken every 15 minutes from the syringe manufacturing process over three shifts. Each sample consists of 5 individual observations. In column G, we calculate the mean of each sample, and in column H, the range.

Figure 8.2 Portion of Excel File *Syringe Samples*

	A	B	C	D	E	F	G	H
1	Syringe Samples							
2								
3	First Shift Data							
4	Sample		Sample Observations				Average	Range
5	1	4.9600	4.9460	4.9500	4.9560	4.9580	4.9540	0.0140
6	2	4.9580	4.9270	4.9350	4.9400	4.9500	4.9420	0.0310
7	3	4.9710	4.9290	4.9650	4.9520	4.9380	4.9510	0.0420
8	4	4.9400	4.9820	4.9700	4.9530	4.9600	4.9610	0.0420
9	5	4.9640	4.9500	4.9530	4.9620	4.9560	4.9570	0.0140

Figure 8.3 Chart of Sample Means for Syringe Data

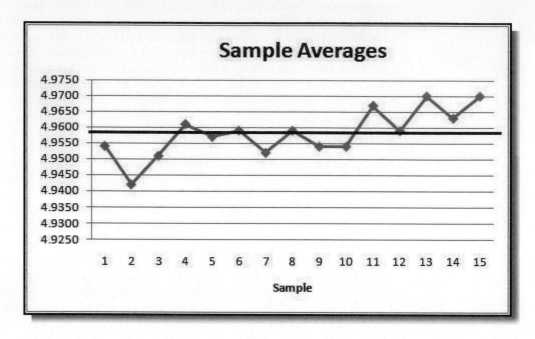

We will work with the first 15 samples (typically, it is recommended that at least 25 to 30 samples be used to construct a control chart, but we will assume that only the first 15 samples are available). After we have calculated the mean and range for each sample, we compute the average mean, $\bar{\bar{x}} = 4.9581$, and the average range, $\bar{R} = 0.0257$. Figures 8.3 and 8.4 show plots of the sample means and ranges. Although

Figure 8.4 Chart of Sample Ranges for Syringe Data

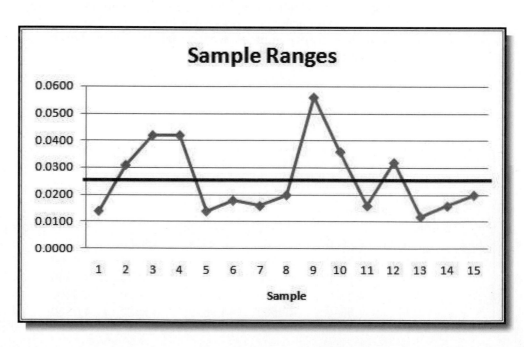

the chart for sample ranges shows some variation, we cannot yet determine statistically whether this variation might be due to some assignable cause or is simply due to chance. The chart for sample means appears to show an increasing trend.

The final step to complete the control charts is to compute control limits. As we explained earlier, control limits are boundaries within which the process is operating in statistical control. Control limits are based on past performance and tell us what values we can expect for $\bar{x}$ or R as long as the process remains stable. If a point falls outside the control limits or if some unusual pattern occurs, then we should be suspicious of a special cause. The upper control limit for the R-chart is given by the formula:

$$UCL_R = D_4 \bar{R} \tag{8.1}$$

and the lower control limit for the R-chart is given by the formula:

$$LCL_R = D_3 \bar{R} \tag{8.2}$$

D_3 and D_4 are constants that depend on the sample size and are found in Table 8.1. The theory is a bit complicated, but suffice it to say that these constants have been determined from the sampling distribution of R so that, for example, $D_4 \bar{R} = \bar{R} + 3s_R$, as we had described earlier.

Because the sample size is 5, $D_4 = 2.114$. Therefore, the upper control limit for the example is $2.114(0.0257) = 0.0543$. In this example, D_3 for a sample size of 5 is 0; therefore, the lower control limit is 0. We then draw and label these control limits on the chart.

SAMPLE SIZE	A_2	D_3	D_4
2	1.880	0	3.267
3	1.023	0	2.574
4	0.729	0	2.282
5	0.577	0	2.114
6	0.483	0	2.004
7	0.419	0.076	1.924
8	0.373	0.136	1.864
9	0.337	0.184	1.816
10	0.308	0.223	1.777
11	0.285	0.256	1.744
12	0.266	0.283	1.717
13	0.249	0.307	1.693
14	0.235	0.328	1.672
15	0.223	0.347	1.653

Table 8.1 Control Chart Factors

PHSTAT NOTE
$\bar{x}$- and R-Charts

From the *PHStat* menu, select *Control Charts,* followed by *R & Xbar Charts.* The dialog box that appears is shown in Figure 8.5. Your worksheet must have already calculated the sample means and ranges. The cell ranges for these data are entered in the appropriate boxes. You must also provide the sample size in the *Data* section and have the option of selecting only the *R*-chart or both the $\bar{x}$- and *R*-charts. *PHStat* will create several worksheets for calculations and the charts.

Figure 8.5 *PHStat* Dialog for *R and XBar Charts*

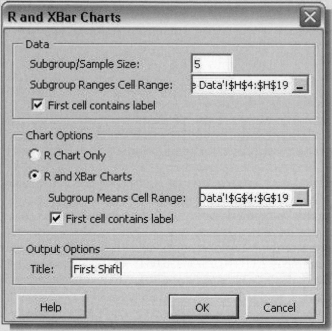

For the $\bar{x}$-chart, the control limits are given by the formulas:

$$UCL_{\bar{x}} = \bar{\bar{x}} + A_2 \overline{R} \qquad (8.3)$$

$$UCL_{\bar{x}} = \bar{\bar{x}} - A_2 \overline{R} \qquad (8.4)$$

Again, the constant A_2 is determined so that $A_2 \overline{R}$ is equivalent to 3 standard errors in the sampling distribution of the mean. For a sample of size 5, $A_2 = 0.577$ from Table 8.1. Therefore, the control limits are:

$$UCL_{\bar{x}} = 4.9581 + (0.577)(0.0257) = 4.973$$

$$LCL_{\bar{x}} = 4.9581 - (0.577)(0.0257) = 4.943$$

We could draw these control limits on the charts to complete the process. *PHStat* includes a routine for constructing $\bar{x}$- and R-charts (see *PHStat Note: $\bar{x}$- and R-Charts*). The charts generated by this routine are shown in Figures 8.6 and 8.7. *PHStat* also creates a worksheet with the calculations using the formulas we have discussed (see Figure 8.8). The next step is to analyze the charts to determine the state of statistical control.

Figure 8.6 *R*-Chart for Syringe Data—First Shift

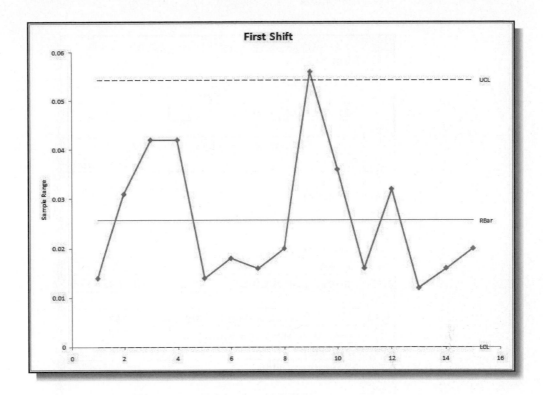

Figure 8.7 *x̄*-Chart for Syringe Data—First Shift

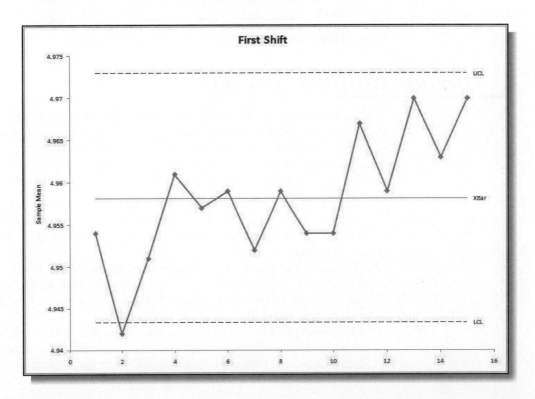

Figure 8.8 *PHStat Calculations* Worksheet for *R & XBar Charts*

	A	B
1	**First Shift**	
2		
3	**Data**	
4	**Sample/Subgroup Size**	5
5		
6	R Chart Intermediate Calculations	
7	RBar	0.025666667
8	D3 Factor	0
9	D4 Factor	2.114
10		
11	**R Chart Control Limits**	
12	**Lower Control Limit**	0
13	**Center**	0.025666667
14	**Upper Control Limit**	0.054259333
15		
16	XBar Chart Intemediate Calculations	
17	Average of Subgroup Averages	4.958133333
18	A2 Factor	0.577
19	A2 Factor * RBar	0.014809667
20		
21	**XBar Chart Control Limits**	
22	**Lower Control Limit**	4.943323667
23	**Center**	4.958133333
24	**Upper Control Limit**	4.972943

ANALYZING CONTROL CHARTS

When a process is in statistical control, the points on a control chart should fluctuate at random between the control limits, and no recognizable patterns should exist. The following checklist provides a set of general rules for examining a control chart to see if a process is in control. These rules are based on the assumption that the underlying distribution of process output—and, therefore, the sampling distribution—is normal.

1. *No points are outside the control limits.* Since the control limits are set at ±3 standard errors from the mean, the probability of a point falling outside the control limits when the process is in control is only 0.0027, under the normality assumption.

2. *The number of points above and below the center line is about the same.* If the distribution is symmetric, as a normal distribution is, we would expect this to occur. If the distribution is highly skewed, we might find a disproportionate amount on one side.

3. *The points seem to fall randomly above and below the center line.* If the distribution is stable, we would expect the same chances of getting a sample above the mean as below. However, if the distribution has shifted during the data collection process, we would expect a nonrandom distribution of sample statistics.

4. *There are no steady upward or downward trends of points moving toward either control limit.* These would indicate a gradual movement of the distribution mean.
5. *Most points, but not all, are near the center line; only a few are close to the control limits.* For a normal distribution, about 68% of observations fall within 1 standard deviation of the mean. If, for instance, we see a high proportion of points near the limits, we might suspect that the data came from two distinct distributions (visualize an inverted normal distribution).

For the syringe data, the R-chart has one point above the upper control limit. In the $\bar{x}$-chart, not only is one point below the lower control limit, but we see a clear upward trend. Thus, we would conclude that the process is not in control, particularly in the ability to maintain a stable average of the syringe length. It is important to keep good records of data—the time at which each sample was taken and the process conditions at that time (who was running the process, where the material came from, etc.). Most of the time, it is easy to identify a logical cause. A common reason for a point falling outside a control limit is an error in the calculation of the sample values of $\bar{x}$ or R. Other possible causes are a sudden power surge, a broken tool, measurement error, or an incomplete or omitted operation in the process. Once in a while, however, they are a normal part of the process and occur simply by chance. When assignable causes are identified, these data should be deleted from the analysis, and new control limits should be computed.

The most common types of other out-of-control conditions are summarized next.

Sudden Shift in the Process Average

When an unusual number of consecutive points fall on one side of the center line, it usually indicates that the process average has suddenly shifted. Typically, this is the result of an external influence that has affected the process; this would be a special cause. In both the $\bar{x}$- and R-charts, possible causes might be a new operator, a new inspector, a new machine setting, or a change in the setup or method. In the R-chart, if the shift is up, the process has become less uniform. Typical causes are carelessness of operators, poor or inadequate maintenance, or possibly a fixture in need of repair. If the shift is down in the R-chart, uniformity of the process has improved. This might be the result of improved workmanship or better machines or materials.

Cycles

Cycles are short, repeated patterns in the chart, having alternative high peaks and low valleys. These are the result of causes that come and go on a regular basis. In the $\bar{x}$-chart, cycles may be the result of operator rotation or fatigue at the end of a shift, different gauges used by different inspectors, seasonal effects such as temperature or humidity, or differences between day and night shifts. In the R-chart, cycles can occur from maintenance schedules, rotation of fixtures or gauges, differences between shifts, or operator fatigue.

Trends

A trend is the result of some cause that gradually affects the quality characteristics of the product and causes the points on a control chart to gradually move up or down from the center line. As a new group of operators gains experience on the job, for example, or as maintenance of equipment improves over time, a trend may occur. In the $\bar{x}$-chart, trends may be the result of improving operator skills, dirt or chip buildup in fixtures, tool wear, changes in temperature or humidity, or aging of equipment. In the R-chart, an increasing trend may be due to a gradual decline in material quality, operator fatigue, gradual loosening of a fixture or a tool, or dulling of a tool. A decreasing trend often is the result of improved operator skill, improved work methods, better purchased materials, or improved or more frequent maintenance.

Hugging the Center Line

Hugging the center line occurs when nearly all the points fall close to the center line. In the control chart, it appears that the control limits are too wide. A common cause of this occurrence is the sample being taken by selecting one item systematically from each of several machines, spindles, operators, and so on. A simple example will serve to illustrate this. Suppose that one machine produces parts whose diameters average 7.508 with variation of only a few thousandths; and a second machine produces parts whose diameters average 7.502, again with only a small variation. Taken together, you can see that the range of variation would probably be between 7.500 and 7.510 and average about 7.505. Now suppose that we sample one part from each machine and compute a sample average to plot on an $\bar{x}$-chart. The sample averages will consistently be around 7.505 since one will always be high and the second will always be low. Even though there is a large variation in the parts taken as whole, the sample averages will not reflect this. In such a case, it would be more appropriate to construct a control chart for each machine, spindle, operator, and so on. An often overlooked cause for this pattern is miscalculation of the control limits, perhaps by using the wrong factor from the table or misplacing the decimal point in the computations.

Hugging the Control Limits

This pattern shows up when many points are near the control limits with very few in between. It is often called a *mixture* and is actually a combination of two different patterns on the same chart. A mixture can be split into two separate patterns. A mixture pattern can result when different lots of material are used in one process or when parts are produced by different machines but fed into a common inspection group.

Quality control practitioners advocate simple rules, based on sound statistical principles, for identifying out-of-control conditions. For example, if eight consecutive points fall on one side of the center line, then you can conclude that the mean has shifted. Why? If the distribution is symmetric, then the probability that the next sample falls above or below the mean is 0.5. Because samples are independent, the probability that eight consecutive samples will fall on one side of the mean is $(0.5)^8 = 0.0039$—a highly unlikely occurrence. Another rule often used to detect a shift is finding 10 of 11 consecutive points on one side of the center line. The probability of this occurring can be found using the binomial distribution:

$$\text{Probability of 10 out of 11 points on one side of center line } =$$

$$\binom{11}{10}(0.5)^{10}(0.5)^1 = 0.00537$$

These examples show the value of statistics and data analysis in common production operations.

Let us return to the syringe data. After examining the first set of charts, a technician was called to adjust the machine prior to the second shift, and 17 more samples were taken. Figure 8.9 shows one of the calculation worksheets created by *PHStat* for developing the control charts. We see that the average range is now 0.0129 (versus 0.0257 in the first set of data) and the average mean is 4.9736 (versus 4.9581). Although the average dispersion appears to have been reduced, the centering of the process has gotten worse, since the target dimension is 4.950. The charts in Figures 8.10 and 8.11 also suggest that the variation has gotten out of control and that the process mean continues to drift upward.

	A	B
1	**Second Shift**	
2		
3	**Data**	
4	**Sample/Subgroup Size**	5
5		
6	R Chart Intermediate Calculations	
7	RBar	0.012882353
8	D3 Factor	0
9	D4 Factor	2.114
10		
11	**R Chart Control Limits**	
12	**Lower Control Limit**	0
13	**Center**	0.012882353
14	**Upper Control Limit**	0.027233294
15		
16	XBar Chart Intemediate Calculations	
17	Average of Subgroup Averages	4.973647059
18	A2 Factor	0.577
19	A2 Factor * RBar	0.007433118
20		
21	**XBar Chart Control Limits**	
22	**Lower Control Limit**	4.966213941
23	**Center**	4.973647059
24	**Upper Control Limit**	4.981080176

Figure 8.9 *PHStat Calculations* Worksheet for Second Shift Data

Figure 8.10 Second Shift *R*-Chart

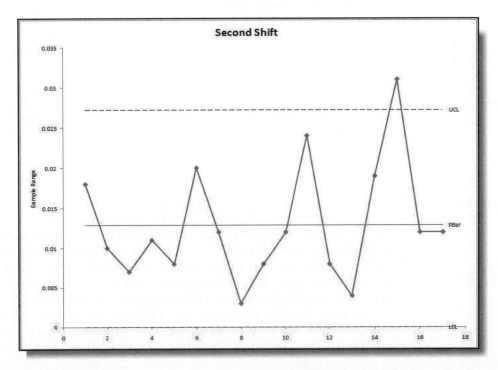

Figure 8.11 Second Shift x̄-Chart

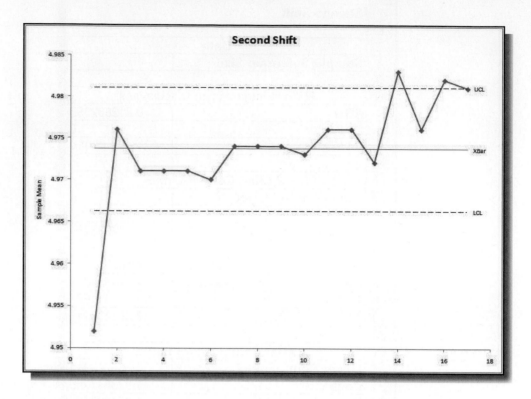

After another adjustment by a technician, the third shift collected another 15 samples. We leave it to you as an exercise to develop the control charts and verify that the *R*-chart appears to be in control, that the variability has stabilized, and that the *x̄*-chart is also in control, although the average length of the syringes is 4.963, slightly above target. In reviewing the process, the maintenance technician discovered that as he tried to move the height adjustment stop down on its threaded shaft, it was difficult to tighten the locknut because of worn threads. As a result, the vibration from the machine loosened the locknut and adjustment cap, resulting in drifts off center of the lengths. When he set the adjustment cap a bit higher (resulting in a slightly higher average length), the threads were good enough to hold the locknut in place, reducing the variation and bringing the process into control. This example shows the value of using control charts to help monitor a process and diagnose quality problems.

CONTROL CHARTS FOR ATTRIBUTES

The most common control chart for attributes is the *p*-chart. A **p-chart** monitors the proportion of nonconforming items. Sometimes it is called a *fraction nonconforming* or *fraction defective* chart. For example, a *p*-chart might be used to monitor the proportion of checking account statements that are sent out with errors, FedEx packages delivered late, hotel rooms not cleaned properly, or surgical infections in a hospital.

As with variables data, a *p*-chart is constructed by first gathering 25 to 30 samples of the attribute being measured. For attributes data, it is recommended that each sample size be at least 100; otherwise, it is difficult to obtain good statistical results.

The size of each sample may vary. It is usually recommended that a constant sample size be used as this makes interpreting patterns in the p-chart easier; however, for many applications this may not be practical or desirable.

The steps in constructing a p-chart are similar to those used for $\bar{x}$- and R-charts. We will first consider the case of a fixed sample size, n. Assume we have k samples, each of size n. For each sample, we compute the fraction nonconforming, p, that is, the number of nonconforming items divided by the number in the sample. The average fraction nonconforming, $\bar{p}$, is computed by summing the total number of nonconforming items in all samples and dividing by the total number of items ($= nk$ if the sample size is constant) in all samples combined. Because the number of nonconforming items in each sample follows a binomial distribution, the standard deviation is:

$$s = \sqrt{\frac{\bar{p}(1 - \bar{p})}{n}} \tag{8.5}$$

Using the principles we described earlier in this chapter, upper and lower control limits are given by:

$$\text{UCL}_p = \bar{p} + 3s \tag{8.6}$$
$$\text{LCL}_p = \bar{p} - 3s \tag{8.7}$$

Whenever LCL_p turns out negative, we use zero as the lower control limit since the fraction nonconforming can never be negative. We may now plot the fraction nonconforming on a control chart just as we did for the averages and ranges and use the same procedures to analyze patterns in a p-chart as we did for $\bar{x}$- and R-charts. That is, we check that no points fall outside of the upper and lower control limits and that no peculiar patterns (runs, trends, cycles, and so on) exist in the chart.

To illustrate a p-chart, suppose that housekeeping supervisors at a hotel inspect 100 rooms selected randomly each day to determine if they were cleaned properly. Any nonconformance, such as a failure to replace used soap or shampoo or empty the wastebasket, results in the room being listed as improperly cleaned. The Excel file *Room Inspection* (shown in Figure 8.12) provides data for 25 days. The total number of nonconforming rooms is 55. Therefore, the average fraction nonconforming, $\bar{p}$, is $55/2{,}500 = 0.022$. This leads to the standard deviation:

$$s = \sqrt{\frac{0.022(1 - 0.022)}{100}} = 0.01467$$

Figure 8.12   Portion of *Room Inspection*

	A	B	C	D
1	Room Inspection Results			
2				
3	Sample	Rooms Inspected	Nonconforming Rooms	Fraction Nonconforming
4	1	100	3	0.03
5	2	100	1	0.01
6	3	100	0	0.00
7	4	100	0	0.00
8	5	100	2	0.02
9	6	100	5	0.05

PHSTAT NOTE
p-Charts

From the *PHStat* menu, select *Control Charts* followed by *p-Chart*. The dialog box, shown in Figure 8.13, prompts you for the cell range for the number of non-conformances and the sample size. This procedure also allows you to have nonconstant sample sizes; if so, you need to enter the cell range of the sample size data. *PHStat* creates several new worksheets for the calculations and the actual chart.

Figure 8.13 *PHStat* Dialog for *p-Charts*

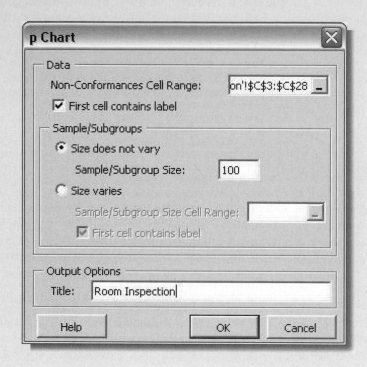

The control limits are computed as:

$$\mathrm{UCL}_p = \bar{p} + 3s = 0.022 + 3(0.01467) = 0.066$$
$$\mathrm{LCL}_p = \bar{p} - 3s = 0.022 - 3(0.01467) = -0.022$$

Because the lower control limit is negative, we use 0.

PHStat has a procedure for constructing *p*-charts (see *PHStat Note: p-Charts*). Figures 8.14 and 8.15 show the *Calculations* worksheet created by the tool and the resulting control chart. Although there is some variation in the proportion of rooms not cleaned properly, the chart appears to be in control, suggesting that this variation is due to common causes within the system (perhaps insufficient training of the housekeeping staff).

Variable Sample Size

In many applications, it is desirable to use all data available rather than a sample. For example, hospitals collect monthly data on the number of infections after surgeries. To monitor the infection rate, a sample would not provide complete information. The

Figure 8.14   *p*-Chart Calculations Worksheet

	A	B
1	**Room Inspection**	
2		
3	Intermediate Calculations	
4	Sum of Subgroup Sizes	2500
5	Number of Subgroups Taken	25
6	Average Sample/Subgroup Size	100
7	Average Proportion of Nonconforming Items	0.022
8	Three Standard Deviations	0.044005
9		
10	**p Chart Control Limits**	
11	**Lower Control Limit**	-0.022005
12	**Center**	0.022
13	**Upper Control Limit**	0.066005

Figure 8.15   *p*-Chart for Room Inspection Data

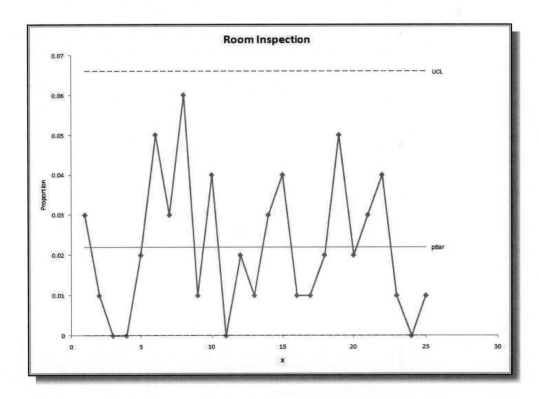

Excel file *Surgery Infections* provides monthly data over a three-year period, a portion of which is shown in Figure 8.16. Because the sample size varies, we must modify the calculation of the standard deviation and control limits. One approach (used by

	A	B	C	D
1	**Surgery Infections**			
2				
3	**Month**	**Surgeries**	**Infections**	**Infection Rate**
4	1	208	1	0.0048
5	2	225	3	0.0133
6	3	201	3	0.0149
7	4	236	1	0.0042
8	5	220	3	0.0136
9	6	244	1	0.0041

Figure 8.16 Portion of *Surgery Infections*

PHStat—see the previous *PHStat* note) is to compute the average sample size, $\bar{n}$, and use this value in the calculation of the standard deviation:

$$s = \sqrt{\frac{\bar{p}(1 - \bar{p})}{\bar{n}}} \tag{8.8}$$

Generally, this is acceptable as long as the sample sizes fall within 25% of the average. If sample sizes vary by a larger amount, then other approaches, which are beyond the scope of this book, should be used. When using this approach, note that because control limits are approximated using the average sample size, points that are actually out of control may not appear so on the chart and nonrandom patterns may be difficult to interpret; thus, some caution should be used. Figure 8.17 shows the control chart constructed using *PHStat*. The chart shows that the infection rate exceeds the upper control limit in month 12, indicating that perhaps some unusual circumstances occurred at the hospital.

Figure 8.17 Control Chart for Surgery Infection Rate Using Average Sample Size

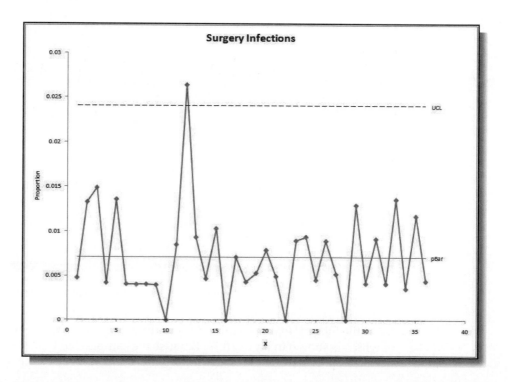

PROCESS CAPABILITY ANALYSIS

The purpose of SPC is to monitor a process over time to maintain a state of statistical control. However, just because a process is in control does not mean that it is capable of meeting specifications on the quality characteristic that is being measured. In the game of golf, for example, you might consistently shoot between 85 and 90; however, this is far from meeting the "specification"—par! **Process capability analysis** involves comparing the distribution of process output to specifications when only common causes (natural variations in materials, machines and tools, methods, operators, and the environment) determine the variation. As such, process capability is meaningless if special causes occur in the process. Therefore, before conducting a process capability analysis, control charts should be used to ensure that all special causes have been eliminated and that the process is in control.

Process capability is measured by the proportion of output that can be produced within design specifications. By collecting data, constructing frequency distributions and histograms, and computing basic descriptive statistics such as the mean and variance, we can better understand the nature of process variation and its ability to meet quality standards.

There are three important elements of process capability: the design specifications, the centering of the process, and the range of variation. Let us examine three possible situations:

1. *The natural variation in the output is smaller than the tolerance specified in the design* (Figure 8.18). The probability of exceeding the specification limits is essentially zero; you would expect that the process will almost always produce output that conforms to the specifications, as long as the process remains centered. Even slight changes in the centering or spread of the process will not affect its ability to meet specifications.
2. *The natural variation and the design specification are about the same* (Figure 8.19). A very small percentage of output might fall outside the specifications. The process should probably be closely monitored to make sure that the centering of the process does not drift and that the spread of variation does not increase.
3. *The range of process variation is larger than the design specifications* (Figure 8.20). The probability of falling in the tails of the distribution outside the specification limits is significant. The only way to improve product quality is to change the process.

Figure 8.18 Capable Process

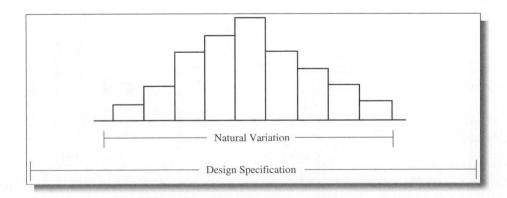

Natural Variation

Design Specification

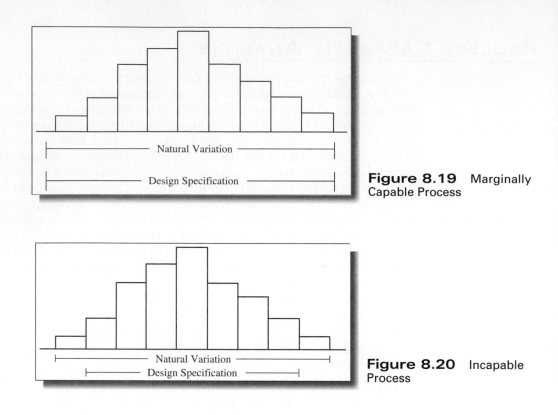

Figure 8.19 Marginally Capable Process

Figure 8.20 Incapable Process

To illustrate process capability analysis, we will use the third shift data in the Excel file *Syringe Samples.* Recall that the specifications of syringe lengths are 4.950 ± 0.030 inches. Figure 8.21 shows the output of Excel's *Descriptive Statistics* and *Histogram* tools applied to these data. The process mean is 4.963667, which is higher

Figure 8.21 Summary Statistics and Histogram for Process Capability Analysis

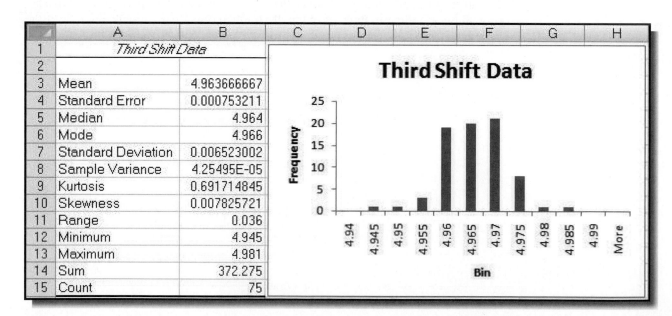

than the target specification for reasons described earlier. We would expect nearly all process output to fall within ±3 standard deviations of the mean. With a standard deviation of 0.006523, this is 4.963667 ± 3(0.006523), or between 4.944 and 4.983. This six standard deviation spread represents the capability of the process. As long as the process remains in control, we would expect the syringe lengths to fall within this range. Thus, we expect a small percentage of syringes to exceed the upper specification of 4.980. As you can see, one of the data points in this sample does exceed this specification.

The relationship between the process capability and the design specifications is often quantified by a measure called the **process capability index**, denoted by C_p (this statistic is unrelated to the C_p statistic that we discussed in the context of regression analysis in Chapter 6). C_p is simply the ratio of the specification width to the process capability.

$$C_p = \frac{\text{USL} - \text{LSL}}{6s} \tag{8.9}$$

For the syringe example, $C_p = (4.98 - 4.92)/(0.039) = 1.54$. A C_p value greater than 1.0 means that the process is capable of meeting specifications; values lower than 1.0 mean that some nonconforming output will always be produced unless the process technology is improved. However, despite the fact that the capability is good, the process is not centered properly on the target value and some nonconforming output will occur.

BASIC CONCEPTS REVIEW QUESTIONS

1. What are some reasons that production output varies? Explain the difference between common causes of variation and special causes of variation.
2. What is the difference between variables and attributes?
3. Describe the meaning of the term *in control*.
4. Describe the steps involved in applying statistical process control.
5. Explain the difference between a run chart and a control chart.
6. What are nominal specifications and tolerances?
7. List the principal rules for examining a control chart to determine if the process is in control.
8. Describe the common types of out-of-control conditions that one might find in a control chart.
9. What is process capability? How is it measured?

SKILL-BUILDING EXERCISES

1. Use *PHStat* to develop control charts for the third shift data in the syringe example. What can you conclude?
2. Using a spreadsheet, compute the probability of finding x consecutive points on one side of the center line if a process is in control, for $x = 2$ to 10. Also find the probabilities of observing x out of n consecutive points on one side of the center line, where n varies from 8 to 12, and x varies from $n - 3$ to n.
3. Build an Excel spreadsheet template for calculating control chart parameters (center line and control limits) for $\bar{x}$- and R-charts that allows you to enter the raw data without having to compute sample averages and ranges. Try to design the template to automatically generate the control charts.
4. Build an Excel spreadsheet template for calculating control chart parameters (center line and control limits) for a p-chart with and without variable sample sizes. Try to design the template to automatically generate the control charts.
5. Build a spreadsheet for computing the process capability index that allows you to enter up to 100 data points and apply it to the third shift syringe data.

PROBLEMS AND APPLICATIONS

1. Find the upper and lower control limits for $\bar{x}$- and R-charts for the width of a chair seat when the sample grand mean (based on 30 samples of 6 observations each) is 27.0 inches, and the average range is 0.35 inches.

2. Suppose that the sample grand mean (based on 25 samples of 10 observations each) for the weight of a package of cereal is 10.1 ounces and the average range is 0.4 ounces. Find the upper and lower control limits for $\bar{x}$- and R-charts.

3. The sample grand mean (based on 30 samples of 8 observations each) for the weight of a can of crabmeat is 0.503 pounds and the average range is 0.068 pounds. Find the upper and lower control limits for $\bar{x}$- and R-charts.

4. If 30 samples of 100 items are tested for nonconformity, and 95 of the 3,000 items are defective, find the upper and lower control limits for a p-chart.

5. If 25 samples of 200 items are run through a battery of tests, and 120 of the 5,000 items are defective, calculate the upper and lower control limits for a p-chart.

6. Suppose that an operation produces output with a standard deviation of 0.35 pounds, with an upper specification limit of 8 pounds and the lower specification limit of 6 pounds. Compute the process capability index.

7. Suppose that the standard deviation of a machining process is 0.12 inch, and that the upper specification limit is 3.60 inches and the lower specification limit is 2.00 inches. What is the process capability index?

For the remaining problems, all data may be found in the Excel workbook *Statistical Quality Control Problems*. Data for each problem are found on a separate worksheet.

8. Compute control limits for $\bar{x}$- and R-charts for the data in the worksheet *Problem 8*. Construct and interpret the charts.

9. Compute control limits for $\bar{x}$- and R-charts for the data in the worksheet *Problem 9*. Construct and interpret the charts.

10. Hunter Nut Company produces cans of mixed nuts, advertised as containing no more than 20% peanuts. Hunter Nut Company wants to establish control limits for their process to ensure meeting this requirement. They have taken 30 samples of 144 cans of nuts from the production process at periodic intervals, inspected each can, and identified the proportion of cans that did not meet the peanut requirement, shown in the worksheet *Problem 10*. Compute the average proportion nonconforming and the upper and lower control limits for this process. Construct the p-chart and interpret the results.

11. A manufacturer of high-quality medicinal soap advertises its product as 99 and 44/100% free of medically offensive pollutants. Twenty-five samples of 100 bars of soap were gathered at the beginning of each hour of production; the numbers of bars not meeting this requirement follows (see the worksheet *Problem 11*). Develop a p-chart for these data and interpret the results.

12. A warehouse double-checks the accuracy of its order fulfillment process each day. The data in the worksheet *Problem 12* shows the number of orders processed each day and the number of inaccurate orders found. Construct a p-chart for these data and interpret the results.

13. Refer to the data in the worksheet *Problem 8*. Suppose that the specification limits for the process are 90 (lower) and 110 (upper). Compute the process capability index and interpret. Eliminate any out-of-control points first, if any are found.

14. Refer to the data in the worksheet *Problem 9*. Suppose that the specification limits for the process are 25 (lower) and 40 (upper). Compute the process capability index and interpret. Eliminate any out-of-control points first, if any are found.

Quality Control Analysis

A manufacturer of commercial and residential lawn mowers has collected data from testing its products. These data are available in the Excel file *Quality Control Case Data*. The worksheet *Blade Weight* provides sample data from the manufacturing process of mower blades; *Mower Test* gives samples of functional performance test results; and *Process Capability* contains additional sample data of mower blade weights. Use appropriate control charts and statistical analyses to provide a formal report to the manufacturing manager about quality issues he should understand or be concerned with.

Part II

Decision Modeling and Analysis

Chapter 9

Building and Using Decision Models

INTRODUCTION

*E*veryone makes decisions. Individuals face personal decisions such as choosing a college or graduate program, making product purchases, selecting the right mortgage, and investing for retirement. Managers in business organizations must determine what products to make and how to price them, where to locate facilities, how many people to hire, where to allocate advertising budgets, whether or not to outsource a business function, and how to schedule production. Developing effective strategies to deal with these types of problems can be a difficult task. Quantitative decision models can greatly assist in these types of decisions. Part II of this book is devoted to the development and application of decision models.

Spreadsheets, in particular, provide a convenient means to manage data, construct models, and analyze them for gaining insight and supporting decisions. Although the early applications of spreadsheets were primarily in accounting and finance, spreadsheets have developed into powerful general-purpose managerial tools for decision modeling and analysis.

In this chapter, we introduce approaches for building decision models, implementing them on spreadsheets, and analyzing them to provide useful business information. The key concepts we will discuss are as follows:

➤ *Understanding the nature of descriptive and prescriptive decision models*

➤ *Excel-based tools for analyzing models and finding optimal solutions*

➤ *Practical knowledge to facilitate the model-building process, including functional relationships, data fitting, and spreadsheet engineering*

➤ *Examples of building both analytical and spreadsheet-based decision models*

➤ *Approaches for incorporating uncertainty in decision models*

➤ *Understanding the effect of assumptions on model complexity and realism*

DECISION MODELS

A **model** is an abstraction or representation of a real system, idea, or object. Models capture the most important features of a problem and present them in a form that is easy to interpret. A **decision model** is one that can be used to understand, analyze, or facilitate making a decision. A model can be a simple picture, a spreadsheet, or a set of mathematical relationships. Decision models generally have three types of inputs:

1. **Data**, which are assumed to be constant for purposes of the model. Some examples would be costs, machine capacities, and intercity distances.
2. **Uncontrollable variables**, which are quantities that can change but cannot be directly controlled by the decision maker. Some examples would be customer demand, inflation rates, and investment returns.
3. **Decision variables**, which are controllable and can be selected at the discretion of the decision maker. Some examples would be production quantities, staffing levels, and investment allocations.

Decision models characterize the relationships among the data, uncontrollable variables, and decision variables and the outputs of interest to the decision maker (see Figure 9.1). A spreadsheet is one way of expressing a decision model through the

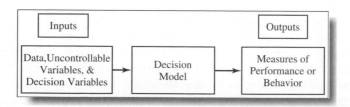

Figure 9.1 Nature of Decision Models

Figure 9.2 *Outsourcing Decision Model*

	A	B
1	Outsourcing Decision Model	
2		
3	Data	
4		
5	Manufactured in-house	
6	Fixed cost	$ 50,000
7	Unit variable cost	$ 125
8		
9	Purchased from supplier	
10	Unit cost	$ 175
11		
12	Model	
13		
14	Demand volume	1500
15		
16	Total manufacturing cost	$ 237,500
17	Total purchased cost	$ 262,500
18	Difference	$ (25,000)
19		
20	Decision	Manufacture

	A	B
1	Outsourcing Decision Model	
2		
3	Data	
4		
5	Manufactured in-house	
6	Fixed cost	50000
7	Unit variable cost	125
8		
9	Purchased from supplier	
10	Unit cost	175
11		
12	Model	
13		
14	Demand volume	1500
15		
16	Total manufacturing cost	=B6+B7*B14
17	Total purchased cost	=B14*B10
18	Difference	=B16-B17
19		
20	Decision	=IF(B18<=0, "Manufacture", "Outsource")

formulas entered in the cells that reflect the relationships among the model components. For any set of inputs, the spreadsheet calculates some output measures of interest. Spreadsheets are ideal vehicles for implementing decision models because of their versatility in managing data, evaluating different scenarios, and presenting results in a meaningful fashion.

Figure 9.2 shows an example of a spreadsheet for evaluating a simple outsourcing decision. Suppose that a manufacturer can produce a part for $125/unit with a fixed cost of $50,000. The alternative is to outsource production to a supplier at a unit cost of $175. The decision depends on the anticipated volume of demand; for high volumes, the cost to manufacture in-house will be lower than outsourcing, because the fixed costs can be spread over a large number of units. For small volumes, it would be more economical to outsource. Knowing the cost of both alternatives and the breakeven point would facilitate the decision. The data consist of the costs associated with manufacturing the product in-house or purchasing it from an outside supplier. The key model input is the demand volume and the outputs are the total manufacturing and purchase cost. Calculating the outputs basically consists of "stepping through" the formulas.

This model can also be expressed mathematically by defining symbols for each component of the model:

F = fixed cost of in-house manufacturing
V = unit variable cost of in-house manufacturing
C = unit cost of outsourcing
D = demand volume

Then the total manufacturing cost can be written as $TMC = F + V \times D$, and the total outsourcing cost as $TOC = C \times D$. Note the correspondence between the spreadsheet formulas and the mathematical model:

$$TMC = B6 + B7*B14$$
$$TOC = B14*B10$$

Thus, if you can write a spreadsheet formula, you can develop a mathematical model!

Mathematical models are easy to manipulate; for example, it is easy to find the breakeven volume by setting $TMC = TOC$ and solving for D:

$$F + V \times D = C \times D$$
$$D = F/(C - V)$$

In contrast, it is more difficult to find the breakeven volume using trial-and-error on the spreadsheet without knowing some advanced tools and approaches. However, spreadsheets have the advantage of allowing you to easily modify the model inputs and calculate the numerical results. We will use both spreadsheets and analytical modeling approaches in our model-building applications, and it is important to be able to "speak both languages."

Models complement decision makers' intuition and often provide insights that intuition cannot. For example, one early application of management science in marketing involved a study of sales operations. Sales representatives had to divide their time between large and small customers and between acquiring new customers and keeping old ones. The problem was to determine how the representatives should best allocate their time. Intuition suggested that they should concentrate on large customers and that it was much harder to acquire a new customer than to keep an old one. However, intuition could not tell whether they should concentrate on the 500 largest or the 5,000 largest customers or how much effort to spend on acquiring customers. Models of sales force effectiveness and customer response patterns provided the insight to make these decisions. However, it is important to understand that all models are only representations of the "real world," and as such, cannot capture every nuance that decision makers face in reality. Decision makers must often modify the policies that models suggest to account for intangible factors that they might not have been able to incorporate into the model.

Decision models take many different forms. Some models are **descriptive;** they simply describe relationships and provide information for evaluation. Regression models that we studied in Chapter 6 are examples of descriptive models; they describe relationships between the dependent and independent variables. The outsourcing model is also descriptive in that it simply allows you to evaluate the cost impact of different demand volumes. Note that demand volume is an uncontrollable input we can vary to evaluate the costs of the decision alternatives. The model itself does not include any decision variables. Descriptive models are used to explain the behavior of systems, to predict future events as inputs to planning processes, and to assist decision makers in making decisions.

Other models, called *optimization models,* are **prescriptive**; they seek to determine an optimal policy, that is, the best course of action that a decision maker should take to maximize or minimize some objective. In a highly competitive world where one percentage point can mean a difference of hundreds of thousands of dollars or more, knowing the best solution can mean the difference between success and failure. To illustrate an example of an optimization model, suppose that an airline has studied the price elasticity for the demand for a round trip between Chicago and Cancun. They discovered that when the price is $600, daily demand is 500 passengers per day, but when the price is $300, demand increases to 1,200 passengers per day. The airplane capacity is 300 passengers, but the airline will add additional flights if demand warrants. The fixed cost associated with each flight is $90,000. The decision is to determine the price that maximizes profit.

Figure 9.3 *Airline Pricing Model* Spreadsheet

	A	B
1	**Airline Pricing Model**	
2		
3	**Data**	
4	Airplane capacity	300
5	Fixed cost	$ 90,000
6	Demand function	
7	slope	-2.33
8	intercept	1900
9		
10	**Model**	
11		
12	Revenue	
13	Unit price	$ 500.00
14	Demand	733
15	Number of flights/day	3
16	Total Revenue	$366,666.67
17	Cost	
18	Fixed Cost	$ 270,000.00
19		
20	Profit	$96,666.67

	A	B
1	**Airline Pricing Model**	
2		
3	**Data**	
4	Airplane capacity	300
5	Fixed cost	90000
6	Demand function	
7	slope	=-7/3
8	intercept	1900
9		
10	**Model**	
11		
12	Revenue	
13	Unit price	500
14	Demand	=B8+B7*B13
15	Number of flights/day	=ROUNDUP(B14/B4,0)
16	Total Revenue	=B13*B14
17	Cost	
18	Fixed Cost	=B5*B15
19		
20	Profit	=B16-B18

To develop the optimization model, we have to first characterize demand as a function of price. Because we are provided with only two data points ($600, 500) and ($300, 1,200), we can assume that demand is a linear function of price and determine the equation of the straight line between them using algebra. Using the basic equation of a straight line, $y = mx + b$, where y is the demand, x is the price, m is the slope, and b is the y-intercept, we calculate the slope and intercept as:

$$m = (500 - 1,200)/(\$600 - \$300) = -7/3 \quad \text{and}$$
$$b = y - mx = 500 - (-7/3)(600) = 1,900$$

Thus, demand $= 1,900 - 7/3$ price. The number of flights/day would be calculated as the demand divided by the airplane capacity (300), rounded up to the next whole number. Now it is easy to calculate daily profit as:

$$\text{Profit} = \text{Demand} \times \text{price} - \text{fixed cost} \times \text{flights/day}$$

Figure 9.3 shows a spreadsheet model for this scenario (Excel file *Airline Pricing Model*). The objective is to find the price that yields the largest profit. In this model, the unit price is a decision variable. The ROUNDUP function is used in cell B15 to ensure that a sufficient number of flights are scheduled to meet demand.

MODEL ANALYSIS

A model helps managers to gain insight into the nature of the relationships among components of a problem, aids intuition, and provides a vehicle for communication. We might be interested in studying the impact of changes in assumptions on model outputs, finding a solution such as a breakeven value, determining the best solution to an optimization model, or evaluating risks associated with decision alternatives. We will discuss some techniques for addressing these objectives; however, the subject of risk analysis will be the focus of the next two chapters. Excel 2007 provides several tools for model analysis—data tables, scenario manager, and goal seek.

What-If Analysis

Spreadsheet models allow you to easily evaluate "what-if" questions—how specific combinations of inputs that reflect key assumptions will affect model outputs. For instance, in the outsourcing decision model described earlier, we might be interested in how different levels of fixed and variable costs affect the total manufacturing cost and the resulting decision. The process of changing key model inputs to determine their effect on the outputs is also often called **sensitivity analysis**. This is one of the most important and valuable approaches to gaining the appropriate insights to make good decisions.

Sensitivity analysis is as easy as changing values in a spreadsheet and recalculating the outputs. However, systematic approaches make this process easier and more

EXCEL NOTE
Creating Data Tables

To create a one-way data table, first create a range of values for some input cell in your model that you wish to vary. The input values must be listed either down a column (column-oriented) or across a row (row-oriented). If the input values are column-oriented, enter the cell reference for the output variable in your model that you wish to evaluate in the row *above* the first value and one cell to the *right* of the column of input values. Reference any other output variable cells to the right of the first formula. If the input values are listed across a row, enter the cell reference of the output variable in the column to the *left* of the first value and one cell *below* the row of values. Type any additional output cell references below the first one. Next, select the range of cells that contains *both* the formulas and values you want to substitute. From the *Data* tab in Excel, select *Data Table* under the *What-If Analysis* menu. In the dialog box (see Figure 9.4), if the input range is column-oriented, type the cell reference

for the input cell in your model in the *Column input cell* box. If the input range is row-oriented, type the cell reference for the input cell in the *Row input cell* box.

To create a two-way data table, type a list of values for one input variable in a column, and a list of input values for the second input variable in a row, starting one row above and one column to the right of the column list. In the cell in the upper left-hand corner immediately above the column list and to the left of the row list, enter the cell reference of the output variable you wish to evaluate. Select the range of cells that contains this cell reference and both the row and column of values. On the *What-If Analysis* menu, click *Data Table*. In the *Row input cell* of the dialog box, enter the reference for the input cell in the model for the input values in the row. In the *Column input cell* box, enter the reference for the input cell in the model for the input values in the column. Then click *OK*.

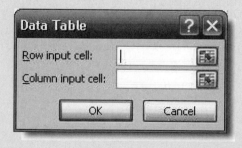

Figure 9.4 *Data Table* Dialog Box

useful. Spreadsheets facilitate sensitivity analysis. Excel 2007 provides several tools—data tables, scenario manager, and goal seek. These can be found within the *What-If Analysis* menu in the *Data* tab.

Data Tables

Data tables summarize the impact of one or two inputs on a specified output. Excel allows you to construct two types of data tables (see *Excel Note: Creating Data Tables*). A **one-way data table** evaluates an output variable over a range of values for a single input variable. **Two-way data tables** evaluate an output variable over a range of values for two different input variables.

We will illustrate the use of data tables to evaluate the impact of cost assumptions in the outsourcing decision model. Suppose we wish to create a one-way data table to evaluate the difference in manufacturing and outsourcing cost and the best decision for varying levels of fixed costs, holding all other model inputs constant. The range of these input values is shown in D4:D11 in Figure 9.5. In cell E3 enter the cell reference for the difference cell (=B18), and in cell F3 enter the cell reference for the decision (=B20). Next, select the data table range, the smallest rectangular block that includes the formula and all the values in the input range (D3:F11). In the Column Input cell of the dialog box, enter B6, the cell that contains the input value of fixed cost. Excel evaluates the difference and decision for each value in the data table as shown in Figure 9.5.

A two-way data table showing the decision for various values of fixed and variable costs is shown in Figure 9.6. Two-way data tables can only evaluate one output variable. In this case, we evaluate the decision in cell B20 of the model and reference this in cell D16. In the *Data Table* dialog box, the row input cell is B7 and the column

Figure 9.5 One-Way Data Table

	A	B	C	D	E	F
1	**Outsourcing Decision Model**					
2				Fixed Costs	Difference	Decision
3	**Data**		Column input cell		$ (25,000)	Manufacture
4				$ 30,000	$ (45,000)	Manufacture
5	**Manufactured in-house**			$ 40,000	$ (35,000)	Manufacture
6	Fixed cost	$	50,000	$ 50,000	$ (25,000)	Manufacture
7	Unit variable cost	$	125	$ 60,000	$ (15,000)	Manufacture
8				$ 70,000	$ (5,000)	Manufacture
9	**Purchased from supplier**			$ 80,000	$ 5,000	Outsource
10	Unit cost	$	175	$ 90,000	$ 15,000	Outsource
11				$ 100,000	$ 25,000	Outsource
12	**Model**					
13						
14	Demand volume		1500			
15						
16	Total manufacturing cost	$	237,500			
17	Total purchased cost	$	262,500			
18	Difference	$	(25,000)			
19						
20	Decision	Manufacture				

Figure 9.6 Two-Way Data Table

	D	E	F	G	H	I
15	Fixed Costs			Variable costs		
16	Manufacture	100	120	130	140	150
17	$ 30,000	Manufacture	Manufacture	Manufacture	Manufacture	Manufacture
18	$ 40,000	Manufacture	Manufacture	Manufacture	Manufacture	Outsource
19	$ 50,000	Manufacture	Manufacture	Manufacture	Manufacture	Outsource
20	$ 60,000	Manufacture	Manufacture	Manufacture	Outsource	Outsource
21	$ 70,000	Manufacture	Manufacture	Outsource	Outsource	Outsource
22	$ 80,000	Manufacture	Manufacture	Outsource	Outsource	Outsource
23	$ 90,000	Manufacture	Outsource	Outsource	Outsource	Outsource
24	$ 100,000	Manufacture	Outsource	Outsource	Outsource	Outsource

input cell is B6. The result shows those combination of inputs for which manufacturing or outsourcing is more economical.

Scenario Manager

The Excel *Scenario Manager* tool allows you to create **scenarios**—sets of values that are saved and can be substituted automatically on your worksheet (see the Excel Note on using the *Scenario Manager*). For example, suppose that the fixed and variable costs, as well as the demand volume are uncertain but that the supplier cost is fixed. Through discussions with key managers, you have determined best-case, worst-case, and most-likely scenarios for these inputs:

	FIXED COST	UNIT VARIABLE COST	DEMAND VOLUME
Best case	$40,000	$120	1800
Worst case	$60,000	$140	1000
Most likely case	$55,000	$125	1500

Using the *Scenario Manager*, the summary report shown in Figure 9.7 was created. This indicates that among these three scenarios, only the worst case results in a decision to outsource the part.

Goal Seek

If you know the result that you want from a formula, but are not sure what input value the formula needs to get that result, use the *Goal Seek* feature in Excel. *Goal Seek* works only with one variable input value. If you want to consider more than one input value or wish to maximize or minimize some objective, you must use the *Solver* add-in, which will be described later. For example, in the outsourcing decision model, you might be interested in finding the breakeven point. The breakeven point would be the value of demand volume for which total manufacturing cost would equal total purchased cost, or equivalently, for which the difference is zero. Therefore, you seek to find the value of demand volume in cell B14 that yields a value of zero in cell B18. In the *Goal Seek* dialog box (see *Excel Note: Using Goal Seek*), enter B18 for the *Set cell*, enter 0 in the *To value* box, and B14 in the *By changing cell* box. The *Goal Seek* tool determines that the breakeven demand is 1000.

Figure 9.7 Scenario Summary for Outsourcing Model

		Current Values:	Best case	Worst case	Most likely case
Scenario Summary					
Changing Cells:					
	$B$6	$ 50,000	$ 40,000	$ 60,000	$ 55,000
	$B$7	$ 125	$ 120	$ 140	$ 125
	$B$14	1500	1800	1000	1500
Result Cells:					
	$B$20	Manufacture	Manufacture	Outsource	Manufacture
	$B$18	$ (25,000)	$ (59,000)	$ 25,000	$ (20,000)

Notes: Current Values column represents values of changing cells at time Scenario Summary Report was created. Changing cells for each scenario are highlighted in gray.

EXCEL NOTE
Using the *Scenario Manager*

The Excel *Scenario Manager* is found under the *What-if Analysis* menu in the *Data Tools* group on the *Data* tab. When the tool is started, the *Add Scenario* dialog opens (see Figure 9.8). Enter the name of the scenario in the *Scenario name* box. In the *Changing cells* box, enter the references for the cells that you want to specify in

Figure 9.8 *Add Scenario* Dialog Box

your scenario. After all scenarios are added, they can be selected by clicking on the name of the scenario and then the *Show* button (see Figure 9.9). Excel will change all values of the cells in your spreadsheet to correspond to those defined by the scenario in order for you to evaluate the results. You can also create a summary report on a new worksheet by clicking the *Summary* button on the *Scenario Manager* dialog.

Figure 9.9 *Scenario Manager* Dialog Box

Model Optimization

What-if analyses are useful approaches for descriptive models; however, the purpose of optimization models is to find the *best* solution. For some models, analytical solutions—closed-form mathematical expressions—can be obtained using such techniques as calculus. In most cases, however, an algorithm is needed to provide the solution. An **algorithm** is a systematic procedure that finds a solution to a problem. Researchers have developed algorithms to solve many types of optimization problems. However, we will not be concerned with the detailed mechanics of these algorithms; our focus will be on the use of the algorithms to solve the models we develop.

Microsoft Excel contains an add-in called *Solver* that allows you to find optimal solutions to optimization problems formulated as spreadsheet models. *Solver* was developed and is maintained by Frontline Systems, Inc. (www.solver.com). Frontline Systems also supports a more powerful version of *Solver, Premium Solver,* an educational version of which is also packaged with this book. We suggest that you install it and use it because it is a more robust product than the Excel-supplied version (*Standard Solver*), and we will use it exclusively in the last two chapters of this book. For now, we will use the standard version.

We will illustrate the use of *Solver* for the airline pricing model. *Solver* can be found in the *Analysis* group under the *Data* tab. (If *Solver* is not activated, click the

EXCEL NOTE
Using *Goal Seek*

On the *Data* tab, in the *Data Tools* group, click *What-If Analysis*, and then click *Goal Seek*. The dialog box shown in Figure 9.10 will be shown. In the *Set cell* box, enter the reference for the cell that contains the formula that you want to resolve. In the *To value* box, type the formula result that you want. In the *By changing cell* box, enter the reference for the cell that contains the value that you want to adjust.

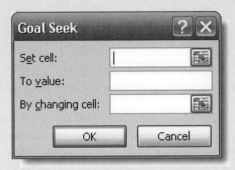

Figure 9.10 *Goal Seek* Dialog Box

Office button, choose *Excel Options*, and select *Add-ins*. Click the *Go* button to manage Excel Add-ins and check the box for *Solver Add-in*.) When *Solver* is started, the *Solver Parameters* dialog box appears as shown in Figure 9.11. The *Target Cell* is the cell in the spreadsheet that you wish to optimize; in this case the profit function in cell B20. Because we want to find the largest profit, we click the radio button for *Max*. *Changing Cells* are another name for decision variables. In this example we wish to find the unit price (cell B13) that results in the maximum profit. Clicking the *Solve* button displays the solution found in the spreadsheet:

$$\text{Unit price} = \$428.57 \text{ and Profit} = \$115,714.28$$

Figure 9.11 *Solver Parameters* Dialog Box

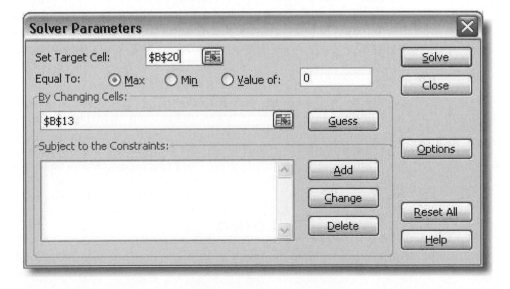

Figure 9.12 *Solver Results* Dialog

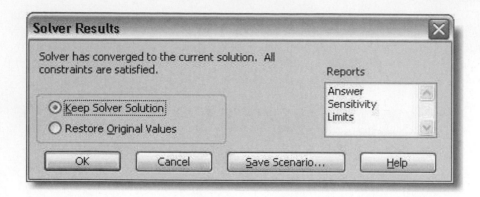

The *Solver Results* box (Figure 9.12) provides several reports, but these are not important at this point. We will describe these and other features of *Solver* in Chapter 13.

Although the airline pricing model did not, most optimization models have **constraints**—limitations, requirements, or other restrictions that are imposed on any solution, such as "do not exceed the allowable budget" or "ensure that all demand is met." For instance, a consumer products company manager would probably want to ensure that a specified level of customer service is achieved with the redesign of the distribution system. The presence of constraints makes identifying the best solution considerably more difficult, and we will address constrained optimization problems in the last two chapters of the book.

If possible, we would like to ensure that an algorithm such as the one *Solver* uses finds an optimal solution. However, some models are so complex that it is impossible to solve them optimally in a reasonable amount of computer time because of the extremely large number of computations that may be required or because they are so complex that an optimal solution cannot be guaranteed. In these cases, analysts use **heuristics**—solution procedures that generally find good solutions without guarantees of finding an optimal solution. (The term *heuristic* stems from a Greek word meaning "to discover.") Researchers have developed powerful heuristics to obtain good solutions to extremely difficult optimization problems. Another practical reason for using heuristics is that a manager might be satisfied with a solution that is good but not necessarily optimal, particularly when:

- inexact or limited data used to estimate uncontrollable quantities in models may contain more error than that of a nonoptimal solution,
- the assumptions used in a model make it an inaccurate representation of the real problem, making having the "best" solution pointless, or
- anything better than the current solution will suffice, so long as it can be obtained at a reasonable cost and in a reasonable amount of time.

TOOLS FOR MODEL BUILDING

Building decision models is more of an art than a science; however, there are many approaches that can facilitate the development of useful models.

Logic and Business Principles

Building good decision models requires a solid understanding of basic business principles in all functional areas, such as accounting, finance, marketing, and operations, knowledge of business practice and research, and logical skills. For example, suppose that you wish to create a model to compute the profit associated with production and sale of a product. A fundamental business principle is:

$$Profit = Revenue - Cost \qquad (9.1)$$

Using your knowledge and experience, you can expand *Revenue* and *Cost* terms as follows:

$$Revenue = (Unit\ price)(Quantity\ sold)$$
$$Cost = [Fixed\ cost + (Unit\ cost)(Quantity\ produced)]$$

Thinking more about this, you might realize that *Quantity sold* is related to both *Quantity produced* and the demand for the product. Specifically, the quantity sold must be equal to the smaller of the demand or the quantity produced:

$$Quantity\ sold = Min(Quantity\ produced, Demand)$$

Therefore, the final model is:

$$Profit = (Unit\ price)Min(Quantity\ produced, Demand)$$
$$- [Fixed\ cost + (Unit\ cost)(Quantity\ produced)]$$

Many business ventures are evaluated on the basis of financial criteria such as *net present value (NPV)*. This will be used in several examples and problems in the remainder of this book; more complete discussions can be found in basic finance texts. See *Excel Note: Net Present Value and the NPV Function* for details about implementing it on spreadsheets. The more you learn about fundamental theory in business, the better prepared you will be to develop good models.

EXCEL NOTE
Net Present Value and the NPV Function

Net present value (also called **discounted cash flow**) measures the worth of a stream of cash flows, taking into account the time value of money. That is, a cash flow of F dollars t time periods in the future is worth $F/(1 + i)^t$ dollars today, where i is the **discount rate.** The discount rate reflects the opportunity costs of spending funds now versus achieving a return through another investment, as well as the risks associated with not receiving returns until a later time. The sum of the present values of all cash flows over a stated time horizon is the net present value:

$$NPV = \sum_{t=0}^{n} \frac{F_t}{(1 + i)^t} \qquad (9.2)$$

where F_t = cash flow in period t. A positive NPV means that the investment will provide added value since the projected return exceeds the discount rate.

The Excel function NPV(*rate, value1, value2, . . .*) calculates the net present value of an investment by using a discount rate and a series of future payments (negative values) and income (positive values). *Rate* is the rate of discount over the length of one period (i), and *value1, value2, . . .* are 1 to 29 arguments representing the payments and income. The values must be equally spaced in time and are assumed to occur at the end of each period. The NPV investment begins one period before the date of the *value1* cash flow and ends with the last cash flow in the list. The NPV calculation is based on future cash flows. If the first cash flow (such as an initial investment or fixed cost) occurs at the beginning of the first period, then it must be added to the NPV result and *not* included in the function arguments.

Common Mathematical Functions

Understanding different functional relationships is instrumental in model building. For example, in the airline pricing model, we developed a linear function relating price and demand using two data points. Common types of mathematical functions used in models include:

Linear: $y = mx + b$. Linear functions show steady increases or decreases over the range of x.

Logarithmic: $y = ln(x)$. Logarithmic functions are used when the rate of change in a variable increases or decreases quickly and then levels out, such as with diminishing returns to scale.

Polynomial: $y = ax^2 + bx + c$ (second order—quadratic function), $y = ax^3 + bx^2 + dx + e$ (third order—cubic function), etc. A second-order polynomial is parabolic in nature and has only one hill or valley; a third-order polynomial has one or two hills or valleys. Revenue models that incorporate price elasticity are often polynomial functions.

Power: $y = ax^b$. Power functions define phenomena that increase at a specific rate. Learning curves that express improving times in performing a task are often modeled with power functions having $a > 0$ and $b < 0$.

Exponential: $y = ab^x$. Exponential functions have the property that y rises or falls at constantly increasing rates. For example, the perceived brightness of a lightbulb grows at a decreasing rate as the wattage increases. In this case, a would be a positive number and b would be between 0 and 1. The exponential function is often defined as $y = ae^x$, where $b = e$, the base of natural logarithms (approximately 2.71828).

Data Fitting

For many applications, functional relationships used in decision models are derived from the analysis of data. The Excel *Trendline* tool (described in Chapter 6) provides a convenient method for determining the best fitting functional relationship. Figure 9.13

Figure 9.13   Chart of Crude Oil Prices

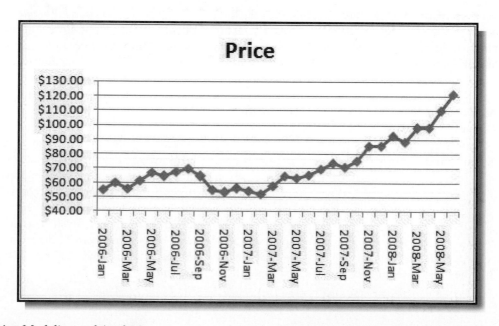

Figure 9.14   Polynomial Fit of Crude Oil Prices

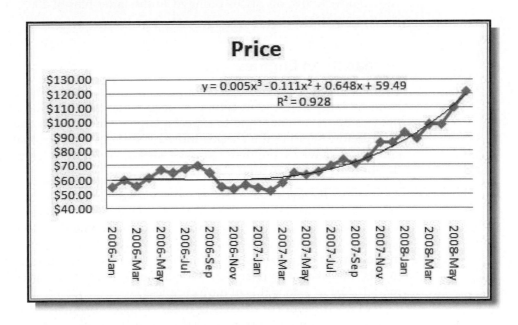

shows a chart of historical data on crude oil prices on the first Friday of each month from January 2006 through June 2008 (data are in the Excel file *Crude Oil Prices* in the *Decision Model Files* folder). Using the *Trendline* tool, we can try to fit the various functions to these data (here x represents the number of months starting with January 2006). The results are:

Exponential: $y = 50.49e^{0.021x}$ $R^2 = 0.664$
Logarithmic: $y = 13.02ln(x) + 39.60$ $R^2 = 0.382$
Polynomial (second order): $y = 0.130x^2 - 2.399x + 68.01$ $R^2 = 0.905$
Polynomial (second order): $y = 0.005x^3 - 0.111x^2 + 0.648x + 59.49$ $R^2 = 0.928$
Power: $y = 45.96x^{.0169}$ $R^2 = 0.397$

The best-fitting model is the third-order polynomial, shown in Figure 9.14.

Of course, the proper model to use depends on the scope of the data. As the chart shows, crude oil prices were relative stable until early 2007 and then began to increase rapidly. By including the early data, the long-term functional relationship might not adequately express the short-term trend. For example, fitting a model to only the data beginning with January 2007 yields the best-fitting models:

Exponential: $y = 50.56e^{0.044x}$ $R^2 = 0.969$
Polynomial (second order): $y = 0.121x^2 + 1.232x + 53.48$ $R^2 = 0.968$
Linear: $y = 3.548x + 45.76$ $R^2 = 0.944$

The difference in prediction can be significant. For example, to predict the price 6 months after the last data point ($x = 36$) yields $172.24 for the third-order polynomial fit with all the data, and $246.45 for the exponential model with only the recent data. Thus, the analysis must be careful to select the proper amount of data for the analysis. The question now becomes one of choosing the best assumptions for the model. Is it reasonable to assume that prices would increase exponentially or perhaps at a slower rate, such as with the linear model fit? Or, would they level off and start falling? Clearly, factors other than historical trends would enter into this choice. As we now know, oil prices plunged in the latter half of 2008; thus, all predictive models are risky.

Spreadsheet Engineering

First and foremost, spreadsheets should be accurate. Spreadsheet errors can be disastrous. A large investment company once made a $2.6 billion error. They notified holders of one mutual fund to expect a large dividend; fortunately, they caught the error before sending the checks. Industry surveys estimate that more than 90% of spreadsheets with more than 150 rows were incorrect by at least 5%. So what can we do about it? There are three basic things:

1. *Improve the design and format of the spreadsheet itself.* After the inputs, outputs, and key relationships are well understood, you should sketch a logical design of the spreadsheet. For example, you might want the spreadsheet to resemble a financial statement to make it easier for managers to read. It is good practice to separate the model inputs from the model itself and to reference the input cells in the model formulas; that way, any changes in the inputs will be automatically reflected in the model. We have done this in the examples.

Another useful approach is to break complex formulas into smaller pieces. This reduces typographical errors, makes it easier to check your results, and also makes the spreadsheet easier to read for the user. For example, we could write one long formula for profit using the equation we developed earlier in this section. However, setting up the model in a form similar to a financial statement and breaking out the revenue and costs makes the calculations much clearer, especially to a manager who will be using the model.

2. *Improve the process used to develop a spreadsheet.* If you sketched out a conceptual design of the spreadsheet, work on each part individually before moving on to the others to ensure that it is correct. As you enter formulas, check the results with simple numbers (such as 1) to determine if they make sense, or use inputs with known results. Be careful in using the *Copy* and *Paste* commands in Excel, particularly with respect to relative and absolute addresses. Use the function wizard (the f_x button on the toolbar) to ensure that you are entering the correct values in the correct fields of the function. Use cell and range names to simplify formulas and make them more user-friendly. For example, suppose that the unit price is stored in cell B13 and quantity sold in cell B14. Suppose you wish to calculate revenue in cell C15. Instead of writing the formula =B13*B14, you could define the name of cell B13 in Excel as "UnitPrice" and the name of cell B14 as "QuantitySold." Then in cell C15, you could simply write the formula =UnitPrice*QuantitySold. (In this book, however, we will use cell references so that you can more easily trace formulas in the examples.)

3. *Inspect your results carefully and use appropriate tools available in Excel.* The *Data Validation* tool can signal errors if an input value is not the right type. For example, it does not make sense to input a Quantity Produced that is not a whole number. With this tool, you may define validation criteria for model inputs and pop up an error message if the wrong values or type of data are entered. The Excel *Auditing* tool also helps you to validate the logic of formulas by visually showing the relationships between input data and cell formulas. We encourage you to learn how to use these tools.

Modeling Examples

In this section, we present several examples that apply the modeling principles we have discussed.

Gasoline Consumption

Automobiles have different fuel economies (mpg), and commuters drive different distances to work or school. Suppose that a state Department of Transportation (DOT) is interested in measuring the average monthly fuel consumption of commuters in a certain city. The DOT might sample a group of commuters and collect information on the number of miles driven per day, number of driving days per month, and the fuel economy of their cars.

We can develop a model for the amount of gasoline consumed by first applying a bit of logic. If a commuter drives m miles per day and d days/month, then the total number of miles driven per month is $m \times d$. If a car gets f miles per gallon in fuel economy, then the number of gallons consumed per month must be $(m \times d)/f$. Notice that the dimensions of this expression are (miles/day)(days/month)/(miles/gallon) = gallons/month. Consistency in dimensions is an important validation check for model accuracy.

Revenue Model

Suppose that a firm wishes to determine the best pricing for one of its products to maximize revenue over the next year. A market research study has collected data that estimate the expected annual sales for different levels of pricing as shown in Figure 9.15. Plotting these data on a scatter chart suggests that sales and price have

Figure 9.15 Price-Sales Data and Trendline

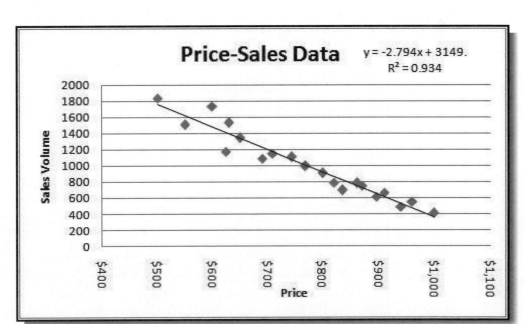

Figure 9.16 *Moore Pharmaceuticals* Spreadsheet Model

	A	B	C	D	E	F
1	**Moore Pharmaceuticals**					
2						
3	**Data**					
4						
5	Market size	2,000,000				
6	Unit (monthly Rx) revenue	$ 130.00				
7	Unit (monthly Rx) cost	$ 40.00				
8	Discount rate	9%				
9						
10	Project Costs					
11	R&D	$ 700,000,000				
12	Clinical Trials	$ 150,000,000				
13	Total Project Costs	$ 850,000,000				
14						
15	**Model**					
16						
17	Year	1	2	3	4	5
18	Market growth factor		3.00%	3.00%	3.00%	3.00%
19	Market size	2,000,000	2,060,000	2,121,800	2,185,454	2,251,018
20	Market share growth rate		20.00%	20.00%	20.00%	20.00%
21	Market share	8.00%	9.60%	11.52%	13.82%	16.59%
22	Sales	160,000	197,760	244,431	302,117	373,417
23						
24	Annual Revenue	$ 249,600,000	$ 308,505,600	$ 381,312,922	$ 471,302,771	$ 582,530,225
25	Annual Costs	$ 76,800,000	$ 94,924,800	$ 117,327,053	$ 145,016,237	$ 179,240,069
26	Profit	$ 172,800,000	$ 213,580,800	$ 263,985,869	$ 326,286,534	$ 403,290,156
27	Cumulative Net Profit	$ (677,200,000)	$ (463,619,200)	$ (199,633,331)	$ 126,653,203	$ 529,943,358
28						
29	Net Present Value	$ 185,404,860				

	A	B	C	D	E	F
1	**Moore Pharmaceuticals**					
2						
3	**Data**					
4						
5	Market size	2000000				
6	Unit (monthly Rx) revenue	130				
7	Unit (monthly Rx) cost	40				
8	Discount rate	0.09				
9						
10	Project Costs					
11	R&D	700000000				
12	Clinical Trials	150000000				
13	Total Project Costs	=B11+B12				
14						
15	**Model**					
16						
17	Year	1	2	3	4	5
18	Market growth factor		0.03	0.03	0.03	0.03
19	Market size	=B5	=B19*(1+C18)	=C19*(1+D18)	=D19*(1+E18)	=E19*(1+F18)
20	Market share growth rate		0.2	0.2	0.2	0.2
21	Market share	0.08	=B21*(1+C20)	=C21*(1+D20)	=D21*(1+E20)	=E21*(1+F20)
22	Sales	=B19*B21	=C19*C21	=D19*D21	=E19*E21	=F19*F21
23						
24	Annual Revenue	=B22*$B$6*12	=C22*$B$6*12	=D22*$B$6*12	=E22*$B$6*12	=F22*$B$6*12
25	Annual Costs	=B22*$B$7*12	=C22*$B$7*12	=D22*$B$7*12	=E22*$B$7*12	=F22*$B$7*12
26	Profit	=B24-B25	=C24-C25	=D24-D25	=E24-E25	=F24-F25
27	Cumulative Net Profit	=B26-B13	=B27+C26	=C27+D26	=D27+E26	=E27+F26
28						
29	Net Present Value	=NPV(B8,B26:F26)-B13				

a linear relationship. Adding a linear trendline confirms a strong correlation between the variables. Thus, the linear model, Sales = −2.794 × Price + 3149 is a good predictor of sales for any pricing decision between $500 and $1000. Because revenue equals price × sales, a model for total revenue is:

$$\text{Total revenue} = \text{Price} \times (-2.794 \times \text{Price} + 3149)$$
$$= -2.794 \times \text{Price}^2 + 3149 \times \text{Price}$$

This revenue function is a second-order polynomial (quadratic).

New Product Development

The decision to launch a new product always involves careful financial analysis. In the pharmaceutical industry, for example, the process of research and development is a long and arduous process; total development expenses can approach one billion dollars. Suppose that Moore Pharmaceuticals has discovered a potential drug breakthrough in the laboratory and needs to decide whether to go forward to conduct clinical trials and seek FDA approval to market the drug. Total R&D costs will be expected to reach $700 million, and the cost of clinical trials will be about $150 million. The current market size is estimated to be 2 million people and is expected to grow at a rate of 3% each year. In the first year, Moore estimates gaining an 8% market share, which is anticipated to grow by 20% each year. It is difficult to estimate beyond 5 years as new competitors are expected to be entering the market. A monthly prescription is anticipated to generate a revenue of $130 while incurring variable costs of $40. A discount rate of 9% is assumed. The company needs to know how long it will take to recover its fixed expenses and the net present value over the first five years.

Figure 9.16 shows a spreadsheet model for this situation (Excel file *Moore Pharmaceuticals*). Many models used in business involve multiple time periods and cannot be formulated as simple equations, thus requiring spreadsheets to represent them. You can see that the design and implementation of this spreadsheet incorporates fundamental business concepts and basic logic. The results show that the product becomes profitable in the fourth year, resulting in a net present value of about $185 million. However, the model is based on some rather tenuous assumptions about the market size and market share growth rates. Failure to estimate these accurately might lead to a poor decision to move forward with the drug.

MODELS INVOLVING UNCERTAINTY

In the examples we have seen thus far, all the data—particularly the uncontrollable inputs—are assumed to be known and constant. In many situations, this assumption may be far from reality because uncontrollable inputs usually exhibit random behavior. Some examples would be customer demand, arrivals to ATM machines, and returns on investments. We often assume such variables to be constant in order to simplify the model and the analysis. However, many situations dictate that randomness be explicitly incorporated into our models. This is usually done by specifying probability distributions for the appropriate uncontrollable inputs. Models that include randomness are called *probabilistic*, or *stochastic*, models. These types of models help to evaluate risks associated with undesirable consequences and to find optimal decisions under uncertainty.

Recall that in Chapter 3 we described how to sample randomly from probability distributions and to generate certain random variates using Excel tools and functions. We will use these techniques to show how to incorporate uncertainty into decision models.

Newsvendor Model

Banana Republic, a division of Gap, Inc., was trying to build a name for itself in fashion circles as parent Gap shifted its product line to basics such as cropped pants, jeans, and khakis. In one recent holiday season, the company had bet that blue would be the top-selling color in stretch merino wool sweaters. They were wrong; as the company president noted, "The number 1 seller was moss green. We didn't have enough."[1]

This situation describes one of many practical situations in which a one-time purchase decision must be made in the face of uncertain demand. Department store buyers must purchase seasonal clothing well in advance of the buying season, and candy shops must decide on how many special holiday gift boxes to assemble. The general scenario is commonly known as the *newsvendor problem*: A street newsvendor sells daily newspapers and must make a decision of how many to purchase. Purchasing too few results in lost opportunity to increase profits, but purchasing too many will result in a loss since the excess must be discarded at the end of the day.

We will first develop a general model for this problem then illustrate it with an example. Let us assume that each item costs $C to purchase and is sold for $R. At the end of the period, any unsold items can be disposed of at $S each (the salvage value). Clearly, it makes sense to assume that $R > C > S$. Let D be the number of units demanded during the period and Q be the quantity purchased. Note that D is an uncontrollable input while Q is a decision variable. If demand is known, then the optimal decision is obvious: Choose $Q = D$. However, if D is not known in advance, we run the risk of over-purchasing or under-purchasing. If $Q < D$, then we lose the opportunity of realizing additional profit (since we assume that $R < C$), and if $Q > D$, we incur a loss (because $C > S$).

Notice that we cannot sell more than the minimum of the actual demand and the amount produced. Thus, the quantity sold at the regular price is the smaller of D and Q. Also, the surplus quantity is the larger of 0 and $Q - D$. The net profit is calculated as:

$$\text{Net profit} = R \times \text{Quantity Sold} + S \times \text{Surplus Quantity} - C \times Q$$

To illustrate this model, let us suppose that a small candy store makes Valentine's Day gift boxes for $12.00 and sells them for $18.00. In the past, at least 40 boxes have been sold by Valentine's Day, but the actual amount is uncertain, and in the past, the owner has often run short or made too many. After the holiday, any unsold boxes are discounted 50% and are eventually sold. Figure 9.17 shows a spreadsheet model that computes the profit for any input values of purchase quantity and demand (Excel file *Newsvendor Model*). An Excel data table can be used to evaluate the profit for various combinations of the inputs. This table shows, for example, that higher purchase quantities have the potential for higher profits, but are also at risk for yielding lower profits than smaller purchase quantities that provide safer and more stable profit expectations.

Monte Carlo Simulation

The basic question in the newsvendor problem is to determine how many boxes of candy to make to maximize the expected profit. To address this question we need

---

[1]Louise Lee, "Yes, We Have a New Banana," *BusinessWeek*, May 31, 2004, 70–72.

Figure 9.17 *Newsvendor Model*

	A	B	C	D	E	F	G	H	I	J	K	L	M	N
1	Newsvendor Model			Demand					Purchase Quantity					
2				$ 237.00	40	41	42	43	44	45	46	47	48	49
3	Data			40	$240.00	$237.00	$234.00	$231.00	$228.00	$225.00	$222.00	$219.00	$216.00	$213.00
4				41	$240.00	$246.00	$243.00	$240.00	$237.00	$234.00	$231.00	$228.00	$225.00	$222.00
5	Selling price	$ 18.00		42	$240.00	$246.00	$252.00	$249.00	$246.00	$243.00	$240.00	$237.00	$234.00	$231.00
6	Cost	$ 12.00		43	$240.00	$246.00	$252.00	$258.00	$255.00	$252.00	$249.00	$246.00	$243.00	$240.00
7	Discount price	$ 9.00		44	$240.00	$246.00	$252.00	$258.00	$264.00	$261.00	$258.00	$255.00	$252.00	$249.00
8				45	$240.00	$246.00	$252.00	$258.00	$264.00	$270.00	$267.00	$264.00	$261.00	$258.00
9	Model			46	$240.00	$246.00	$252.00	$258.00	$264.00	$270.00	$276.00	$273.00	$270.00	$267.00
10				47	$240.00	$246.00	$252.00	$258.00	$264.00	$270.00	$276.00	$282.00	$279.00	$276.00
11	Demand	41		48	$240.00	$246.00	$252.00	$258.00	$264.00	$270.00	$276.00	$282.00	$288.00	$285.00
12	Purchase Quantity	44		49	$240.00	$246.00	$252.00	$258.00	$264.00	$270.00	$276.00	$282.00	$288.00	$294.00
13														
14	Quantity Sold	41												
15	Surplus Quantity	3												
16														
17	Profit	$237.00												

	A	B
1	Newsvendor Model	
2		
3	Data	
4		
5	Selling price	18
6	Cost	12
7	Discount price	9
8		
9	Model	
10		
11	Demand	41
12	Purchase Quantity	44
13		
14	Quantity Sold	=MIN(B11,B12)
15	Surplus Quantity	=MAX(0,B12-B11)
16		
17	Profit	=B14*B5+B15*B7-B12*B6

to make an assumption about the demand for the number of boxes. Assume that demand has a normal distribution with a mean of 43, a standard deviation of 4 boxes, and a lower limit of 40. We could use the Excel function *NORMINV (probability, mean, standard_deviation)* to generate random values of the demand by replacing the input in cell B11 of Figure 9.17 with the function $= MAX(40, ROUND(NORMINV(RAND(), 43, 4),0))$ as described in Chapter 3. We use the MAX function to ensure that the demand is at least 40 as noted in the problem statement, and the ROUND function to ensure whole numbers. Whenever the F9 key is pressed, the worksheet will be recalculated and the value of demand will change randomly.

We can use Monte Carlo sampling techniques that we introduced in Chapter 3 to analyze this model as a basis for a decision. For a specific purchase quantity in the newsvendor model, we can randomly generate a demand and compute the profit, and then repeat this for some number of trials. **Monte Carlo simulation** is the process of generating random values for uncertain inputs in a model, computing the output variables of interest, and repeating this process for many trials in order to understand the distribution of the output results.

Figure 9.18 Monte Carlo Simulation of *Newsvendor Model*

	A	B	C	D	E	F
1	**Newsvendor Model**				**Demand**	**Profit**
2				Trial	40	$ 228
3	**Data**			1	40	$ 228
4				2	40	$ 228
5	Selling price	$ 18.00		3	50	$ 264
6	Cost	$ 12.00		4	40	$ 228
7	Discount price	$ 9.00		5	43	$ 255
8				6	42	$ 246
9	**Model**			7	40	$ 228
10				8	44	$ 264
11	Demand	40		9	43	$ 255
12	Purchase Quantity	44		10	45	$ 264
13				11	46	$ 264
14	Quantity Sold	40		12	45	$ 264
15	Surplus Quantity	4		13	40	$ 228
16				14	42	$ 246
17	Profit	$ 228.00		15	40	$ 228
18				16	40	$ 228
19				17	46	$ 264
20				18	43	$ 255
21				19	40	$ 228
22				20	45	$ 264
23				Average	43	$ 246.45

Monte Carlo simulation can easily be accomplished for the newsvendor model using a data table as shown in Figure 9.18. Construct a data table by listing the number of trials down a column (here we used 20 trials), and referencing the value of demand and profit in cells E2 and F2 (that is, the formula in cell E2 is =B11, and in cell F2, =B17). Select the range of the table (D2:F22)—and here's the trick—in the *Column Input Cell* field in the *Data Table* dialog box, enter any blank cell in the spreadsheet. This is done because the trial number does not relate to any parameter in the model; we simply want to repeat the spreadsheet recalculation each time, knowing that the demand will change each time because of the use of the RAND function in the demand formula.

As you can see from the figure, each trial has a randomly generated demand and associated profit. The average profit is $246.45, using a purchase quantity of 44. You may then change the purchase quantity and rerun the Monte Carlo simulation, seeking to identify the purchase quantity that yields the largest average profit. *Caution: You must manually recalculate the AVERAGE function as it is not updated as the data table changes!*

Although the use of a data table illustrates the value of incorporating uncertainty into a decision model, it is impractical to apply to more complex problems. For example, in the *Moore Pharmaceuticals* model, many of the model parameters such as the initial market size, project costs, market size growth factors, and market share growth rates may all be uncertain. In addition, we need to be able to capture and save the results of each trial, and it would be useful to construct a histogram of the results in order to conduct further analyses. Fortunately, sophisticated software approaches that easily perform these functions are available. In the next chapter, you will learn how to use *Crystal Ball* software to perform large-scale Monte Carlo simulation. *Crystal Ball* also has a powerful distribution fitting capability, which we discuss next.

Fitting Probability Distributions to Data

In the absence of data, uncertainty in data inputs can be specified judgmentally by choosing a probability distribution that has the shape that would most reasonably represent the analyst's understanding about the uncertain variable. For example, a normal distribution is symmetric, with a peak in the middle. Exponential data are very positively skewed, with no negative values. A triangular distribution has a limited range and can be skewed in either direction. However, for many decision models, empirical data may be available, either in historical records or collected through special efforts. For example, maintenance records might provide data on machine failure rates and repair times, or observers might collect data on service times in a bank or post office. This provides a factual basis for choosing the appropriate probability distribution to model the input variable.

Goodness-of-fit tests provide statistical evidence to test hypotheses about the *nature* of the distribution. Three methods commonly used methods are the chi-square test, which is similar in nature to the approach used in the test for independence of categorical variables, Anderson–Darling test, and Kolmogorov–Smirnov test. These approaches test the hypotheses:

H_0: *the sample data come from a specified distribution* (*e.g., normal*)

H_1: *the sample data do not come from the specified distribution*

As with any hypothesis test, you can disprove the null hypothesis but cannot statistically *prove* that data come from the specified distribution. However, if you cannot reject the null, then you at least have some measure of faith that the hypothesized distribution fits the data rather well. The chi-square goodness-of-fit test breaks down the hypothesized distribution into areas of equal probability and compares the data points within each area to the number that would be expected for that distribution. The Kolmogorov–Smirnov test compares the cumulative distribution of the data with the theoretical distribution and bases its conclusion on the largest vertical distance between them. The Anderson–Darling method is similar, but puts more weight on the differences between the tails of the distributions. Special software programs, such as *Crystal Ball*, automate this process (see *Crystal Ball Note: Distribution Fitting*).

CRYSTAL BALL NOTE
Distribution Fitting

Crystal Ball, an Excel add-in that is used in Monte Carlo simulation that we will study in the next chapter, provides a useful data-fitting procedure. Not only does it determine the closeness of each fit using one of the standard goodness-of-fit tests, but it also determines the set of parameters for each distribution that best describes the characteristics of the data. For the newsvendor example, suppose that the store owner kept records for the past 20 years on the number of boxes sold (Excel file *Historical Candy Sales*). Figure 9.19 shows these data in a spreadsheet along with the *Crystal Ball* menu bar.

To fit distributions to data, first select a data cell and click *Define Assumption* from the *Define* group in the menu bar. Click on the *Fit* button in the *Distribution Gallery* that pops up. In the next dialog, specify the range of the data, which distributions to fit, and the ranking method for choosing the best fit

Figure 9.19   *Crystal Ball* Menu Options (Excel 2007)

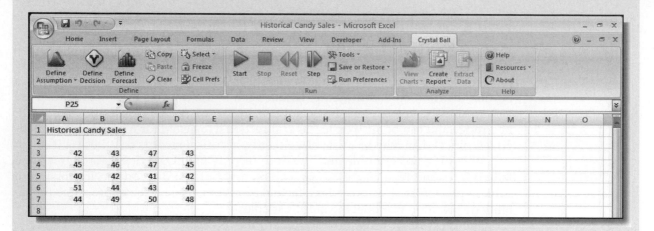

(see Figure 9.20). Click on *OK*. The fitted distributions appear in the *Comparison Chart* dialog (Figure 9.21); the shaded values indicate the highest rankings. The *Next* and *Previous* buttons beneath the *Comparison Chart* will allow you to scroll through the fitted probability distributions. Each probability distribution is shown super-imposed over the data. The parameters corresponding to the fitted distributions are also shown at the right. In the next chapter, we will learn how to use this to automatically enter the chosen distribution into a *Crystal Ball* simulation model.

Figure 9.20   *Fit Distribution* Dialog Box

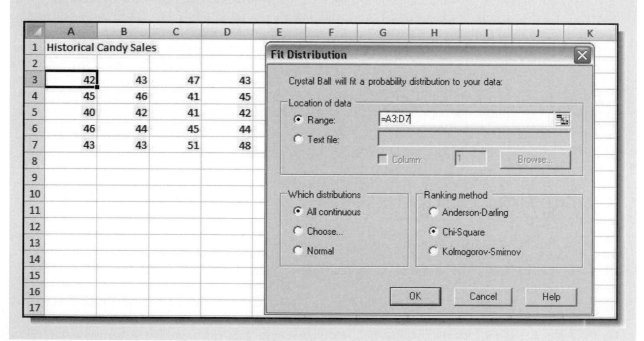

Figure 9.21 Comparison Chart and Distribution Fitting Results

Comparison Chart

Edit View Preferences Help

20 Values Split View 20 Displayed

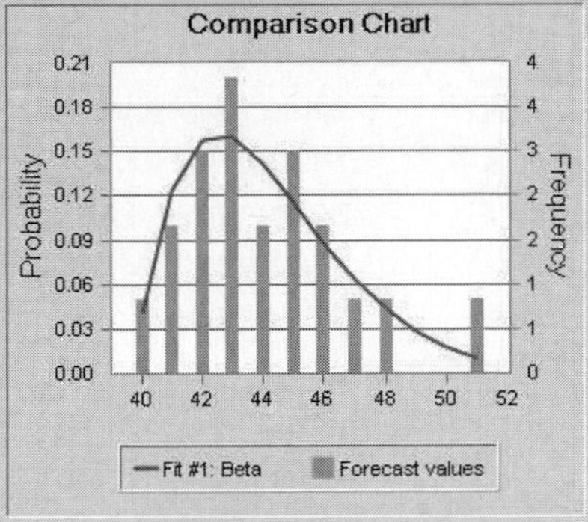

Ranked by: Chi-Square

Distribution	A-D	Chi-Square	K-S	Parameters
Lognormal	.3074	.4000	.1474	Mean=44.05,Std. Dev.=2.62
Weibull	.2975	.4000	.1459	Location=38.93,Scale=6.03,Shape=2.1049
Gamma	.1947	.4000	.1002	Location=38.87,Scale=1.35,Shape=3.82877
Beta	.2196	.4000	.1028	Minimum=39.63,Maximum=62.07,Alpha=2.12717,Beta=8.67416
Logistic	.2965	.4000	.1419	Mean=43.86,Scale=1.48
Max Extreme	.1947	.4000	.1062	Likeliest=42.85,Scale=2.07
Normal	.3718	1.2000	.1532	Mean=44.05,Std. Dev.=2.67
Student's t	.4605	1.2000	.1378	Midpoint=44.05,Scale=2.91,Deg. Freedom=11.70098
BetaPERT	.6842	1.6000	.1882	Minimum=39.63,Likeliest=40.65,Maximum=62.07
Triangular	1.1984	2.4000	.2080	Minimum=38.46,Likeliest=43.00,Maximum=53.51
Min Extreme	.9788	4.8000	.1719	Likeliest=45.44,Scale=3.00
Pareto	1.5391	6.4000	.2668	Location=39.80,Shape=10.02603
Uniform	2.4402	6.8000	.3089	Minimum=39.42,Maximum=51.58
Exponential	7.8162	60.0000	.5780	Rate=0.02

<< Previous Next >> Accept Cancel Help

Model Assumptions, Complexity, and Realism

Models cannot capture every detail of the real problem, and managers must understand the limitations of models and their underlying assumptions. **Validity** refers to how well a model represents reality. One approach for judging the validity of a model is to identify and examine the assumptions made in a model to see how they agree with one's perception of the real world; the closer the agreement, the higher the validity. A "perfect" model corresponds to the real world in every respect; unfortunately, no such model has ever existed, and never will exist in the future, because it is impossible to include every detail of real life in one model. To add more realism to a model generally requires more complexity and analysts have to know how to balance these.

To illustrate this, consider the following. The Nemesh Printing Company (NPC) receives orders of various sizes it classifies as either small or large. Typically, NPC receives between 5 and 10 small orders and 1 and 6 large orders each month. Revenues for small orders range between \$3000 and \$15,000, and revenues for large orders range between \$15,000 and \$50,000. Returns and allowances typically range between 5% and 10% of revenues. Marginal costs are 35% for small orders and 65% for large orders. Fixed costs amount to \$55,000 each month.

Figure 9.22 shows a basic Monte Carlo model for this scenario (Excel file *Nemesh Printing Company*). We used uniform probability distributions to express the number of orders, revenues, and returns and allowances by applying the formula described in Chapter 3:

$$U = a + (b - a) \times \text{RAND}()$$

where U is a uniform random variate between a and b. For the number of orders, we need to make a modification to ensure the correct integer values over the ranges specified. For instance, the formula for the number of small orders is =TRUNC(5 + 6*RAND()). We use a value one larger than the range $(b - a)$ so that the result of 5 + 6*RAND() will be a number greater than 5 and less than 11. When

Figure 9.22 *Nemesh Printing Company* Model

	A	B
1	**Nemesh Printing Company**	
2		
3	**Data**	
4		
5	Number of small orders	8
6	Number of large orders	2
7	Revenue - small orders	$ 10,898
8	Revenue - large orders	$ 35,253
9	Returns and allowances	5.03%
10	Marginal cost - small orders	35%
11	Marginal cost - large orders	65%
12	Fixed costs	$ 55,000
13		
14	**Model**	
15		
16	Revenue	
17	Small orders	$ 87,182
18	Large orders	$ 70,506
19	Total revenue	$ 157,688
20	Costs	
21	Small orders	$ 30,514
22	Large orders	$ 45,829
23	Returns and allowances	$ 7,930
24	Fixed costs	$ 55,000
25	Total costs	$ 139,273
26		
27	Net profit	$ 18,416

	A	B
1	**Nemesh Printing Company**	
2		
3	**Data**	
4		
5	Number of small orders	=TRUNC(5+6*RAND())
6	Number of large orders	=TRUNC(1+6*RAND())
7	Revenue - small orders	=3000+12000*RAND()
8	Revenue - large orders	=15000+35000*RAND()
9	Returns and allowances	=5%+5%*RAND()
10	Marginal cost - small orders	0.35
11	Marginal cost - large orders	0.65
12	Fixed costs	55000
13		
14	**Model**	
15		
16	Revenue	
17	Small orders	=B5*B7
18	Large orders	=B6*B8
19	Total revenue	=SUM(B17:B18)
20	Costs	
21	Small orders	=B17*B10
22	Large orders	=B18*B11
23	Returns and allowances	=B9*B19
24	Fixed costs	=B12
25	Total costs	=SUM(B21:B24)
26		
27	Net profit	=B19-B25

the fractions are truncated, the resulting integer values will be between 5 and 10. Because the revenues are real numbers, we do not have to modify those formulas.

Examine the model closely. What assumptions are evident? In computing the revenue, the model assumes that *each* of the orders has the same unit revenue. While this provides a rough estimate of the total revenue that might be realized, it is far from realistic. It is unlikely that each of the number of small or large orders used in the data inputs will have the same revenue. Because the revenues will vary over the ranges specified, a more realistic model would use independent values of revenues for each order. This requires a complete redesign of the model, shown in Figure 9.23 (Excel file *Nemesh Printing Company Revised Model*). In this model, we created tables for the revenue associated with each of the possible numbers of orders. In columns F and I, we accumulate the revenues. Then we use the VLOOKUP function (see the Excel note in Chapter 3) in cells B17 and B18 to extract the cumulative revenue associated with the randomly generated number of orders in cells B5 and B6. We will examine the impact of these different assumptions on the results in the next chapter. Suffice it to say that building realistic models requires some thought and creativity, and a good working knowledge of the capabilities of Excel.

Figure 9.23 Nemesh Printing Company Revised Model

Top spreadsheet (formula view):

	A	B	C	D	E	F	G	H	I
1	Nemesh Printing Company								
2									
3	Data								
4									
5	Number of small orders	=TRUNC(5+6*RAND())							
6	Number of large orders	=TRUNC(1 + 6*RAND())							
7	Revenue - small orders	=3000+12000*RAND()							
8	Revenue - large orders	=15000+35000*RAND()							
9	Returns and allowances	=5% + 5%*RAND()							
10	Marginal cost - small orders	0.35							
11	Marginal cost - large orders	0.65							
12	Fixed costs	55000							
13									
14	Model								
15									
16	Revenue			Small order	Revenue		Cumulative	Large Order	Revenue	Cumulative
17	Small orders	=VLOOKUP(B5,D17:F26,3)		1	=3000+12000*RAND()	=E17	1	=15000+35000*RAND()	=H17
18	Large orders	=VLOOKUP(B6,G17:I22,3)		2	=3000+12000*RAND()	=F17+E18	2	=15000+35000*RAND()	=I17+H18
19	Total revenue	=SUM(B17:B18)		3	=3000+12000*RAND()	=F18+E19	3	=15000+35000*RAND()	=I18+H19
20	Costs			4	=3000+12000*RAND()	=F19+E20	4	=15000+35000*RAND()	=I19+H20
21	Small orders	=B17*B10		5	=3000+12000*RAND()	=F20+E21	5	=15000+35000*RAND()	=I20+H21
22	Large orders	=B18*B11		6	=3000+12000*RAND()	=F21+E22	6	=15000+35000*RAND()	=I21+H22
23	Returns and allowances	=B9*B19		7	=3000+12000*RAND()	=F22+E23			
24	Fixed costs	=B12		8	=3000+12000*RAND()	=F23+E24			
25	Total costs	=SUM(B21:B24)		9	=3000+12000*RAND()	=F24+E25			
26				10	=3000+12000*RAND()	=F25+E26			
27	Net profit	=B19-B25							

Bottom spreadsheet (value view):

	A	B	C	D	E	F	G	H	I
1	Nemesh Printing Company								
2									
3	Data								
4									
5	Number of small orders	7							
6	Number of large orders	4							
7	Revenue - small orders	$ 4,267.86							
8	Revenue - large orders	$ 36,757.44							
9	Returns and allowances	6.29%							
10	Marginal cost - small orders	35%							
11	Marginal cost - large orders	65%							
12	Fixed costs	$ 55,000							
13									
14	Model								
15									
16	Revenue			Small order number	Revenue	Cumulative	Large Order Number	Revenue	Cumulative
17	Small orders	$ 65,102		1	$ 9,232.75	$ 9,232.75	1	$ 18,198.93	$ 18,198.93
18	Large orders	$ 91,512		2	$ 6,451.34	$ 15,684.09	2	$ 16,060.28	$ 34,259.21
19	Total revenue	$ 156,614		3	$ 14,397.81	$ 30,081.90	3	$ 20,384.20	$ 54,643.41
20	Costs			4	$ 13,870.57	$ 43,952.47	4	$ 36,868.26	$ 91,511.67
21	Small orders	$ 22,786		5	$ 4,749.41	$ 48,701.88	5	$ 15,063.72	$ 106,575.40
22	Large orders	$ 59,483		6	$ 4,195.00	$ 52,896.88	6	$ 46,184.44	$ 152,759.83
23	Returns and allowances	$ 9,843		7	$ 12,204.95	$ 65,101.83			
24	Fixed costs	$ 55,000		8	$ 3,158.95	$ 68,260.78			
25	Total costs	$ 147,111		9	$ 14,026.93	$ 82,287.71			
26				10	$ 6,808.90	$ 89,096.61			
27	Net profit	$ 9,502							

BASIC CONCEPTS REVIEW QUESTIONS

1. What is a decision model, and what are the three types of inputs common to decision models?
2. Explain the difference between descriptive and prescriptive (optimization) models.
3. Describe how to use Excel data tables, scenario manager, and goal seek tools to analyze decision models.
4. Explain the purpose of *Solver* and what type of decision model it is used for.
5. Describe the reasons why a manager might use a heuristic instead of an optimization algorithm.
6. Summarize the important knowledge that you need to successfully build good decision models.
7. Explain basic spreadsheet engineering approaches for implementing decision models in Excel.
8. What approaches can you use to incorporate uncertainty into decision models?
9. What does validity mean? What issues must an analyst consider in building realistic models?
10. Provide some examples of how you might use decision models in your personal life or in the context of current or prior work experience.

SKILL-BUILDING EXERCISES

1. Develop a one-way data table for the airline pricing model for airplane capacities ranging from 200 to 360 in increments of 20.
2. Develop a two-way data table for the airline pricing model for airplane capacities ranging from 200 to 360 in increments of 20, and fixed costs ranging from $60,000 to $120,000 in increments of $10,000.
3. Implement the gasoline consumption model developed in this chapter on a spreadsheet. Survey five of your fellow students and obtain data on miles driven per day on a routine trip (work or school), number of days per month, and fuel economy. Use the *Scenario Manager* to define scenarios for each individual and create a Scenario Summary.
4. Implement the simple revenue model developed in this chapter on a spreadsheet and use *Solver* to find the price that maximizes total revenue.
5. Implement the model we developed in this chapter for profit: *Profit = (Unit price)[Min(Quantity produced, Demand)] − [Fixed cost + (Unit cost)(Quantity produced)]* on a spreadsheet using sound spreadsheet design principles. Use the following values: unit price = $40, unit cost = $24, fixed cost = $400,000, demand = 50,000, quantity produced = 40,000.

6. Use the Moore Pharmaceuticals spreadsheet model and the goal seek tool to determine the value of the total fixed project costs that would result in a net present value of zero.
7. Replicate the Monte Carlo simulation of the newsvendor model in Figure 9.17 for purchase quantities ranging from 40 to 50 and find the value that yields the largest average profit.
8. Apply Monte Carlo simulation to the Nemesh Printing Company example (revised model) by constructing a data table for profit and recalculating the spreadsheet for 50 trials, similar to the way we applied it to the newsvendor model. What is the average profit achieved? What percent of the time was the profit negative?
9. Generate 20 samples from a normal distribution with mean 15 and standard deviation 3 using the formula = NORMINV(RAND(), 15,3). Use *Crystal Ball* to find the best distribution fit for these data. Repeat with 200 samples. What conclusions can you reach?

PROBLEMS AND APPLICATIONS

1. A supermarket has been experiencing long lines during peak periods of the day. The problem is noticeably worse on certain days of the week, and the peak periods are sometimes different according to the day of the week. There are usually enough workers on the job to open all cash registers. The problem is knowing when to call some of the workers stocking shelves up to the front to work the checkout counters. How might decision models help the supermarket? What data would be needed to develop these models?

2. Four key marketing decision variables are price (P), advertising (A), transportation (T), and product quality (Q). Consumer demand (D) is influenced by these variables. The simplest model for describing demand in terms of these variables is:

$$D = k - pP + aA + tT + qQ$$

where k, p, a, t, and q are constants. Discuss the assumptions of this model. Specifically, how does each variable affect demand? How do the variables influence each other? What limitations might this model have? How can it be improved?

3. *Total marketing effort* is a term used to describe the critical decision factors that affect demand: price, advertising, distribution, and product quality. Define the variable x to represent total marketing effort. A typical model that is used to predict demand as a function of total marketing effort is based on the power function:

$$D = ax^b$$

Suppose that a is a positive number. Different model forms result from varying the constant b. Sketch the graphs of this model for $b = 0, b = 1, 0 < b < 1, b < 0$, and $b > 1$. (We encourage you to use Excel to do this.) What does each model tell you about the relationship between demand and marketing effort? What assumptions are implied? Are they reasonable? How would you go about selecting the appropriate model?

4. A manufacturer is preparing to set the price on a new action game. Demand is thought to depend on the price and is represented by the model:

$$D = 2000 - 3P$$

The accounting department estimates that the total costs can be represented by:

$$C = 5,000 + 4D$$

a. Develop a model for the total profit and implement it on a spreadsheet.
b. Develop a one-way data table to evaluate profit as a function of price (choose a price range that is reasonable and appropriate).
c. Use *Solver* to find the price that maximizes profit.

5. The Radio Shop sells two popular models of portable sport radios, model A and model B. The sales of these products are not independent of each other (in economics, we call these substitutable products, because if the price of one increases, sales of the other will increase). The store wishes to establish a pricing policy to maximize revenue from these products. A study of price and sales data shows the following relationships between the quantity sold (N) and prices (P) of each model:

$$N_A = 19.5 - 0.6P_A + 0.25P_B$$
$$N_B = 30.1 + 0.08P_A - 0.5P_B$$

a. Construct a model for the total revenue and implement it on a spreadsheet.
b. Develop two-way data table to estimate the optimal prices for each product in order to maximize the total revenue.
c. Use *Solver* to find the optimal prices.

6. A forest fire is burning down a narrow valley 3 miles wide at a speed of 40 feet per minute. The fire can be contained by cutting a firebreak through the forest across the valley. It takes 30 seconds for one person to clear one foot of the firebreak. The value of lost timber is $4,000 per square mile. Each person hired is paid $12 per hour, and it costs $30 to transport and supply each person with the appropriate equipment.

a. Develop a model for determining how many people should be sent to contain the fire and for determining the best location for the firebreak (draw a picture first!).

b. Implement your model on a spreadsheet and find the optimal solution using *Solver*.

7. The Hyde Park Surgery Center specializes in high-risk cardiovascular surgery. The center needs to forecast its profitability over the next three years to plan for capital growth projects. For the first year, the hospital anticipates serving 1500 patients, which is expected to grow by 8% per year. Based on current reimbursement formulas, each patient provides an average billing of $150,000, which will grow by 3% each year. However, because of managed care, the center collects only 35% of billings. Variable costs for supplies and drugs are calculated to be 12% of billings. Fixed costs for salaries, utilities, and so on, will amount to $20,000,000 in the first year and are assumed to increase by 6% per year. Develop a spreadsheet model to calculate the net present value of profit over the next three years. Use a discount rate of 7%.

8. Each year, Kurbe Marketing Research (KMR) conducts hundreds of research studies for a variety of clients. Throughout the year, the demand for telephone research varies. KMR conducts most of its research projects in-house, but if it cannot handle all of the work, it subcontracts some of the telephone interviewing at a lower profit margin. Costs include the fixed costs of facilities and supervision, general overhead ($125,000 per month), the costs of operating computer-assisted telephone interviewing stations ($3,000 per month for each), and the cost of telephone interviewers ($1,500 per month per operator). The key decision variable is the number of interviewing stations to have. If the number of stations is high, costs would increase, but the amount of subcontracted work would decrease. If the number is low, subcontracting would increase while costs would decrease. Each station can handle $9,000 of research demand per month.

KMR receives 15% of the revenue for all work subcontracted to an outside vendor. The table below provides the average monthly demands for the past several years.

Month	Demand
January	$940,000
February	$820,000
March	$575,000
April	$860,000
May	$695,000
June	$320,000
July	$840,000
August	$700,000
September	$245,000
October	$770,000
November	$150,000
December	$170,000

a. Develop a spreadsheet for computing the total annual profit from both in-house and subcontracting work for a variable number of interviewing stations.

b. Make a recommendation for the optimal number of stations to have.

c. Examine the sensitivity of your solution if demand estimates vary by 6% in either direction. Does this change your recommendation?

9. Financial analysts often use the following model to characterize changes in stock prices:

$$P_t = P_0 e^{(\mu - 0.5\sigma^2)t + \sigma Z \sqrt{t}}$$

where

P_0 = current stock price
P_t = price at time t
μ = mean (logarithmic) change of the stock price per unit time
σ = (logarithmic) standard deviation of price change
Z = standard normal random variable

This model assumes that the logarithm of a stock's price is a normally distributed random variable (see the discussion of the lognormal distribution and note that the first term of the exponent is the mean of the lognormal distribution). Using historical data, one can estimate values for μ and σ. Suppose that the average daily change for a stock is $0.003227, and the standard deviation is 0.026154. Develop a spreadsheet to simulate the price of the stock over the next 30 days, if the current price is $53. Use the Excel function NORMSINV(RAND( )) to generate values for Z. Construct a chart showing the movement in the stock price.

10. The Miller–Orr model in finance addresses the problem of managing its cash position by purchasing or selling securities at a transaction cost in order to lower or raise its cash position. That is, the firm needs to have enough cash on hand to meet its obligations, but does not want to maintain too high a cash balance because it loses the opportunity for earning higher interest by investing in other securities. The Miller–Orr model assumes that the firm will maintain a minimum cash balance, m, a maximum cash balance, M, and an ideal level, R, called the return point. Cash is managed using a decision rule that states that whenever the cash balance falls to m, $R - m$ securities are sold to bring the balance up to the return point. When the cash balance rises to M, $M - R$ securities are purchased to reduce the cash balance back to the return point. Using some advanced mathematics, the return point and maximum cash balance levels are shown to be:

$$R = m + Z$$
$$M = R + 2Z$$

where

$$Z = \left(\frac{3C_0 \sigma^2}{4r} \right)^{1/3}$$

σ^2 = variance of the daily cash flows
r = average daily rate of return corresponding to the premium associated with securities

For example, if the premium is 4%, $r = 0.04/365$. To apply the model, note that we do not need to know the actual demand for cash, only the daily variance. Essentially, the Miller–Orr model determines the decision rule that minimizes the expected costs of making the cash-security transactions and the expected opportunity costs of maintaining the cash balance based on the variance of the cash requirements. Suppose that the daily requirements are normally distributed with a mean of 0 and variance of $60,000. Assume a transaction cost equal to $35, interest rate premium of 4%, and required minimum balance of $7500. Develop a spreadsheet implementation for this model. Apply Monte Carlo simulation to simulate the cash balance over the next year (365 days). Your simulation should apply the decision rule that if the cash balance for the current day is less than or equal to the minimum level, sell securities to bring the balance up to the return point. Otherwise, if the cash balance exceeds the upper limit, buy enough securities (i.e., subtract an amount of cash) to bring the balance back down to the return point. If neither of these conditions hold, there is no transaction and the balance for the next day is simply the current value plus the net requirement. Show the cash balance results on a line chart.

11. The admissions director of an engineering college has $500,000 in scholarships each year from an endowment to offer to high-achieving applicants. The value of each scholarship offered is $25,000 (thus, 20 scholarships are offered). The benefactor who provided the money would like to see all of it used each year for new students. However, not all students accept the money; some take offers from competing schools. If they wait until the end of the admissions deadline to decline the scholarship, it cannot be offered to someone else, as any other good students would already have committed to other programs. Consequently, the admissions director offers more money than available in anticipation that a percentage of offers will be declined. If more than 10 students accept the offers, the college is committed to honoring them, and the additional amount has to come out of the dean's budget. Based on prior history, the percentage of applicants that accept the offer has varied from about 60% to 85%. Develop a spreadsheet model for this situation, and apply whatever analysis tools you deem appropriate to help the admissions director make a decision on how many scholarships to offer. Explain your results in a business memo to the director, Mr. P. Woolston.

12. J&G Bank receives a large number of credit card applications each month, an average of 30,000 with a standard deviation of 4,000, normally distributed. Approximately 60% of them are approved, but this typically varies between 55% and 65%. Each customer charges a total $5,000, normally distributed, with a standard deviation of $400, to his or her credit card each month. Approximately 85% pay off their balance in full, and the remaining incur finance charges. The average finance charge has recently varied from 3.25% to 3.75% per month. The bank also receives income from fees charged for late payments and annual fees associated with the credit cards. This is a percentage of total monthly charges, and has varied between 6.8% and 7.2%. It costs the bank $15 per application, whether it is approved or not. The monthly maintenance cost for credit card customers is normally distributed with a mean of $10 and standard deviation of $1.50. Finally, losses due to charge-offs of customers' accounts range between 4.8% and 5.2%.

   a. Using average values for all uncertain inputs, develop a spreadsheet model to calculate the bank's total monthly profit.

   b. Modify the spreadsheet to incorporate the uncertainty associated with the uncontrollable inputs.

   c. Perform a Monte Carlo simulation to obtain 25 random samples of profit values using a data table. Summarize your results.

13. Each worksheet in the Excel file *LineFit Data* contains a set of data that describes a functional relationship between the dependent variable *y* and the independent variable *x*. Construct a line chart of each data set, and use the *Add Trendline* tool to determine the best-fitting functions to model these data sets.

14. The data in the Excel file *Real Estate Sales* represent the annual sales of 100 branch offices of a national real estate company. Use the *Crystal Ball* distribution fitting procedure to find the best-fitting distribution for these data. Use the Kolmogorov–Smirnov statistic to rank the fit.

15. The data in the Excel file *Technical Support Data* represent a sample of the number of minutes spent resolving customer support issues for a computer manufacturer.

Use the *Crystal Ball* distribution fitting procedure to find the best-fitting distribution for these data. Use the Anderson–Darling statistic to rank the fit.

16. Suppose that a 30-year-old MBA graduate takes a job with a company that provides a 401k retirement plan and would like to retire at age 60. The company matches a portion of the employee's contributions based on the organization's financial performance. Assume that the employee's starting salary is $50,000, that he or she contributes 6% per year to the plan, and that the employer makes a matching contribution of 50%. Further assume that the employee's salary is expected to increase 3% per year, and that the investment will return 5% per year.
    a. Build a spreadsheet model to compute the retirement balance at the end of each year and the final amount at retirement. What assumptions did you make in the formulas you used to calculate the return on investment each year? Provide at least one alternative approach to calculating the return on investment for each year. Which approach is the most realistic?
    b. Recognizing that neither the salary increase nor the investment return will be constant each year, what modifications must you make to the model to incorporate this more realistic assumption?

CASE

An Inventory Management Decision Model

Inventories represent a considerable investment for every organization; thus, it is important that they be managed well. Excess inventories can indicate poor financial and operational management. On the other hand, not having inventory when it is needed can also result in business failure. The two basic inventory decisions that managers face are *how much* to order or produce for additional inventory, and *when* to order or produce it to minimize total inventory cost, which consists of the cost of holding inventory and the cost of ordering it from the supplier.

Holding costs, or *carrying costs*, represent costs associated with maintaining inventory. These costs include interest incurred or the opportunity cost of having capital tied up in inventories; storage costs such as insurance, taxes, rental fees, utilities, and other maintenance costs of storage space; warehousing or storage operation costs, including handling, recordkeeping, information processing, and actual physical inventory expenses; and costs associated with deterioration, shrinkage, obsolescence, and damage. Total holding costs are dependent on how many items are stored and for how long they are stored. Therefore, holding costs are expressed in terms of *dollars associated with carrying one unit of inventory for one unit of time.*

Ordering costs represent costs associated with replenishing inventories. These costs are not dependent on how many items are ordered at a time, but on the number of orders that are prepared. Ordering costs include overhead, clerical work, data processing, and other expenses that are incurred in searching for supply sources, as well as costs associated with purchasing, expediting, transporting, receiving, and inspecting. It is typical to assume that the ordering cost is constant and is expressed in terms of *dollars per order.*

For a manufacturing company that you are consulting for, managers are unsure about making inventory decisions associated with a key engine component. The annual demand is estimated to be 15,000 units and is assumed to be constant throughout the year. Each unit costs $80. The company's accounting department estimates that its opportunity cost for holding this item in stock for one year is 18% of the unit value. Each order placed with the supplier costs $220. The company's policy is to place a fixed order for Q units whenever the inventory level reaches a predetermined reorder point that provides sufficient stock to meet demand until the supplier's order can be shipped and received.

As a consultant, your task is to develop and implement a decision model to help them arrive at the best decision. As a guide, consider the following:

1. Define the data, uncontrollable inputs, and decision variables that influence total inventory cost.
2. Develop mathematical functions that compute the annual ordering cost and annual holding cost based on average inventory held throughout the year in order to arrive at a model for total cost.
3. Implement your model on a spreadsheet.
4. Use data tables to find an approximate order quantity that results in the smallest total cost.
5. Use *Solver* to verify your result.
6. Conduct what-if analyses to study the sensitivity of total cost to changes in the model parameters.
7. Explain your results and analysis in a memo to the vice president of operations.

Chapter 10

Risk Analysis and Monte Carlo Simulation

INTRODUCTION

*W*e introduced the basic ideas of Monte Carlo simulation in Chapter 9. Monte Carlo simulation is the process of repeatedly generating inputs to a decision model based on sampling from their assumed probability distributions, calculating the outputs, and analyzing the results. Of particular interest are the distributions of the output variables, which characterize the likelihood that certain output values will be achieved. Understanding this distribution provides the basis for assessing risk in making decisions. **Risk** is simply the probability of occurrence of an undesirable outcome. For example, we could answer such questions as: What is the probability that we will incur a financial loss? What is the probability that we will run out of inventory? What are the chances that a project will be completed on time?

Risk analysis is an approach for developing "a comprehensive understanding and awareness of the risk associated with a particular variable of interest (be it a payoff measure, a cash flow profile, or a macroeconomic forecast)."[1] Hertz and Thomas present a simple scenario to illustrate the concept of risk analysis:

> The executives of a food company must decide whether to launch a new packaged cereal. They have come to the conclusion that five factors are the determining variables: advertising and promotion expense, total cereal market, share of market for this product, operating costs, and new capital investment. On the basis of the "most likely" estimate for each of these variables, the picture looks very bright—a healthy 30% return, indicating a significantly positive expected net present value. This future, however, depends on each of the "most likely" estimates coming true in the actual case. If each of these "educated guesses" has, for example, a 60% chance of being correct, there is only an 8% chance that all five will be correct ($0.60 \times 0.60 \times 0.60 \times 0.60 \times 0.60$) if the factors are assumed to be independent. So the "expected" return, or present value measure, is actually dependent on a rather unlikely coincidence. The decision maker needs to know a great deal more about the other values used to make each of the five estimates and about what he stands to gain or lose from various combinations of these values.[2]

Thus, risk analysis seeks to examine the impacts of uncertainty in the estimates and their potential interaction with one another on the output variable of interest. Hertz and Thomas also note that the challenge to risk analysts is to frame the output of risk analysis procedures in a manner that makes sense to the manager and provides clear insight into the problem, suggesting that simulation has many advantages.

In this chapter we provide the details on how to apply Monte Carlo simulation with the Excel add-in, *Crystal Ball*. *Crystal Ball* is a powerful commercial package that is used by many of the *Fortune* 1000 companies. Specifically, we will discuss the following:

➤ *Capabilities of* Crystal Ball *software*
➤ *Using* Crystal Ball *for conducting risk analysis simulations*
➤ *Using* Crystal Ball *tools for analyzing risk analysis simulations*
➤ *Practical applications of risk analysis simulation*

---

[1] David B. Hertz and Howard Thomas, *Risk Analysis and Its Applications* (Chichester, UK: John Wiley & Sons, Ltd., 1983), 1.
[2] *Ibid.*, 24.

MONTE CARLO SIMULATION USING CRYSTAL BALL

Crystal Ball is an Excel add-in that performs Monte Carlo simulation for risk analysis. We have also seen that *Crystal Ball* includes a forecasting module (Chapter 7) and distribution fitting capability (Chapter 9). Start *Crystal Ball* by selecting it from the Windows *Start* menu. Click the *Crystal Ball* tab in the Excel menu bar to display the menus (as shown in Chapter 9 in Figure 9.19).

To use *Crystal Ball*, you must perform the following steps:

1. Develop the spreadsheet model.
2. Define *assumptions* for uncertain variables, that is, the probability distributions that describe the uncertainty.
3. Define the *forecast cells*, that is, the output variables of interest.
4. Set the number of trials and other run preferences.
5. Run the simulation.
6. Interpret the results.

To illustrate this process, we will use a financial spreadsheet model (Excel file *New Store Financial Model*) as the basis for discussion.

A Financial Risk Analysis Simulation

Figure 10.1 shows the *New Store Financial Model* spreadsheet. To set the context for this model, think of any retailer that operates many stores throughout the country, such as Old Navy, Hallmark Cards, or Radio Shack, to name just a few. The retailer is often seeking to open new stores and has developed the model to evaluate the profitability of a new site of a certain size that would be leased for five years. If you examine the model closely, you will see that the key assumptions in the model might be based on historical data (e.g., cost of merchandise as a percent of sales and operating expenses), current economic forecasts (e.g., inflation rate), or judgmental estimates based on preliminary market research (e.g., first-year sales revenue and annual growth rates). These assumptions represent the "most likely" estimates, and as a deterministic model, the spreadsheet shows that the new store will appear to be quite profitable by the end of five years. However, the model does not provide any information about what might happen if these variables do not attain these most likely values, and considerable uncertainty exists about their true values. We might be interested in such questions about risk as the following:

- What are the chances that the store would not be profitable by the third year?
- How likely is it that cumulative profits over five years would not exceed $100,000?
- What profit are we likely to realize with a probability of at least 0.70?

Defining Model Inputs

As we saw in Chapter 9, the inputs to a decision model are either constant values, or are uncertain and characterized by some probability distribution. We can characterize these probability distributions either by fitting historical data using distribution-fitting techniques described in Chapter 9, or when a sufficient amount of historical data is not available, using judgment. We can draw upon the properties of common probability distributions and typical applications that we discussed in Chapter 3 to help choose a representative distribution (see the sections entitled *Continuous Probability Distributions* and *Other Useful Distributions* in

Figure 10.1 New Store Financial Analysis Spreadsheet Model

	A	B	C	D	E	F	G
1	**New Store Financial Analysis Model**						
2							
3	**Model Assumptions**		*Year 1*	*Year 2*	*Year 3*	*Year 4*	*Year 5*
4	Annual Growth Rate			20%	12%	9%	5%
5	Sales Revenue		$ 800,000				
6							
7	Cost of Merchandise (% of sales)	30%					
8	Operating Expenses						
9	Labor Cost	$ 200,000					
10	Rent Per Square Foot	$ 28					
11	Other Expenses	$ 325,000					
12							
13	Inflation Rate	2%					
14	Store Size (square feet)	5,000					
15	Total Fixed Assets	$ 300,000					
16	Depreciation period (straight line)	5					
17	Discount Rate	10%					
18	Tax Rate	34%					
19							
20	**Model Outputs**	Year	*1*	*2*	*3*	*4*	*5*
21	Sales Revenue		$ 800,000	$ 960,000	$ 1,075,200	$ 1,171,968	$ 1,230,566
22	Cost of Merchandise		$ 240,000	$ 288,000	$ 322,560	$ 351,590	$ 369,170
23	Operating Expenses						
24	Labor Cost		$ 200,000	$ 204,000	$ 208,080	$ 212,242	$ 216,486
25	Rent Per Square Foot		$ 140,000	$ 142,800	$ 145,656	$ 148,569	$ 151,541
26	Other Expenses		$ 325,000	$ 331,500	$ 338,130	$ 344,893	$ 351,790
27	Net Operating Income		$ (105,000)	$ (6,300)	$ 60,774	$ 114,674	$ 141,579
28	Depreciation Expense		$ 60,000	$ 60,000	$ 60,000	$ 60,000	$ 60,000
29	Net Income Before Tax		$ (165,000)	$ (66,300)	$ 774	$ 54,674	$ 81,579
30	Income Tax		$ (56,100)	$ (22,542)	$ 263	$ 18,589	$ 27,737
31	Net After Tax Income		$ (108,900)	$ (43,758)	$ 511	$ 36,085	$ 53,842
32	Plus Depreciation Expense		$ 60,000	$ 60,000	$ 60,000	$ 60,000	$ 60,000
33	Annual Cash Flow		$ (48,900)	$ 16,242	$ 60,511	$ 96,085	$ 113,842
34	Discounted Cash Flow		(44,454.55)	13,423.14	45,462.69	65,627.36	70,687.05
35	Cumulative Discounted Cash Flow		(44,454.55)	(31,031.40)	14,431.28	80,058.65	150,745.70

Chapter 3). For example, if we are modeling machine failure times, then an exponential or Weibull distribution might be appropriate to use.

Very often, uniform or triangular distributions are used in the absence of good data. These distributions depend on simple parameters that one can easily identify based on managerial knowledge and judgment. For example, to define the uniform distribution, we need to know only the smallest and largest possible values that the variable might assume. For the triangular distribution, we also include the most likely value. In the construction industry, for instance, experienced foremen can easily tell you the fastest, most likely, and slowest times it would take to perform a task such as framing a house, taking into account possible weather and material delays, labor absences, and so on.

Assumptions

In *Crystal Ball,* uncertain inputs are called **assumptions**. Suppose that the new business development manager for the firm has identified the following distributions and parameters for these variables:

- First-year sales revenue: normal, mean = $800,000, standard deviation = $70,000
- Annual growth rate, year 2: lognormal, mean = 20%, standard deviation = 8%

- Annual growth rate, year 3: lognormal, mean = 12%, standard deviation = 4%
- Annual growth rate, year 4: lognormal, mean = 9%, standard deviation = 2%
- Annual growth rate, year 5: lognormal, mean = 5%, standard deviation = 1%
- Cost of merchandise: uniform between 27% and 33%
- Labor cost: triangular, minimum = $175,000, most likely = $200,000, maximum = $225,000
- Rent per square foot: uniform between $26 and $30
- Other expenses: triangular, minimum = $310,000, most likely = $325,000, maximum = $350,000
- Inflation rate: triangular, minimum = 1%, most likely = 2%, maximum = 5%

To define them in *Crystal Ball,* first select the cell corresponding to the uncertain input. Assumption cells must contain a value; they cannot be defined for formula, nonnumeric, or blank cells. From the *Define* group in the *Crystal Ball* tab, click *Define Assumption. Crystal Ball* displays a gallery of probability distributions from which to choose and prompts you for the parameters. For example, let us define the distribution for the first-year sales revenue. First, click on cell C5 and then select *Define Assumption. Crystal Ball* displays the distribution gallery shown in Figure 10.2. You can select one of the folders on the left of the dialog box to display all distributions or previously defined favorites. Since we assume that this variable has a normal distribution, we click on the normal distribution then the *OK* button (or simply double-click the distribution). A dialog box is then displayed, prompting you for the parameters associated with this distribution.

Figure 10.2 *Crystal Ball Distribution Gallery*

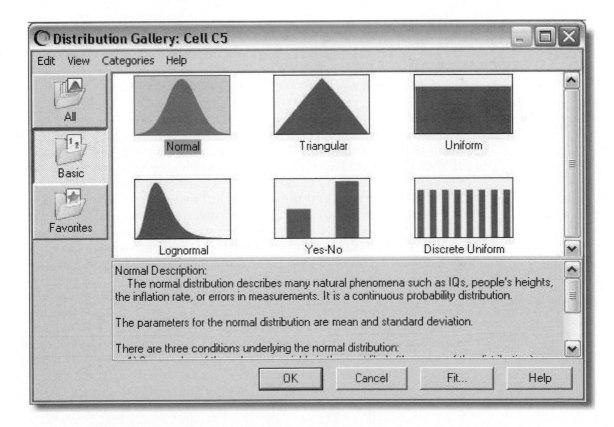

Figure 10.3 First-Year Sales Revenue Normal Distribution Assumption

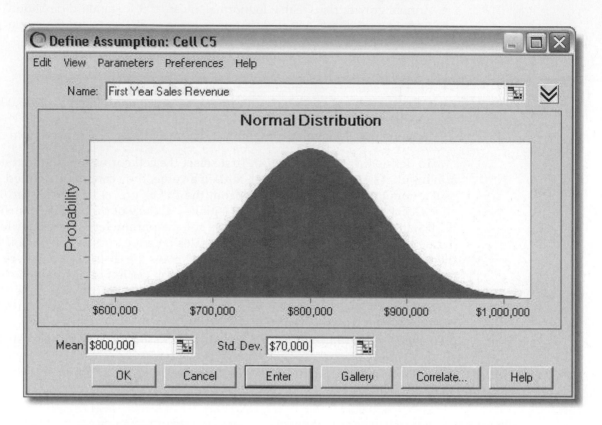

We suggest that you first enter a clear, descriptive name for your assumptions in the top box. *Crystal Ball* will automatically use text in cells immediately to the left or above assumption cells, but these may not be the correct ones for your application.

Crystal Ball anticipates the default parameters based on the current values in the spreadsheet model. For example, with a normal distribution, the default mean is the assumption cell value, and the standard deviation is assumed to be 10% of the mean. Therefore, in our example, we need to change the standard deviation to $70,000. Clicking on *Enter* fixes these values and rescales the picture to allow you to see what the distribution looks like (this feature is quite useful for flexible families of distributions such as the triangular, gamma, or beta). Figure 10.3 shows the completed *Define Assumption* screen with the correct parameters for the first-year sales' revenue. After an assumption is defined, the spreadsheet cell is shaded in green (default color) so that you may easily identify the assumptions.

We repeat this process for each of the probabilistic assumptions in the model. Figure 10.4, for example, shows the lognormal distribution for the annual growth rate in year 2. Figures 10.5 and 10.6 show the dialog boxes for the uniform distribution assumption for cost of merchandise and triangular assumption for inflation rate, respectively. We encourage you to work through this example and enter all the assumptions for this model. *Crystal Ball* also has various options for specifying input information and customizing the views of the assumptions (see *Crystal Ball Note: Customizing Define Assumption*).

Figure 10.4 Year 2 Annual Growth Rate Assumption

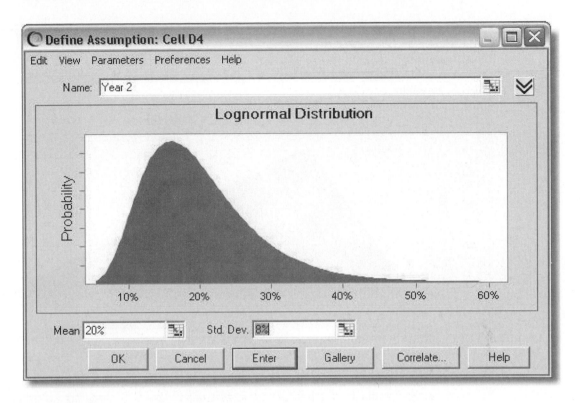

Figure 10.5 Cost of Merchandise Assumption

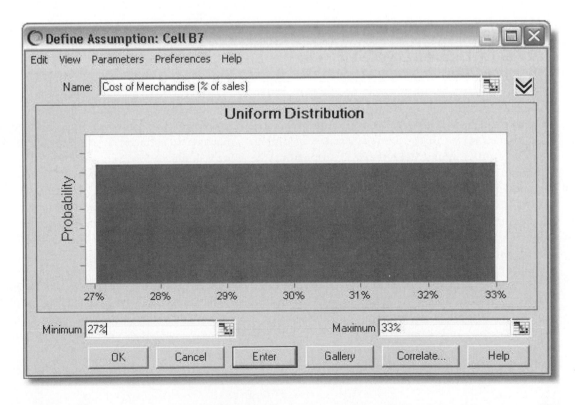

Figure 10.6 Inflation Rate Assumption

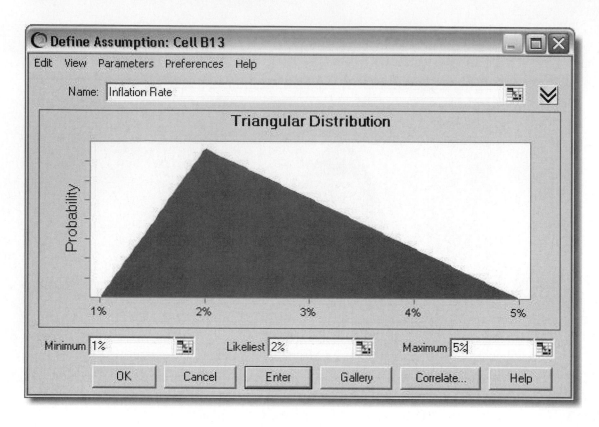

CRYSTAL BALL NOTE
Customizing *Define Assumption*

For most continuous distributions, you have several options on how to specify the distribution in the *Parameters* menu in the *Define Assumption* dialog. For example, with the normal distribution, the default is to enter the mean and standard deviation; however, you can also define the distribution by its 10th and 90th percentiles, the mean and the 90th percentile, and several other ways. This option is useful when only percentile information is available or when specific parameters such as the mean and standard deviation are unknown. As a practical illustration, suppose that you are interviewing a construction manager to identify the distribution of the time it takes to complete a task. Although a beta distribution is often appropriate in such applications, it would be very difficult to define the

parameters for the beta distribution from judgmental information. However, it would be easy for the manager to estimate the 10th and 90th percentiles for task times.

Crystal Ball also allows you to customize the charts by adding gridlines, 3D effects, "marker lines" that show various statistical information, such as the mean, median, etc., as well as formatting the *x*-axis and other options. Marker lines are especially useful for distributions in which the mean or standard deviation are not input parameters. These options can be invoked by selecting *Chart* from the *Preferences* menu in the *Define Assumption* dialog. Figure 10.7 shows a marker line for the mean in the inflation rate assumption (the axis was also reformatted to show two decimal places).

Figure 10.7 Example of Customized Assumption Preferences

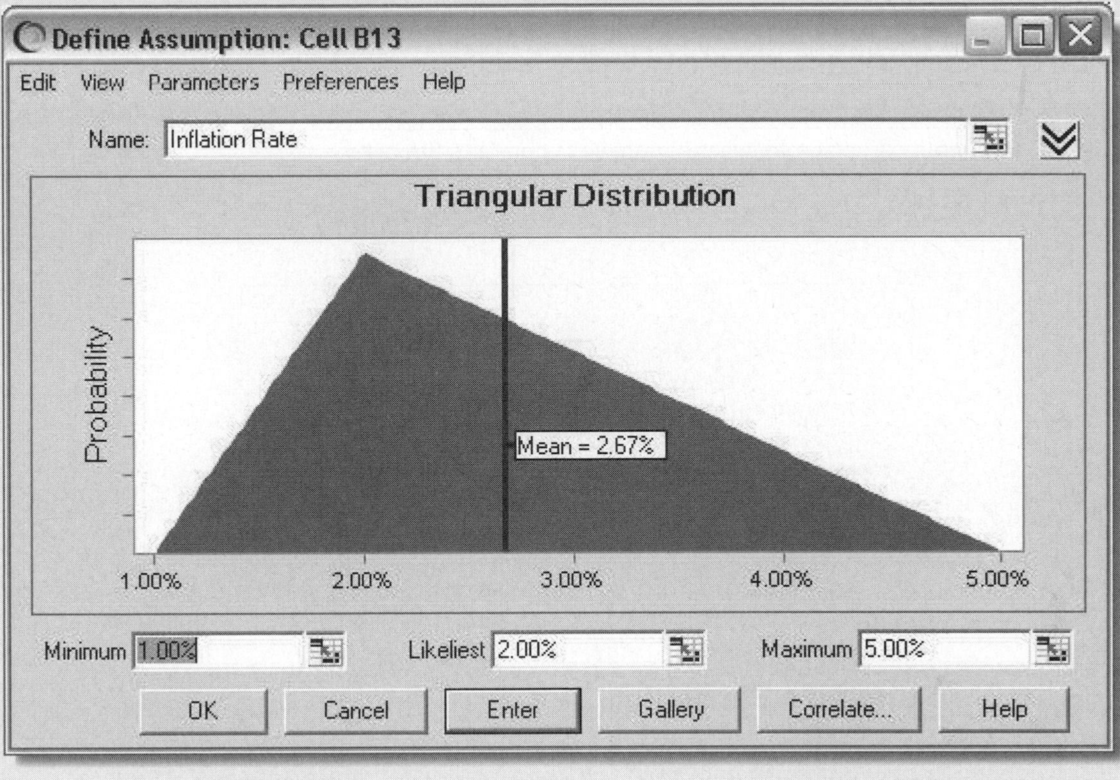

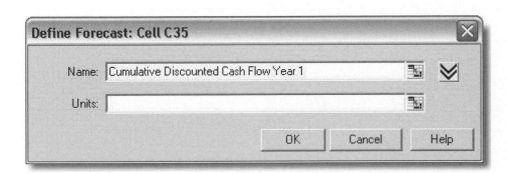

Forecasts

The output variables in which we are interested are called **forecast cells**. In our example these are the cumulative discounted cash flows in cells C35:G35. Instead of choosing them individually, highlight the entire range (*Crystal Ball* will cycle through each of the cells, prompting for input), and select *Define Forecast* from the *Define* group. In the *Define Forecast* dialog, enter a name for the forecast and unit of measure, if desired. Figure 10.8 shows this for their first-year cumulative discounted cash flow. After a forecast is defined, the spreadsheet cell is shaded in blue to indicate the forecast and distinguish it from the assumptions.

Figure 10.8 Define Forecast Dialog

The Excel *Cut, Copy,* and *Paste* commands apply only to Excel data and formulas and *do not* work for *Crystal Ball* data. *Crystal Ball* provides editing commands to let you copy, paste, or clear assumptions or forecasts from cells. These are useful, for example, when you have several assumptions with the same distributional properties. These commands are found in the *Define* group (see Figure 10.9). The *Select* button in the *Define* group also allows you to select some or all assumptions and forecasts.

Figure 10.9 *Crystal Ball Define* Group Options

Crystal Ball provides commands for selecting, copying, pasting, and clearing assumptions and forecasts (see *Crystal Ball Note: Working with Assumptions and Forecasts*).

Running a Simulation

Prior to running a simulation, you need to set some specifications for the simulation. To do this, select the *Run Preferences* item from the *Run* group. The first dialog box has several tabs as shown in Figure 10.10. The first tab *Trials,* allows you to choose the

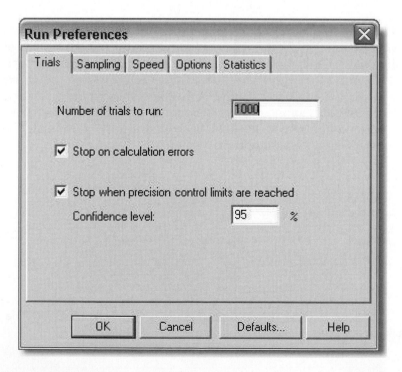

Figure 10.10 *Run Preferences* Dialog

number of times that *Crystal Ball* will generate assumptions for the assumption cells in the model and recalculate the entire spreadsheet. Because Monte Carlo simulation is essentially statistical sampling, the larger the number of trials you use, the more precise will be the result. Unless the model is extremely complex, a large number of trials will not unduly tax today's microprocessors, so we recommend that you use at least the default value of 1,000 trials. You should use a larger number of trials as the number of assumption cells in your model increases so that *Crystal Ball* can generate representative samples from all distributions for assumptions. For the new store example, we will set the number of trials to 10,000.

In the *Sampling* tab, you can choose to use the same sequence of random numbers for generating the random variates in the assumption cells; this will guarantee that the same assumption values will be used each time you run the model. This is useful when you wish to change a controllable variable in your model and compare results for the same assumption values. If you check the box "*Use same sequence of random numbers*," you may specify an *Initial seed value*. As long as you use the same number, the assumptions generated will be the same for all simulations.

Crystal Ball has two types of sampling methods: Monte Carlo and Latin Hypercube. Monte Carlo sampling selects random variates independently over the entire range of possible values. With Latin Hypercube sampling, *Crystal Ball* divides each assumption's probability distribution into intervals of equal probability and generates an assumption value randomly within each interval. Latin Hypercube sampling results in a more even distribution of forecast values because it samples the entire range of the distribution in a more consistent manner, thus achieving more accurate forecast statistics (particularly the mean) for a fixed number of Monte Carlo trials. However, Monte Carlo sampling is more representative of reality and should be used if you are interested in evaluating the model performance under various "what-if" scenarios. We recommend leaving the default values in the *Speed, Options,* and *Statistics* tabs of the *Run Preferences* dialog.

The last step is to run the simulation by clicking the *Start* button in the *Run* group and watch *Crystal Ball* go to work! You may choose the *Step* run option if you wish to see the results of individual trials. This is a useful option in the early stages of model building and to debug your spreadsheets. You may stop and reset the simulation (which clears all statistical results) using the appropriate buttons.

Saving *Crystal Ball* Runs

When you save your spreadsheet in Excel, any assumptions and forecasts that you defined for *Crystal Ball* are also saved. However, this does not save the results of a *Crystal Ball* simulation. To save a *Crystal Ball* simulation, select *Save or Restore* from the *Run* group and then click on *Save Results*. Doing so allows you to save any customized chart settings and other simulation results and recall them without rerunning the simulation. To retrieve a *Crystal Ball* simulation, choose *Restore Results* from the *Save or Restore* menu.

Analyzing Results

The principal output results provided by *Crystal Ball* are the *forecast chart, percentiles summary,* and *statistics summary.* The forecast chart is automatically displayed when the simulation ends, and the percentiles and statistics summaries can be selected from the *View* menu in the forecast chart (selecting *Split View* first allows you to display the forecast chart and/or the percentiles and statistics in one

Figure 10.11   Forecast Chart for Cumulative Discounted Cash Flow Year 5

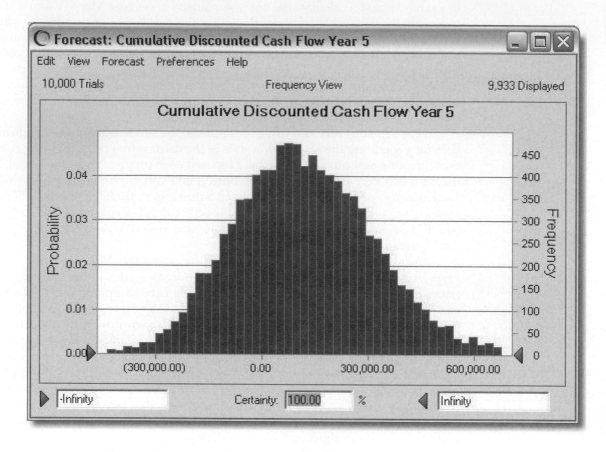

window). Figure 10.11 shows the forecast chart for the five-year cumulative discounted cash flow. The forecast chart is simply a histogram of the outcome variable that includes all values within a default value of 2.6 standard deviations of the mean, which represents approximately 99% of the data. The actual number of values displayed is shown in the upper right corner of the chart.

Just below the horizontal axis at the extremes of the distribution are two small triangles called *endpoint grabbers.* The range values of the variable at these positions are given in the boxes at the bottom left and right corners of the chart; in Figure 10.11, this shows that the grabbers are positioned at minus and plus infinity. The percent of data values between the grabbers is displayed in the *Certainty* box at the lower center of the chart. A **certainty level** is a probability interval that states the probability of the forecast falling within the specified range of the grabbers.

Questions involving risk can now be answered by manipulating the endpoint grabbers or by changing the range and certainty values in the boxes. Several options exist.

1. *You may move an endpoint grabber by clicking on the grabber and dragging it along the axis.*   As you do, the distribution outside of the middle range changes color, the range value corresponding to the grabber changes to reflect its current position, and the certainty level changes to reflect the new percentage between the grabbers.

2. *You may type in specific values in the range boxes.*   When you do, the grabbers automatically move to the appropriate positions and the certainty level changes

Figure 10.12 Probability of Positive Discounted Cash Flow

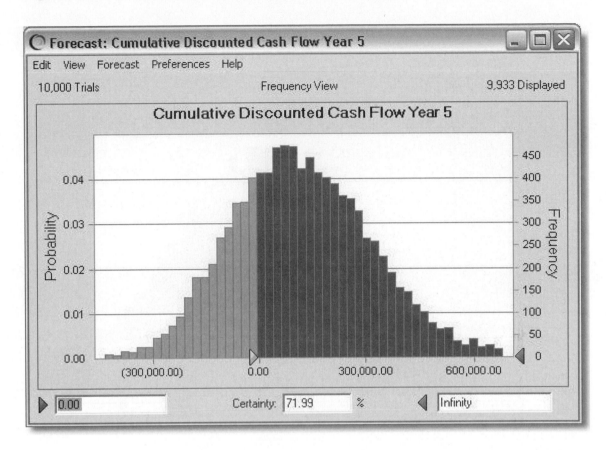

to reflect the new percentage of values between the range values. For example, suppose you wanted to determine the percentage of values greater than $0, or, equivalently, the probability that the cash flow is positive by the fifth year. If you enter this in the left range box and press the enter key, the grabber will automatically move to that position, the portion of the histogram to the left of 0 will change color, and the certainty level will change to reflect the percentage of the distribution between the grabbers. This is illustrated in Figure 10.12, which shows a 71.99% chance of a positive cash flow in year 5.

3. *You may specify a certainty level.*   If the endpoint grabbers are free (as indicated by a dark gray color), the certainty range will be centered around the median. For example, Figure 10.13 shows a 90% certainty level. The range centered about the median is from −$185,564 to $465,478. You may anchor an endpoint grabber by clicking on it. When anchored, the grabber will be a lighter color. (To free an anchored grabber, click on it again.) If a grabber is anchored and you specify a certainty level, the free grabber moves to a position corresponding to this level. Finally, you may cross over the grabbers and move them to opposite ends to determine certainty levels for the tails of the distribution.

We caution you that a certainty range is *not* a confidence interval. It is simply a probability interval. Confidence intervals, as discussed in Chapter 4, depend on the sample size and relate to the sampling distribution of a statistic.

The forecast chart may be customized to change its appearance from the *Preferences* menu in the forecast chart. We encourage you to experiment with these

Figure 10.13 A 90% Probability Interval around the Median

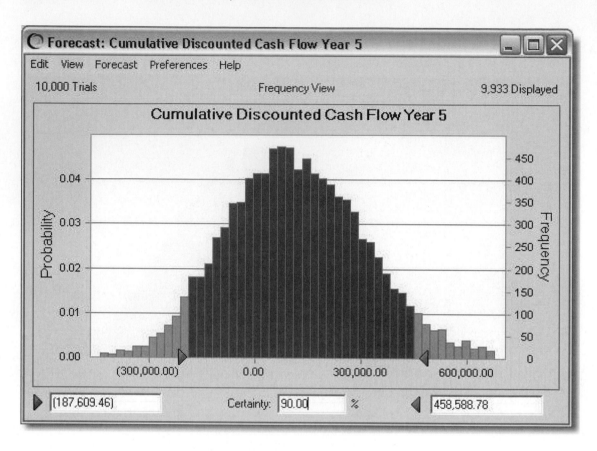

options. You may also fit the forecast data to a distribution by selecting *Fit Probability Distribution* from the *Forecast* menu in the chart.

The percentiles summary is shown in Figure 10.14. This is essentially the cumulative probability distribution of the forecast. For example, we see that the chance that the total profit will be less than $219,586 is 70%.

The statistics summary, shown in Figure 10.15, provides a summary of key descriptive statistical measures. The mean standard error is reported on the last line of the statistics report and defines the standard deviation for the sampling distribution of the mean as discussed in Chapter 3. We may use this to construct a confidence interval for the mean using the formula given in Chapter 4:

$$\bar{x} \pm z_{\alpha/2}\left(s/\sqrt{n}\right)$$

Because a *Crystal Ball* simulation will generally have a large number of trials, we may use the standard normal *z*-value instead of the *t*-distribution. Thus, for the year 5 cumulative discounted cash flow results, a 95% confidence interval for the mean would be:

$$\$120,823.22 \pm 1.96(\$1987.73) \text{ or } [\$116,927, \$124,719]$$

To reduce the size of the confidence interval, we would need to run the simulation for a larger number of trials. For most risk analysis applications, however, the mean is less important than the actual distribution of outcomes.

Forecast: Cumulative Discounted Cash Flow Year 5

Edit View Forecast Preferences Help

10,000 Trials Percentiles View Sheet1!G35

Percentile	Forecast values
0%	(548,560.30)
10%	(126,656.46)
20%	(48,862.16)
30%	10,993.43
40%	62,512.88
50%	110,247.81
60%	163,240.10
70%	219,586.60
80%	283,925.09
90%	376,228.23
100%	1,116,772.12

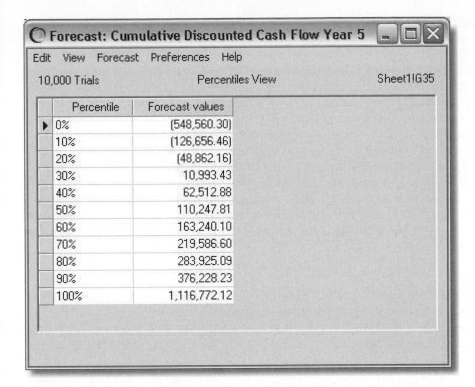

Figure 10.14 Percentiles Summary

Forecast: Cumulative Discounted Cash Flow Year 5

Edit View Forecast Preferences Help

10,000 Trials Statistics View Sheet1!G35

Statistic	Forecast values
Trials	10,000
Mean	120,823.22
Median	110,257.63
Mode	---
Standard Deviation	198,773.28
Variance	39,510,816,518.79
Skewness	0.2901
Kurtosis	3.28
Coeff. of Variability	1.65
Minimum	(548,560.30)
Maximum	1,116,772.12
Mean Std. Error	1,987.73

Figure 10.15 Statistics Summary

Crystal Ball also provides a variety of charts to help you analyze the results of a simulation. These can be selected from the *View Charts* option in the *Analyze* group. *Assumption Charts* show the sampled random variates superimposed over the theoretical distributions chosen as assumptions. They allow you to compare the effects of different settings, such as number of trials, on the simulated values. Larger samples generate smoother curves that conform more closely to the theoretical distributions. The *Forecast Charts* option allows you to open or close selected forecast charts.

An important reason for using simulation for risk analysis is the ability to conduct sensitivity analyses to understand the impacts of individual variables or their distributional assumptions on forecasts. A somewhat naïve way to investigate the impact of assumptions on forecast cells is to freeze, or hold, certain assumptions constant in the model and compare the results with a base case simulation. The *Freeze* command in the *Define* group allows you to temporarily disable certain assumptions from a simulation and conduct this type of sensitivity analysis.

The uncertainty in a forecast is the result of the combined effect of the uncertainties of all assumptions as well as the formulas used in the model. An assumption might have a high degree of uncertainty yet have little effect on the forecast because it is not weighted heavily in the model formulas. For instance, a spreadsheet formula for a forecast might be defined as:

$$0.9(Assumption \ 1) \ + \ 0.1(Assumption \ 2)$$

In the model, the forecast is nine times as sensitive to changes in the value of Assumption 1 as it is to changes in the value of Assumption 2. Thus, even if Assumption 2 has a much higher degree of uncertainty, as specified by the variance of its probability distribution, it would have a relatively minor effect on the uncertainty of the forecast.

The *Sensitivity Chart* feature allows you to determine the influence that each assumption has individually on a forecast. The sensitivity chart displays the rankings of each assumption according to their impact on a forecast cell as a bar chart. See *Crystal Ball Note: Sensitivity Charts*. A sensitivity chart provides three benefits:

1. It tells which assumptions are influencing forecasts the most and which need better estimates.
2. It tells which assumptions are influencing forecasts the least and can be ignored or discarded altogether.
3. By understanding how assumptions affect your model, you can develop more realistic spreadsheet models and improve the accuracy of your results.

A *Contribution to Variance* sensitivity chart for the cash flow example is shown in Figure 10.16. The assumptions are ranked from top to bottom, beginning with the assumption having the highest sensitivity. Positive values indicate a direct relationship between the assumption and forecast, while negative values reflect an inverse relationship. The percentages represent the contribution that each assumption has on the variance of the forecast. For example, we see that First Year Sales Revenue accounts for about 66% of the variance, Year 2 annual growth rate accounts for about 19%, and most other assumptions have a much smaller effect. This means that if you want to reduce the variability in the forecast the most, you would need to obtain better information about first year's sales revenue and use an assumption that has less variation.

If a simulation has multiple related forecasts, the *Overlay Charts* feature allows you to superimpose the frequency data from selected forecasts on one chart in order to compare differences and similarities that might not be apparent. You may select the forecasts that you wish to display from the *Choose Forecasts* dialog that appears when creating a new chart. Figure 10.17 shows an overlay chart for the cumulative distribution of cash flows for years 1 and 5. This chart makes it clear that the variance in year 5 is much larger than that in year 1. In the overlay chart, you may also view

Figure 10.16 Sensitivity Chart for Cumulative Discounted Cash Flow Year 5

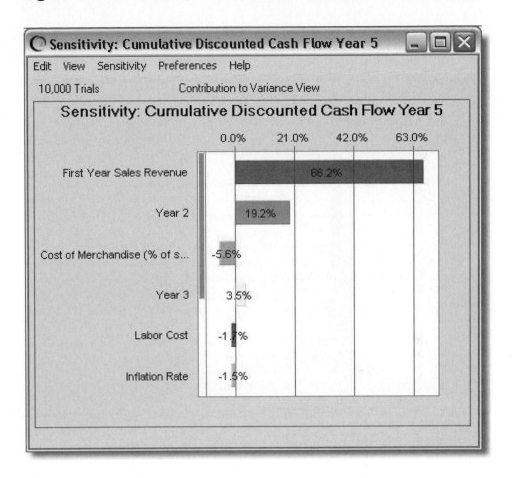

CRYSTAL BALL NOTE
Sensitivity Charts

To open a sensitivity chart, select *Sensitivity Charts* from the *View Charts* menu in the *Analyze* group. Click the *New* button to create a new chart, or check the box corresponding to a previously defined chart and click the *Open* button. For a new chart, check the box corresponding to the forecast you wish to create a sensitivity chart for in the *Choose Forecast* dialog that appears. Two types of charts are available: *Rank Correlation View* and *Contribution to Variance View* (default). The *Contribution to Variance View* addresses the question "What percentage of the variance in the target forecast is due to a particular assumption?" For the *Rank Correlation View, Crystal Ball* computes rank correlation coefficients between each assumption and forecast. Rank correlation uses the ranking of assumption values rather than the actual numbers. These correlation coefficients provide a measure of the degree to which assumptions and forecasts change together. Positive coefficients indicate that an increase in the assumption is associated with an increase in the forecast; negative coefficients imply the reverse. The larger the absolute value of the correlation coefficient, the stronger is the relationship. This chart may be displayed by selecting *Rank Correlation Chart* from the *View* menu. In addition, choosing *Sensitivity Data* from the *View* menu displays numerical results for both charts instead of graphical views.

Figure 10.17 Overlay Chart

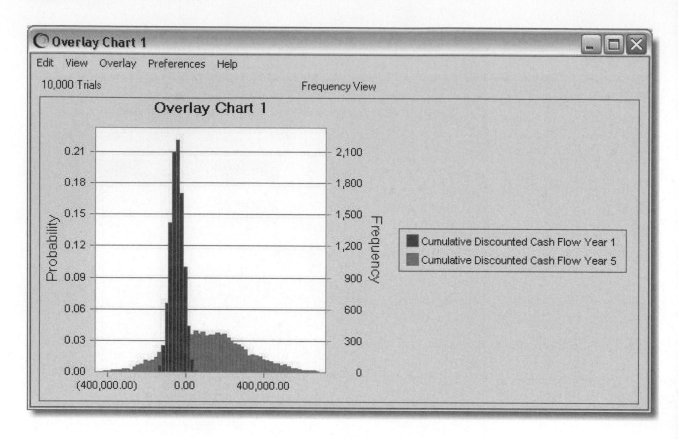

comparative statistics and percentiles from the *View* menu, and fit probability distributions to the forecasts from the *Overlay* menu.

If a simulation has multiple forecasts that are related to one another (such as over time), you can view the certainty ranges of all forecasts on a single chart, called a *Trend Chart*, which can be selected from the *View Charts* menu. Figure 10.18 shows a trend chart for the five-year cumulative discounted cash flows in our example. The trend chart displays certainty ranges in a series of patterned bands centered on the medians. For example, the band representing the 90% certainty range shows the range of values into which a forecast has a 90% chance of falling. From the trend chart in Figures 10.18, we see that although the median cash flow increases over time, so does the variation, indicating that the uncertainty also increases with time.

Finally, *Crystal Ball* can create *Scatter Charts*, which show correlations, dependencies, and other relationships between pairs of variables (forecasts and/or assumptions) plotted against each other. See the *Help* files for further information.

Crystal Ball Reports and Data Extraction

The *Analyze* group in the menu bar has two other options: *Create Report* and *Extract Data*. *Create Report* allows you to build customized reports for all or a subset of assumptions and output information that we described. The (default) full report contains a summary of run preferences and run statistics, the most recent views of forecast charts that you may have analyzed, forecast statistics and percentiles,

Figure 10.18 Trend Chart

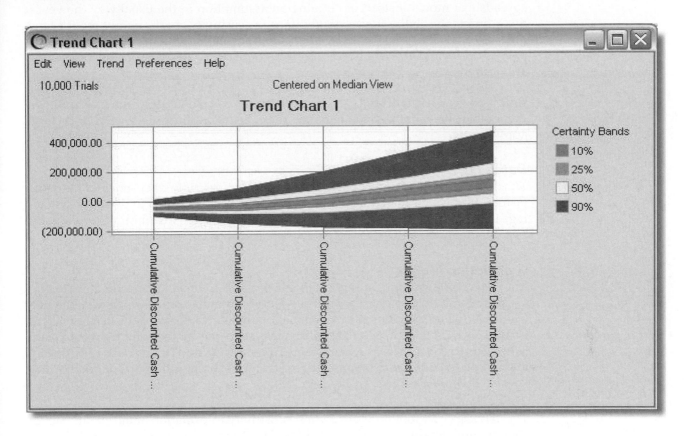

definitions of all assumptions, and any other charts that you may have created, such as sensitivity, overlay, etc. The report can be created in the current or new workbook from the *Options* tab of the *Create Report Preferences* dialog. You also have the option to display an image of the charts, or display them as Excel charts, which allows you to edit them using Excel chart commands. The report data may be edited, copied, and pasted into Word documents, or printed.

In addition, *Crystal Ball* allows you to extract selected data to an Excel worksheet for further analysis. In the dialog box that appears after clicking on *Extract Data* from the *Analyze* group, you may select various types of data to extract:

- Statistics
- Percentiles
- Chart Bins—the intervals in the forecast chart along with their probability and frequency of occurrence
- Sensitivity Data—sensitivity data for all pairs of assumptions and forecasts
- Trial Values—the generated assumption and forecast values for each simulation trial (be careful with this one, especially if you use 10,000 trials!)

Crystal Ball Functions and Tools

A very useful feature of *Crystal Ball* is the ability to use its functions within Excel. A complete list may be found by clicking the *Insert Function* button (f_x) on the Excel menu bar and selecting the *Crystal Ball* category. The random variate generation functions (CB.uniform, CB.normal, CB.exponential, and so on) may be used in a spreadsheet model instead of defining an assumption in the usual way (however, this restricts the ability to use assumptions in sensitivity charts, output reports, and other features of *Crystal Ball*). These functions are often useful to embed within complex Excel formulas. We will see examples of using these functions in some of the applications later in this chapter.

Crystal Ball has a set of modeling and analysis tools to complement the basic simulation capability. These can be found in the *Tools* menu under the *Run* group on the toolbar. We will briefly summarize them here and illustrate several of them in applications.

Batch Fit

The *Batch Fit* tool fits probability distributions to multiple data series in the same fashion that we discussed in Chapter 9 for fitting distributions to data. The advantage of this tool is that it eliminates the necessity to fit each distribution individually. The only requirement is that the data must be in adjacent rows or columns.

Correlation Matrix

In *Crystal Ball* each random input variable is assumed to be independent of the others; that is, random variates are sampled independently from each assumption distribution during the course of the simulation. In many situations this is not realistic because assumptions would naturally be related to one another, and we would want to explicitly model such dependencies between variables. The *Correlation Matrix* tool allows you to define correlations between groups of assumptions in a model.

Tornado Chart

The *Tornado Chart* tool provides *a priori* sensitivity information about the impact of each model variable on a target forecast. The tornado chart differs from the sensitivity chart in that it tests each assumption independently while freezing the other variables at their base values. This is useful in quickly prescreening variables in a model to determine which variables are the most important candidates to define as assumptions before building a simulation model. Those having little effect on a forecast might not need to be defined as assumptions and kept as constants in the model, thus simplifying it.

Bootstrap Tool

The classical approach for confidence intervals assumes that the sampling distribution of the mean is normal. However, if the sampling distribution is not normally distributed, such a confidence interval is not valid. Also, if we wanted to develop a confidence interval for the median, standard deviation, or maximum forecast value, for example, we may not know the sampling distribution of these statistics. A statistical technique called **bootstrapping** analyzes sample statistics empirically by repeatedly sampling the data and creating distributions of the statistics. This approach allows you to estimate the sampling distribution of any statistic, even an unconventional one such as the minimum or maximum endpoint of a forecast. The *Bootstrap* tool does this.

Decision Table

The *Decision Table* tool runs multiple simulations to test different values for one or two decision variables. We will illustrate this tool for an example later in this chapter.

Scenario Analysis

The *Scenario Analysis* tool runs a simulation and then sorts and matches all the resulting values of a target forecast with their corresponding assumption values. This allows you to investigate which combination of assumption values gives you a particular result.

Two-dimensional Simulation

The *Two-dimensional Simulation* tool allows you to distinguish between uncertainty in assumptions due to limited information or to data and variability—assumptions that change because they describe a population with different values. Theoretically, you can eliminate uncertainty by gathering more information; practically, it is usually impossible or cost prohibitive. Variability is inherent in the system, and you cannot eliminate it by gathering more information. Separating these two types of assumptions lets you more accurately characterize risk. The tool runs an outer loop to simulate the uncertainty values and then freezes them while it runs an inner loop to simulate variability. The process repeats for some small number of outer simulations, providing a portrait of how the forecast distribution varies due to the uncertainty.

APPLICATIONS OF MONTE CARLO SIMULATION

In this section, we present several additional examples of Monte Carlo simulation using *Crystal Ball*. These serve to illustrate the wide range of applications to which the approach may be applied and also various features of *Crystal Ball*.

Newsvendor Model

In Chapter 9, we developed the newsvendor model to analyze a single-period purchase decision using a naïve approach for Monte Carlo simulation. Here we will apply *Crystal Ball* and illustrate how to incorporate the distribution fitting and *Decision Table* tools into the analysis. To model the probability distribution of demand, we will use the historical candy sales introduced in Chapter 9 to illustrate the distribution fitting capability of *Crystal Ball*.

Figure 10.19 shows the newsvendor model spreadsheet (Excel file *Newsvendor CB Model*) to which we have added the historical sales data. Cell B11 represents the assumption cell for demand. We can have *Crystal Ball* automatically define an assumption cell for demand based on fitting the data. Because demand should be an integer value, we must make a slight modification to the spreadsheet. Rather than define the assumption cell as B11, we will define the assumption in cell C11 and then copy and round the result to a whole number in cell B11. First, click on cell C11, then *Define Assumption*, and then the *Fit* button. Proceed to fit the data. When the *Comparison Chart* pops up identifying the Beta distribution as the best fit, click the *Accept* button. This will display the *Define Assumption* dialog for cell C11 with the proper parameters for the Beta distribution automatically entered. Click *OK* and the demand assumption will be defined. Now enter the formula =ROUND(C11,0) into cell B11. This will enter a whole number based on the Beta assumption in the model. Next, define the profit in cell B17 as a forecast. You may now run *Crystal Ball*.

The forecast chart, shown in Figure 10.20, looks somewhat odd. However, recall that if demand exceeds the purchase quantity, then sales are limited to

Figure 10.19 *Crystal Ball* Implementation of Newsvendor Model

	A	B	C	D	E	F	G
1	Newsvendor Model			*Historical Candy Sales*			
2							
3	Data			42	43	47	43
4				45	46	41	45
5	Selling price	$ 18.00		40	42	41	42
6	Cost	$ 12.00		46	44	45	44
7	Discount price	$ 9.00		43	43	51	48
8							
9	Model						
10							
11	Demand	0	0				
12	Purchase Quantity	44					
13							
14	Quantity Sold	0					
15	Surplus Quantity	44					
16							
17	Profit	$ (132.00)					

Figure 10.20 Newsvendor Simulation Results for Purchase Quantity = 44

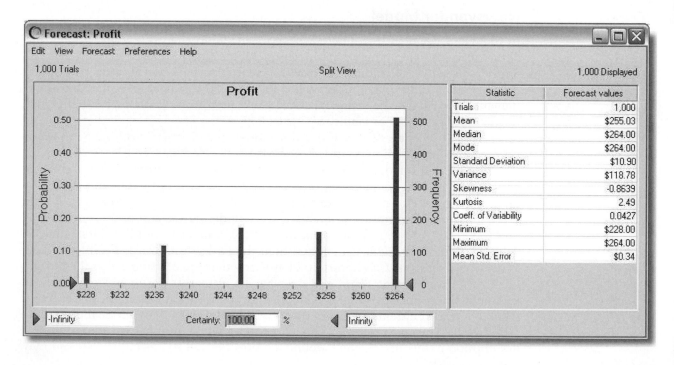

Figure 10.21 *Define Decision Variable* Dialog

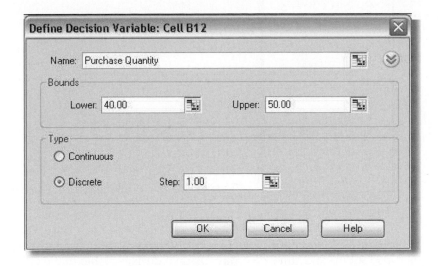

the number purchased; hence, the large spike at the right of the distribution. Although we could repeat the simulation for different values of purchase quantity, we will use the *Decision Table* tool to automatically run simulations for a range of values.

First, define the purchase quantity in cell B12 as a decision variable in *Crystal Ball*. Click on cell B12 and then on *Define Decision* in the *Define* group. Enter 40 in the *Lower* bound box, and 50 in the *Upper* bound box. Because this should be a whole number, choose *Discrete* with a *Step* size of 1 under *Type* as shown in Figure 10.21. Next, select the *Decision Table* tool. The first two dialog boxes that appear ask you to select the forecast and decision variable to evaluate; in this example, we have only one option for each to choose. In the third (Options) dialog box, set the number of trials for each simulation and click the *Run* button. The tool will run a simulation for each of the 11 decision variable values defined. When they are completed, a new worksheet will be displayed that shows the mean values of profit for each value of the decision variable (Figure 10.22). By selecting one of these cells and clicking on the *Forecast Charts* button, you can display the forecast chart for that simulation run. You may also select a range of cells and display trend or overlay charts. We see that a purchase quantity of 45 provides the largest average profit. The *Decision Table* tool is

Figure 10.22 *Decision Table* Tool Results

	A	B	C	D	E	F	G	H	I	J	K	L
1	Trend Chart / Overlay Chart / Forecast Charts	Purchase Quantity (40)	Purchase Quantity (41)	Purchase Quantity (42)	Purchase Quantity (43)	Purchase Quantity (44)	Purchase Quantity (45)	Purchase Quantity (46)	Purchase Quantity (47)	Purchase Quantity (48)	Purchase Quantity (49)	Purchase Quantity (50)
2		$240.00	$245.65	$249.92	$253.28	$255.23	$255.97	$254.85	$253.10	$251.79	$249.14	$245.89
3		1	2	3	4	5	6	7	8	9	10	11

SALES	PROBABILITY
40	0.02
41	0.05
42	0.07
43	0.09
44	0.05
45	0.07
46	0.05
47	0.02
48	0.02
49	0.00
50	0.00
51	0.02

Table 10.1 Probability Distribution of Candy Sales

useful whenever you wish to identify the best values of a decision variable. You may also set up and run a two-dimensional decision table that evaluates all combinations of two decision variables.

An alternative approach to sampling from a fitted distribution is to sample from the empirical distribution of the data. Table 10.1 shows a probability distribution of sales based on the frequencies of the sample data. We may use the Custom distribution in *Crystal Ball* to define an assumption based on this probability distribution (see *Crystal Ball Note: Using the Custom Distribution*).

While sampling from empirical data is easy to do, it does have some drawbacks. First, the empirical data may not adequately represent the true underlying population because of sampling error. Second, using an empirical distribution precludes sampling values outside the range of the actual data. Therefore, it is usually advisable to fit a distribution and use it for the assumption.

CRYSTAL BALL NOTE
Using the Custom Distribution

The Custom Distribution can be used to model any type of discrete probability function that has a series of single values, discrete ranges, or continuous ranges. To define the demand in cell C11 as an assumption having this distribution, choose Custom Distribution from the distribution gallery. You may select the different options from the *Parameters* menu, with the default being *Weighted Values* (that is, a discrete probability distribution). Enter the value and probability for each outcome in the appropriate fields in the dialog. With the Custom Distribution, you may either enter the data manually or load the data from a worksheet, which is typically easier to do. To do this, click the *More* button (the downward pointing arrow) to the right of the name field in the *Custom Distribution* dialog box, which will expand the dialog box, then click on the *Load* button. *Crystal Ball* will prompt you for the location of the data; simply enter the cell reference of the data range. The range of the data must be so that the outcomes or outcome range columns are first and the probabilities are on the right. The data in Table 10.1 are set up in this fashion. The result is shown in Figure 10.23.

You can also create distributions that have continuous ranges associated with discrete probabilities. To do this, choose *Continuous Ranges* from the *Parameters* menu in the *Custom Distribution* dialog box. Other options are also available; see the *Help* files for further information.

Figure 10.23 Completed Custom Distribution Example

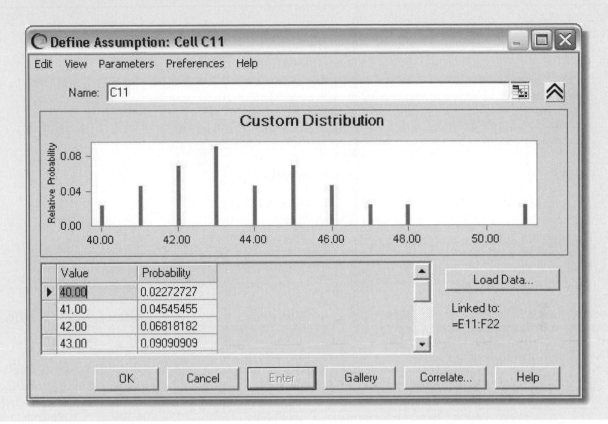

Overbooking Model

An important operations decision for service businesses such as hotels, airlines, and car rental companies is the number of reservations to accept to effectively fill capacity with the knowledge that some customers may not use their reservations nor tell the business. If a hotel, for example, holds rooms for customers who do not show up, they lose revenue opportunities. (Even if they charge a night's lodging as a guarantee, rooms held for additional days may go unused.) A common practice in these industries is to overbook reservations. When more customers arrive than can be handled, the business usually incurs some cost to satisfy them (by putting them up at another hotel, or for most airlines, providing extra compensation such as ticket vouchers). Therefore, the decision becomes how much to overbook to balance the costs of overbooking against the lost revenue for underuse.

We will illustrate how a simulation model can help in making this decision. Figure 10.24. shows a spreadsheet simulation model (Excel file *Overbooking Model*) for a popular resort hotel that has 300 rooms and is usually fully booked. The hotel charges $120 per room. About 4% of the time a reservation is cancelled by the 6:00 p.m. deadline with no penalty incurred by the customer. The hotel has estimated that the average overbooking cost is $100.

In the model section of the spreadsheet, cell B12 represents the decision variable of how many reservations to accept. Rather than use the *Define Assumption* option in *Crystal Ball*, we will show how to incorporate *Crystal Ball* functions into this model.

Figure 10.24 Overbooking Model

	A	B
1	Hotel Overbooking Model	
2		
3	Data	
4		
5	Rooms available	300
6	Price	$120
7	Probability of Cancellation	0.04
8	Overbooking cost	$100
9		
10	Model	
11		
12	Reservation limit	300
13	Customer demand	284
14	Reservations made	284
15	Cancellations	15
16	Customer arrivals	285
17	Overbooked customers	0
18		
19	Net revenue	$34,200

	A	B
1	Hotel Overbooking Model	
2		
3	Data	
4		
5	Rooms available	300
6	Price	120
7	Probability of Cancellation	0.04
8	Overbooking cost	100
9		
10	Model	
11		
12	Reservation limit	300
13	Customer demand	=CB.Poisson(320)
14	Reservations made	=MIN(B12,B13)
15	Cancellations	=CB.Binomial(B7,B14)
16	Customer arrivals	=B12-B15
17	Overbooked customers	=MAX(0,B16-B5)
18		
19	Net revenue	=MIN(B16,B5)*B6-B17*B8

In cell B13, we define the customer demand by a Poisson distribution with a mean of 320 using the function CB.Poisson(320). The number of reservations made, therefore, is the smaller of the customer demand and the reservation limit. Because each reservation has a constant probability of being cancelled, the number of cancellations (cell B15) can be modeled using a binomial distribution with n = number of reservations made and p = probability of cancellation, using the function CB.binomial(B7, B14). If the actual number of customer arrivals exceeds the room capacity, overbooking occurs. Net revenue and overbooked customers are defined as forecast cells.

Figure 10.25 shows forecast charts for accepting 310 reservations. There is about a 21% chance of overbooking at least one customer. Can you explain why the Net Revenue forecast chart looks the way it does? An exercise at the end of this chapter asks you to use the *Decision Table* tool in *Crystal Ball* to automatically run simulations for a range of values of the number of reservations to accept.

Cash Budgeting

Cash budgeting[3] is the process of projecting and summarizing a company's cash inflows and outflows expected during a planning horizon, usually 6 to 12 months. The cash budget also shows the monthly cash balances and any short-term borrowing used to cover cash shortfalls. Positive cash flows can increase cash, reduce outstanding loans, or be used elsewhere in the business; negative cash flows can reduce cash available or be offset with additional borrowing. Most cash budgets are based on sales forecasts. With the inherent uncertainty in such forecasts, Monte Carlo simulation is an appropriate tool to analyze cash budgets.

[3] Adapted from Douglas R. Emery, John D. Finnerty, and John D. Stowe, *Principles of Financial Management* (Upper Saddle River, NJ: Prentice Hall, 1998), 652–654.

Figure 10.25 Overbooking Model Results

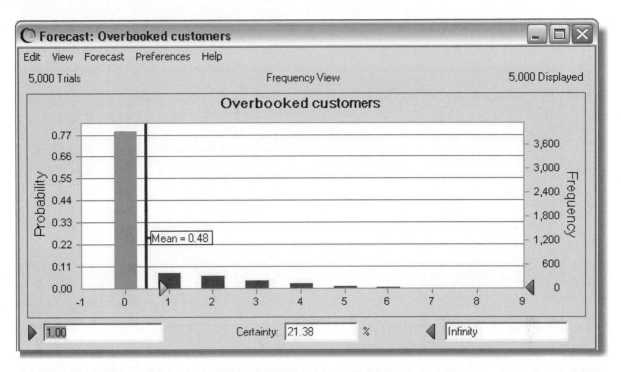

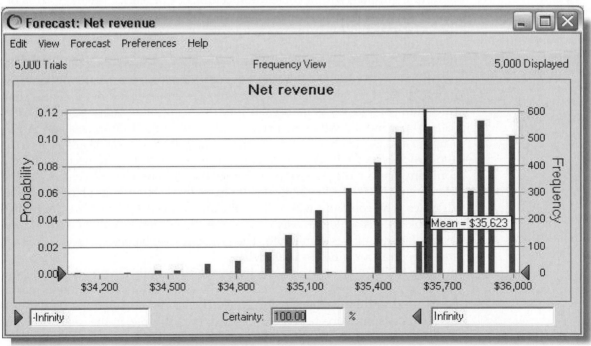

Figure 10.26 shows an example of a cash budget spreadsheet (Excel file *Cash Budget*). The budget begins in April (thus, sales for April and subsequent months are uncertain). These are assumed to be normally distributed with a standard deviation of 10% of the mean. In addition, we assume that sales in adjacent months are correlated

Figure 10.26 Cash Budget Model

	A	B	C	D	E	F	G	H	I	J	K
1	Cash Budgeting										
2	Desired Minimum Balance	$ 100,000									
3			February	March	April	May	June	July	August	September	October
4		Sales	$400,000	$500,000	$600,000	$700,000	$800,000	$800,000	$700,000	$600,000	$500,000
5	Cash Receipts										
6	Collections (current)	20%			$120,000	$140,000	$160,000	$160,000	$140,000	$120,000	
7	Collections (previous month)	50%			$250,000	$300,000	$350,000	$400,000	$400,000	$350,000	
8	Collections (2nd month previous)	30%			$120,000	$150,000	$180,000	$210,000	$240,000	$240,000	
9	Total Cash Receipts				$490,000	$590,000	$690,000	$770,000	$780,000	$710,000	
10											
11	Cash Disbursements										
12	Purchases				$420,000	$480,000	$480,000	$420,000	$360,000	$300,000	
13	Wages and Salaries				$ 72,000	$ 84,000	$ 96,000	$ 96,000	$ 84,000	$ 72,000	
14	Rent				$ 10,000	$ 10,000	$ 10,000	$ 10,000	$ 10,000	$ 10,000	
15	Cash Operating Expenses				$ 30,000	$ 30,000	$ 30,000	$ 30,000	$ 25,000	$ 25,000	
16	Tax Installments				$ 20,000			$ 30,000			
17	Capital Expenditure						$150,000				
18	Mortgage Payment					$ 60,000					
19	Total Cash Disbursements				$552,000	$664,000	$766,000	$586,000	$479,000	$407,000	
20											
21	Ending Cash Balance										
22	Net Cash Flow				$ (62,000)	$ (74,000)	$ (76,000)	$184,000	$301,000	$303,000	
23	Beginning Cash Balance				$150,000	$100,000	$100,000	$100,000	$122,000	$423,000	
24	Available Balance				$ 88,000	$ 26,000	$ 24,000	$284,000	$423,000	$726,000	
25	Monthly Borrowing				$ 12,000	$ 74,000	$ 76,000	$ -	$ -	$ -	
26	Monthly Repayment				$ -	$ -	$ -	$162,000	$ -	$ -	
27	Ending Cash Balance			$150,000	$100,000	$100,000	$100,000	$122,000	$423,000	$726,000	
28	Cumulative Loan Balance			$ -	$ 12,000	$ 86,000	$162,000	$ -	$ -	$ -	

with one another, with a correlation coefficient of 0.6. On average, 20% of sales is collected in the month of sale, 50% in the month following the sale, and 30% in the second month following the sale. However, these figures are uncertain, so a uniform distribution is used to model the first two values (15% to 20% and 40% to 50%, respectively) with the assumption that all remaining revenues are collected in the second month following the sale. Purchases are 60% of sales and are paid for one month prior to the sale. Wages and salaries are 12% of sales and are paid in the same month as the sale. Rent of $10,000 is paid each month. Additional cash operating expenses of $30,000 per month will be incurred for April through July, decreasing to $25,000 for August and September. Tax payments of $20,000 and $30,000 are expected in April and July, respectively. A capital expenditure of $150,000 will occur in June, and the company has a mortgage payment of $60,000 in May. The cash balance at the end of March is $150,000, and managers want to maintain a minimum balance of $100,000 at all times. The company will borrow the amounts necessary to ensure that the minimum balance is achieved. Any cash above the minimum will be used to pay off any loan balance until it is eliminated. The available cash balances in row 24 of the spreadsheet are the *Crystal Ball* forecast cells.

Crystal Ball allows you to specify correlation coefficients to define dependencies between assumptions. *Crystal Ball* uses the correlation coefficients to rearrange the generated random variates to produce the desired correlations. This can be done only after assumptions have been defined. For example, in the cash budget model, if the sales in April are high, then it would make sense that the sales in May would be high also. Thus, we might expect a positive correlation between these variables.

The *Crystal Ball Correlation Matrix* tool is used to specify the correlations between sales assumptions (see *Crystal Ball Note: Correlation Matrix Tool*). Figure 10.27 shows the trend chart for the monthly cash balances. We see that there is a high likelihood that cash balances will be negative for the first three months before increasing. Viewing the forecast chart and statistics for individual months will provide the

Figure 10.27 Cash Balance Trend Chart

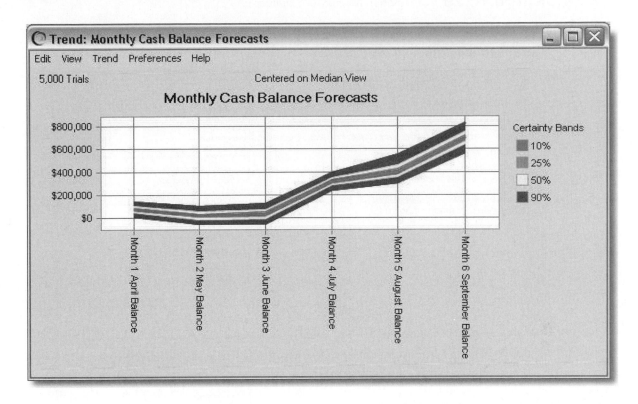

details on the distribution of cash balances and the likelihood of requiring loans. For example, the forecast chart for April shows that the probability that the balance will not exceed the minimum of $100,000 and require an additional loan is about 0.70, actually worsens for May and June, and is zero by July.

CRYSTAL BALL NOTE
Correlation Matrix Tool

In *Crystal Ball*, you may enter correlations between any pair of assumptions by clicking on one of them, selecting *Define Assumption*, and then choosing the *Correlate* button in the *Define Assumption* dialog. This brings up the *Define Correlation* dialog, which allows you to choose the assumption to correlate with and enter the value of the correlation coefficient. The dialog displays a scatter chart of how the assumptions would be related to each other. Instead of manually entering the correlation coefficients this way, you can use the *Correlation Matrix* tool to define a matrix of correlations between assumptions in one simple step. This saves time and effort when building your spreadsheet model, especially for models with many correlated assumptions.

Choose the *Correlation Matrix* from the *Tools* menu in the *Run* group. This brings up a dialog box in which you select the assumptions to correlate. In the cash budget example, we wish to correlate sales, so choose all assumptions except the current and previous month's collections as shown in Figure 10.28. The correlation matrix is either an upper or lower triangular matrix with ones along the diagonal and correlation coefficients entered in the blank cells. The tool allows you to select the type of matrix and whether to create in a new worksheet or the existing worksheet; we recommend using the existing worksheet simply to document the input data. The example is shown in Figure 10.29. If you enter inconsistent correlations, *Crystal Ball* tries to adjust the correlations so they don't conflict.

Figure 10.28 *Correlation Matrix* Tool Dialog

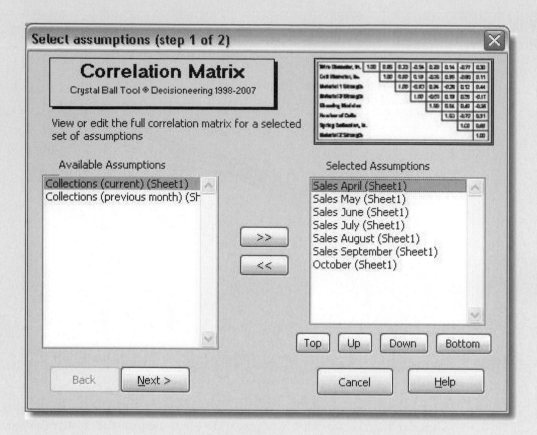

Figure 10.29 Correlation Matrix for Cash Budget Example

	Sales April (Sheet1)	Sales May (Sheet1)	Sales June (Sheet1)	Sales July (Sheet1)	Sales August (Sheet1)	Sales September (Sheet1)	October (Sheet1)
Sales April (Sheet1)	1.000	0.600					
Sales May (Sheet1)		1.000	0.600				
Sales June (Sheet1)			1.000	0.600			
Sales July (Sheet1)				1.000	0.600		
Sales August (Sheet1)					1.000	0.600	
Sales September (Sheet1)						1.000	0.600
October (Sheet1)							1.000

New Product Development Model

We will revisit the Moore Pharmaceuticals model that we developed in Chapter 9. Suppose that analysts have made the following assumptions:

R&D costs: Triangular($500, $700, $800) in millions of dollars
Clinical trials costs: Triangular($135, $150, $160) in millions of dollars
Initial market size: Normal(2000000, 250000)
Initial market share: Uniform(6%, 10%)

Figure 10.30 shows the distribution of the forecast for net present value of profit based on these assumptions. As you can see, there is about a 70% chance that the NPV will be positive. This is a somewhat risky decision. The large variability in the forecast makes it difficult to gain an accurate projection of profitability and make a good decision. The Sensitivity Chart provides an assessment of the factors that contribute to the variance of the forecast. As the chart in Figure 10.31 shows, the market size assumption contributes about 57% of the variability in the forecast followed by the market share assumption. Thus, the company would be advised to conduct better market research to try to obtain more accurate estimates of these values and use assumptions with less variability.

Crystal Ball has a tool that can analyze the sensitivity of each assumption or other model inputs to the forecast prior to conducting a simulation, the *Tornado*

Figure 10.30 Net Present Value of Profit for New Product Development Model

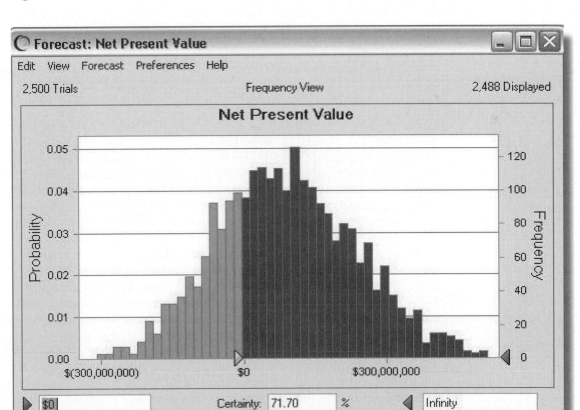

Figure 10.31   Sensitivity Chart for Moore's Pharmaceuticals

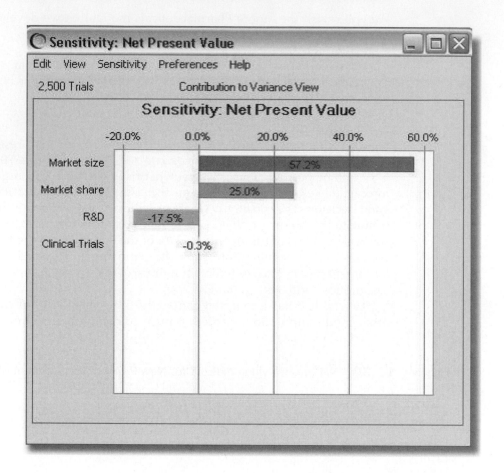

Chart tool (see *Crystal Ball Note: Tornado Charts*). The *Tornado Chart* tool is useful for measuring the sensitivity of variables that you have defined in *Crystal Ball* or quickly prescreening the variables in your model to determine which ones are good candidates to define as assumptions or decision variables. For example, suppose that we wanted to understand the impact of the four variables—R&D cost, clinical trials cost, initial market size, and initial market share—prior to defining any *Crystal Ball* assumptions. In the *Tornado Chart* tool dialog, use the *Add Range* button to select these cells in step 2 of the input process and run the tool. The tool creates two charts, a tornado chart and a spider chart. Figure 10.32 shows the tornado chart and data table that is created (we have edited the variable names in the table). In the data table, the tool examines values ±10% away from the base case and evaluates the NPV forecast while holding all other variables constant. We see that the initial market size has the most impact while clinical trials cost has the least. The spider chart in Figure 10.33 shows these results as rates of change; the larger the slope, the higher the impact. While these charts show similar insights as the sensitivity chart, running this before developing the *Crystal Ball* model might have led you to not bother trying to determine an appropriate assumption for clinical trials cost, as it would have little impact on the forecast. Instead, it could have been left constant with little change in the final results. For models with large numbers of potential assumptions, this is a useful tool.

Figure 10.32 Tornado Chart Results

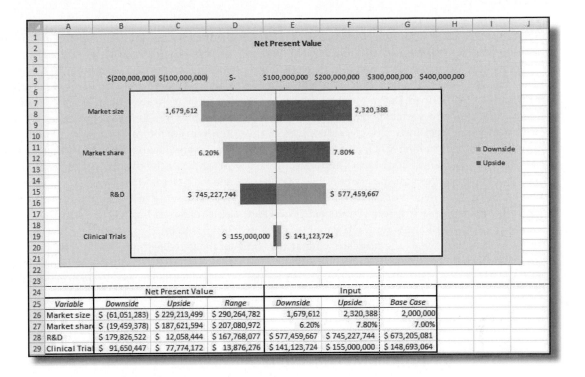

	Net Present Value			Input		
Variable	Downside	Upside	Range	Downside	Upside	Base Case
Market size	$ (61,051,283)	$ 229,213,499	$ 290,264,782	1,679,612	2,320,388	2,000,000
Market share	$ (19,459,378)	$ 187,621,594	$ 207,080,972	6.20%	7.80%	7.00%
R&D	$ 179,826,522	$ 12,058,444	$ 167,768,077	$ 577,459,667	$ 745,227,744	$ 673,205,081
Clinical Trial	$ 91,650,447	$ 77,774,172	$ 13,876,276	$ 141,123,724	$ 155,000,000	$ 148,693,064

Figure 10.33 Spider Chart Results

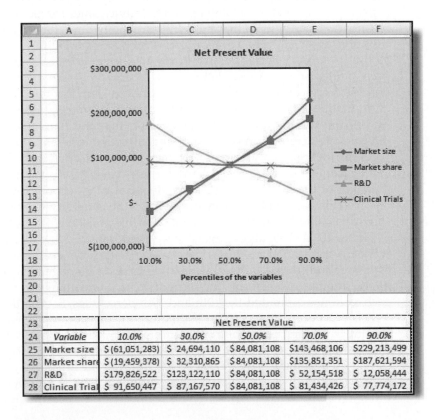

	Net Present Value				
Variable	10.0%	30.0%	50.0%	70.0%	90.0%
Market size	$ (61,051,283)	$ 24,694,110	$ 84,081,108	$143,468,106	$229,213,499
Market share	$ (19,459,378)	$ 32,310,865	$ 84,081,108	$135,851,351	$187,621,594
R&D	$179,826,522	$123,122,110	$ 84,081,108	$ 52,154,518	$ 12,058,444
Clinical Trial	$ 91,650,447	$ 87,167,570	$ 84,081,108	$ 81,434,426	$ 77,774,172

CRYSTAL BALL NOTE
Tornado Charts

Select *Tornado Charts* from the *Tools* menu in the *Run* group. The first dialog box that opens asks you to select the forecast you wish to analyze. In the next box, click the appropriate button to add assumptions or other variables defined in your model. Use the *Add Range* button to add variables that are not defined as assumptions or decisions. The default values in the third box can be left alone. Click on *Start*.

The *Tornado Chart* tool tests the range of each variable at percentiles you specify and then calculates the value of the forecast at each point. The tornado chart illustrates the swing between the maximum and minimum forecast values for each variable, placing the variable that causes the largest swing at the top and the variable that causes the smallest swing at the bottom. The top variables have the most effect on the forecast, and the bottom variables have the least effect

on the forecast. While analyzing one variable, the tool freezes the other variables at their base values. This measures the effect each variable has on the forecast cell while removing the effects of the other variables.

The bars next to each variable represent the forecast value range across the variable tested. Next to the bars are the values of the variables that produced the greatest swing in the forecast values. The bar colors indicate the direction of the relationship between the variables and the forecast. For variables that have a positive effect on the forecast, the upside of the variable (dark shading in Figure 10.32, actually shown in blue) is to the right of the base case and the downside of the variable (light shading in Figure 10.32, actually shown in red) is to the left side of the base case. For variables that have a reverse relationship with the forecast, the bars are reversed.

Project Management

Project management is concerned with scheduling the activities of a project involving interrelated activities. An important aspect of project management is identifying the expected completion time of the project. Activity times can be deterministic or probabilistic. We often assume that probabilistic activity times have a beta or triangular distribution, especially when times are estimated judgmentally. Analytical methods, such as the Program Evaluation and Review Technique (PERT), allow us to determine probabilities of project completion times by assuming that the expected activity times define the critical path and invoking the Central Limit theorem to make an assumption of normality of the distribution of project completion time. However, this assumption may not always be valid, and we will explore this later. Simulation can provide a more realistic characterization of the project completion time and the associated risks. We will illustrate risk analysis in project management through the following example.

Becker Consulting has been hired to assist in the evaluation of new software. The manager of the Information Systems department is responsible for coordinating all of the activities involving consultants and the company's resources. The activities shown in Table 10.2 have been defined for this project, which is depicted graphically in Figure 10.34. The target project completion date is 150 working days. Because this is a new application, no historical data on activity times are available and must be estimated judgmentally. The Information Systems manager has determined the most likely time for each activity but, recognizing the uncertainty in the times to complete each task, has estimated the 10th and 90th percentiles. Setting the parameters in this way is a common approach for estimating the distribution since managers typically cannot estimate the absolute minimum or maximum times but can reasonably determine a time that might be met or exceeded 10% of the time. With only these estimates, a triangular distribution is an appropriate assumption.

Figure 10.35 shows a spreadsheet designed to simulate the project completion time (Excel file *Becker Consulting Project Management Model*). For those activity times

Table 10.2
Activity and Time Estimate List

ACTIVITY		PREDECESSORS	10TH PERCENTILE	MOST LIKELY	90TH PERCENTILE
A	Select steering committee	—	15	15	15
B	Develop requirements list	—	40	45	60
C	Develop system size estimates	—	10	14	30
D	Determine prospective vendors	—	2	3	5
E	Form evaluation team	A	5	7	9
F	Issue request for proposal	B,C,D,E	4	5	8
G	Bidders conference	F	1	1	1
H	Review submissions	G	25	30	50
I	Select vendor short list	H	3	5	10
J	Check vendor references	I	3	7	10
K	Vendor demonstrations	I	20	30	45
L	User site visit	I	3	4	5
M	Select vendor	J,K,L	3	3	3
N	Volume sensitive test	M	10	13	20
O	Negotiate contracts	M	10	14	28
P	Cost-benefit analysis	N,O	2	2	2
Q	Obtain board of directors approval	P	5	5	5

Figure 10.34 Project Network Structure

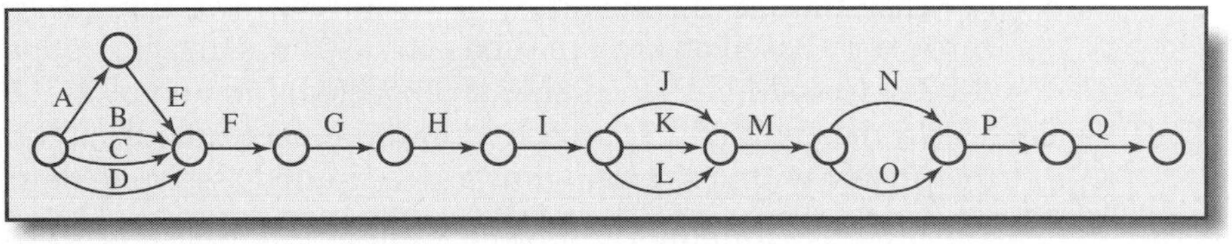

Figure 10.35 Becker Consulting Project Management Spreadsheet

	A	B	C	D	E	F	G	H	I	J	K
1	Becker Consulting Project Management Model										
2											
3		10th	Most	90th	Activity	Early	Early	Latest	Latest		On Critical
4	Activity	Percentile	Likely	Percentile	Time	Start	Finish	Start	Finish	Slack	Path?
5	A	15	15	15	15.00	0.00	15.00	27.32	42.32	27.32	0
6	B	40	45	60	49.32	0.00	49.32	0.00	49.32	0.00	1
7	C	10	14	30	19.19	0.00	19.19	30.12	49.32	30.12	0
8	D	2	3	5	3.43	0.00	3.43	45.88	49.32	45.88	0
9	E	5	7	9	7.00	15.00	22.00	42.32	49.32	27.32	0
10	F	4	5	8	5.86	49.32	55.18	49.32	55.18	0.00	1
11	G	1	1	1	1.00	55.18	56.18	55.18	56.18	0.00	1
12	H	25	30	50	36.49	56.18	92.67	56.18	92.67	0.00	1
13	I	3	5	10	6.29	92.67	98.96	92.67	98.96	0.00	1
14	J	3	7	10	6.57	98.96	105.53	124.54	131.11	25.58	0
15	K	20	30	45	32.15	98.96	131.11	98.96	131.11	0.00	1
16	L	3	4	5	4.00	98.96	102.96	127.11	131.11	28.15	0
17	M	3	3	3	3.00	131.11	134.11	131.11	134.11	0.00	1
18	N	10	13	20	14.72	134.11	148.83	137.71	152.43	3.60	0
19	O	10	14	28	18.32	134.11	152.43	134.11	152.43	0.00	1
20	P	2	2	2	2.00	152.43	154.43	152.43	154.43	0.00	1
21	Q	5	5	5	5.00	154.43	159.43	154.43	159.43	0.00	1
22											
23			Project completion time		159.43						

that are not constant, we define the cell for the activity time as a *Crystal Ball* assumption using the triangular distribution. After selecting the triangular distribution in the *Crystal Ball* gallery, click on the *Parameters* menu in the dialog box. This provides a list of alternative ways to input the data. Select the *10th percentile, likeliest,* and *90th percentile* option. *Crystal Ball* determines the appropriate minimum (a) and maximum (b) values for the triangular distribution based on the percentile information. To facilitate data input, we may use cell references in the input boxes instead of inputting the values. Then we may use the *Copy Data* and *Paste Data* commands under the *Define* menu to copy the *Crystal Ball* assumptions to the other appropriate cells (remember not to use the Excel *Copy* and *Paste* commands).

The project completion time depends on the specific time for each activity. To find this, we compute the activity schedule and slack for each activity. Activities A, B, C, and D have no immediate predecessors and, therefore, have early start times of 0. The early start time for each other activity is the maximum of the early finish times for the activity's immediate predecessor. Early finish times are computed as the early start time plus the activity time. The early finish time for the last activity, Q, represents the earliest time the project can be completed, that is, the minimum project completion time. This is defined as the forecast cell for the *Crystal Ball* simulation.

To compute late start and late finish times, we set the late finish time of the terminal activity equal to the project completion time. The late start time is computed by subtracting the activity time from the late finish time. The late finish time for any other activity, say X, is defined as the minimum late start of all activities to which activity X is an immediate predecessor. Slack is computed as the difference between the late finish and early finish. The critical path consists of activities with zero slack. Based on the expected activity times, the critical path consists of activities B-F-G-H-I-K-M-O-P-Q and has an expected duration of about 160 days.

In the analytical approach found in most textbooks, probabilities of completing the project within a certain time are computed assuming that:

1. the distribution of project completion times is normal (by applying the Central Limit theorem);
2. the expected project completion time is the sum of the expected activity times along the critical path, which is found using the expected activity times; and
3. the variance of the distribution is the sum of the variances of those activities along the critical path, which is found using the expected activity times. If more than one critical path exists, use the path with the largest variance.

Using the minimum and maximum values for the triangular distribution computed by *Crystal Ball* from the percentile data input, we may use the formulas presented in Chapter 3 to compute the variance for each activity, as shown in Figure 10.36. Thus, for this example, the variance of the critical path is 281.88 (found by adding the variances of those activities with zero slack). With the normality assumption, the probability that the project will be completed within 150 days is found by computing the z-value:

$$z = \frac{150 - 159.43}{16.789} = -0.562$$

From Table A.1 in the appendix at the end of the book, this corresponds to a probability of approximately 0.29.

Variations in actual activity times may yield different critical paths than the one resulting from expected times. This may change both the mean and variance of the actual project completion time, resulting in an inaccurate assessment of risk. Simulation can easily address these issues. Figure 10.37 shows the *Crystal Ball* forecast and statistics

Figure 10.36 Critical Path Mean and Variance Calculations

a	m	b	mean	variance
15	15	15	15	0
33.63	45	69.31	49.3133	55.369872
4.07	14	39.51	19.1933	55.704406
0.95	3	6.34	3.43	1.2336167
3.38	7	10.62	7	2.1840667
2.73	5	9.89	5.87333	2.2314056
1	1	1	1	0
17.58	30	61.89	36.49	87.07235
0.67	5	13.21	6.29333	6.7612389
0.03	7	12.68	6.57	6.6907167
10.66	30	55.79	32.15	85.441017
2.19	4	5.81	4	0.5460167
3	3	3	3	0
6.62	13	24.55	14.7233	13.766439
4.48	14	36.48	18.32	44.999467
2	2	2	2	0
5	5	5	5	0
	On critical path:		159.44	281.87535
		standard deviation		16.789144

Figure 10.37 Project Completion Time Results

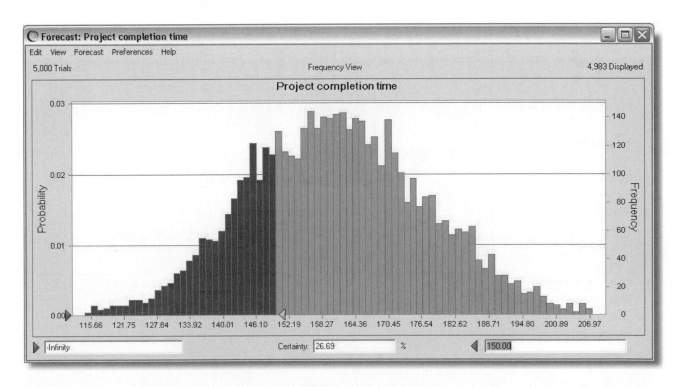

charts for 5,000 trials. The mean and variance are quite close to those predicted; in fact, fitting the normal distribution to the forecast in an overlay chart results in a very good fit. Whereas the analytical approach computed the probability of completing the project within 150 days as 0.29, analysis of the forecast chart shows this to be somewhat smaller, about 0.26. Note, however, that the structure of the project network is quite "linear," resulting in few options for critical paths. For projects that have many more parallel paths, the results of the simulation and analytical models may differ significantly more.

In this example, the simulation results provide estimates of key statistics such as the mean, median, maximum value, and so on. If we run the model again, these values will change slightly because of the different random assumptions that would be generated. A logical question is how accurate are these estimates? For example, suppose that Becker Consulting is concerned about the maximum project time. From this simulation, an estimate of the maximum project time is about 207 days. To better understand the variability, we can use the *Bootstrap* tool in *Crystal Ball* (see *Crystal Ball Note: Bootstrap Tool*). The *Bootstrap* tool uses a simple technique that estimates the reliability or accuracy of forecast statistics or other sample data. Classical methods used in the past relied on mathematical formulas to describe the accuracy of sample statistics. In contrast, bootstrapping analyzes sample statistics empirically by repeatedly sampling the data and creating sampling distributions of the different statistics. The term *bootstrap* comes from the saying, "to pull oneself up by one's own bootstraps," since this method uses the distribution of statistics themselves to analyze the statistics' accuracy.

We used 100 bootstrap samples with 500 trials per sample. Figure 10.38 shows results obtained using the *Bootstrap* tool. Row 2 in the spreadsheet shows the statistics

Figure 10.38 Bootstrap Statistics Results

	A	B Mean	C Median	D Mode	E Standard Deviation	F Variance	G Skewness	H Kurtosis	I Coeff. of Variability	J Mean Std. Error	K Minimum	L Maximum	M Range
1													
2	Project completion time	160.93	160.51	0.00	16.32	266.77	0.15	2.82	0.10	0.73	117.62	210.74	93.11
3													
4	Correlations:												
5	Mean	1.000	0.747	0.085	0.037	0.037	-0.176	-0.103	-0.086	0.037	0.213		
6	Median		1.000	0.162	0.050	0.050	-0.424	-0.026	-0.041	0.050	0.120		
7	Mode			1.000	-0.119	-0.119	0.023	0.015	-0.123	-0.119	0.092		
8	Standard Deviation				1.000	1.000	-0.206	-0.232	0.989	1.000	-0.448		
9	Variance					1.000	-0.206	-0.232	0.989	1.000	-0.448		
10	Skewness						1.000	0.443	-0.183	-0.206	0.342		
11	Kurtosis							1.000	-0.217	-0.232	-0.037		
12	Coeff. of Variability								1.000	0.989	-0.486		
13	Mean Std. Error									1.000	-0.448		
14	Minimum										1.000		
15	Maximum												
16	Range												

Figure 10.39 Forecast Chart for Maximum Project Completion Time

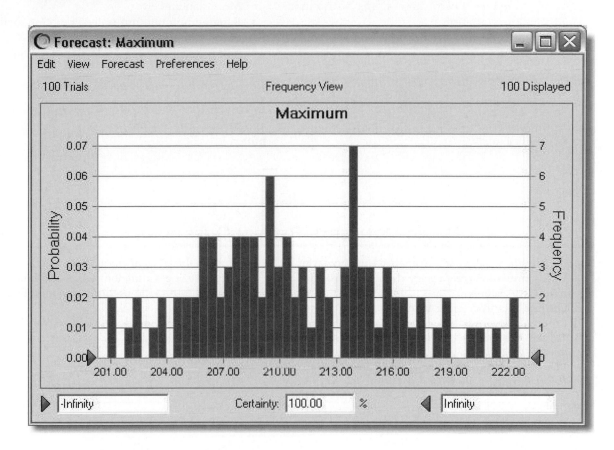

of the sampling distribution of the cash flow. The results spreadsheet also displays a correlation matrix showing the correlations between the various statistics. High correlation between certain statistics, such as between the mean and the standard deviation, usually indicates a highly skewed distribution. The forecast chart constructed for each statistic visually conveys the accuracy of the statistic. A narrow and symmetric distribution is better than a wide and skewed distribution. A small standard error of the statistic and coefficient of variability are quantitative indicators of reliable statistical estimates. Figure 10.39 shows the sampling distribution for the maximum project completion time. Although the maximum project completion time has a mean of about 210 days, we see that this might reasonably vary from about 200 to 220 days.

CRYSTAL BALL NOTE
Bootstrap Tool

Choose *Bootstrap* from the *Tools* menu in the *Run* group. The first dialog box asks you to select the forecast you wish to analyze. In the second box, you must choose one of two alternative methods:

1. *One-simulation method*, which simulates the model data once and repeatedly samples with replacement

2. *Multiple-simulation method*, which repeatedly simulates the model and creates a sampling distribution from each simulation; this method is more accurate but is slower

The *Bootstrap* tool constructs sampling distributions for the following statistics:

- Mean
- Median
- Standard deviation
- Variance
- Skewness
- Kurtosis
- Coefficient of variability

- Mean standard error
- Range minimum (multiple-simulation method only)
- Range maximum (multiple-simulation method only)
- Range width (multiple-simulation method only)

In the third dialog box, set the number of samples and number of trials per sample and click the *Start* button.

BASIC CONCEPTS REVIEW QUESTIONS

1. Explain the concept of risk analysis and how Monte Carlo simulation can provide useful information for making decisions.
2. Outline the process for using *Crystal Ball* to run a Monte Carlo simulation.
3. Explain the terms *assumption, forecast,* and *decision* as used in *Crystal Ball.*
4. What information can you extract from a *Crystal Ball* forecast chart?
5. Explain the types of charts that *Crystal Ball* provides for analyzing simulation results and how they may be used to provide useful information to a decision maker.
6. Summarize the modeling and analysis tools available in *Crystal Ball* and what information they provide.

SKILL-BUILDING EXERCISES

1. Open the *New Store Financial Model* spreadsheet and set up and run the *Crystal Ball* simulation as described in the chapter. Find the probabilities of a positive cumulative discounted cash flow for each of years 1 through 5. Create a spreadsheet summary of the simulation statistics and percentiles for each year.
2. Open the *New Store Financial Model* spreadsheet and set up and run the *Crystal Ball* simulation using *Crystal Ball* functions instead of defining assumptions in the model.
3. Open the *New Store Financial Model* spreadsheet and use the *Correlation Matrix* tool to define the following correlations and run the simulation. How do the results compare to the uncorrelated model?

	YEAR 2	YEAR 3	YEAR 4	YEAR 5
Year 2	1.00	0.60		
Year 3		1.00	0.70	
Year 4			1.00	0.80
Year 5				1.00

4. Run the *Tornado Chart* tool for the *New Store Financial Model* spreadsheet and explain the results.
5. Set up the demand assumption for the newsvendor model using the Custom Distribution for the empirical data of candy sales. Compare the differences in the results with using the fitted distribution.
6. Modify the Nemesh Printing Company (revised) model from Chapter 9 to replace the Excel-based formulas with *Crystal Ball* assumptions (hint: use the discrete uniform distribution for the number of small and large orders). Run the model to find the distribution of total revenue, total costs, and net profit.
7. Use the *Decision Table* tool to find the best overbooking policy for the overbooking model developed in this chapter.
8. For the best policy for the overbooking model identified in exercise 6, run the model for 500 trials using both Monte Carlo and Latin Hypercube sampling. Compare the assumption charts and forecast charts and explain the differences you observe.
9. Open the *Cash Budget Model* spreadsheet and remove the correlations from the model and rerun the simulation. Compare the forecast charts and statistics

with the correlated model. What differences do you observed?

10. Apply the *Bootstrap* tool to the Moore Pharmaceuticals model to find the sampling distribution of the mean net present value.

11. Run the Becker Consulting Project Management model to examine the Sensitivity Chart. What impact do the assumptions have on the project completion time?

PROBLEMS AND APPLICATIONS

Data for problems 3 and 15 can be found in the Excel file *Chapter 10 Problem Data*.

1. For the Outsourcing Decision Model used in Chapter 9, suppose that the demand volume is lognormally distributed with a mean of 1500 and standard deviation of 500. What is the distribution of the cost differences between manufacturing in-house and purchasing? What decision would you recommend? Define both the cost difference and decision as forecast cells. Because *Crystal Ball* forecast cells must be numeric, replace the formula in cell B20 with =IF(B18<=0, 1, 0); that is, "1" represents manufacturing and "0" represents outsourcing.

2. For the Airline Pricing Model in Chapter 9, suppose that the fixed cost is triangular with a minimum of $80,000, most likely value of $90,000, and maximum value of $95,000. Also assume that the values of the slope and intercept in the demand function are uniformly distributed plus or minus 5% around their current values. Find the distribution of profit for a unit price of $500. Use the *Decision Table* tool to find the best price between $400 and $600 in steps of $25.

3. Using the generic profit model developed in the section *Logic and Business Principles* in Chapter 9, develop a financial simulation model for a new product proposal and construct a distribution of profits under the following assumptions: Price is fixed at $1,000. Unit costs are unknown and follow the distribution.

UNIT COST	PROBABILITY
$400	.20
$600	.40
$700	.25
$800	.15

Demand is also variable and follows the following distribution:

DEMAND	PROBABILITY
120	.25
140	.50
160	.25

Fixed costs are estimated to follow the following distribution:

FIXED COSTS	PROBABILITY
$45,000	.20
$50,000	.50
$55,000	.30

Implement your model using *Crystal Ball* to determine the best production quantity to maximize the average profit. Would you conclude that this product is a good investment?

4. For the Hyde Park Surgery Center scenario described in Problem 7 in Chapter 9, suppose that the following assumptions are made. The number of patients served the first year is uniform between 1300 and 1700; the growth rate for subsequent years is triangular with parameters (5%, 8%, 9%), and the growth rate for year 2 is independent of the growth rate for year 3; average billing is normal with mean of $150,000 and standard deviation $10,000; and the annual increase in fixed costs is uniform between 5% and 7%, and independent of other years. Find the distribution of the NPV of profit over the 3-year horizon and analyze the sensitivity and trend charts. Summarize your conclusions.

5. Refer back to the college admission director scenario (Problem 11 in Chapter 9). Develop a spreadsheet model and apply *Crystal Ball* tools to make a recommendation on how many scholarships to offer.

6. For the J&G Bank scenario (Problem 12 in Chapter 9), use *Crystal Ball* to analyze the profitability of the credit card product and summarize your results to the manager of the credit card division. Use whatever *Crystal Ball* tools you deem appropriate to analyze your results and provide a complete and useful report.

7. Review the MBA's retirement planning situation in Problem 16 of Chapter 9. Modify the spreadsheet you developed for part (b) to include the assumptions that the annual raise is triangular with a minimum of 1%, most likely value of 3%, and maximum value of 5%, and that the investment return each year is triangular with minimum of −8%, most likely value of 5%, and maximum value of 9%. Use *Crystal Ball* to find the distribution of the ending retirement fund balance under these assumptions. How do the results compare with the base case?

8. A local pharmacy orders 15 copies of a monthly magazine. Depending on the cover story, demand for the magazine varies. Historical records suggest that the probability distribution of demand is Poisson with a mean of 10. The pharmacy purchases the magazines for $2.25 and sells them for $5.00. Any magazines left over at the end of the month are donated to hospitals and other health care facilities. Is the 15-copy order quantity the most profitable? Conduct a simulation analysis using *Crystal Ball* to answer this question.

9. The manager of an apartment complex has observed that the number of units rented during any given month varies between 30 and 40 (use a triangular distribution with minimum 30, most likely 34, maximum 40). Rent is $500 per month. Operating costs average $15,000 per month but vary following a normal distribution with mean $15,000. Operating costs are assumed to be normal with a standard deviation of $300. Use *Crystal Ball* to estimate the 80%, 90%, and 95% confidence intervals for the profitability of this business.
   a. What is the probability that monthly profit will be positive?
   b. What is the probability that monthly profit will exceed $4,000?
   c. Compare the 80%, 90%, and 95% certainty ranges.
   d. What is the probability that profit will be between $1,000 and $3,000?

10. A garage band wants to hold a concert. The expected crowd is normally distributed with mean of 3,000 and standard deviation 400 (minimum of 0). The average expenditure on concessions is also normally distributed with mean $15, standard deviation $3, and minimum 0. Tickets sell for $10 each, and the band's profit is 80% of the gate, along with concession sales, minus a fixed cost of $10,000. Develop a spreadsheet model and simulate 1,000 trials using *Crystal Ball* to identify the mean profit, the minimum observed profit, maximum observed

profit, and the probability of achieving a positive profit. Develop and interpret a confidence interval for the mean profit for the 1000-trial simulation.

11. Develop a *Crystal Ball* model for a three-year financial analysis of total profit based on the following data and information. Sales volume in the first year is estimated to be 100,000 units and is projected to grow at a rate that is normally distributed with a mean of 7% per year and a standard deviation of 4%. The selling price is $10, and the price increase is normally distributed with a mean of $0.50 and standard deviation of $0.05 each year. Per-unit variable costs are $3, and annual fixed costs are $200,000. Per-unit costs are expected to increase by an amount normally distributed with a mean of 5% per year and standard deviation of 2%. Fixed costs are expected to increase following a normal distribution with a mean of 10% per year and standard deviation of 3%. Based on 5,000 simulation trials, find the average three-year cumulative profit, and explain the percentile report. Generate and explain a trend chart showing net profit by year.

12. MasterTech is a new software company that develops and markets productivity software for municipal government applications. In developing their income statement, the following formulas are used:

$$\text{Gross profit} = \text{Net sales} - \text{Cost of sales}$$

$$\text{Net operating profit} = \text{Gross profit} - \text{Administrative expenses} - \text{Selling expenses}$$

$$\text{Net income before taxes} = \text{Net operating profit} - \text{Interest expense}$$

$$\text{Net income} = \text{Net income before taxes} - \text{Taxes}$$

Net sales are uniformly distributed between $600,000 and $1,200,000. Cost of sales is normally distributed, with mean of $540,000 and standard deviation of $20,000. Selling expenses has a fixed component that is uniform between $75,000 and $110,000. The variable component is estimated to be 7% of net sales. Administrative expenses are normal, with a mean of $50,000 and standard deviation of $3,500. Interest expenses are $10,000. The company is taxed at a 50% rate. Develop a risk profile of net income using *Crystal Ball* and write a report to management.

13. A plant manager is considering investing in a new $30,000 machine. Use of the new machine is expected to generate a cash flow of about $8,000 per year for each of the next five years. However, the cash flow is uncertain, and the manager estimates that the actual cash flow will be normally distributed with a mean of $8,000 and a standard deviation of $500. The discount rate is set at 8% and assumed to remain constant over the next five years. The company evaluates capital investments using net present value. How risky is this investment? Develop an appropriate simulation model and conduct experiments and statistical output analysis to answer this question.

14. Koehler Vision Associates (KVA) specializes in laser-assisted corrective eye surgery. Prospective patients make appointments for prescreening exams to determine their candidacy for the surgery, and if they qualify, the $300 charge is applied as a deposit for the actual procedure. The weekly demand averages 175, but anywhere between 10% and 20% of prospective patients fail to show up or cancel their exam at the last minute. Patients that do not show up do not pay the pre-screening fee. KVA can handle 125 patients per week and is considering overbooking its appointments to reduce the lost revenue associated with cancellations. However, any patient that is overbooked may spread unfavorable comments about the company; thus, the over-booking cost is estimated to be $125, the value of a referral. Develop a *Crystal Ball* model to determine the best level of overbooking to maximize the net profit

(revenue less overbooking costs). Assume that the demand is uniform between 110 and 160 per week.

15. A software development project consists of six activities. The activities and their predecessor relationships are given in the following table. The Excel file *Chapter 10 Problem Data* provides a sample of times in weeks for activities A, B, E, and F from 40 past projects.

Activity		Activity Times		Predecessors
A	Requirements analysis	See Excel file		None
B	Programming	See Excel file		A
C	Hardware	Constant	3 weeks	A
D	User training	Constant	12 weeks	A
E	Implementation	See Excel file		B,C
F	Testing	See Excel file		E

Assume that activity start time is zero if there are no predecessor activities. Fit distributions to the sample data and use *Crystal Ball* to find the mean project completion time, minimum and maximum project completion times, skewness of the completion time distribution, and probability of completing the project in 14, 15, 16, or 17 weeks.

16. A stock broker calls on potential clients from referrals. For each call, there is a 15% chance that the client will decide to invest with the firm. Sixty percent of those interested are not found to be qualified based on the brokerage firm's screening criteria. The remaining are qualified. Of these, half will invest between $2,000 and $10,000, 25% will invest between $10,000 and $25,000, 15% will invest between $25,000 and $50,000, and the remainder will invest between $50,000 and $100,000. The commission schedule is as follows:

Transaction Amount	Commission
Up to $25,000	$60 + 0.5% of the amount
$25,001 to $50,000	$85 + 0.4% of the amount
$50,001 to $100,000	$135 + 0.3% of the amount

The broker keeps half the commission. How many calls per month must the broker make each month to have at least a 75% chance of making at least $5,000?

17. The director of a nonprofit ballet company in a medium-sized U.S. city is planning its next fundraising campaign. In recent years, the program has found the following percentages of donors and gift levels:

Gift Level	Amount	Number of Gifts
Benefactor	$10,000	1–3
Philanthropist	$5,000	3–7
Producer's Circle	$1,000	16–25
Director's Circle	$500	31–40
Principal	$100	5–7% of solicitations
Soloist	$50	5–7% of solicitations

The company has set a financial goal of $150,000. How many prospective donors must they contact for donations at the $100 level or below to have a 95% chance of meeting this goal? Assume that the number of gifts at each level follow a uniform distribution. (At the high levels, these must be discrete numbers!)

18. Arbino Mortgage Company obtains business from direct marketing letters signed by a fictitious loan officer named Jackson Smith. The percentages of interested responses from Jackson Smith letters are triangular with parameters 1%, 1.2%, and 2%. Of those responding, the percentage of respondents that actually close on a loan is also triangular with parameters 10%, 15%, and 18%. The average loan fee for a Jackson Smith loan is $3,500 with a standard deviation of $650. Other loan requests are obtained from two other sources: referrals and repeat customers, and unsolicited customers obtained from other advertising (billboard ads, Google searches, and so on). Fees for referrals and repeat customers average $2,600 with a standard deviation of $500 (these are less in order to provide an incentive for future business), and unsolicited customers' loan fees are the same as Jackson Smith loans. The company has 15 loan officers. Each loan officer will close an average of one loan per month from referrals and repeat customers, and about one loan every four months from unsolicited customers (use judgment to define a reasonable uniform distribution around these averages). The company is moving to a new office and will double in size. This requires them to cover additional overhead expenses. The general manager wants to be 90% certain that the office will close at least $600,000 in total loan fees each month from all sources. The principal question is how many Jackson Smith letters should be sent each month to ensure this. Develop a simulation model to help identify the best decision.

CASE

The Bellin Project

Jennifer Bellin has been put in charge planning her company's annual leadership conference. The dates of the conference have been determined by her company's executive team. The Excel file *Chapter 10 Problem Data* contains information about the activities, predecessors, and activity times in the worksheet *Bellin Case*. Assume a BetaPERT distribution for the activity times. How far in advance of the conference must Jennifer begin the project activities to ensure a 95% chance of completing the project by the scheduled date? How sensitive are the results to the distributional assumption? For example, suppose a triangular or normal distribution was used to model the activity times. How do the results change? Are the differences significant?

Chapter

11

Decisions, Uncertainty, and Risk

INTRODUCTION

A world of difference exists between building, analyzing, and solving decision models and making decisions. Decision models such as Monte Carlo simulations can provide insight about the impacts of potential decisions or, in the case of prescriptive models, recommendations as to the best course of action to take. However, people make decisions, and their decisions often have significant economic or human resource consequences that cannot always be predicted accurately. Therefore, understanding the philosophy of decision making and how to deal with uncertainty and risk is vitally important to being a good manager.

Managers make many kinds of decisions. Some of these decisions are repetitive, perhaps on an annual or quarterly basis. These might include selecting new employees from a pool of applicants or selecting specific business improvement projects from a pool of proposals. These types of decisions have little financial impact and would not be considered very risky. Other decisions, such as deciding where to locate a plant or warehouse or determining whether to pursue a major investment opportunity, commit the firm to spending large sums of money. These usually are one-time decisions and generally involve a higher level of risk because of uncertain data and imperfect information.

In this chapter we present various approaches for evaluating and making decisions. Specifically, we will discuss:

➤ *Approaches to making decisions without uncertainty and risk*

➤ *Concepts of uncertainty and risk, and decision strategies for making decisions with uncertain consequences*

➤ *The appropriateness of expected value decision making and examples in which it can be used*

➤ *The concepts of expected value of perfect information and portfolio risk*

➤ *Using decision trees for modeling and analyzing decision problems*

➤ *Basic concepts of utility theory and its application in making decisions*

DECISION MAKING WITHOUT UNCERTAINTY AND RISK

Many decisions involve making a choice among one or more alternatives without consideration of uncertainty and risk. The first steps in making a decision are to understand the alternative decisions that can be made, and identify a rational criterion (or multiple criteria) by which to evaluate the impact of the decision. **Decision alternatives** represent the choices that a decision maker can make. They might be a simple set of decisions, such as locating a factory from five potential sites or choosing one of three corporate health plan options. Other situations require a more complex sequence of decisions. For example, in new product introduction, a marketing manager might have to decide on whether to test-market a new product then, based on the test-market results, decide to conduct further tests, begin a full-scale marketing effort, or drop the product from further consideration. In either case the manager must list the options that are available. Generating viable alternatives might involve some prescreening (perhaps using optimization models). For instance, a company might develop, solve, and perform sensitivity analysis on an optimization model to generate potential plant location sites based on total distribution costs. However, making the final decision would involve many qualitative factors such as labor supply, tax incentives, environmental regulations, future uncertainties, and so on. Managers must ensure that they have considered all possible options so that the "best" one will be included in the list. This often requires a good deal of creativity to define unusual options that might not normally be considered. Managers must put aside the tendency to jump right into the process of finding a solution to consider creative alternatives. After the alternatives are defined, we must select the criteria on which to evaluate them. Decision criteria might be to maximize discounted net profits or social benefits, or to minimize costs, environmental impact, or some measure of loss.

Decisions Involving a Single Alternative

Decisions involving a single alternative, such as whether to outsource a product or invest in a research and development project, generally are straightforward to make. The choice is either to accept or reject the alternative based on some (usually financial) criterion. For example, in the outsourcing decision model in Chapter 9, the decision could be made by finding the breakeven volume and determining whether the anticipated volume is larger or smaller than the breakeven point.

One common criterion used to evaluate capital investments is the **internal rate of return (IRR)**. This is the discount rate that makes the total present value of all cash flows sum to zero:

$$\sum_{t=0}^{n} \frac{F_t}{(1 + \text{IRR})^t} = 0 \qquad (11.1)$$

IRR is often used to compare a project against a predetermined *hurdle rate*, a rate of return required by management to accept a project, which is often based on the rate of return required by shareholders. If IRR is greater than the hurdle rate, the project should be accepted; otherwise, it should be rejected. The internal rate of return can be computed in Excel (see *Excel Note: Using the IRR Function*).

To illustrate this, consider the model for Moore Pharmaceuticals in Chapter 9 (Figure 9.16). The model calculated the net profit for each of five years and the NPV. The company's hurdle rate has been determined to be 10%. While the model showed a positive NPV of over $185 million, it is not clear whether this project will meet the required hurdle rate. Using the data in Figure 9.16, the IRR is found to be 15.96%, and therefore, the company should continue to develop the drug.

Decisions Involving Non-Mutually Exclusive Alternatives

If several non-mutually exclusive alternatives are being considered, various ranking criteria, such as return on investment (ROI) or more general benefit/cost ratios, provide a basis for evaluation and selection. ROI is computed as:

$$\text{ROI} = \frac{\text{Annual Revenue} - \text{Annual Costs}}{\text{Initial Investment}} \qquad (11.2)$$

EXCEL NOTE
Using the IRR Function

The Excel function for internal rate of return is IRR(*values, guess*). *Values* represents the series of cash flows (at least one of which must be positive and one of which must be negative). *Guess* is a number believed close to the value of IRR that is used to facilitate the mathematical algorithm used to find the solution. Occasionally, the function might not converge to a solution; in those cases, you should try a different value for *guess*. In most cases, the value of *guess* can be omitted from the function.

Proposals are typically ranked in order of highest ROI first and are selected until total investment exceeds a budget. Benefit/cost analysis is based on examining the ratios of expected benefits to expected costs. Benefit/cost analysis is often used in evaluating social projects where benefits generally cannot be quantified, for example, municipal projects such as parks and urban renewal. Ratios greater than 1.0 generally indicate that a proposal should be adopted if sufficient resources exist to support it.

Decisions Involving Mutually Exclusive Alternatives

Many decisions involve a choice among several mutually exclusive alternatives. Some examples would be decisions about purchasing automobiles, choosing colleges, selecting mortgage instruments, investing money, introducing new products, locating plants, and choosing suppliers, to name just a few. Decision criteria might be to maximize discounted net profits or social benefits, or to minimize costs or some measure of loss.

When only one alternative can be selected from among a small set of alternatives, the best choice can usually be identified by evaluating each alternative using the criterion chosen. For example, in the newsvendor model we analyzed in Chapter 10, we used a *Crystal Ball* decision table to evaluate the average profit and identify the best purchase quantity based on this criterion. When the set of alternatives is very large or infinite, then an optimization model and a solution technique such as *Solver* can help to identify the best choice. The airline pricing model we developed in Chapter 9 is one example. Chapters 13 and 14 will deal with more complex optimization models.

For decisions involving multiple criteria, simple scoring models are often used. A **scoring model** is a quantitative assessment of a decision alternative's value based on a set of attributes. Usually, each attribute is characterized by several levels, and a score is assigned to each level that reflects the relative benefits of that level. For example, in evaluating new product ideas, some attributes might include product development time, the competitive environment, and return on investment. A simple scoring model for these attributes might be the one shown in Figure 11.1.

| 1. Product development time | |
LEVEL	SCORE
a. Less than 6 months	+5
b. 6 months to 1 year	+3
c. 1–2 years	0
d. 2–3 years	−3
e. more than 2 years	−5
2. Competitive environment	
a. None	+5
b. Few minor competitors	+3
c. Many minor competitors	0
d. Few major competitors	−3
e. Many major competitors	−5
3. Return on investment	
a. 30% or more	+5
b. 25–30%	+3
c. 15–25%	0
d. 10–15%	−3
e. below 15%	−5

Figure 11.1 Scoring Model Example

For each decision alternative, a score would be assigned to each attribute (which might be weighted), and the overall score would be used as a basis for selection.

DECISIONS INVOLVING UNCERTAINTY AND RISK

Uncertainty is imperfect knowledge of what will happen; *risk* is associated with the consequences of what actually happens. Even though uncertainty may exist, there may be no risk. For example, the change in the stock price of Google on the next day of trading is uncertain. This uncertainty has no impact if you don't own Google stock. However, if you do, then you bear the risk associated with the possibility of losing money. Thus, risk is an outcome of uncertainty.

The importance of risk in business has long been recognized. The renowned management writer, Peter Drucker, observed in 1974:

> To try to eliminate risk in business enterprise is futile. Risk is inherent in the commitment of present resources to future expectations. Indeed, economic progress can be defined as the ability to take greater risks. The attempt to eliminate risks, even the attempt to minimize them, can only make them irrational and unbearable. It can only result in the greatest risk of all: rigidity.[1]

Consideration of risk is a vital element of decision making. For instance, you would probably not choose an investment simply on the basis of its expected return because, typically, higher returns are associated with higher risk. Therefore, you have to make a trade-off between the benefits of greater rewards and the risks of potential losses. We can see this in the *Crystal Ball* results for the new product development model in Chapter 10. Even though the expected net present value is positive, there is a probability of about 0.4 of realizing a negative net present value. The rational decision would probably be not to continue to develop the drug.

Decisions involving uncertainty and risk have been studied for many years. A large body of knowledge has been developed that helps to explain the philosophy associated with making decisions and also provide techniques for incorporating uncertainty and risk in making decisions.

Making Decisions with Uncertain Information

Many decisions involve a choice from among a small set of decisions with uncertain consequences. We may characterize such decisions by defining three things:

1. the decision alternatives,
2. the outcomes that may occur once a decision is made, and
3. the payoff associated with each decision and outcome.

Outcomes, often called **events**, may be quantitative or qualitative. For instance, in selecting the size of a new factory, the future demand for the product would be uncertain. The demand might be expressed quantitatively in sales units or dollars. On the other hand, suppose that you are planning a spring break vacation to Florida in January; you might define uncertain weather-related outcomes qualitatively: sunny and warm, sunny and cold, rainy and warm, or rainy and cold, etc. The payoff

---

[1]P.F. Drucker, *The Manager and the Management Sciences in Management: Tasks, Responsibilities, Practices* (London: Harper and Row, 1974).

is a measure of the value of making a decision and having a particular outcome occur. This might be a simple estimate made judgmentally, or a value computed from a complex spreadsheet model. Payoffs are often summarized in a **payoff table**, a matrix whose rows correspond to decisions and whose columns correspond to events. The following example illustrates these concepts.

Many young families face the decision of choosing a mortgage instrument. Suppose the Durr family is considering purchasing a new home and would like to finance $150,000. Three mortgage options are available, a 1-year adjusted-rate mortgage (ARM) at a low interest rate, a 3-year ARM at a slightly higher rate, and a 30-year fixed mortgage at the highest rate. However, both ARMs are sensitive to interest rate changes and the rates may change resulting in either higher or lower interest charges; thus the potential changes in interest rates are the uncertain outcomes. As the family anticipates staying in the home for at least five years, they are interested in the total interest costs they might incur; these represent the payoffs associated with their choice and the future change in interest rates and can easily be calculated using a spreadsheet. The payoff table is:

DECISION	OUTCOME		
	RATES RISE	RATES STABLE	RATES FALL
1-year ARM	$61,134	$46,443	$40,161
3-year ARM	$56,901	$51,075	$46,721
30-year fixed	$54,658	$54,658	$54,658

Clearly, no decision is best for all outcome scenarios. The best decision clearly depends on what outcome occurs. If rates rise, then the 30-year fixed would be the best decision. If rates remain stable or fall, then the 1-year ARM is best. Of course, you cannot predict the outcome with certainty, so the question is how to choose one of the options. Not everyone views risk in the same fashion. Most individuals will weigh their potential losses against potential gains. For example, if they choose the 1-year ARM mortgage instead of the fixed-rate mortgage, they risk losing money if rates rise; however, they would clearly save a lot if rates remain stable or fall. Would the potential savings be worth the risk?

Evaluating risk should take into account both the magnitude of potential gains and losses, as well as their probabilities of occurrence, if this can be assessed. For instance, suppose that you are offered a chance to win a $40,000 car in a charity raffle for $100 in which only 1,000 tickets are sold. Although the probability of losing is 0.999, most individuals would not view this to be very risky because the loss of only $100 would not be viewed as catastrophic (ignoring the charitable issues!).

Decision Strategies

We will discuss several quantitative approaches that model different risk behaviors for making decisions involving uncertainty.

Average Payoff Strategy

Because the future events are unpredictable, we might simply assume that each one is as likely to occur as the others. This approach was proposed by the French mathematician Laplace, who stated the *principle of insufficient reason*: If there is no reason for one state of nature to be more likely than another, treat them as equally likely. Under this assumption, which is called the **Laplace**, or **average payoff strategy**, we evaluate each decision by simply averaging the payoffs. We then select the decision with the best average payoff, depending on whether the objective is to maximize or minimize the result. For the mortgage selection problem we have the following:

DECISION	OUTCOME			
	RATES RISE	RATES STABLE	RATES FALL	AVERAGE PAYOFF
1-year ARM	$61,134	$46,443	$40,161	$49,246
3-year ARM	$56,901	$51,075	$46,721	$51,566
30-year fixed	$54,658	$54,658	$54,658	$54,658

Based on this criterion, we choose the decision having the smallest average payoff, or the 1-year ARM. If the payoffs were revenues or profits, then clearly we should choose the largest average payoff.

Aggressive Strategy

An aggressive decision maker might seek the option that holds the promise of maximizing his or her potential return or minimizing the potential loss. This type of decision maker would first ask the question, "What is the best that could result from each decision?" and then choose the decision that corresponds to the "best of the best." For our example, this is summarized below:

DECISION	OUTCOME			
	RATES RISE	RATES STABLE	RATES FALL	BEST PAYOFF
1-year ARM	$61,134	$46,443	$40,161	$40,161
3-year ARM	$56,901	$51,075	$46,721	$46,721
30-year fixed	$54,658	$54,658	$54,658	$54,658

Because our goal is to minimize costs, we would choose the 1-year ARM. For a minimization objective, this strategy is also often called a **minimin strategy**; that is, we choose the decision that minimizes the minimum payoff. For a maximization objective, the best payoff would be the largest value, and we would choose the decision corresponding to the largest of these, also called a **maximax strategy**. Aggressive decision makers are often called speculators, particularly in financial arenas because they increase their exposure to risk in hopes of increasing their return.

Conservative Strategy

A conservative decision maker, on the other hand, might take a more pessimistic attitude and ask, "What is the worst thing that might result from my decision?" and then select the decision that represents the "best of the worst." For the mortgage decision problem, the largest costs for each option are as follows:

DECISION	OUTCOME			
	RATES RISE	RATES STABLE	RATES FALL	WORST PAYOFF
1-year ARM	$61,134	$46,443	$40,161	$61,134
3-year ARM	$56,901	$51,075	$46,721	$56,901
30-year fixed	$54,658	$54,658	$54,658	$54,658

In this case we want to choose the decision that has the smallest worst payoff, or the 30-year fixed mortgage. Thus, no matter what the future holds, a cost of $54,658 is guaranteed. Such a strategy is also known as a **minimax strategy** because we seek the decision that corresponds to the minimum value of the largest cost. For a maximization objective, the worst payoff would be the minimum value over all outcomes and we would want to choose the largest; this would be called a **maximin strategy**. Conservative decision makers are often called **risk averse** and are willing to forgo potential returns in order to reduce their exposure to risk.

Opportunity Loss Strategy

A fourth approach that underlies decision choices for many individuals is to consider the *opportunity loss* associated with a decision. Opportunity loss represents the "regret" that people often feel after making a nonoptimal decision ("I should have bought that stock years ago!"). In our example, suppose we chose the 30-year fixed mortgage and later find out that the interest rates rose. We could not have done any better by selecting a different decision; in this case, the opportunity loss is zero. However, if the interest rates remained stable, the best decision *would have been* to choose the one-year ARM. By choosing the 30-year fixed instrument, the investor lost a total of $54,658 − $46,443 = $8215. This represents the opportunity loss associated with making the wrong decision. If the rates fell, the best decision would have been the 1-year ARM also, and choosing the 30-year fixed mortgage would result in a larger opportunity loss of $54,658 − $40,161 = $14,497.

In general, the opportunity loss associated with any decision and event is the difference between the *best* decision for that particular outcome and the payoff for the decision that was chosen. *Opportunity losses can only be nonnegative values!* Thus, you need to be careful when computing these and pay attention to whether the objective is to minimize or maximize the payoff, especially if some payoffs are negative.

Once opportunity losses are computed, the decision strategy is similar to a conservative strategy. The decision maker would select the decision that minimizes the largest opportunity loss. For these reasons, this is also called a **minimax regret strategy**. This is summarized below:

	OUTCOME			
DECISION	RATES RISE	RATES STABLE	RATES FALL	MAX OPPORTUNITY LOSS
1-year ARM	$6,476	$—	$—	$6,476
3-year ARM	$2,243	$4,632	$6,560	$6,560
30-year fixed	$—	$8,215	$14,497	$14,497

Using this strategy, we would choose the 1-year ARM. This ensures that, no matter what outcome occurs, we will never be further than $6,476 away from the least cost we could have incurred. *Different criteria, different decisions.* Which criterion best reflects your personal values?

Risk and Variability

A serious drawback of the average payoff strategy is that it neglects to consider the actual outcomes that can occur. For any decision (with the trivial exception of equal payoffs), the average outcome will *never occur*. For instance, choosing the 1-year ARM will never result in the average payoff of $49,246; the cost to the Durr family will *either be* $61,134, $46,443, or $40,161. Choosing the 3-year ARM results in an average cost of $51,566, but the possible outcomes vary only between $46,721 and

$56,901. Therefore, while the averages are fairly similar, note that the one-year ARM has a larger variation in the possible outcomes.

In financial investment analysis, risk is often measured by the standard deviation. For example, *Fortune* magazine evaluates mutual fund risk using the standard deviation, because it measures the tendency of a fund's monthly returns to vary from their long-term average (as *Fortune* stated in the 1999 Investor's Guide, ". . . standard deviation tells you what to expect in the way of dips and rolls. It tells you how scared you'll be."). For example, a mutual fund's return might have averaged 11% with a standard deviation of 10%. Thus, about two-thirds of the time the annualized monthly return was between 1% and 21%. By contrast, another fund's average return might be 14%, but have a standard deviation of 20%. Its returns would have fallen in a range of −6% to 34% and, therefore, is more risky.

The statistical measure of coefficient of variation, which is the ratio of the standard deviation to the mean, provides a relative measure of risk to return. The smaller the coefficient of variation, the smaller the relative risk is for the return provided. The reciprocal of the coefficient of variation, called **return to risk**, is often used because it is easier to interpret. That is, if the objective is to maximize return, a higher return-to-risk ratio is often considered better. A related measure in finance is the **Sharpe ratio**, which is the ratio of a fund's excess returns (annualized total returns minus Treasury bill returns) to its standard deviation. *Fortune* noted that, for example, although both the American Century Equity-Income fund and Fidelity Equity-Income fund had 3-year returns of about 21% per year, the Sharpe ratio for the American Century fund was 1.43 compared with 0.98 for Fidelity. If several investment opportunities have the same mean but different variances, a rational (risk-averse) investor will select the one that has the smallest variance.[2] This approach to formalizing risk is the basis for modern portfolio theory, which seeks to construct minimum-variance portfolios. As *Fortune* noted, "It's not that risk is always bad. . . . It's just that when you take chances with your money, you want to be paid for it."

Standard deviations may not tell the complete story about risk, however. It is also important to consider the skewness and kurtosis of the distribution of outcomes. For example, both a negatively and positively skewed distribution may have the same standard deviation, but clearly if the objective is to achieve high return, the negatively skewed distribution will have higher probabilities of larger returns. Higher kurtosis values indicate that there is more area in the tails of the distribution than for distributions with lower kurtosis values. This indicates a greater potential for extreme and possibly catastrophic outcomes.

Applying these concepts to the mortgage example, we may compute the standard deviation of the outcomes associated with each decision:

DECISION	STANDARD DEVIATION
1-year ARM	$10,763.80
3-year ARM	$5,107.71
30-year fixed	$—

Based solely on the standard deviation, the 30-year fixed mortgage has no risk at all while the 1-year ARM appears to be the riskiest. Although based only on three data points, the three-year ARM is fairly symmetric about the mean while the one-year ARM is positively skewed—most of the variation around the average is driven by

[2]David G. Luenberger, *Investment Science* (New York: Oxford University Press, 1998).

the upside potential (i.e., lower costs), not the downside risk of higher costs. Although none of the formal decision strategies chose the 3-year ARM, viewing risk from this perspective might lead to this decision. For instance, a conservative decision maker who is willing to tolerate a moderate amount of risk might choose the 3-year ARM over the 30-year fixed because the downside risk is relatively small (and is smaller than the 1-year ARM) and the upside potential is much larger. The larger upside potential associated with the one-year ARM might even make this decision attractive. Thus, it is important to understand that making decisions under uncertainty cannot be done using only simple rules, but careful evaluation of risk versus rewards. This is why top executives make the big bucks!

EXPECTED VALUE DECISION MAKING

The average payoff strategy is not appropriate for one-time decisions because, as we noted, average payoffs don't occur and one must evaluate risk. However, for decisions that occur on a repeated basis, an average payoff strategy can be used. For example, real estate development, day trading, and pharmaceutical research projects all fall into this scenario. Drug development is a good example. The cost of research and development projects in the pharmaceutical industry is generally in the hundreds of millions of dollars and could approach $1 billion. Many projects never make it to clinical trials or might not get approved by the Food and Drug Administration. Statistics indicate that seven of ten products fail to return the cost of the company's capital. However, large firms can absorb such losses because the return from one or two blockbuster drugs can easily offset these losses. On an average basis, drug companies make a net profit from these decisions. This leads to the notion of expected value decision making. The newsvendor model that we discussed in Chapter 10 is a classic example of expected value decision making. We based the optimal purchase quantity on the average profit obtained by Monte Carlo simulation.

For decisions with a finite set of outcomes, we will assume that we know or can estimate the probabilities. If historical data on past occurrences of events are available, then we can estimate the probabilities objectively. If not, managers will often be able to estimate the likelihood of events from their experience and good judgment. The **expected monetary value (EMV) approach** selects the decision based on the best expected payoff. Define $V(D_i, S_j)$ to be the payoff associated with choosing decision i and having outcome S_j subsequently occur. Suppose that $P(S_j) = $ the probability that outcome S_j occurs, and $n = $ the number of outcomes. The expected payoff for each decision alternative D_i is:

$$E(D_i) = \sum_{j=1}^{n} P(S_j)V(D_i, S_j)$$ (11.3)

To illustrate this, let us consider a simplified version of the typical revenue management process that airlines use. At any date prior to a scheduled flight, airlines must make a decision as to whether to reduce ticket prices to stimulate demand for unfilled seats. If the airline does not discount the fare, empty seats might not be sold and the airline will lose revenue. If the airline discounts the remaining seats too early (and could have sold them at the higher fare), they would lose profit. The decision depends on the probability p of selling a full-fare ticket if they choose not to discount the price. Because an airline makes hundreds or thousands of such decisions each day, the expected value approach is appropriate.

PHSTAT NOTE
Using the *Expected Monetary Value* Tool

Click *PHStat* from the Excel *Add-Ins* tab and select *Decision Making* then *Expected Monetary Value* from the menu. The dialog box is shown in Figure 11.2; you need only specify the number of actions (alternatives) and events. *PHStat* creates a worksheet in which you must enter your data (see Figure 11.3). Note that the events correspond to rows and decisions to columns, which is the opposite of the way it is presented in this text and most other books. You may customize the worksheet to change the row and column labels in

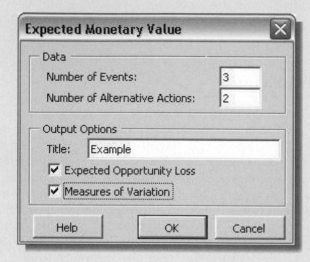

Figure 11.2 *PHStat* Dialog for Expected Monetary Value

	A	B	C	D
1	**Example**			
2				
3	**Probabilities & Payoffs Table:**			
4		P	A1	A2
5	E1			
6	E2			
7	E3			

Figure 11.3 *PHStat* EMV Template

the Probabilities & Payoffs Table for your specific problem. After you enter the data, the expected values and other statistical information are automatically computed. The tool also computes the opportunity loss table and the expected opportunity loss for each decision alternative. EVPI is the expected value of perfect information, and is discussed in the text.

Assume that only two fares are available: full and discount. Suppose that a full-fare ticket is $560, the discount fare is $400, and $p = 0.75$. For simplification, assume that if the price is reduced, then any remaining seats would be sold at that price. The expected value of not discounting the price is $0.25(0) + 0.75(\$560) = \$420$. Because this is higher than the discounted price, the airline should not discount at this time. In reality, airlines constantly update the probability p based upon information they collect and analyze in a database. When the value of p drops below the breakeven point: $\$400 = p(\$560)$ or $p = 0.714$, then it is beneficial to discount. It can also work in reverse; if demand is such that the probability that a higher-fare ticket would be sold, then the price may be adjusted upward. This is why published fares constantly change and why you may receive last-minute discount offers or may pay higher prices if you wait too long to book a reservation. Other industries such as cruise lines use similar decision strategies.

   PHStat provides tools for computing EMV and other decision information (see *PHStat Note: Using the Expected Monetary Value Tool*). A completed example is shown in Figure 11.4 (Excel file *Day Trading Example*) for a situation in which a day trader wants to decide on investing $100 in either a Dow index fund or a NASDAQ index fund. The probabilities and anticipated payoffs are shown in the Probabilities & Payoffs Table. We see that the best expected value decision would be to invest in the NASDAQ fund, which has a higher EMV as well as a better return-to-risk ratio, which indicates a higher comparative risk, despite a larger standard deviation. Of course, this analysis assumes that the probabilities and payoffs will be constant, at least over the near term.

Opportunity Loss and Expected Value of Perfect Information

The *PHStat Expected Monetary Value* tool also computes an Opportunity Loss Table, shown in Figure 11.5 for the day trading example. This can also be done independently using the *PHStat* tool *Opportunity Loss* from the *Decision Making* options. The **expected opportunity loss,** shown in row 35 of Figure 11.5, represents the average additional amount the investor would have achieved by making the right decision

Figure 11.4 Expected Monetary Value Results for *Day Trading Example*

	A	B	C	D
1	**Day Trading Decisions**			
2				
3	**Probabilities & Payoffs Table:**			
4		**Probability**	**Dow Index Fund**	**NASDAQ Index Fund**
5	Dow Up/NASDAQ Up	0.42	$ 4.00	$ 5.00
6	Dow Up/NASDAQ Unchanged	0.04	$ 3.00	$ -
7	Dow Up/NASDAQ Down	0.10	$ 1.00	$ (3.00)
8	Dow Unchanged/NASDAQ Up	0.25	$ -	$ 4.00
9	Dow Unchanged/NASDAQ Unchanged	0.02	$ -	$ -
10	Dow Unchanged/NASDAQ Down	0.05	$ -	$ (2.00)
11	Dow Down/NASDAQ Up	0.05	$ (2.00)	$ 1.00
12	Dow Down/NASDAQ Unchanged	0.03	$ (3.00)	$ -
13	Dow Down/NASDAQ Down	0.04	$ (6.00)	$ (8.00)
14				
15	**Statistics for:**		**Dow Index Fund**	**NASDAQ Index Fund**
16	**Expected Monetary Value**		1.47	2.43
17	**Variance**		6.9291	12.3051
18	**Standard Deviation**		2.63231837	3.507862597
19	**Coefficient of Variation**		1.790692769	1.443564855
20	**Return to Risk Ratio**		0.558443088	0.692729528

Figure 11.5 Opportunity Loss Table for *Day Trading Example*

22	Opportunity Loss Table:				
23		Optimum	Optimum	Alternatives	
24		Action	Profit	Dow Index Fund	NASDAQ Index Fund
25	Dow Up/NASDAQ Up	NASDAQ Index Fund	5	1	0
26	Dow Up/NASDAQ Unchanged	Dow Index Fund	3	0	3
27	Dow Up/NASDAQ Down	Dow Index Fund	1	0	4
28	Dow Unchanged/NASDAQ Up	NASDAQ Index Fund	4	4	0
29	Dow Unchanged/NASDAQ Unchanged	Dow Index Fund	0	0	0
30	Dow Unchanged/NASDAQ Down	Dow Index Fund	0	0	2
31	Dow Down/NASDAQ Up	NASDAQ Index Fund	1	3	0
32	Dow Down/NASDAQ Unchanged	NASDAQ Index Fund	0	3	0
33	Dow Down/NASDAQ Down	Dow Index Fund	-6	0	2
34				**Dow Index Fund**	**NASDAQ Index Fund**
35			Expected Opportunity Loss	1.66	0.7
36					EVPI

instead of a wrong one. To find the expected opportunity loss, we create an opportunity loss table as we discussed earlier in this chapter. For example, if the event "Dow Up/NASDAQ Up" occurs, the best decision would have been to invest in the NASDAQ index fund and no opportunity loss would be incurred; if the Dow fund was chosen, the opportunity loss would be $1. Once the opportunity loss table is constructed, the expected opportunity loss for each action is found by weighting the values by the event probabilities. We see that the NASDAQ index fund has the smallest expected opportunity loss. *It will always be true that the decision having the best expected value will also have the minimum expected opportunity loss.*

The minimum expected opportunity loss is called the **expected value of perfect information (EVPI)**. EVPI represents the maximum improvement in the expected return that can be achieved if the decision maker is able to acquire—before making a decision—perfect information about the future event that will take place. For example, if we know with certainty that both the Dow and NASDAQ will rise, then the optimal decision would be to choose the NASDAQ index fund and get a return of $5 instead of only $4. Likewise, if we know that the second event will occur, we should choose the Dow index fund, and so on. By weighting these best outcomes by their probabilities of occurrence, we can compute the expected return under the assumption of having perfect information:

Expected Return with Perfect Information = 0.42($5) + 0.04($3) + 0.1($1)
  + 0.25($4) + 0.02($0) + 0.05($0) + 0.05($1) + 0.03($0) + 0.04(−$6) = $3.13

Because the expected value without having the perfect information is only $2.43, we would have increased our average return by $3.13 − $2.43 = $0.70 This is the expected value of perfect information. EVPI is often used to assess the value of acquiring less-than-perfect information; one would never want to pay more than the EVPI for any information about the future event, no matter how good.

Analysis of Portfolio Risk

Concepts of expected value decisions and risk may be applied to the analysis of portfolios. A **portfolio** is simply a collection of assets, such as stocks, bonds, or other investments, that are managed as a group. In constructing a portfolio, one usually seeks to maximize expected return while minimizing risk. The expected return of a two-asset portfolio is computed as:

$$E[P] = w_X E[X] + w_Y E[Y] \tag{11.4}$$

where w_X = fraction of portfolio for asset X and w_Y − fraction of portfolio for asset Y.

Risk depends on the correlation among assets in a portfolio. This can be seen by examining the formula for the standard deviation of a two-asset portfolio:

$$\sigma_P = \sqrt{w_X^2 \sigma_X^2 + w_Y^2 \sigma_Y^2 + 2 w_X w_Y \sigma_{XY}} \tag{11.5}$$

Here, σ_{XY} is the covariance between X and Y. Note that if the covariance is zero, the variance of the portfolio is simply a weighted sum of the individual variances. If the covariance is negative, the standard deviation of the portfolio is smaller, while if the covariance is positive, the standard deviation of the portfolio is larger. Thus, if two investments (such as two technology stocks) are positively correlated, the overall risk increases, for if the stock price of one falls, the other would generally fall also. However, if two investments are negatively correlated, when one increases, the other decreases, reducing the overall risk. This is why financial experts advise diversification.

The *PHStat* tool *Covariance and Portfolio Analysis*, found under the *Decision Making* menu, can be used to perform these calculations for a two-asset portfolio. In the dialog that appears, enter the number of outcomes and check the box "Portfolio Management Analysis." *PHStat* creates a template in which you enter the probabilities and outcomes similar to those in the *Expected Monetary Value* tool. Figure 11.6 shows the output for equal proportions of the index funds in the day trading example. Note that the expected return is $1.95, the covariance is positive, and the portfolio risk is 2.83. Using the tool, it is easy to change the weighting and experiment with different values. For instance, a portfolio consisting of only the Dow index fund has a portfolio risk of 2.63 while yielding an expected return of $1.47, and a portfolio consisting of only the NASDAQ index fund has a portfolio risk of 3.51 while yielding an expected return of $2.43. You may also use *Solver* (see Chapter 9) to find the best weights to minimize the portfolio risk (an exercise at the end of this chapter asks you to do this).

The "Flaw of Averages"

One might think that any decision model with probabilistic inputs can be easily analyzed by simply using average values for the inputs. Let's see what happens if we do this for the newsvendor model. If we find the average of the historical candy sales, we obtain 44.05, or rounded to a whole number, 44. Using this value for demand and purchase quantity, the model predicts a profit of $264 (see Figure 11.7). However, if we construct a data table to evaluate the profit for each of the historical values (also shown in Figure 11.7), we see that the average profit is $255. Dr. Sam Savage, a strong proponent of spreadsheet modeling, coined the term "the flaw of averages" to describe this phenomenon. Basically what this says is that the evaluation of a model output using the average value of the input is not necessarily equal to the average value of the outputs when evaluated with each of the input values, or in mathematical terms, $f(E[X]) \neq E[f(X)]$, where f is a function, X is a random variable, and $E[\ ]$ signifies expected value. The reason this occurs in the newsvendor example is because the quantity sold is limited by the smaller of the demand and purchase

Figure 11.6 *PHStat* Output for Portfolio Risk Analysis

	A	B	C	D
1	Day Trading Portfolio			
2				
3	Probabilities & Outcomes:	Probability	Dow Index Fund	NASDAQ Index Fund
4		0.42	$ 4.00	$ 5.00
5		0.04	$ 3.00	$ -
6		0.10	$ 1.00	$ (3.00)
7		0.25	$ -	$ 4.00
8		0.02	$ -	$ -
9		0.05	$ -	$ (2.00)
10		0.05	$ (2.00)	$ 1.00
11		0.03	$ (3.00)	$ -
12		0.04	$ (6.00)	$ (8.00)
13				
14	Weight Assigned to X	0.5		
15				
16	Statistics			
17	E(X)	1.47		
18	E(Y)	2.43		
19	Variance(X)	6.9291		
20	Standard Deviation(X)	2.63231837		
21	Variance(Y)	12.3051		
22	Standard Deviation(Y)	3.5078626		
23	Covariance(XY)	6.3479		
24	Variance(X+Y)	31.93		
25	Standard Deviation(X+Y)	5.65066368		
26				
27	Portfolio Management			
28	Weight Assigned to X	0.5		
29	Weight Assigned to Y	0.5		
30	Portfolio Expected Return	1.95		
31	Portfolio Risk	2.82533184		

quantity, so even when demand exceeds the purchase quantity, the profit is limited. Using averages in models can conceal risk, and this is a common error among users of quantitative models.

DECISION TREES

A useful approach to structuring a decision problem involving uncertainty is to use a graphical model called a **decision tree**. Decision trees consist of a set of **nodes** and **branches**. Nodes are points in time at which events take place. The event can be a selection of a decision from among several alternatives, represented by a **decision node**, or an outcome over which the decision maker has no control, an **event node**. Event nodes are conventionally depicted by circles, while decision nodes are expressed by squares. Many decision makers find decision trees useful because *sequences* of decisions and outcomes over time can be modeled easily.

Figure 11.7 *Newsvendor Model* with Average Demand

	A	B	C	D	E	F
1	**Newsvendor Model**			**Historical Candy Sales**		
2					$ 264.00	
3	**Data**			42	$ 246.00	
4				45	$ 264.00	
5	Selling price	$ 18.00		40	$ 228.00	
6	Cost	$ 12.00		46	$ 264.00	
7	Discount price	$ 9.00		43	$ 255.00	
8				43	$ 255.00	
9	**Model**			46	$ 264.00	
10				42	$ 246.00	
11	Demand	44	44	44	$ 264.00	
12	Purchase Quantity	44		43	$ 255.00	
13				47	$ 264.00	
14	Quantity Sold	44		41	$ 237.00	
15	Surplus Quantity	0		41	$ 237.00	
16				45	$ 264.00	
17	Profit	$ 264.00		51	$ 264.00	
18				43	$ 255.00	
19				45	$ 264.00	
20				42	$ 246.00	
21				44	$ 264.00	
22				48	$ 264.00	
23				Average	$ 255.00	

New Drug Development Model

To illustrate the application of decision analysis techniques using decision trees, we will consider the process of developing a new drug (you might recall the basic financial model we developed for the Moore Pharmaceuticals example in Chapter 9). Suppose that the company has spent $300 million to date in research expenses. The next decision is whether or not to proceed with clinical trials. The cost of clinical trials is estimated to be $250 million, and the probability of a successful outcome is 0.3. After clinical trials are completed, the company may seek approval from the Food and Drug Administration. This is estimated to cost $25 million and there is a 60% chance of approval. The market potential has been identified as either large, medium, or small with the following characteristics:

	MARKET POTENTIAL EXPECTED REVENUES (MILLIONS OF $)	PROBABILITY
Large	$4,500	0.6
Medium	$2,200	0.3
Small	$1,500	0.1

The CD-ROM accompanying this book contains an Excel add-in called *TreePlan*, which allows you to construct decision trees and perform calculations within an Excel worksheet (see *TreePlan Note: Constructing Decision Trees in Excel*). The disk contains further documentation and examples of how to use *TreePlan*, and we encourage you to experiment with this software.

TREEPLAN NOTE
Constructing Decision Trees in Excel

To use *TreePlan* within Excel, first select the upper left corner of the worksheet where you wish to draw the tree. Note that *TreePlan* writes over existing values in the spreadsheet; therefore, begin your tree to the *right* of the area where your data is stored, and do not subsequently add or delete rows or columns in the tree-diagram area. Double-click the file *tree172a* and allow Excel to enable macros. *TreePlan Academic Version* will appear in the *Add-Ins* tab of Excel 2007; click on this to start *TreePlan*. *TreePlan* then prompts you with a dialog box with three options; choose *New Tree* to begin a new tree. *TreePlan* will then draw a default initial decision tree with its upper left corner at the selected cell as shown in Figure 11.8.

Expand a tree by adding or modifying branches or nodes in the default tree. To change the branch labels or probabilities, click on the cell containing the label or probability and type the new label or probability. To modify the structure of the tree (e.g., add or delete branches or nodes in the tree), select the node or branch in the tree to modify and select *TreePlan* from the *Add-Ins* tab. *TreePlan* will then present a dialog box showing the available commands. For example, to add an event node to the top branch of the tree in Figure 11.8, select the terminal node at the end of that branch (F3) and press Ctrl-T. *TreePlan* then presents the dialog box shown in Figure 11.9. To add an event node to the branch, change the selected terminal node to an event node by selecting *Change to event node* in the dialog box, select the number of branches (two), and press Enter. *TreePlan* then redraws the tree with a chance node in place of the terminal node, as shown in Figure 11.10. The dialog boxes presented by *TreePlan* vary depending on what is selected. For instance, if you select an event node, a different dialog box appears, allowing you options to add a branch, insert another event, change it to a decision, and so on. When building large trees, the *Copy subtree* option allows you to copy a

Figure 11.8 *TreePlan* Initial Decision Tree Structure

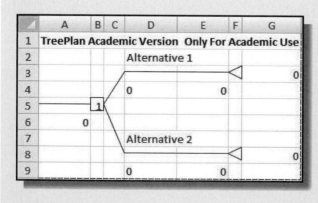

Figure 11.9 *TreePlan* Dialog

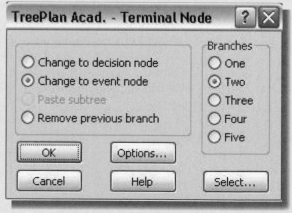

selected node and everything to the right of it. You can then select a terminal node and choose *Paste subtree*.

The *Options* button in a *TreePlan* dialog box allows you to use expected values (default) or an exponential utility function and to select whether the objective is to maximize (default) or minimize. Exponential utility functions convert cash flows into "utilities," or scaled payoffs for individuals with risk-averse attitudes.

Figure 11.10 Expanded Decision Tree

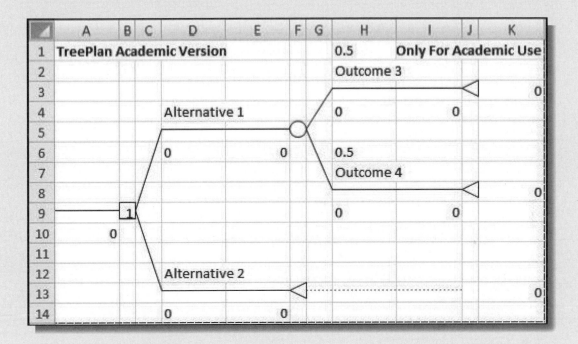

A decision tree for this situation constructed in *TreePlan* is shown in Figure 11.11 (Excel file *Drug Development Decision Tree*). In *TreePlan* there are two ways of specifying the terminal values (payoffs). First, a value for the terminal value at the end of a path can simply be entered in the appropriate cell. A second method is to enter values or formulas for partial cash flows associated with each branch, as we have done here. *TreePlan* will accumulate the cash flows at the end of the tree. This approach is particularly useful when we wish to examine the sensitivity of the decision to specific values associated with decisions or events.

A decision tree is evaluated by "rolling back" the tree from right to left. When we encounter an event node, we compute the expected value of all events that emanate from the node since each branch will have an associated probability. For example, the rollback value of the top-right event node in Figure 11.11 is found by taking the expected value of the payoffs associated with market potential (note that payoffs in parentheses are negative values):

$$\$3{,}925 \times 0.6 + 1{,}625 \times 0.3 + \$925 \times 0.1 = \$2{,}935$$

Likewise, the expected value corresponding to the branch "Seek FDA approval" is computed as:

$$\$2{,}935 \times 0.6 + (\$575) \times 0.4 = \$1{,}531$$

Figure 11.11   New Drug Development Decision Tree

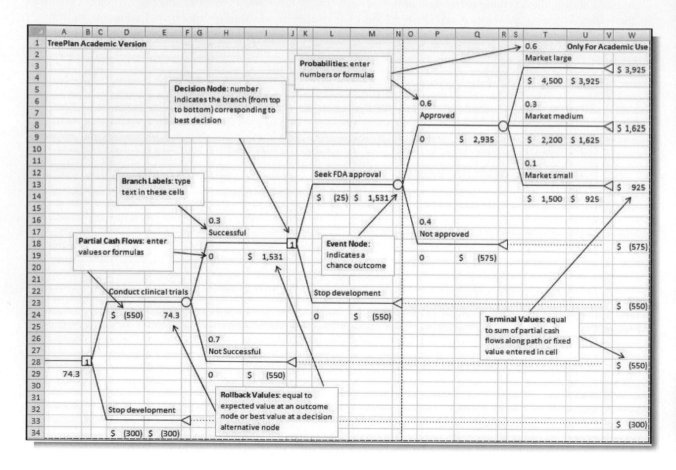

When we encounter a decision node (for instance, whether or not to seek FDA approval), we take the best of the expected values of all nodes that follow. The best decision here is to seek approval because the rollback value is $1,531 versus a negative $550 if the company decides to stop development. *TreePlan* designates the best choice by entering the number of the branch in the decision node square.

A **decision strategy** is a specification of an initial decision and subsequent decisions to make after knowing what events occur. For example, the best strategy is to conduct clinical trials, and if successful, seek FDA approval and market the drug. The expected net revenue is calculated as $74.3 million. Another strategy would be to conduct clinical trials, and even if successful, stop development.

Decision Trees and Risk

The decision tree approach is an example of expected value decision making, or essentially, the average payoff strategy in which the outcomes are weighted by their probabilities of occurrence. Thus, in the drug development example, if the company's portfolio of drug development projects has similar characteristics, then pursuing further development is justified on an expected value basis. However, this approach does not explicitly consider risk.

From a classical decision analysis perspective, we may summarize the company's decision as the following payoff table:

	Unsuccessful Clinical Trials	Successful Clinical Trials; No FDA Approval	Successful Trials and Approval; Large Market	Successful Trials and Approval; Medium Market	Successful Trials and Approval; Small Market
Develop drug	$(550)	$(575)	$3,925	$1,625	$925
Stop development	$(300)	$(300)	$(300)	$(300)	$(300)

If we apply the aggressive, conservative, and opportunity loss decision strategies to these data (note that the payoffs are profits as opposed to costs, so it is important to use the correct rule as discussed earlier in the chapter), we obtain:

Aggressive strategy (maximax):

	Maximum
Develop drug	$3,925
Stop development	$(300)

The decision that maximizes the maximum payoff is to develop the drug.

Conservative strategy (maximin):

	Minimum
Develop drug	$(575)
Stop development	$(300)

The decision that maximizes the minimum payoff is to stop development.

Opportunity loss:

	Unsuccessful Clinical Trials	Successful Clinical Trials; No FDA Approval	Successful Trials and Approval; Large Market	Successful Trials and Approval; Medium Market	Successful Trials and Approval; Small Market	Maximum
Develop drug	$250	$275	$—	$—	$—	$275
Stop development	$—	$—	$4,225	$1,925	$1,225	$4,225

The decision that minimizes the maximum opportunity loss is to develop the drug.

As we noted, however, we must evaluate risk by considering both the magnitude of the payoffs and their chances of occurrence. The aggressive, conservative, and opportunity loss rules do not consider the probabilities of the outcomes.

Each decision strategy has an associated payoff distribution, called a **risk profile**. Risk profiles show the possible payoff values that can occur and their probabilities.

For example, consider the strategy of pursuing development. The possible outcomes that can occur and their probabilities are:

TERMINAL OUTCOME	NET REVENUE	PROBABILITY
Market large	$3,925	0.108
Market medium	$1,625	0.054
Market small	$925	0.018
FDA not approved	($575)	0.120
Clinical trials not successful	($550)	0.700

The probabilities are computed by multiplying the probabilities on the event branches along the path to the terminal outcome. For example, the probability of getting to "Market large" is $0.3 \times 0.6 \times 0.6 = 0.108$. Thus, we see that the probability that the drug will not reach the market is 0.82 and the company will incur a loss of over $500 million. On the other hand, if they decide not to pursue clinical trials, the loss would only be $300 million, the cost of research to date. If this were a one-time decision, what decision would you make if you were a top executive of this company?

Sensitivity Analysis in Decision Trees

We may use Excel data tables to investigate the sensitivity of the optimal decision to changes in probabilities or payoff values. We will illustrate this using the airline revenue management scenario we discussed earlier in this chapter.

Figure 11.12 shows the decision tree (Excel file *Revenue Management Decision Tree*) for deciding whether or not to discount the fare with a data table for varying the probability of success with two output columns, one providing the expected value from cell A10 in the tree, and the second column providing the best decision. The formula

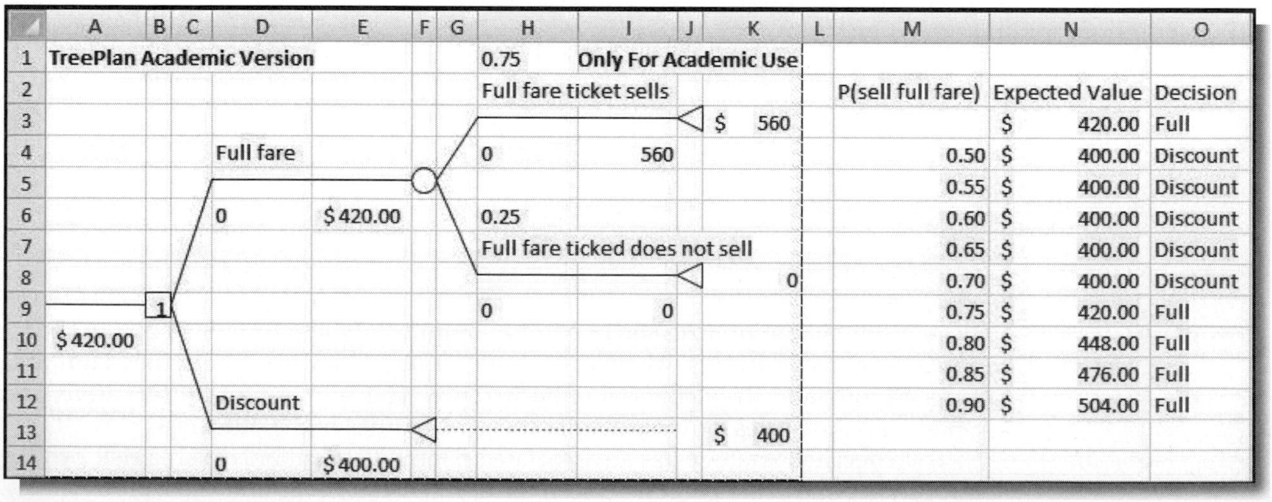

Figure 11.12 Revenue Management Decision Tree and Data Table

in cell O3 is = IF(B9 = 1,"Full","Discount"). However, we must first modify the worksheet prior to constructing the data table so that probabilities will always sum to 1. To do this, enter the formula = 1 − H1 in cell H6 corresponding to the probability of not selling the full fare ticket. From the results, we see that if the probability of selling the full fare ticket is 0.7 or less, then the best decision is to discount the price. Two-way data tables may also be used in a similar fashion to study simultaneous changes in model parameters.

UTILITY AND DECISION MAKING

A typical charity raffle involves selling one thousand $100 tickets to win a $40,000 automobile. Based on the expected payoff, a simple decision tree would show this to be a poor gamble. Nevertheless, many people would take this chance because the financial risk is low. On the other hand, if the ticket cost $5,000 and the probability of winning was increased to 0.50, yielding a much higher expected value, most people would *not* take the chance because of the personal monetary risk involved.

An approach for assessing risk attitudes quantitatively is called **utility theory**. This approach quantifies a decision maker's relative preferences for particular outcomes. We can determine an individual's utility function by posing a series of decision scenarios. This is best illustrated with an example; we will use a personal investment problem to do this.

Suppose that you have $10,000 to invest and are expecting to buy a new car in a year, so you can tie the money up for only 12 months. You are considering three options: a bank CD paying 4%, a bond mutual fund, and a stock fund. Both the bond and stock funds are sensitive to changing interest rates. If rates remain the same over the coming year, the share price of the bond fund is expected to remain the same, and you expect to earn $840. The stock fund would return about $600 in dividends and capital gains. However, if interest rates rise, you can anticipate losing about $500 from the bond fund after taking into account the drop in share price, and likewise expect to lose $900 from the stock fund. If interest rates fall, however, the yield from the bond fund would be $1,000 and the stock fund would net $1,700. Table 11.1 summarizes the payoff table for this decision problem.

The decision could result in a variety of payoffs, ranging from a profit of $1,700 to a loss of $900. The first step in determining a utility function is to rank-order the payoffs from highest to lowest. We arbitrarily assign a utility of 1.0 to the highest payoff and a utility of zero to the lowest:

PAYOFF, x	UTILITY, $U(x)$
$1,700	1.0
$1,000	
$840	
$600	
$400	
−$500	
−$900	0.0

Decision/Event	Rates Rise	Rates Stable	Rates Fall
Bank CD	$400	$400	$400
Bond fund	-$500	$840	$1,000
Stock fund	-$900	$600	$1,700

Table 11.1
Investment Return
Payoff Table

Figure 11.13 Decision Tree Lottery for Determining the Utility of $1000

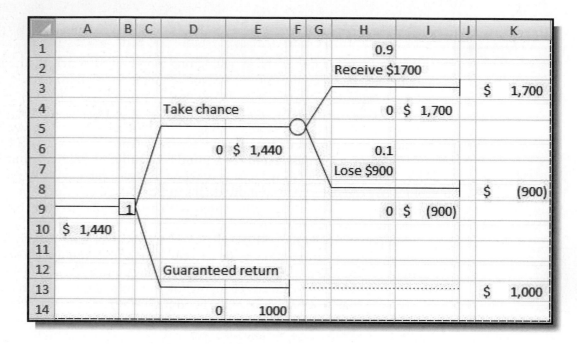

Next, for each payoff between the highest and lowest, we present you with the following situation: Suppose you have the opportunity of achieving a *guaranteed return of* x, or taking a chance of receiving $1,700 (the highest payoff) with probability p or losing $900 (the lowest payoff) with probability $1 - p$. What value of p would make you indifferent to these two choices? Let us start with x = $1,000. This is illustrated in the simple decision tree in Figure 11.13. Because this is a relatively high value, you decide that p would have to be at least .9 to take this risk. This represents the utility of a payoff of $1,000, denoted as U($1,000).

We repeat this process for each payoff, the probabilities p that you select for each scenario from your utility function. Suppose this process results in the following:

Payoff, x	Utility, U(x)
$1,700	1.0
$1,000	0.90
$840	0.85
$600	0.80
$400	0.75
-$500	0.35
-$900	0.0

If we compute the expected value of each of the gambles for the chosen values of p, we see that they are higher than the corresponding payoffs. For example, for the payoff of $1,000 and the corresponding $p = .9$, the expected value of taking the gamble is:

$$0.9(\$1,700) + 0.1(-\$900) = \$1,440$$

This is larger than accepting $1,000 outright. We can interpret this to mean that you require a risk premium of $1,440 − $1,000 = $440 to feel comfortable enough to risk losing $900 if you take the gamble. In general, the **risk premium** is the amount an individual is willing to forgo to avoid risk. This indicates that you are a *risk-averse individual*, that is, relatively conservative.

Another way of viewing this is to find the *breakeven probability* at which you would be indifferent to receiving the guaranteed return and taking the gamble. This probability is found by solving the equation:

$$1,700p − 900(1 − p) = 1,000$$

resulting in $p = 19/26 = 0.73$. Because you require a higher probability of winning the gamble, it is clear that you are uncomfortable taking the risk.

If we graph the utility versus the payoffs, we can sketch a utility function as shown in Figure 11.14. This utility function is generally *concave downward*. This type of curve is characteristic of risk-averse individuals. Such decision makers avoid risk, choosing conservative strategies and those with high return-to-risk values. Thus, a gamble must have a higher expected value than a given payoff to be preferable, or equivalently, a higher probability of winning than the breakeven value.

Other individuals might be risk taking. What would their utility functions look like? As you might suspect, they are *concave upward*. These individuals would take a gamble that offers higher rewards even if the expected value is less than a certain

Figure 11.14   Example of a Risk-Averse Utility Function

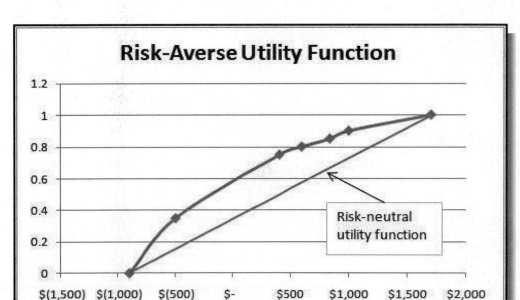

payoff. An example of a utility function for a risk-taking individual in this situation would be as follows:

Payoff, x	Utility, $U(x)$
$1,700	1.0
$1,000	0.6
$840	0.55
$600	0.45
$400	0.40
−$500	0.1
−$900	0.0

A skill-building exercise will ask you to chart this to see its shape. For the payoff of $1,000, this individual would be indifferent between receiving $1,000 and taking a chance at $1,700 with probability 0.6 and losing $900 with probability 0.4. The expected value of this gamble is:

$$0.6(\$1,700) + 0.4(-\$900) = \$660$$

Since this is considerably less than $1,000, the individual is taking a larger risk to try to receive $1,700. Note that the probability of winning is less than the breakeven value. Risk takers generally prefer more aggressive strategies.

Finally, some individuals are risk neutral; they prefer neither taking risks nor avoiding them. Their utility function would be linear and would correspond to the breakeven probabilities for each gamble. For example, a payoff of $600 would be equivalent to the gamble if:

$$\$600 = p(\$1,700) + (1 - p)(-\$900)$$

Solving for p we obtain $p = 15/26$, or .58, which represents the utility of this payoff. The decision of accepting $600 outright or taking the gamble could be made by flipping a coin. These individuals tend to ignore risk measures and base their decisions on the average payoffs.

A utility function may be used instead of the actual monetary payoffs in a decision analysis by simply replacing the payoffs with their equivalent utilities then computing expected values. The expected utilities and the corresponding optimal decision strategy then reflect the decision maker's preferences toward risk. For example, if we use the average payoff strategy (because no probabilities of events are given) for the data in Table 11.1, the best decision would be to choose the stock fund. However, if we replace the payoffs in Table 11.1 with the (risk-averse) utilities that we defined, and again use the average payoff strategy, the best decision would be to choose the bank CD as opposed to the stock fund, as shown in the table below.

Decision/Event	Rates Rise	Rates Stable	Rates Fall	Average Utility
Bank CD	0.75	0.75	0.75	0.75
Bond fund	0.35	0.85	0.9	0.70
Stock fund	0	0.80	1.0	0.60

If assessments of event probabilities are available, these can be used to compute the expected utility and identify the best decision.

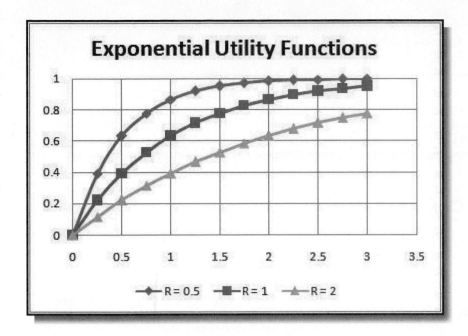

Figure 11.15 Examples of Exponential Utility Functions

Exponential Utility Functions

It can be rather difficult to compute a utility function, especially for situations involving a large number of payoffs. Because most decision makers are risk averse, we may use an exponential utility function to approximate the true utility function. The exponential utility function is:

$$U(x) = 1 - e^{-x/R} \qquad (11.6)$$

where e is the base of the natural logarithm (2.71828 . . .) and R is a shape parameter. Figure 11.15 shows several examples of $U(x)$ for different values of R. Notice that all of these functions are concave, and that as R increases, the functions become flatter, indicating more tendency toward risk neutrality.

One approach to estimating a reasonable value of R is to find the maximum payoff $\$R$ for which the decision maker is willing to take an equal chance on winning $\$R$ or losing $\$R/2$. The smaller the value of R, the more risk averse is the individual. For instance, would you take a bet on winning \$100 versus losing \$50? How about winning \$10,000 versus losing \$5000? Most people probably would not worry about taking the first gamble, but might definitely think twice about the second. Finding one's maximum comfort level establishes the utility function. For the example problem, suppose that $R = \$400$. The utility function would be $U(x) = 1 - e^{-x/400}$, resulting in the following utility values:

Payoff, x	Utility, $U(x)$
\$1,700	0.9857
\$1,000	0.9179
\$840	0.8775
\$600	0.7769
\$400	0.6321
−\$500	−2.4903
−\$900	−8.4877

Using these values in the payoff table, we find that the bank CD remains the best decision as shown in the table below.

Decision/Event	Rates Rise	Rates Stable	Rates Fall	Average Utility
Bank CD	0.6321	0.6321	0.6321	0.6321
Bond fund	−2.4903	0.8775	0.9179	−0.2316
Stock fund	−8.4877	0.7769	0.9857	−2.2417

BASIC CONCEPTS REVIEW QUESTIONS

1. Describe the types of decisions that do not involve uncertainty and risk and common approaches used for analyzing them.
2. Explain what internal rate of return is and how it is used in decision making.
3. How do you compute return on investment?
4. What are the characteristics of a scoring model?
5. Describe how to model decisions involving uncertainty.
6. Summarize the decision strategies that model different risk behaviors for making decisions involving uncertainty.
7. Explain how the standard deviation, coefficient of variation, skewness, and kurtosis of the distribution of outcomes provide information about risk.

8. Provide examples of when expected value decision making is appropriate and when it is not.
9. Explain the concept of expected value of perfect information.
10. What are the implications of the "flaw of averages"?
11. Explain the components of a decision tree and how optimal decisions are computed.
12. What is a risk profile and how can it be used in conjunction with the solution of a decision tree?
13. Explain the concept of utility theory and how utilities can be assessed.
14. What do risk-averse and risk-taking utility functions look like?

SKILL-BUILDING EXERCISES

1. An investment requires an initial cash outlay of $100,000 and additional outlays of $50,000 at the end of each of the first three years. This investment is expected to result in incomes of $40,000 at the end of the first year, $70,000 at the end of the second year, $90,000 at the end of the third year, and $120,000 at the end of the fourth year. Calculate the internal rate of return using the IRR function.
2. Develop a spreadsheet for the Durr family mortgage example and use it to compute the information needed to implement the average payoff, aggressive, conservative, and opportunity loss strategies.
3. Develop a spreadsheet for the new drug development payoff table example and use it to compute the information needed to implement the average payoff, aggressive, conservative, and opportunity loss strategies.
4. Use *Solver* to find the best weights to minimize risk for the day trading portfolio example.
5. Illustrate the "flaw of averages" using the overbooking model developed in Chapter 10.
6. Set up a decision tree for the airline revenue management example shown in Figure 11.12 using *TreePlan*

and find the breakeven probability to within 0.01 using a data table.
7. Perform a sensitivity analysis using a data table on the probability of FDA approval for the new drug development decision tree example to identify the breakeven probability that would change the decision strategy.
8. Using a two-way data table, perform a sensitivity analysis on both the probability of a successful clinical trial and probability of FDA approval to determine how they impact the decision strategy.
9. Generate an Excel chart for the utility function of a risk-taking individual in the text example.
10. Consider the investment return decision problem in Table 11.1. Find the decisions that would result from the aggressive, conservative, and opportunity loss strategies. Compare these decisions with those that would result from using the risk-averse utilities provided in the chapter instead of the payoff values, and the exponential utility function values that were calculated using $R = \$400$.

PROBLEMS AND APPLICATIONS

Data for selected problems can be found in the Excel file *Chapter 11 Problem Data* to facilitate your problem-solving efforts.

1. A company has estimated that a proposed $10,000 investment will generate $3250 for each of the next four years. Their required rate of return is 9%. Using internal rate of return as a decision criterion, determine whether or not this proposal should be accepted.

2. An e-commerce firm is developing a new application. Financial analysts have estimated the expenses and revenues over the next five years:

MONTH	DEVELOPMENT EXPENSE	OPERATING EXPENSE	REVENUE
Initial investment	$50,000.00	$—	$—
Year 1	$10,000.00	$10,000.00	$5,000.00
Year 2	$—	$10,000.00	$15,000.00
Year 3	$—	$10,000.00	$25,000.00
Year 4	$—	$10,000.00	$50,000.00
Year 5	$—	$10,000.00	$50,000.00

The company's discount rate is 8%. Compute the IRR for net profit and make a recommendation on whether or not to pursue the project. Then use a data table to evaluate the impact of changing the initial investment in increments of $5,000 between $30,000 and $70,000. What might this mean with regard to the company's decision?

3. The finance committee at Olson, Inc. has 13 proposed new technology projects under consideration. Estimated initial investments, external revenues, and internal cost reductions are:

PROJECT	INVESTMENT	REVENUE IMPACT	COST REDUCTION
A1	$175,600.00	$—	$200,358.00
A2	$126,512.00	$422,580.00	$(103,420.00)
A3	$198,326.00	$415,625.00	$(226,413.00)
B4	$421,618.00	$—	$486,312.00
B5	$322,863.00	$—	$456,116.00
B6	$398,810.00	$—	$508,213.00
B7	$212,506.00	$—	$356,067.00
C8	$813,620.00	$416,283.00	$386,229.00
C9	$850,418.00	$583,260.00	$398,014.00
D10	$522,615.00	$916,426.00	$(155,106.00)
D11	$486,283.00	$816,420.00	$(103,210.00)
D12	$683,407.00	$758,420.00	$(75,896.00)
D13	$722,813.00	$950,128.00	$(120,063.00)

Note that a positive cost reduction increases revenue, while a negative cost reduction implies additional costs to the firm. Using ROI, determine which projects should be selected for further consideration by the committee. How would the committee rank the projects using a benefit/cost analysis?

4. The cost accountant of a large truck fleet is evaluating options for dealing with a large volume of flat tires. Currently, the company repairs tires on the open market by having the driver take the flat tire to the nearest tire dealer. Last year, this cost an average of $30 per flat tire. The volume of flat tires experienced per year was 10,000 last year, and the expected rate of growth in flat tires is 10% per year. However, some feel that flat tire growth will be as low as 5%; others as high as 15%. A complicating factor is that the cost to repair a tire grows an average of 3% per year.

The company has two alternatives. A tire dealer has offered to fix all the company's flat tires for a fixed rate of $36 per tire over a three-year period. The other alternative is for the company to go into the tire repair business for themselves. This option is expected to require an investment in equipment of $200,000, with a salvage value of $50,000 after three years. It would require an overhead expense of $40,000 per year in the first year, $45,000 the second year, and $50,000 the third year. The variable cost for fixing a flat tire is $12 per tire for the three-year period of analysis. Compare the net present costs using a discount rate of 8% over three years for each of these three options under conditions of tire growth rate ranging from 5% to 15% per year in increments of 1%. What is the best option under each scenario? What risk factors should be considered?

5. Your neighbor's son or daughter is starting to look at college choices and he has asked you to help make a good choice. You have suggested that a scoring model would be useful in helping to narrow down the choices. Develop a scoring model on a spreadsheet by defining a set of attributes and levels of those attributes that would be appropriate for evaluating a prospective college. Use your own retrospective experience in developing the model.

6. Slaggert Systems is considering becoming certified to the ISO 9000 series of quality standards. Becoming certified is expensive, but the company could lose a substantial amount of business if its major customers suddenly demand ISO certification and the company does not have it. At a management retreat, the senior executives of the firm developed the following payoff table, indicating the net present value of profits over the next five years. What decision should they make under the average payoff, aggressive, conservative, and opportunity loss decision strategies?

	CUSTOMER RESPONSE	
	STANDARDS REQUIRED	STANDARDS NOT REQUIRED
Become certified	$550,000	$480,000
Stay uncertified	$300,000	$520,000

7. The DoorCo Corporation is a leading manufacturer of garage doors. All doors are manufactured in their plant in Carmel, Indiana, and shipped to distribution centers or major customers. DoorCo recently acquired another manufacturer of garage doors, Wisconsin Door, and is considering moving its wood door operations to the Wisconsin plant. A key consideration in this decision is the transportation and production costs at the two plants and the new construction and relocation costs. Complicating matters is the fact that marketing is predicting a decline in the demand for wood doors. The company developed three scenarios:
    1. Demand falls slightly, with no noticeable effect on production.
    2. Demand and production decline 20%.
    3. Demand and production decline 45%.

The table below shows the total costs under each decision and scenario.

	SLIGHT DECLINE	20% DECLINE	40% DECLINE
Stay in Carmel	$982,000	$830,000	$635,000
Move to Wisconsin	$993,000	$832,000	$629,000

    a. What decision should DoorCo make using the average payoff, aggressive, conservative, and opportunity loss decision strategies discussed in this chapter?

    b. Suppose the probabilities of the three scenarios are estimated to be 0.15, 0.45, and 0.40, respectively. Construct a decision tree and compute the rollback values to find the best expected value decision.

8. Suppose that a car rental agency offers insurance for a week that will cost $10 per day. A minor fender bender will cost $1,200, while a major accident might cost $10,000. Without the insurance, you would be personally liable for any damages. What should you do? Clearly, there are two decision alternatives: take the insurance, or do not take the insurance. The uncertain consequences, or events that might occur, are that you would not be involved in an accident, that you would be involved in a fender bender, or that you would be involved in a major accident. Assume that you researched insurance industry statistics and found out that the probability of major accident is 0.06%, and that the probability of a fender bender is 0.15%. What is the expected value decision? Would you choose this? Why or why not? What would be some alternate ways to evaluate risk?

9. An investor can invest in three highly speculative opportunities. The returns and standard deviations are given here.

	EXPECTED RETURN	STANDARD DEVIATION
Investment A	$50,000	$25,000
Investment B	$40,000	$24,000
Investment C	$30,000	$10,000

Based on the return to risk, which of these is the best investment?

10. An information system consultant is bidding on a project that involves some uncertainty. Based on past experience, if all went well (probability 0.1), the project would cost $1.2 million to complete. If moderate debugging were required (probability 0.7), the project would probably cost $1.4 million. If major problems were encountered (probability 0.2), the project could cost $1.8 million. Assume that the firm is bidding competitively, and the expectation of successfully gaining the job at a bid of $2.2 million is 0, at $2.1 million is 0.1, at $2.0 million is 0.2, at $1.9 million is 0.3, at $1.8 million is 0.5, at $1.7 million is 0.8, and at $1.6 million is practically certain.

    a. Calculate the expected monetary value for the given bids.

    b. What is the best bidding decision?

11. Mountain Ski Sports, a chain of ski equipment shops in Colorado, purchases skis from a manufacturer each summer for the coming winter season. The most popular intermediate model costs $150 and sells for $260. Any skis left over at the end of the winter are sold at the store's half-price sale (for $130). Sales over the

years are quite stable. Gathering data from all its stores, Mountain Ski Sports developed the following probability distribution for demand:

DEMAND	PROBABILITY
150	0.05
175	0.15
200	0.40
225	0.30
250	0.10

The manufacturer will take orders only for multiples of 20, so Mountain Ski is considering the following order sizes: 160, 180, 200, 220, and 240.

a. Construct a payoff table for Mountain Ski's decision problem of how many pairs of skis to order. What is the best decision from an expected value basis?
b. Find the expected value of perfect information.
c. What is the expected demand? Is the optimal order quantity equal to the expected demand? Why?

12. Bev's Bakery specializes in sourdough bread. Early each morning, Bev must decide how many loaves to bake for the day. Each loaf costs $0.75 to make and sells for $2.85. Bread left over at the end of the day can be sold the next for $1.00. Past data indicate that demand is distributed as follows:

NUMBER OF LOAVES	PROBABILITY
15	0.05
16	0.05
17	0.10
18	0.10
19	0.20
20	0.40
21	0.05
22	0.05

a. Construct a payoff table and determine the optimal quantity for Bev to bake each morning.
b. What is the optimal quantity for Bev to bake if the unsold loaves cannot be sold to the day-old store at the end of the day (so that unsold loaves are a total loss)?

13. An investor is considering a two-asset portfolio. Stock A has an expected return of $4.50 per share with a standard deviation of $1.00, while stock B has an expected return of $3.75 with a standard deviation of $0.75. The covariance between the two stocks is −0.35. Find the portfolio risk if:
a. the stocks are weighted equally in the portfolio,
b. the amount of stock A is one-fourth as much as stock B, and
c. the amount of stock B is one-fourth as much as stock A.

14. A patient arrives at an emergency room complaining of abdominal pain. The ER physician must decide on whether to operate or to place the patient under observation for a non-appendix-related condition. If an appendectomy is performed immediately, the doctor runs the risk that the patient does not have appendicitis. If it is delayed and the patient does indeed have appendicitis, the appendix might perforate, leading to a more severe case and possible complications. However, the patient might recover without the operation.

a. Construct a decision tree for the doctor's dilemma.

b. How might payoffs be determined?

c. Would utility be a better measure of payoff than actual costs? If so, how might utilities be derived for each path in the tree?

15. Midwestern Hardware must decide how many snow shovels to order for the coming snow season. Each shovel costs $12.00 and is sold for $19.95. No inventory is carried from one snow season to the next. Shovels unsold after February are sold at a discount price of $10.00. Past data indicate that sales are highly dependent on the severity of the winter season. Past seasons have been classified as mild or harsh and the following distribution of regular price demand have been tabulated:

MILD WINTER		HARSH WINTER	
NO. OF SHOVELS	PROBABILITY	NO. OF SHOVELS	PROBABILITY
250	0.5	1,500	0.2
300	0.4	2,500	0.4
350	0.1	3,000	0.4

Shovels must be ordered from the manufacturer in lots of 200. Construct a decision tree to illustrate the components of the decision model, and find the optimal quantity for Midwestern to order if the forecast calls for a 70% chance of a harsh winter.

16. Perform a sensitivity analysis of the Midwestern Hardware scenario (Problem 15). Find the optimal order quantity and optimal expected profit for probabilities of a harsh winter ranging from 0.2 to 0.8 in increments of 0.2. Plot optimal expected profit as a function of the probability of a harsh winter.

17. Dean Kuroff started a business of rehabbing old homes. He recently purchased a circa-1800 Victorian mansion and converted it into a three-family residence. Recently, one of his tenants complained that the refrigerator was not working properly. Since Dean's cash flow was not extensive, he was not excited about purchasing a new refrigerator. He is considering two other options: purchase a used refrigerator or repair the current unit. He can purchase a new one for $400, and it will easily last three years. If he repairs the current one, he estimates a repair cost of $150, but he also believes that there is only a 30% chance that it will last a full three years and he will end up purchasing a new one anyway. If he buys a used refrigerator for $200, he estimates that there is a 0.6 probability that it will last at least three years. If it breaks down, he will still have the option of repairing it for $150 or buying a new one. Develop a decision tree for this situation and determine Dean's optimal strategy.

18. Drilling decisions by oil and gas operators involve intensive capital expenditures made in an environment characterized by limited information and high risk. A well site is dry, wet, or gushing. Historically, 50% of all wells have been dry, 30% wet, and 20% gushing. The value (net of drilling costs) for each type of well is as follows:

Dry	−$80.000
Wet	$100,000
Gushing	$200,000

Wildcat operators often investigate oil prospects in areas where deposits are thought to exist by making geological and geophysical examinations of the area before obtaining a lease and drilling permit. This often includes recording shock waves from detonations by a seismograph and using a magnetometer to measure the intensity of the Earth's magnetic effect to detect rock formations below the surface. The cost of doing such studies is approximately $15,000. Of course, one may choose to drill in a location based on "gut feel" and avoid the cost of the study. The geological and geophysical examination classify an area into one of three categories: no structure (NS), which is a bad sign; open structure (OS), which is an "OK" sign; and closed structure (CS), which is hopeful. Historically, 40% of the tests have resulted in NS, 35% resulted in OS, and 25% resulted in CS readings. After the result of the test is known, the company may decide not to drill. The following table shows probabilities that the well will actually be dry, wet, or gushing based on the classification provided by the examination (in essence, the examination cannot accurately predict the actual event):

	DRY	WET	GUSHING
NS	0.73	0.22	0.05
OS	0.43	0.34	0.23
CS	0.23	0.372	0.398

    a. Construct a decision tree of this problem that includes the decision of whether or not to perform the geological tests.

    b. What is the optimal decision under expected value when no experimentation is conducted?

    c. Find the overall optimal strategy by rolling back the tree.

19. Many automobile dealers advertise lease options for new cars. Suppose that you are considering three alternatives:

    1. Purchase the car outright with cash.

    2. Purchase the car with 20% down and a 48-month loan.

    3. Lease the car.

      Select an automobile whose leasing contract is advertised in a local paper. Using current interest rates and advertised leasing arrangements, perform a decision analysis of these options. Make, but clearly define, any assumptions that may be required.

20. Consider the car rental insurance scenario in Problem 8. Use the approach described in this chapter to develop your personal utility function for the payoffs associated with this decision. Determine the decision that would result using the utilities instead of the payoffs. Is the decision consistent with your choice?

21. A college football team is trailing 14–0 late in the game. The team is getting close to making a touchdown. If they can score now, hold the opponent, and score one more time, they can tie or win the game. The coach is wondering whether to go for an extra-point kick or a two-point conversion now, and what to do if they can score again.

    a. Develop a decision tree for the coach's decision. Develop a utility function to represent the final score for each path in the tree.

    b. Estimate probabilities for successful kicks or two-point conversions. (You might want to do this by some group brainstorming or by calling on experts, such as your school's coach or a sports journalist.) Using the probabilities and utilities from part (a), determine the optimal strategy.

    c. Perform a sensitivity analysis on the probabilities to evaluate alternative strategies (such as when the starting kicker is injured).

CASE

The Sandwich Decision

A national restaurant chain has developed a new specialty sandwich. Initially, it faces two possible decisions: introduce the sandwich nationally at a cost of $200,000 or evaluate it in a regional test market at a cost of $30,000. If it introduces the sandwich nationally, the chain might find either a high or low response to the idea. Probabilities of these events are estimated to be 0.6 and 0.4, respectively. With a high response, gross revenues of $700,000 (at net present value) are expected; with a low response, the figure is $150,000. If it starts with a regional marketing strategy, it might find a low response or a high response at the regional level with probabilities 0.3 and 0.7, respectively. This may or may not reflect the national market potential. In any case, the chain next needs to decide whether to remain regional, market nationally, or drop the product. If the regional response is high and it remains regional, the expected revenue is $200,000. If it markets nationally (at an additional cost of $200,000), the probability of a high

national response is 0.9 with revenues of $700,000 ($150,000 if the national response is low). If the regional response is low and it remains regional, the expected revenue is $100,000. If it markets nationally (at an additional cost of $200,000), the probability of a high national response is 0.05 with revenues of $700,000 ($150,000 if the national response is low).

a. Using *TreePlan*, construct a decision tree and determine the optimal strategy.
b. Conduct sensitivity analyses for the probability estimates using both one- and two-way data tables as appropriate.
c. Develop the risk profile associated with the optimal strategy.
d. Evaluate the risk associated with this decision, considering that it is a one-time decision.
e. Summarize all your results, including your recommendation and justification for it, in a formal report to the executive in charge of making this decision.

Chapter 12

Queues and Process Simulation Modeling

INTRODUCTION

*M*any production and service operations involve the flow of some type of entity through a system over time. The entity might be a customer, a physical object, or a piece of information. Some common examples are people being served at a driver's license bureau, jobs being processed in a factory, messages moving through a communication system, and calls being processed at a call center. Most of these situations involve *queuing,* or waiting for service to occur. Decisions regarding such operations often involve determining the best design configuration or operating policies of the system to reduce customer wait time or to improve the efficiency of the operation.

Modeling these situations is considerably more difficult than modeling other types of problems, primarily because the sequence of events over time must be explicitly taken into account. While some simple situations are amenable to analytical models, most practical systems are modeled and analyzed using process simulation. **Process simulation** is an approach for modeling the logical sequence of events as they take place.

In this chapter we introduce basic concepts of queuing systems and process simulation modeling using a simple Excel-based software package, *SimQuick.*[1] The key concepts that we will discuss are listed here:

➤ *Fundamental concepts of queues and analytical models for queuing analysis*

➤ *Process-driven simulation models for queuing problems*

➤ *Other applications of process simulation*

➤ *An introduction to system dynamics and continuous simulation models*

QUEUES AND QUEUING SYSTEMS

Waiting lines occur in many important business operations as well as in everyday life. Most service systems, such as fast-food restaurants, banks, gasoline stations, and technical support telephone hotlines involve customer waiting. In these systems,

---

[1] David Hartvigsen, *SimQuick, Process Simulation with Excel*, 2nd Ed. (Upper Saddle River, NJ: Pearson Prentice Hall, 2004).

customers arrive at random times, and service times are rarely predictable. Managers of these systems would be interested in how long customers have to wait, the length of waiting lines, use of the servers, and other measures of performance. The important issue in designing such systems involves the trade-off between customer waiting and system cost, usually determined by the number of servers. A design that balances the average demand with the average service rate will cause unacceptable delays. The decision is difficult because the marginal return for increasing service capacity declines. For example, a system that can handle 99% of expected demand will cost much more than a system designed to handle 90% of expected demand. In this section we discuss the basic components of waiting line models and illustrate a process-driven simulation model for a simple case.

Basic Concepts of Queuing Systems

The analysis of waiting lines, called *queuing theory,* applies to any situation in which customers arrive to a system, wait, and receive service. Queuing theory had its origins in 1908 with a Danish telephone engineer, A.K. Erlang, who began to study congestion in the telephone service of the Copenhagen Telephone Company. Erlang developed mathematical formulas to predict waiting times and line lengths. Over the years, queuing theory has found numerous applications in telecommunications and computer systems and has expanded to many other service systems. The objectives of queuing theory are to improve customer service and reduce operating costs. As consumers began to differentiate firms by their quality of service, reducing waiting times has become an obsession with many firms. Many restaurant and department store chains take waiting seriously—some have dedicated staffs that study ways to speed up service.

All queuing systems have three elements in common:

1. *Customers waiting for service.* Customers need not be people but can be machines awaiting repair, airplanes waiting to take off, subassemblies waiting for a machine, computer programs waiting for processing, or telephone calls awaiting a customer service representative.
2. *Servers providing the service.* Again, servers need not be only people, such as clerks, customer service representatives, or repairpersons; servers may be airport runways, machine tools, repair bays, ATMs, or computers.
3. A *waiting line* or *queue.* The queue is the set of customers waiting for service. In many cases, a queue is a physical line, as you experience in a bank or grocery store. In other situations, a queue may not even be visible or in one location, as with computer jobs waiting for processing or telephone calls waiting for an open line.

To understand the operation of a queuing system, we need to describe the characteristics of the customer, server, and queue, and how the system is configured.

Customer Characteristics

Customers arrive to the system according to some *arrival process,* which can be deterministic or probabilistic. Examples of deterministic arrivals would be parts feeding from an automated machine to an assembly line or patients arriving at appointed times to a medical facility. Most arrival processes, such as people arriving at a supermarket, are probabilistic. We can describe a probabilistic arrival process by a probability distribution representing the number of arrivals during a specific time interval, or by a distribution that represents the time between successive arrivals.

Many models assume that arrivals are governed by a **Poisson process**. This means that:

1. customers arrive one at a time, independently of each other and at random;
2. past arrivals do not influence future arrivals; that is, the probability that a customer arrives at any point in time does not depend on when other customers arrived (sometimes we say that the system has *no memory*); and
3. the probability of an arrival does not vary over time (the arrival rate is **stationary**).

One way to validate these assumptions is to collect empirical data about the pattern of arrivals. We can observe and record the actual times of individual arrivals in order to determine the probability distribution and check if the arrival rate is constant over time. We can also observe if customers arrive individually or in groups and whether they exhibit any special behavior, such as not entering the system if the line is perceived as too long.

If the arrival pattern is described by a Poisson process, then the Poisson probability distribution with mean arrival rate λ (customers per unit time) can be used to describe the probability that a particular number of customers arrives during a specified time interval. For example, an arrival rate of two customers per minute means that on the average, customers arrive every half minute, or every 30 seconds. Thus, an equivalent way of expressing arrivals is to state the **mean interarrival time** between successive customers. If λ is the mean arrival rate, the mean interarrival time, t, is simply $1/\lambda$. One useful result is that if the number of arrivals follows a Poisson process with mean λ, then the time between arrivals has an exponential distribution with a mean rate $1/\lambda$. This fact will be very useful when simulating queuing systems later in this chapter.

Sometimes, the arrival rate is not stationary (that is, customers arrive at different rates at different times). For instance, the demand for service at a quick-service restaurant is typically low in the mid-morning and mid-afternoon and peaking during the breakfast, lunch, and dinner hours. Individual customers may also arrive singly and independently (telephone calls to a mail order company) or in groups (a pallet-load of parts arriving at a machine center, or patrons at a movie theater).

The **calling population** is the set of potential customers. In many applications, the calling population is assumed to be infinite; that is, an unlimited number of possible customers can arrive to the system. This would be the case with telephone calls to a mail order company or shoppers at a supermarket. In other situations the calling population is finite. One example would be a factory in which failed machines await repair.

Once in line, customers may not always stay in the same order as they arrived. It is common for customers to **renege**, or leave a queue, before being served if they get tired of waiting. In queuing systems with multiple queues, customers may **jockey**, or switch lines if they perceive another to be moving faster. Some customers may arrive at the system, determine that the line is too long, and decide not to join the queue. This behavior is called **balking**.

Service Characteristics

Service occurs according to some service process. The time it takes to serve a customer may be deterministic or probabilistic. In the probabilistic case, the service time is described by some probability distribution. In many queuing models, we make the assumption that service times follow an exponential distribution with a mean service rate μ, the average number of customers served per unit of time. Thus, the average service time is $1/\mu$. One reason that the exponential distribution describes many realistic service phenomena is that it has a useful property—the probability of small service times is large. For example, the probability that t exceeds the mean is only 0.368. This means that we see a large number of short service times and a few long ones. Think of your own experience in grocery stores. Most customers' service times are relatively short; however, every once in a while you see a shopper with a large number of groceries.

The exponential distribution, however, does not seem to be as common in modeling service processes as the Poisson is in modeling arrival processes. Analysts have found that many queuing systems have service time distributions that are not exponential, and may be constant, normal, or some other probability distribution.

Other service characteristics include nonstationary service times (taking orders and serving dinner might be longer than for breakfast), and service times that depend on the type of customer (patients at an emergency room). The service process may include one or several servers. The service characteristics of multiple servers may be identical or different. In some systems certain servers may only service specific types of customers. In many systems, such as restaurants and department stores, managers vary the number of servers to adjust to busy or slack periods.

Queue Characteristics

The order in which customers are served is defined by the **queue discipline**. The most common queue discipline is first come, first served (FCFS). In some situations, a queue may be structured as last come, first served (LCFS); just think of the in-box on a clerk's desk. At an emergency room, patients are usually serviced according to a priority determined by triage, with the more critical patients served first.

System Configuration

The customers, servers, and queues in a queuing system can be arranged in various ways. Three common queuing configurations are as follows:

1. One or more parallel servers fed by a single queue—this is the typical configuration used by many banks and airline ticket counters.
2. Several parallel servers fed by their own queues—most supermarkets and discount retailers use this type of system.
3. A combination of several queues in series—this structure is common when multiple processing operations exist, such as in manufacturing facilities.

Performance Measures

A queuing model provides measures of system performance, which typically are as follows:

- The quality of the service provided to the customer
- The efficiency of the service operation and the cost of providing the service

Various numerical measures of performance can be used to evaluate the quality of the service provided to the customer. These include the following:

- Waiting time in the queue
- Time in the system (waiting time plus service time)
- Completion by a deadline

The efficiency of the service operation can be evaluated by computing such measures as the following:

- Average queue length
- Average number of customers in the system (queue plus in service)

- Throughput—the rate at which customers are served
- Server use—percentage of time servers are busy
- Percentage of customers who balk or renege

Usually we can use these measures to compute an operating cost in order to compare alternative system configurations. The most common measures of queuing system performance, called the **operating characteristics** of the queuing system, and the symbols used to denote them are shown here:

$$L_q = \text{average number in the queue}$$
$$L = \text{average number in the system}$$
$$W_q = \text{average waiting time in the queue}$$
$$W = \text{average time in the system}$$
$$P_0 = \text{probability that the system is empty}$$

A key objective of waiting line studies is to minimize the total expected cost of the waiting line system. The total system cost consists of service costs and waiting costs. As the level of service increases (e.g., as the number of checkout counters in a grocery store increases), the cost of service increases. Simultaneously, customer waiting time will decrease, and consequently, expected waiting cost will decrease. Waiting costs are difficult to measure because they depend on customers' perceptions of waiting. It is difficult to quantify how much revenue is lost because of long lines. However, managerial judgment, tempered by experience, can provide estimates of waiting costs. If waiting occurs in a work situation, such as workers waiting in line at a copy machine, the cost of waiting should reflect the cost of productive resources lost while waiting. Although we usually cannot eliminate customer waiting completely without prohibitively high costs, we can minimize the total expected system cost by balancing service and waiting costs.

ANALYTICAL QUEUING MODELS

Many analytical models have been developed for predicting the characteristics of waiting line systems. Analytical models of queuing behavior depend on some key assumptions about the arrival and service processes, the most common being Poisson arrivals and exponential services. In applying these models, as well as for simulation purposes, the unit of time that we use in modeling both arrival and service processes can be arbitrary. For example, a mean arrival rate of 2 customers per minute is equivalent to 120 customers per hour. We must be careful, however, to express the arrival rate and service rate in the same time units.

Except for some special cases, queuing models in general are rather difficult to formulate and solve even when the distribution of arrivals and departures is known. We present analytical results for the simplest case, the single-server model.

Single-Server Model

The most basic queuing model assumes Poisson arrivals with mean arrival rate λ, exponential service times with mean service rate μ, a single server, and a FCFS queue discipline. For this model, the operating characteristics are:

$$\text{Average number in the queue} = L_q = \frac{\lambda^2}{\mu(\mu - \lambda)} \tag{12.1}$$

$$\text{Average number in the system} = L = \frac{\lambda}{\mu - \lambda} \qquad (12.2)$$

$$\text{Average waiting time in the queue} = W_q = \frac{\lambda}{\mu(\mu - \lambda)} \qquad (12.3)$$

$$\text{Average time in the system} = W = \frac{1}{\mu - \lambda} \qquad (12.4)$$

$$\text{Probability that the system is empty} = P_0 = 1 - \lambda/\mu \qquad (12.5)$$

Note that these formulas are valid only if $\lambda < \mu$. If $\lambda \geq \mu$ (that is, the rate of arrivals is at least as great as the service rate), the numerical results become nonsensical. In practice, this means that the queue will never "average out" but will grow indefinitely (we will discuss this further in the section on simulation). It should be obvious that when $\lambda > \mu$, the server will not be able to keep up with the demand. However, it may seem a little strange that this will occur even when $\lambda = \mu$. You would think that an equal arrival rate and service rate should result in a "balanced" system. This *would* be true in the deterministic case when both arrival and service rates are constant. However, when *any* variation exists in the arrival or service pattern, the queue will eventually build up indefinitely. The reason is that individual arrival times and service times vary in an unpredictable fashion even though their averages may be constant. As a result, there will be periods of time in which demand is low and the server is idle. This time is lost forever, and the server will not be able to make up for periods of heavy demand at other times. This also explains why queues form when $\lambda < \mu$.

To illustrate the use of this model, suppose that customers arrive at an airline ticket counter at a rate of two customers per minute and can be served at a rate of three customers per minute. Note that the average time between arrivals is 1/2 minute per customer and the average service time is 1/3 minute per customer. Using the queuing formulas, we have:

$$\text{Average number in the queue} = L_q = \frac{\lambda^2}{\mu(\mu - \lambda)} = \frac{2^2}{3(3 - 2)} = 1.33 \text{ customers}$$

$$\text{Average number in the system} = L = \frac{\lambda}{\mu - \lambda} = \frac{2}{3 - 2} = 2.00 \text{ customers}$$

$$\text{Average waiting time in the queue} = W_q = \frac{\lambda}{\mu(\mu - \lambda)} = \frac{2}{3(3 - 2)} = 0.67 \text{ minutes}$$

$$\text{Average time in the system} = W = \frac{1}{\mu - \lambda} = \frac{1}{3 - 2} = 1.00 \text{ minutes}$$

$$\text{Probability that the system is empty} = P_0 = 1 - \lambda/\mu = 1 - \frac{2}{3} = 0.33$$

These results indicate that on the average, 1.33 customers will be waiting in the queue. In other words, if we took photographs of the waiting line at random times, we would find an average of 1.33 customers waiting. If we include any customers in service, the average number of customers in the system is 2. Each customer can expect to wait an average of 0.67 minutes in the queue, and spend an average of 1 minute in the system. About one-third of the time, we would expect to see the system empty and the server idle.

The analytical formulas provide long-term expected values for the operating characteristics; they do not describe short-term dynamic behavior of system performance. In a real waiting line system, we typically see large fluctuations around the averages, and in systems in which the system begins empty, it may take a very long time to reach these expected performance levels. Simulation provides information about the dynamic behavior of waiting lines that analytical models cannot. In addition, simulation is not constrained by the restrictive assumptions necessary to obtain a mathematical solution. Thus, simulation has some important advantages over analytical approaches.

Analytical queuing models provide steady-state values of operating characteristics. By **steady state**, we mean that the probability distribution of the operating characteristics does not vary with time. This means that no matter when we observe the system, we would expect to see the same average values of queue lengths, waiting times, and so on. However, this usually does not happen in practice, even if the average arrival rate and average service rate are constant over time. To understand this, think of an amusement park that opens at 10:00 A.M. When it opens, there are no customers in the system and, hence, no queues at any of the rides. As customers arrive, it will take some time for queues to build up. For the first hour or so, the lines and waiting times for popular rides grow longer and eventually level off. This is called the **transient period**. Thus, if we are interested in how long it takes to reach steady state or understand the dynamic behavior during the transient period, we must resort to other methods of analysis, such as simulation.

Little's Law

MIT Professor John D.C. Little has made many contributions to the field of management science. He is most famous for recognizing a simple yet powerful relationship among operating characteristics in queuing systems. Little's Law, as it has become known, is very simple:

$$\text{For any steady-state queuing system}, L = \lambda W \tag{12.6}$$

This states that the average number of customers in a system is equal to the mean arrival rate times the average time in the system. An intuitive explanation of this result can be seen in the following way. Suppose that you arrive at a queue and spend W minutes in the system (waiting plus service). During this time, more customers will arrive at a rate λ. Thus, when you complete service, a total of λW customers will have arrived after you. This is precisely the number of customers that remain in the system when you leave, or L. Using similar arguments, we can also show that *for any steady-state queuing system, $L_q = \lambda W_q$*. This is similar to the first result and states that the average length of the queue equals the mean arrival rate times the average waiting time.

These results provide an alternative way of computing operating characteristics instead of using the formulas provided earlier. For example, if L is known, then we may compute W by L/λ. Also, W_q can be computed as L_q/λ. Two other general relationships that are useful are:

$$L = L_q + \lambda/\mu \tag{12.7}$$

and

$$W = W_q + 1/\mu \tag{12.8}$$

The first relationship states that the average number in the system is equal to the average queue length plus λ/μ. This makes sense if you recall that the probability that the system is empty is $P_0 = 1 - \lambda/\mu$. Thus, λ/μ is the probability that at least one customer is in the system. If there is at least one customer in the system, then the server must be busy. The term λ/μ simply represents the expected number of customers in service.

The second relationship states that the average time in the system is equal to the average waiting time plus the average service time. This makes sense because the time spent in the system for any customer consists of the waiting time plus the time in service.

PROCESS SIMULATION CONCEPTS

Process simulation is used routinely in business to address complex operational problems. Building a process simulation model involves first describing how the process operates, normally using some type of graphical flowchart that describes all process steps and logical decisions that route entities to different locations. Second, all key inputs such as how long it takes to perform each step of the process and resources needed to perform process tasks must be identified. Typically the activity times in a process are uncertain and described by probability distributions. The intent is for the model to duplicate the real process so that "what-if" questions can easily be evaluated without having to make time-consuming or costly changes to the real process. Once the model is developed, the simulation process repeatedly samples from the probability distributions of the input variables to drive the flow of entities.

To understand the logic behind a process simulation model, we will use a single-server queue. Consider the sequence of activities that each customer undergoes:

1. Customer arrives.
2. Customer waits for service if the server is busy.
3. Customer receives service.
4. Customer leaves the system.

In order to compute the waiting time, we need to know the time a customer arrived and the time service began; the waiting time is the difference. Similarly, to compute the server idle time, we need to know if the arrival time of the next customer is greater than the time at which the current customer completes service. If so, the idle time is the difference. To find the number in the queue, we note that when a customer arrives, then all prior customers who have not completed service by that time must still be waiting. We can make three other observations:

1. If a customer arrives and the server is idle, then service can begin immediately upon arrival.
2. If the server is busy when a customer arrives, then the customer cannot begin service until the previous customer has completed service.
3. The time that a customer completes service equals the time service begins plus the actual service time.

These observations provide all the information we need to run a small manual simulation. Table 12.1 shows such a simulation. We assume that the system opens at time 0 and that the arrival times and service times have been generated by some random

CUSTOMER	ARRIVAL TIME	SERVICE TIME	START TIME	END TIME	WAITING TIME	SERVER IDLE TIME
1	3.2	3.7	3.2	6.9	0.0	3.2
2	10.5	3.5	10.5	14.0	0.0	3.6
3	12.8	4.3	14.0	18.3	1.2	0.0
4	14.5	3.0	18.3	21.3	3.8	0.0
5	17.2	2.8	21.3	24.1	4.1	0.0
6	19.7	4.2	24.1	28.3	4.4	0.0
7	28.7	2.8	28.7	31.5	0.0	0.4
8	29.6	1.3	31.5	32.8	1.9	0.0
9	32.7	2.1	32.8	34.9	0.1	0.0
10	36.9	4.8	36.9	41.7	0.0	2.0

Table 12.1
Manual Process Simulation of a Single-Server Queue

mechanism and are known. We can use the logic above to complete the last four columns. For example, the first customer arrives at time 3.2 (the server is idle from time 0 until this event). Because the queue is empty, customer 1 immediately begins service and ends at time 6.9. The server is idle until the next customer arrives at time 10.5 and completes service at time 14.0. Customer 3 arrives at time 12.8. Because customer 2 is still in service, customer 3 must wait until time 14.0 to begin service, incurring a waiting time of 1.2. You should verify the calculations for the remaining customers in this simulation.

This logic is not difficult to implement in an Excel spreadsheet, and an exercise at the end of the chapter asks you to do this. However, modeling and simulation of queuing systems as well as other types of process simulation models is facilitated by special software applications, one of which we introduce in the next section.

PROCESS SIMULATION WITH *SIMQUICK*

Many different commercial software packages are available for process simulation. Although these are very powerful, they can take considerable time to learn and master. We will use an Excel-based academic package, *SimQuick,* that is quite similar in nature to more sophisticated commercial software to learn how to perform process simulation. Read *SimQuick Note: Getting Started with SimQuick* before continuing.

SIMQUICK NOTE
Getting Started with *SimQuick*

SimQuick allows you to run simple simulations in an Excel spreadsheet. To launch *SimQuick,* simply double-click the Excel file *SimQuick-v2* on the CD-ROM accompanying this book. *SimQuick* contains hidden Excel macros, so please allow macros to run in Excel 2007. This can be done by clicking on *Excel Options* from the *Office* button, clicking the *Trust Center Settings* button, and checking the appropriate box in the *Macro Settings* list. Figure 12.1 shows the *SimQuick* control panel. The control panel has several buttons that are used for entering information in the model.

SimQuick uses three types of building blocks for simulation models: *objects, elements,* and *statistical distributions.* Objects represent the entities that move in a process (customers, parts, messages, and so on). Elements are stationary in a process and consist of five types:

1. *Entrances*—where objects enter a process

2. *Buffers*—places where objects can be stored (inventory storage, queues of people or parts, and so on)

3. *Work Stations*—places where work is performed on objects (machines, service personnel, and so on)

4. *Decision Points*—where an object goes in one of two or more directions (outcomes of processing activities, routings for further processing, and so on)

5. *Exits*—places where objects leave a process according to a specified schedule

Simulation models are created by specifying elements and their properties.

Statistical distributions are limited to one of the following:

- Normal: Nor(*mean, standard deviation*)
- Exponential: Exp(*mean*)
- Uniform: Uni(*lower, upper*)
- Constant
- Discrete: Dis(i), where i is the reference to table i of the worksheet *Discrete Distributions* (click the *Other Features* button to find this)

To save a *SimQuick* model, save the Excel worksheet under a different name. The appendix at the end of this chapter provides a short *SimQuick* reference manual.

Figure 12.1 *SimQuick* Control Panel

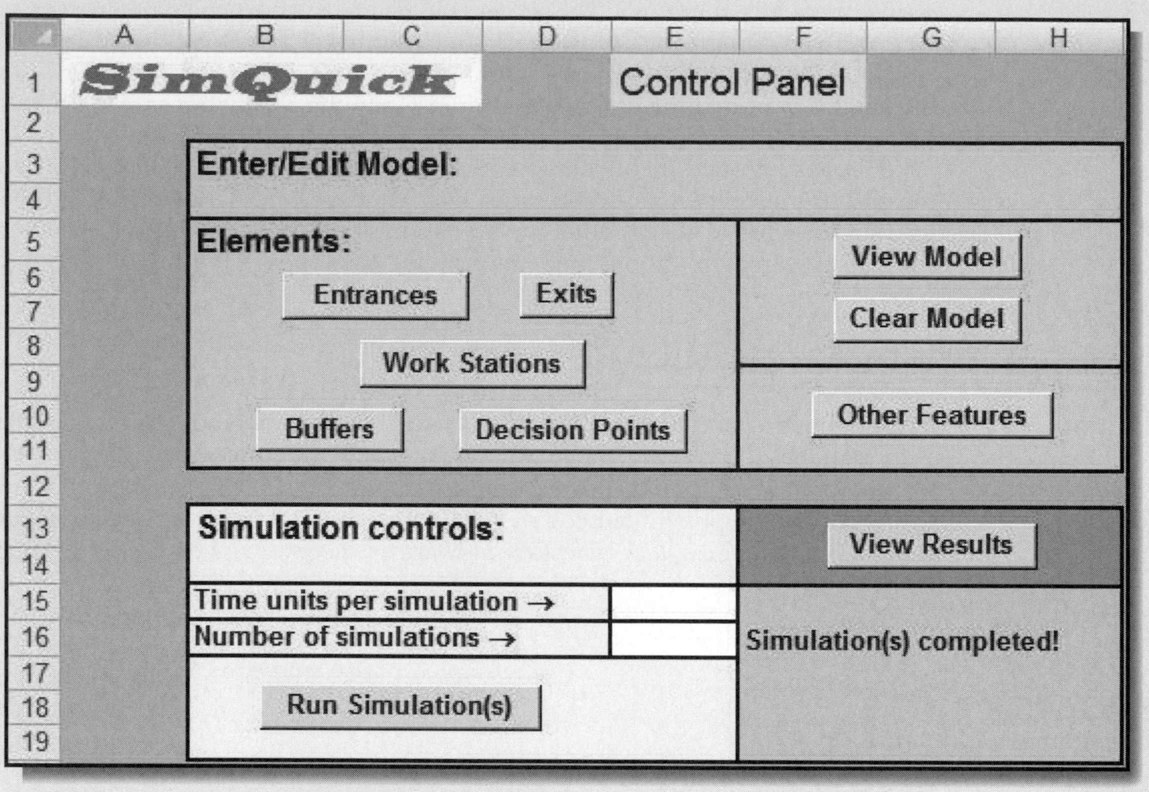

A Queuing Simulation Model

To illustrate simulation modeling and analysis for queues, we will begin with a simple single-server system and compare the results with the analytical solution we presented earlier. Dan O'Callahan operates a car wash and is in charge of finance, accounting, marketing, and analysis; his son is in charge of production. During the "lunch hour," which Dan defines as the period from 11 A.M. to 1 P.M., customers arrive randomly at an average of 15 cars per hour (or one car every 4 minutes). A car takes an average of 3 minutes to wash (or 20 cars per hour), but this fluctuates quite a bit due to variations in hand-prepping. Dan doesn't understand how a line could possibly pile up when his son can work faster than the rate at which cars arrive. Although customers complain a bit, they do not leave if they have to wait. Dan is particularly interested in understanding the waiting time, the number waiting, and how long his son is actually busy before considering improving his facility.

The first step is to draw a flowchart that represents the flow of objects through the process using the five element structures in *SimQuick*. This is shown in Figure 12.2. Note that a buffer is used to represent cars leaving the system and will allow us to count the number of customers served. The *Exit* element is used only when objects leave the model according to a specified schedule instead of when they are ready to leave.

To build a *SimQuick* model, click on the *Entrances* button in the control panel. This brings up a new worksheet that prompts you for information. Fill out one table (working from left to right) for each entrance in the model. In the Name cell, enter the name

Figure 12.2 Process Flow Map for Car Wash System

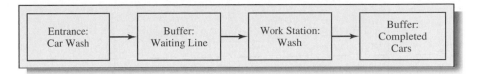

of the Entrance block. Next, enter the time between arrivals and number of objects per arrival; this may be one of the statistical distributions, or a custom schedule, which may be accessed from the *Other Features* button. In many queuing models, we assume that the number of arrivals in a fixed time period (λ) is Poisson distributed. If this is the case, then it is true that the time between arrivals ($1/\lambda$) has an exponential distribution. Therefore, we specify the time between arrivals to be Exp(4), using minutes for simulated time. Finally, specify where the objects go after entering the system; this is the next block in the process map that we labeled Waiting Line. Figure 12.3 shows the completed worksheet. Click on the *Return to Control Panel* button.

Next, click on *Buffers* to input information for the waiting line. In the table that appears, enter the name of the buffer, its capacity—that is, the maximum number of objects that can be stored in the buffer, use the word *Unlimited* to represent an infinite capacity—the initial number of objects in the buffer at the start of the simulation, and the output destinations of this process block. You have the option to move objects in different group sizes (think of a pallet of machined parts that would be transferred to a new work station as a group). Figure 12.4 shows the completed *Buffers* worksheet.

The next block in the process flow map is the work station labeled Wash. Click on the *Work Stations* button in the Control Panel. Enter the name and working time using one of the statistical distributions. In this example we will assume that car washing time is exponentially distributed with a mean of $1/\mu = 3$ minutes (note that μ represents the number of completed services per unit time in the analytical solution). The

Figure 12.3 *Entrances* Worksheet

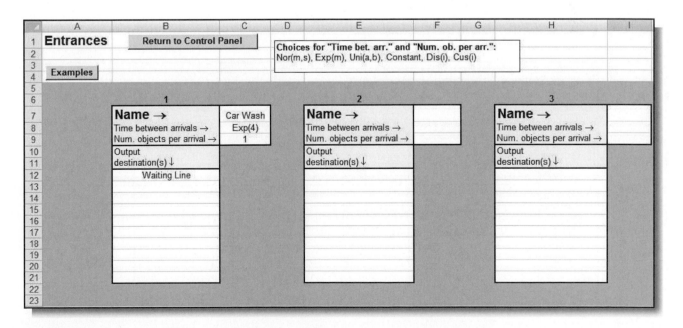

Figure 12.4 *Buffers* Worksheet

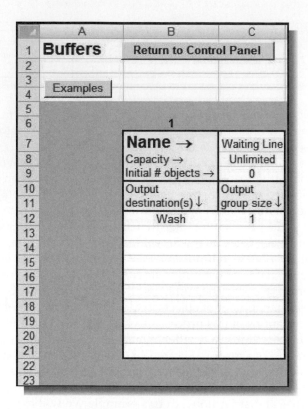

Resource fields are not needed in this example; we will discuss them later. Figure 12.5 shows the completed *Work Stations* worksheet.

Next, click on the *Buffers* button again and enter the data for the last block in the process flow map, Completed Cars. Because this is the last block in the process flow map, there is no output destination. Return to the Control Panel and enter the time

Figure 12.5 *Work Stations* Worksheet

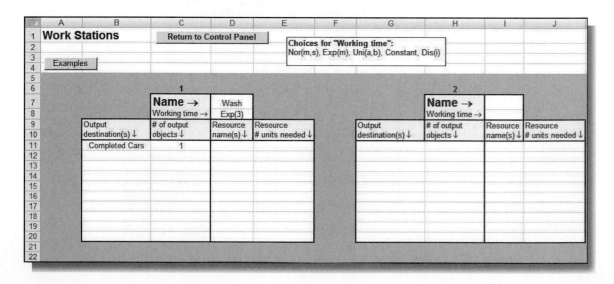

units per simulation and number of simulations. The time units can be any time interval, even fractions, but must be used consistently in all statistical distributions. Because we used minutes in the time between arrivals and service times, we should specify the simulation duration as minutes. Because we are simulating the system from 11:00 A.M. to 1:00 P.M., enter 120 as the time units per simulation. The number of simulations is the number of replications to run the simulation (analogous to the number of trials in *Crystal Ball*). For reasonable statistical results, let us run the simulation 30 times. As with risk analysis simulations, as the number of uncertain inputs increases, the simulation should be run for a larger number of trials.

The complete model is available on the CD-ROM as the Excel file *SimQuick Car Wash Simulation*. You may click on the *View Model* button for a summary of your inputs. To run the model, click on the *Run Simulation(s)* button (you may press the Esc key at any time to abort). When the simulation has completed, click on the *View Results* button. *SimQuick* creates a spreadsheet with statistical information for each model element, including overall means and detailed statistics for each run. Figure 12.6 shows a portion of the simulation results. For the Car Wash Entrance element, we have information on two things. First, we know the number of objects that entered and could not enter (this is applicable for models in which the following buffer had a finite capacity and might have blocked incoming objects). Second, we know the service level,

Figure 12.6 Portion of *SimQuick* Car Wash Simulation Results

	A	B	C	D	E	F	G	H
1	**Simulation Results**			Return to Control Panel				
2								
3	**Element**	**Element**	**Statistics**	**Overall**	**Simulation Numbers**			
4	**types**	**names**		**means**	**1**	**2**	**3**	**4**
5								
6	Entrance(s)	Car Wash	Objects entering process	29.33	19	27	30	31
7			Objects unable to enter	0.00	0	0	0	0
8			Service level	1.00	1.00	1.00	1.00	1.00
9								
10	Work Station(s)	Wash	Final status	NA	Working	Working	Working	Working
11			Final inventory (int. buff.)	0.00	0	0	0	0
12			Mean inventory (int. buff.)	0.00	0.00	0.00	0.00	0.00
13			Mean cycle time (int. buff.)	0.00	0.00	0.00	0.00	0.00
14			Work cycles started	27.60	19	27	27	31
15			Fraction time working	0.65	0.37	0.45	0.54	0.85
16			Fraction time blocked	0.00	0.00	0.00	0.00	0.00
17								
18	Buffer(s)	Waiting Line	Objects leaving	27.60	19	27	27	31
19			Final inventory	1.73	0	0	3	0
20			Minimum inventory	0.00	0	0	0	0
21			Maximum inventory	4.63	3	2	4	6
22			Mean inventory	1.23	0.27	0.20	0.61	1.43
23			Mean cycle time	4.89	1.72	0.89	2.71	5.54
24								
25		Completed Cars	Objects leaving	0.00	0	0	0	0
26			Final inventory	26.73	18	26	26	30
27			Minimum inventory	0.00	0	0	0	0
28			Maximum inventory	26.73	18	26	26	30
29			Mean inventory	13.23	10.91	14.13	12.58	14.99
30			Mean cycle time	Infinite	Infinite	Infinite	Infinite	Infinite

which is defined as the percentage of objects entering the process divided by the total number that tried to enter.

For the Wash Work Station, which represents the server in the queuing system, the statistics are defined as follows:

- Final status: status of the work station when the simulation ends
- Final inventory (int. buff.), Mean inventory (int. buff.), and Mean cycle time (int. buff.): Work stations have small internal buffers with enough room to hold one object after it has completed processing. In some models it might not be able to pass along an object to the next buffer if it is full or another work station if it is working. In such a case, the work station is called *blocked*. These statistics provide information on the levels in this internal buffer.
- Work cycles started: the number of times the work station has started processing
- Fraction time working: use of the work station
- Fraction time blocked: fraction of time that the work station was waiting to pass an object on to the next element

We see that the mean fraction of time the car wash was busy is 0.65. However, note that the variability over different simulation runs is quite high, ranging from 0.37 to 0.85 over the first four runs.

The buffer statistics provide information about the waiting line or the buffer representing completed cars. The results summarized are as follows:

- Objects leaving: number of objects that left the buffer
- Final inventory: "Inventory" refers to the number of objects in the buffer. Final inventory is the number remaining at the end of the simulation
- Minimum inventory, Maximum inventory, Mean inventory: statistics on the number of objects during the simulation
- Mean cycle time: mean time that an object spends in the buffer

In this example, the waiting line itself had a mean number of cars of 1.23 over the 30 simulations. Again, this number varied considerably over the different simulation runs. The mean cycle time is the average time in the queue.

Using the results in the spreadsheet, it would be rather straightforward to perform additional statistical analyses, such as computing the minimum and maximum values of individual statistics, standard deviations, and histograms to better understand the variability in the 30 simulated runs.

How do the simulation results compare with the analytical results? Table 12.2 shows some comparisons. The simulated averages appear to be significantly different from the analytical results. Recall that the analytical results provide steady state averages for the behavior of the queuing system. Each of our simulated runs was for only a two-hour period, and each simulation began with an empty system. Early arrivals to the system would not expect to wait very long, but when a short-term random surge of customers arrives, the queue begins to build up. As the length of the simulation increases, the number in the queue averaged over all customers begins to level off, reaching steady state. However, during the lunch period, the car wash would probably not run long enough to reach steady state; therefore, the analytical

	ANALYTICAL RESULTS	SIMULATION MEANS (30 RUNS)
Average number in the queue	$L_q = 2.25$ customers	1.23 customers
Average waiting time in the queue	$W_q = 0.15$ hours $= 9$ minutes	4.89 minutes
Probability that the system is empty	$P_0 = 0.25$	$1 - 0.65 = 0.35$

Table 12.2
Analytical Results versus Simulation Statistics

Table 12.3 Mean Queuing Statistics as a Function of Simulation Run Time

| | | | SIMULATION RUN TIME | | | | | ANALYTICAL RESULT |
| | | | 120 | 480 | 1,000 | 5,000 | 10,000 | |
ELEMENT TYPE	NAME	STATISTICS						
Work Station	Wash	Fraction time working	0.65	0.74	0.74	0.75	0.76	$1 - P_0 = 0.75$
Buffer	Waiting Line	Mean inventory	1.23	1.97	2.49	2.12	2.44	$L_q = 2.25$
		Mean cycle time	4.89	7.94	0.97	8.48	9.71	$W_q = 9$

results will never present an accurate picture of the system behavior. Thus, the differences are not surprising. However, from a practical perspective, the simulation provided information about the system behavior during this short time period that the analytical formulas could not. Dan might have made some poor decisions based on the analytical results alone.

Table 12.3 shows how these simulation statistics change as the number of time units per simulation increases. Note that as the simulation run time increases, the statistical results tend to converge toward the steady state analytical results. Therefore, if you are interested in obtaining steady state results from a simulation, then a long run time must be chosen.

In the following sections, we present a variety of additional examples of process simulation models. These serve not only to illustrate the variety of problems to which simulation can be applied, but also to demonstrate some of the additional features available in *SimQuick*.

Queues in Series with Blocking

In many manufacturing processes, the output from one work station is sent directly to the queue of another work station. Because of space limitations, these queues may have limited capacity. If they fill up, the work stations in front of them become blocked, meaning that they cannot process a new unit because they are unable to transfer the unit to the next queue. A flow process map for such a situation is shown in Figure 12.7. Assume that orders enter the system with a time between arrivals that is exponentially distributed with a mean of 0.4 minutes. The processing time at work station 1 is exponential with a mean of 0.25 minutes, and for work station 2, exponential with a mean of 0.5 minutes. The queue for work station 1 has a capacity of 4, while the queue for work station 2 has a capacity of 2. If an arriving order cannot enter the production process because the queue for work station 1 is full, then it is subcontracted to another manufacturer. To model these capacities in *SimQuick*, enter the capacities in the buffer tables as shown in Figure 12.8.

Figure 12.7 Flow Process Map for Serial Queues and Work Stations

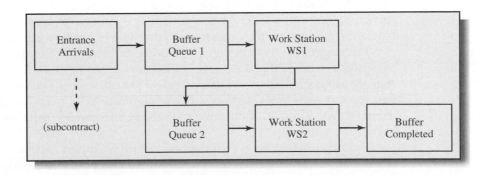

Figure 12.8 Queue Capacities in Buffer Tables

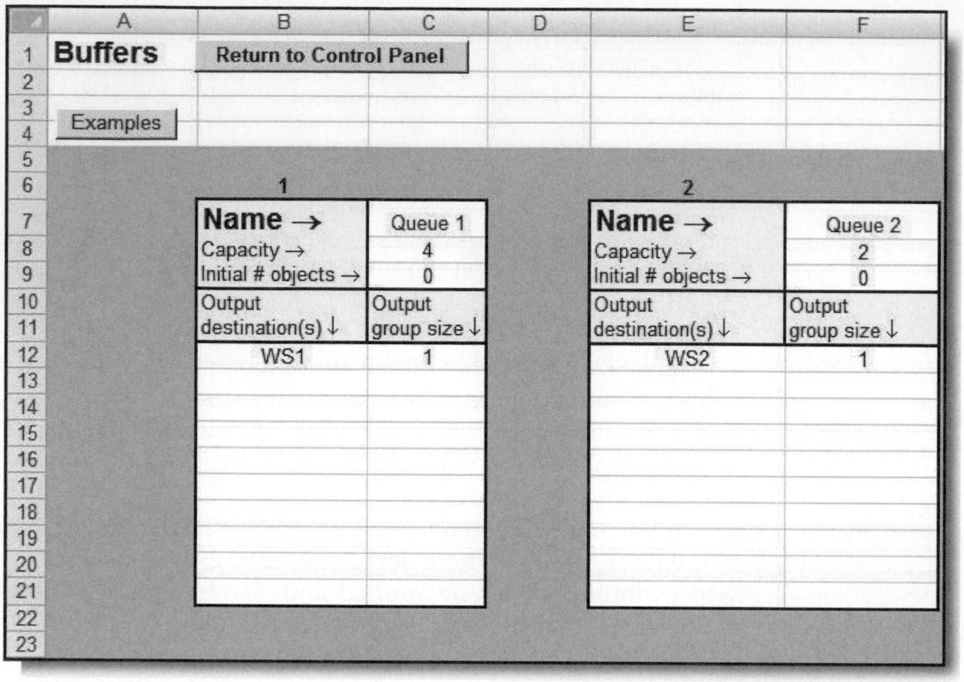

Figure 12.9 shows the *SimQuick* results for 30 replications of a 24-hour (1,440 minutes) simulation in the Excel model file *SimQuick Serial Queues*. In the Entrance statistics, we see that the service level is 75%, or equivalently, 25% of the arriving objects are unable to enter the system and, hence, are subcontracted. In the Work Station statistics, we see that work station 1 is blocked 42% of the time because the following queue is full.

Grocery Store Checkout Model with Resources

A small grocery store has two checkout lines and a floating bagger, who bags groceries at one of the two lines as needed. Customers enter the queue for the checkout lines, arriving on average every 9 minutes (exponentially distributed). Checkout time is exponential with a mean of 2 minutes. If line 1 is open, a customer will go there first. If line 1 is occupied and line 2 is free, the customer will go to line 2 instead. If both lines are occupied, the customer will wait in the queue until one of them finishes service. The bagger can only work one line at a time and takes an average of 6 minutes to bag, again exponentially distributed. If a checker finishes scanning all of a customer's items and the bagger is working at the other line, we will assume that the customer and checker will wait for the bagger. In such cases the checker is blocked from processing another customer who might be waiting in line.

The bagger is an example of a **resource** in *SimQuick*. Resources are defined and assigned to work stations. A work station cannot start processing unless all resources that are assigned to it are available. If more than one work station competes for a limited resource, the work station with the higher priority gets it. The priority is determined by the number of the table in the *SimQuick* model (the lower the number, the higher the priority).

The process flow map of the grocery store operation and the distribution of processing times are shown in Figure 12.10. In the *SimQuick* Excel model *SimQuick Grocery*

	B	C	D
3	Element	Statistics	Overall
4	names		means
5			
6	Arrivals	Objects entering process	2696.73
7		Objects unable to enter	918.43
8		Service level	0.75
9			
10	WS1	Final status	NA
11		Final inventory (int. buff.)	0.30
12		Mean inventory (int. buff.)	0.42
13		Mean cycle time (int. buff.)	0.23
14		Work cycles started	2694.47
15		Fraction time working	0.47
16		Fraction time blocked	0.42
17			
18	WS2	Final status	NA
19		Final inventory (int. buff.)	0.00
20		Mean inventory (int. buff.)	0.00
21		Mean cycle time (int. buff.)	0.00
22		Work cycles started	2692.23
23		Fraction time working	0.93
24		Fraction time blocked	0.00
25			
26	Queue1	Objects leaving	2694.47
27		Final inventory	2.27
28		Minimum inventory	0.00
29		Maximum inventory	4.00
30		Mean inventory	2.15
31		Mean cycle time	1.15
32			
33	Queue2	Objects leaving	2692.23
34		Final inventory	1.30
35		Minimum inventory	0.00
36		Maximum inventory	2.00
37		Mean inventory	1.49
38		Mean cycle time	0.80
39			
40	Completed	Objects leaving	0.00
41		Final inventory	2691.30
42		Minimum inventory	0.00
43		Maximum inventory	2691.30
44		Mean inventory	1340.78
45		Mean cycle time	Infinite

Figure 12.9 Simulation Results for Serial Queue Model

Base Case, the bagger is defined as a resource by clicking the button *Other Features* in the control panel then choosing *Resources.* In the Resources worksheet, enter the name of the resource (Bagger) and the number of resources available (1) as shown in Figure 12.11. For the work stations associated with the bagging processes, the resource Bagger is assigned to each of them (see Figure 12.12). Because only one bagger is available, the resource can be used by only one of the bagging processes at any one time.

Figure 12.13 shows the *SimQuick* results for 30 runs of 600 time units each. The key statistic for the work stations is the fraction time blocked, which represents the time the checker must wait for a bagger after servicing a customer and before he or she can service another waiting customer. For Checkout lines 1 and 2, these values are, respectively, 0.19 and 0.11. Note that the fraction of time working is higher for Checkout 1 because that line has a higher priority for entering customers. In the buffer statistics, there are an average of 0.16 customers waiting (mean inventory) for checkout, with an average waiting time of 1.39 (mean cycle time). However, for the bag buffers, the mean waiting times are 5.13 and 7.60, respectively, which is probably an unacceptable service level for a grocery store.

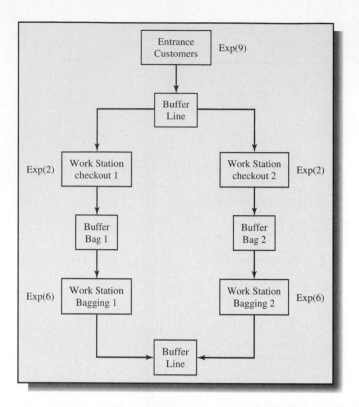

Figure 12.10 Process Flow Map of Grocery Store Operation

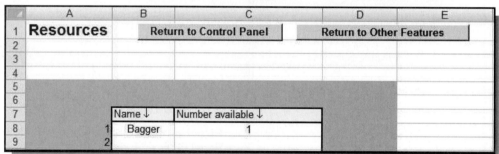

Figure 12.11 *SimQuick Resources* Worksheet

	A	B	C	D	E
1	**Resources**	Return to Control Panel		Return to Other Features	
2					
3					
4					
5					
6					
7		Name ↓	Number available ↓		
8	1	Bagger	1		
9	2				

Figure 12.12 Assigning Resources to Work Stations

	L	M	N	O	P	Q	R	S	T
5									
6		**3**					**4**		
7		**Name →**	Bagging 1				**Name →**	Bagging 2	
8		Working time →	Exp(6)				Working time →	Exp(6)	
9	Output	# of output	Resource	Resource		Output	# of output	Resource	Resource
10	destination(s) ↓	objects ↓	name(s) ↓	# units needed ↓		destination(s) ↓	objects ↓	name(s) ↓	# units needed ↓
11	Finish	1	Bagger	1		Finish	1	Bagger	1
12									
13									
14									
15									
16									
17									
18									
19									
20									

Figure 12.13 *SimQuick* Results for Grocery Store Simulation Model

	A	B	C	D
3	**Element**	**Element**	**Statistics**	**Overall**
4	**types**	**names**		**means**
5				
6	Entrance(s)	Customers	Objects entering process	69.00
7			Objects unable to enter	0.00
8			Service level	1.00
9				
10	Work Station(s)	Checkout 1	Final status	NA
11			Final inventory (int. buff.)	0.17
12			Mean inventory (int. buff.)	0.19
13			Mean cycle time (int. buff.)	2.46
14			Work cycles started	48.57
15			Fraction time working	0.17
16			Fraction time blocked	0.19
17				
18		Checkout 2	Final status	NA
19			Final inventory (int. buff.)	0.13
20			Mean inventory (int. buff.)	0.11
21			Mean cycle time (int. buff.)	3.03
22			Work cycles started	20.10
23			Fraction time working	0.07
24			Fraction time blocked	0.11
25				
26		Bagging 1	Final status	NA
27			Final inventory (int. buff.)	0.00
28			Mean inventory (int. buff.)	0.00
29			Mean cycle time (int. buff.)	0.00
30			Work cycles started	47.93
31			Fraction time working	0.47
32			Fraction time blocked	0.00
33				
34		Bagging 2	Final status	NA
35			Final inventory (int. buff.)	0.00
36			Mean inventory (int. buff.)	0.00
37			Mean cycle time (int. buff.)	0.00
38			Work cycles started	19.57
39			Fraction time working	0.19
40			Fraction time blocked	0.00

	A	B	C	D
42	Buffer(s)	Line	Objects leaving	68.67
43			Final inventory	0.33
44			Minimum inventory	0.00
45			Maximum inventory	2.60
46			Mean inventory	0.16
47			Mean cycle time	1.39
48				
49		Bag 1	Objects leaving	47.93
50			Final inventory	0.37
51			Minimum inventory	0.00
52			Maximum inventory	1.00
53			Mean inventory	0.41
54			Mean cycle time	5.13
55				
56		Bag 2	Objects leaving	19.57
57			Final inventory	0.30
58			Minimum inventory	0.00
59			Maximum inventory	1.00
60			Mean inventory	0.26
61			Mean cycle time	7.60
62				
63		Finish	Objects leaving	0.00
64			Final inventory	66.73
65			Minimum inventory	0.00
66			Maximum inventory	66.73
67			Mean inventory	32.02
68			Mean cycle time	Infinite
69				
70				
71				
72	**Resource(s)**			
73		Bagger	Mean number in use	0.66

One way of possibly improving the process is to assign the bagger to line 1 exclusively and have the checker in line 2 also bag groceries. To model this (see Excel file *SimQuick Grocery Enhanced*), define the checkout clerk in the second line as a resource, Checker2, in the Resources worksheet. The bagger need not be defined as a resource in this model because the two work stations are no longer competing for it. Then, for the work stations Checkout 2 and Bagging 2, assign Checker 2 as a required resource. In this fashion, Checker 2 may work on only one process at a time. If you examine the results, you will find that the fraction of time blocked has decreased substantially, to 0.07 for line 1 and 0 for line 2; and that the mean waiting times for both the entrance queue and bagging queues have also decreased. The mean waiting time for entering customers is now only 0.27, and for the two bagging operations, has decreased to 2.48 and 1.12, respectively.

Manufacturing Inspection Model with Decision Points

A Decision Point in a *SimQuick* model is a point in a process flow map where the routes for some objects are chosen randomly. Suppose that the output from an electronics manufacturing operation is inspected and that a percentage of the items require adjustment and are reinspected. Assume that units arrive to the inspection process queue (Iqueue) at a constant rate of one every 5 minutes. The inspection process is normal with a mean of 9 minutes and a standard deviation of 2 minutes,

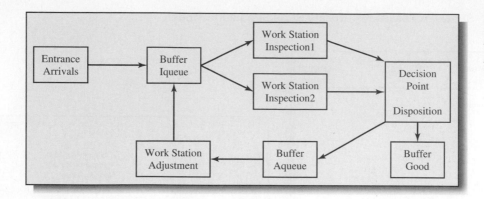

Figure 12.14
Process Flow Map for Inspection and Adjustment Process

and two inspectors are working. Fifteen percent of the units require adjustment and are sent to the queue (Aqueue) of an adjustment process, which takes an average of 30 minutes, with a standard deviation of 5 minutes, normally distributed. The remaining 85% of the units are classified as good and sent on to packaging and distribution.

Figure 12.14 shows the process flow map for this situation. To define a decision point, click on the *Decision Points* button in the control panel. In the *SimQuick* table, enter the name of the output destinations and the percentages of objects that are routed to these destinations. This is shown in Figure 12.15.

Figure 12.15   *SimQuick Decision Points* Table

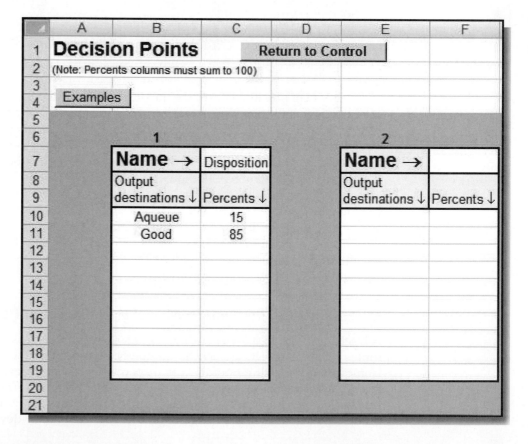

Figure 12.16 *SimQuick* Results for Inspection Adjustment Model

	A	B	C	D
3	**Element**	**Element**	**Statistics**	**Overall**
4	**types**	**names**		**means**
5				
6	Entrance(s)	Arrivals	Objects entering process	80.00
7			Objects unable to enter	0.00
8			Service level	1.00
9				
10	Work Station(s)	Inspection1	Final status	NA
11			Final inventory (int. buff.)	0.00
12			Mean inventory (int. buff.)	0.00
13			Mean cycle time (int. buff.)	0.00
14			Work cycles started	43.83
15			Fraction time working	0.98
16			Fraction time blocked	0.00
17				
18		Inspection2	Final status	NA
19			Final inventory (int. buff.)	0.00
20			Mean inventory (int. buff.)	0.00
21			Mean cycle time (int. buff.)	0.00
22			Work cycles started	43.60
23			Fraction time working	0.97
24			Fraction time blocked	0.00
25				
26		Adjustment	Final status	NA
27			Final inventory (int. buff.)	0.00
28			Mean inventory (int. buff.)	0.00
29			Mean cycle time (int. buff.)	0.00
30			Work cycles started	11.57
31			Fraction time working	0.83
32			Fraction time blocked	0.00

	A	B	C	D
34	Buffer(s)	Iqueue	Objects leaving	87.43
35			Final inventory	3.23
36			Minimum inventory	0.00
37			Maximum inventory	5.10
38			Mean inventory	1.80
39			Mean cycle time	8.25
40				
41		Aqueue	Objects leaving	11.57
42			Final inventory	2.60
43			Minimum inventory	0.00
44			Maximum inventory	3.97
45			Mean inventory	1.40
46			Mean cycle time	46.84
47				
48		Good	Objects leaving	0.00
49			Final inventory	71.27
50			Minimum inventory	0.00
51			Maximum inventory	71.27
52			Mean inventory	34.53
53			Mean cycle time	Infinite
54				
55	Decision Point(s)	Disposition	Objects leaving	85.43
56			Final inventory (int. buff.)	0.00
57			Mean inventory (int. buff.)	0.00
58			Mean cycle time (int. buff.)	0.00

Figure 12.16 shows the results of 30 simulations for 400 time units (see Excel file *SimQuick Inspection Adjustment Model*). We see that the inspections have high utilizations, 98% and 97%. The adjustor is working 83% of the time. The mean waiting time in the inspection queue is 8.25, and an average of 1.8 units wait. In the adjustment queue, the mean waiting time is 46.84, and on average 1.4 units are waiting for adjustment. This process appears to be running adequately. This model might be used to investigate the effect of quality improvements in the manufacturing process or the impact of cross-training inspectors to perform adjustments.

Pull System Supply Chain with Exit Schedules

A supply chain for a particular model of high-definition TV consists of a manufacturer, wholesale supplier, and retail store. Customer demand pulls inventory from the system at a rate that is exponentially distributed with a mean of 1 unit per day. The retail store uses a reorder point inventory system in which six units are ordered whenever the inventory drops to three units. The number of units ordered is called the **lot size** or **order quantity**, and the inventory level that triggers an order is called the **reorder point**. Delivery time from the supplier to the store is normal with a mean of three days and standard deviation of one day. At the supplier's warehouse, inventory is ordered from the manufacturer in lot sizes of 10 with a reorder point of 15. Lead time for delivery from the manufacturer is normal with a mean of two days and a standard deviation of 0.5 days.

A *SimQuick* model of this system (Excel file *SimQuick Pull Supply Chain)* is shown in Figure 12.17. This model introduces a new element, *Exit*. An *Exit* pulls objects

Figure 12.17   Process Flow Map of Pull Supply Chain

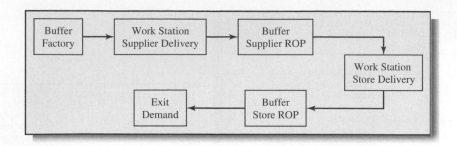

according to a specified "time between departures." Exits are modeled by clicking on the appropriate button in the control panel. You must also specify the number of objects per departure. In this model the *Exit* element, Demand, has a time between departures that is Exp(1), and 1 object per departure. The buffers model the reorder point (ROP) processes. The Store ROP buffer has a capacity and initial number of objects equal to 3, the reorder point at the retail store, and output group size of 1, the demand for each customer. Similarly, the Supplier ROP buffer has a capacity and initial number of objects equal to its reorder point, 15, and an output group size of 6 corresponding to the order quantity it sends to the retail store. The Factory buffer is given an arbitrarily large capacity of 5,000, which also equals the initial number of objects. The output group size from the factory is the order quantity of the supplier, 10. Each time Supplier Delivery obtains one object from Factory, the number of objects in the Factory drops by 10. The Store Delivery work station sends 6 objects at a time (the order quantity) to the Store ROP buffer with a working time defined as Nor(3, 1), and the Supplier Delivery work station sends ten objects at a time to the Supplier ROP buffer with a working time that is Nor(2, 0.5).

Store inventory is modeled by the Store ROP buffer and the internal buffer of the Store Delivery work station. The internal buffer of a work station holds objects completed after one working cycle. Work stations are blocked from working on a new object as long as it has objects in its internal buffer. The Exit element pulls objects from Store ROP, which in turn, pulls objects from the internal buffer of Store Delivery. When the number of objects in the internal buffer falls to zero, the inventory is entirely contained in the Store ROP buffer, and the amount of inventory is less than or equal to the reorder point, 3, which is the capacity of the Store ROP buffer. At this time Store Delivery becomes unblocked and pulls an object from Supplier ROP, which corresponds to placing an order. When Store Delivery is finished, it deposits the order size of 6 units into its internal buffer. Store Delivery is then blocked until it can pass all inventory to the following buffer.

Figure 12.18 shows the mean results for 30 runs of a 365-day simulation (Excel file *SimQuick Pull Supply Chain*). Of particular interest is the service level for the *Exit* element, which is 0.88. This means that only 88% of customer demand can be satisfied (the store is out of stock 12% of the time). Service level is primarily influenced by the reorder point, suggesting that a higher reorder point should be used at the retail store. Other information that we can obtain from the simulation results include the following:

- *Mean number of orders placed.* This is found from the number of work cycles started at the work stations. For example, the supplier placed an average of 33 orders, while the store placed about 54 orders.
- *Mean inventory level.* This is found by adding the mean inventory in the internal buffer plus the mean inventory in the corresponding buffer. Thus, for the

Figure 12.18
Simulation Results for
Pull Supply Chain

	A	B	C	D
3	**Element**	**Element**	**Statistics**	**Overall**
4	**types**	**names**		**means**
5				
6	Work Station(s)	Store Delivery	Final status	NA
7			Final inventory (int. buff.)	1.63
8			Mean inventory (int. buff.)	1.41
9			Mean cycle time (int. buff.)	1.60
10			Work cycles started	54.23
11			Fraction time working	0.45
12			Fraction time blocked	0.55
13				
14		Supplier Delivery	Final status	NA
15			Final inventory (int. buff.)	4.67
16			Mean inventory (int. buff.)	4.57
17			Mean cycle time (int. buff.)	5.15
18			Work cycles started	33.10
19			Fraction time working	0.18
20			Fraction time blocked	0.82
21				
22	Buffer(s)	Factory	Objects leaving	331.00
23			Final inventory	4669.00
24			Minimum inventory	4669.00
25			Maximum inventory	5000.00
26			Mean inventory	4829.74
27			Mean cycle time	5335.02
28				
29		Supplier ROP	Objects leaving	325.40
30			Final inventory	14.60
31			Minimum inventory	8.93
32			Maximum inventory	15.00
33			Mean inventory	14.62
34			Mean cycle time	16.43
35				
36		Store ROP	Objects leaving	322.03
37			Final inventory	2.33
38			Minimum inventory	0.00
39			Maximum inventory	3.00
40			Mean inventory	2.38
41			Mean cycle time	2.71
42				
43	Exit(s)	Demand	Objects leaving process	322.03
44			Object departures missed	44.60
45			Service level	0.88

supplier, the mean amount of inventory held is $4.57 + 14.62 = 20.19$ units. At the store, we have a mean inventory level of $1.41 + 2.38 = 3.79$ units.

Other *SimQuick* Features and Commercial Simulation Software

Two other features of *SimQuick* that we have not illustrated in the examples are discrete distributions and custom schedules. Discrete distributions can be defined by clicking on the *Other Features* button and selecting *Discrete Distributions*. For example, suppose that a more realistic assumption for the working time for the car wash in our first example is:

TIME	PROBABILITY
2	.6
3	.3
4	.1

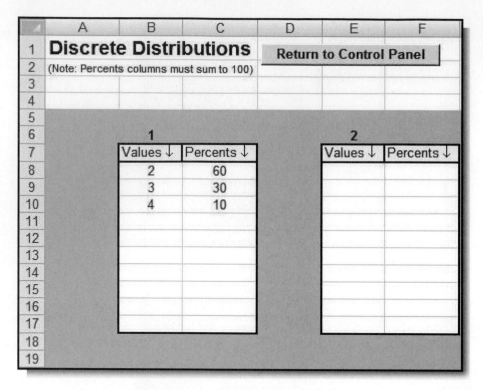

Figure 12.19
Discrete Distributions
Worksheet in *SimQuick*

Car wash times are usually constant but depend on the features that customers order (for instance, underspray, wax, and so on), which would change the amount of time it takes to service the car. In the Discrete Distribution table, enter the values for time and the probabilities expressed as percents as shown in Figure 12.19. Then in the Work Station table, specify Dis(1) as the working time. In general, Dis(*i*) refers to the distribution you specify in table *i* in the Discrete Distributions worksheet.

A Custom Schedule allows a modeler to specify the exact times of arrivals at Entrances and departures at Exits rather than using some probability distribution that generates random values. From the *Other Features* menu, select *Custom Schedules*. In the table enter the times and the quantity that arrive or depart in the appropriate columns as shown in Figure 12.20. Then use Cus(*i*), where *i* represents the table number of the

Figure 12.20 *Custom Schedules* Input Table

	A	B	C	D	E	F
1	**Custom Schedules**		Return to Control Panel			Return to Other Features
2						
3						
4	A Custom Schedule can have at most 1,000 arrivals/departures					
5						
6		1			2	
7		Times ↓	Quantity arriving/departing ↓		Times ↓	Quantity arriving/departing ↓
8						
9						

custom schedule, in the fields for *Time between arrivals* and *Num. objects per arrival* in an entrance element, or in the fields *Time between departures* and *Num. Objects per departure* in an exit element.

SimQuick can model a variety of additional applications, such as manufacturing cells, assembly/disassembly processes, job shops, quality control policies, project management, and more complex inventory situations. For further information about *SimQuick* and additional examples, we recommend that you consult the book by David Hartvigsen, *SimQuick: Process Simulation with Excel*, 2nd edition, Prentice-Hall, 2004.

Although *SimQuick* is a convenient software application for learning the basic concepts of systems simulation, it is limited in its modeling capabilities. More powerful commercial software packages are available that can model and simulate virtually any situation. These include GPSS, ProcessModel, Arena, Extend, and many others. Comprehensive surveys of simulation software are published routinely in *OR/MS Today*, a publication of INFORMS (The Institute for Operations Research and Management Science). A recent survey from the August 2003 issue of the magazine can be found at www.lionhrtpub.com/orms/surveys/Simulation/Simulation.html.

CONTINUOUS SIMULATION MODELING

Many models contain variables that change continuously over time. One example would be a model of an oil refinery. The amount of oil moving between various stages of production is clearly a continuous variable. In other models, changes in variables occur gradually (though discretely) over an extended time period; however, for all intents and purposes, they may be treated as continuous. An example would be the amount of inventory at a warehouse in a production-distribution system over several years. As customer demand is fulfilled, inventory is depleted, leading to factory orders to replenish the stock. As orders are received from suppliers, the inventory increases. Over time, particularly if orders are relatively small and frequent, as we see in just-in-time environments, the inventory level can be represented by a smooth, continuous function.

Continuous variables are often called *state variables*. A continuous simulation model defines equations for relationships among state variables so that the dynamic behavior of the system over time can be studied. To simulate continuous systems, we will decompose time into small increments. The defining equations are used to determine how the state variables change during an increment of time. A specific type of continuous simulation is called **system dynamics**, which dates back to the early 1960s when it was created by Jay Forrester of MIT. System dynamics focuses on the structure and behavior of systems that are composed of interactions among variables and feedback loops. A system dynamics model usually takes the form of an influence diagram that shows the relationships and interactions among a set of variables.

To gain an understanding of system dynamics and how continuous simulation models work, let us develop a model for the cost of medical care and implement it using general Excel features. Doctors and hospitals charge more for services, citing the rising cost of research, equipment, and insurance rates. Insurance companies cite rising court awards in malpractice suits as the basis for increasing their rates. Lawyers stress the need to force professionals to provide their patients with the best care possible and use the courts as a means to enforce patient rights. The medical cost system has received focused attention from those paying for medical care and from government officials.

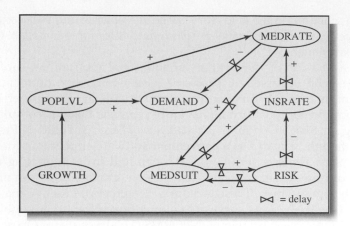

Figure 12.21
Influence Diagram for the
Cost of Medical Services

Let us suppose that we are interested in how medical rates (MEDRATE) are influenced by other factors, specifically:

1. the demand for medical service (DEMAND),
2. insurance rates (INSRATE),
3. population levels (POPLVL),
4. medical-related lawsuits (MEDSUIT), and
5. avoidance of risk by doctors (RISK).

Figure 12.21 shows an influence diagram of how these factors might relate to one another. For example, rates rise as the demand for medical service increases and as insurance rates rise. The demand is influenced by the population level and its growth rate. Also, increasing rates have a negative influence on demand, meaning that as rates rise, the demand will decrease. Insurance rates increase as medical lawsuits increase and drop as doctors avoid taking risks. At the same time, lawsuits increase as medical rates increase but also decline with risk avoidance. Some of these influences do not occur immediately, as noted by the "delay" factors in the figure. It might take about one year before some variables actually influence others.

We may express these relationships quantitatively through a set of equations that describes how each variable changes from one year to the next (that is, year $t - 1$ to year t). At time $t = 0$, we index all variables to 1.0. We will assume that the population level grows each year by a value, GROWTH(t), that is normally distributed with a mean of 0.05 and a standard deviation of 0.03. This is expressed by the equation:

$$POPLVL(t) = POPLVL(t - 1) + GROWTH(t)$$

The demand for medical services increases with the population and decreases with the rate of increase in the cost of medical service, lagged by one year. Thus, demand is computed by the formula:

$$DEMAND(t) = POPLVL(t) - [MEDRATE(t - 1) - MEDRATE(t - 2)]$$

The cost of medical services increases with the change in population level and a portion (80%) of the increase in insurance rates, lagged by one year:

$$MEDRATE(t) = MEDRATE(t - 1) + POPLVL(t) - POPLVL(t - 1)$$
$$+ .8 \times [INSRATE(t - 1) - INSRATE(t - 2)]$$

Insurance rates increase by a fraction (10%) of the previous year's level of lawsuits and decrease with any increases in doctors' adoption of safer practices to avoid risk:

$$INSRATE(t) = INSRATE(t - 1) + .10 \times MEDSUIT(t - 1)$$
$$- [RISK(t - 1) - RISK(t - 2)]$$

Increase in lawsuits is proportional to the increased costs of medical service and inversely proportional to risk avoidance, both lagged by one year:

$$MEDSUIT(t) = MEDSUIT(t - 1) + [MEDRATE(t - 1) - 1]/RISK(t - 1)$$

Finally, the avoidance of risk increases as a proportion (10%) of the increase in the level of lawsuits, based on the previous year:

$$RISK(t) = RISK(t - 1) + .10[MEDSUIT(t - 1) - 1]$$

Figure 12.22 shows a portion (first ten time periods) spreadsheet model for simulating this system (Excel file *Continuous Simulation Model*). Population growth for each year is modeled using the *Crystal Ball* function CB.Normal(0.05,0.03). (Make sure that *Crystal Ball* is loaded to have access to this function in Excel.) The remainder of the simulation model is deterministic because none of the other variables are assumed to be uncertain. Figure 12.23 shows a graph of each of the variables over the 30-year period of the simulation. Based on our assumptions, the population has increased by almost 350%. However, the demand for medical services has not quite reached that level, dampened by a fivefold increase in the cost of medical services. Insurance rates have increased 5 times, and lawsuits have increased 13 times (a compounded rate of 9% per year), while risk avoidance practices have increased an average of more than 10% per year.

System dynamics has been applied to the analysis of material and information flows in logistics systems, sales and marketing problems, social organizations, ecology, and many other fields. System dynamics was quite popular among researchers and practitioners until the early 1970s. The concept was brought back to the attention of business in the 1990s by Peter Senge through his book *The Fifth Discipline* (New York: Currency, 1994), which explores the role and importance of systems thinking in modern organizations and has formed the basis for many contemporary concepts in supply chain management.

Figure 12.22   Portion of Spreadsheet for Continuous Simulation of Medical Rates

	A	B	C	D	E	F	G	H
1	Time	Population	Population	Med. Service	Medical	Insurance	Medical	Risk
2	period	growth	level	demand	rate	rate	lawsuits	avoidance
3	0		1	1	1	1	1	1
4	1	0.098	1.098	1.098	1.098	1	1	1
5	2	0.019	1.117	1.019	1.117	1.1	1.098	1
6	3	0.052	1.170	1.151	1.250	1.210	1.215	1.010
7	4	0.068	1.238	1.105	1.406	1.322	1.463	1.031
8	5	0.043	1.280	1.124	1.538	1.446	1.856	1.078
9	6	0.055	1.335	1.203	1.692	1.586	2.355	1.163
10	7	0.036	1.371	1.217	1.839	1.735	2.950	1.299
11	8	0.046	1.417	1.270	2.006	1.895	3.596	1.494
12	9	0.034	1.451	1.285	2.167	2.060	4.269	1.753
13	10	0.006	1.458	1.296	2.305	2.227	4.935	2.080

Figure 12.23 Dynamic Behavior of Variables in Medical Rate Simulation

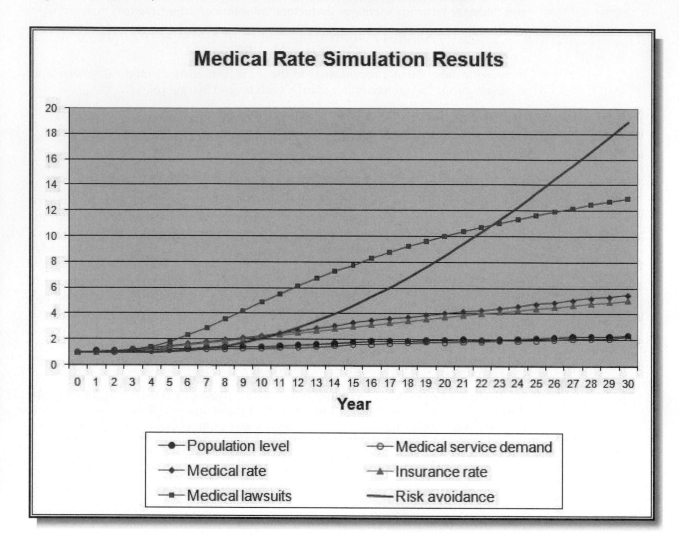

Medical Rate Simulation Results

Legend:
- Population level
- Medical service demand
- Medical rate
- Insurance rate
- Medical lawsuits
- Risk avoidance

X-axis: Year

BASIC CONCEPTS REVIEW QUESTIONS

1. What is process simulation? Explain how process simulation models differ from Monte Carlo simulation models.
2. Provide some examples of waiting lines in everyday life. What decisions must managers of such systems consider?
3. Describe the three elements that all queuing systems have in common.
4. What is a Poisson process?
5. Explain the following terms:
   a. calling population
   b. reneging
   c. jockeying
   d. balking
   e. queue discipline
6. What are the common types of queuing system configurations?
7. List and define the major types of performance measures used in queuing analysis.
8. Why do queues build up when the mean arrival rate equals the mean service rate?
9. Explain the difference between steady state and transient behavior in queuing systems.
10. What is Little's Law, and why is it important?
11. Explain the key features of *SimQuick* and how they are used in simulation models.
12. How does continuous simulation modeling differ from process simulation? Provide some practical examples different from the text where continuous simulation models might be used.

SKILL-BUILDING EXERCISES

1. Develop an Excel worksheet for computing the operating characteristics for the single-server queuing model. Apply your model to verify the calculations for the airline ticket counter example with an arrival rate of two customers per minute and a service rate of three customers per minute.

2. Use the worksheet developed in Exercise 1 to answer the following. An airport ticket counter has a service rate of 180 per hour, exponentially distributed, with Poisson arrivals at the rate of 120 per hour. Find the operating characteristics of this system.

3. Enhance the worksheet developed in Exercise 1 to include Little's Law formulas and verify the calculations.

4. Suppose that the service rate to a waiting line system is ten customers per hour (exponentially distributed). Construct an Excel worksheet to analyze how the average waiting time is expected to change as the arrival rate varies from two to ten customers per hour (exponentially distributed) in increments of 2.

5. Complete the following table for a queuing simulation.

Customer	Arrival Time	Service Time	Start Time	End Time	Waiting Time	Server Idle Time
1	1.1	0.6				
2	2.2	1.1				
3	2.9	3.3				
4	3.7	1.0				
5	4.9	0.8				
6	6.6	4.2				
7	9.4	0.8				
8	10.6	1.3				
9	11.1	0.1				
10	13.3	3.8				

6. Build an Excel model for implementing the process simulation of a single-server queue using the first three columns of data in Table 12.1 (that is, your model should calculate the start time, end time, waiting time, and server idle time). Also apply your model to answer Exercise 5.

7. Conduct a sensitivity analysis of the car wash simulation as the arrival rate and service rate vary. How sensitive are the results to these parameters?

PROBLEMS AND APPLICATIONS

1. Star Savings and Loan is planning to install a drive-through window. Transaction times at the drive-through are expected to average three minutes because customers would use it for simpler transactions. The arrival rate is expected to be ten customers per hour.
   a. Compute the operating characteristics for the drive-through window queuing system.
   b. The demand for drive-through services is expected to increase over the next several years. If you were the bank manager, how high would you let the average arrival rate increase before considering adding a second drive-through window?

2. Michael's Tire Company performs free tire balance and rotation for the life of any tires purchased there on a first-come, first-served basis. One mechanic handles this and can usually complete the job in an average of 20 minutes. Customers arrive at an average rate of two per hour for this service.
   a. How long will a customer expect to wait before being served?
   b. How long will a customer expect to spend at the shop?

c. How many customers will be waiting, on the average?

d. What is the probability that the mechanic will be idle?

3. A college office has one photocopier. Faculty and staff arrive according to a Poisson process at a rate of 5 per hour. Copying times average 8 minutes according to an exponential distribution. What are the operating characteristics for this system?

4. People arrive at a busy airport news kiosk at a rate that is uniformly distributed between 5 and 10 seconds. Most people buy only 1 paper, which takes an average of 6 seconds, but may be as little as 3 or as much as 9 seconds, and 20% buy 2 papers, which takes between 6 and 12 seconds to buy. Develop a *SimQuick* model of this process and interpret the results.

5. People arrive at a self-service cafeteria on average every 15 seconds, exponentially distributed. There are two counters, entrées and drinks. Thirty percent of customers want entrées only, 60% go first to entrées then to drinks, and the remaining get only a drink. Service time at each counter is exponential with a mean of 15 seconds. Develop a *SimQuick* model of this system and interpret the results.

6. A machine shop has a large number of machines that fail regularly. One repairperson is available to fix them. Each machine fails on average every three hours, with time between failures being exponential. Repair time (in minutes) has the distribution:

TIME	PROBABILITY
15	0.1
30	0.2
45	0.3
50	0.4

Develop and run a *SimQuick* model to determine the average time that machines spend waiting for repair and the average percent of time the repairperson is busy.

7. A small factory has two serial work stations: mold/trim and assemble/package. Jobs to the factory arrive at an exponentially distributed rate of 1 every 10 hours. Time at the mold/trim station is exponential, with a mean of 7 hours. Each job proceeds next to the assemble/package station and requires an average of five hours, again exponentially distributed. Using *SimQuick*, estimate the average time in the system, average waiting time at mold/trim, and average waiting time at assembly/packaging.

8. Modify the factory simulation in Problem 7 to include two mold/trim servers (each with the original time distribution). How do your results change?

9. A gasoline station near an outlet mall has two self-serve pumps and one full-serve pump that services only handicapped customers, which is 10% of the calling population. Customers arrive at an exponential rate of 1.6 customers per minute. Service times have discrete distributions as follows:

SELF-SERVE	
Time (minutes)	Probability
2.5	0.1
3.3	0.5
4.0	0.3
5.0	0.1

FULL-SERVE	
Time (minutes)	Probability
3.0	0.2
4.2	0.3
5.3	0.4
6.0	0.06
7.0	0.04

Customers will not wait if four or more cars are waiting in either line. Simulate this system for an eight-hour period. What percentage of customers is lost? How might you improve this system?

10. A small manufacturing facility produces custom-molded plastic parts. The process consists of two sequential operations: mold/trim and assemble/package. Jobs arrive in batches at a mean rate of 5 jobs every 2 hours. For any job, the mold/trim process is exponential, with a mean of 15 minutes, and the assemble/package operation is also exponential, with a mean of 30 minutes. The mold/trim work station has room to store only four jobs awaiting processing. The assemble/package station has room to store only two waiting jobs. Jobs are transported from the first to the second work station in ten minutes. If the mold/trim work station is full when a job arrives, the job is subcontracted. If the assemble/package work station is full when a job is completed at the mold/trim operation, the mold/trim operation cannot process any more jobs until space is freed up in the assemble/package area. Develop a *SimQuick* model to simulate this process, and explain the results.

11. A bank has an inside teller operation with two windows and a drive-through. Teller 2 works both inside and is responsible for the drive-through, while Teller 1 works only inside. Cars arrive to the drive-through uniformly every 5 to 15 minutes. Inside customers arrive uniformly every 10 to 20 minutes. Transaction times at the drive-through are exponential with a mean of 3 minutes. Times for inside transactions are exponential with a mean of 4 minutes. The drive-through has a capacity for only four cars. Develop a *SimQuick* model, and run it for 600 minutes. Explain the results.

12. Prescriptions at a hospital pharmacy arrive at an average rate of 22 per hour, with exponentially distributed times between arrivals. One pharmacist can fill a prescription at a rate of 12 per hour, or one every 5 minutes, exponentially distributed. Develop a *SimQuick* model, and use it to conduct experiments to determine the optimal number of pharmacists to have.

13. A single clerk fills two types of orders. The first type arrives randomly, with a time between arrivals between 6 and 24 minutes; the second type of order has a time between arrivals of 45 to 75 minutes. It takes the clerk anywhere between 5 and 15 minutes to fill an order. Develop a *SimQuick* model to simulate this system and determine the average number of orders waiting, their average waiting time, and the percentage of time the clerk is busy.

14. Computer monitors arrive to a final inspection station with a mean time between arrivals of 5.5 minutes (exponentially distributed). Two inspectors test the units, which take an average of 10 minutes (again exponential). However, only 90% of the units pass the test; the remaining 10% are routed to an adjustment station with a single worker. Adjustment takes an average of 30 minutes, after which the units are routed back to the inspection station. Construct a *SimQuick* model of

this process and determine the average waiting time at each station, number of units in each queue, and percent use of the workers.

15. A retail store uses a reorder point inventory system for a popular MP3 player that orders 40 units whenever the reorder point of 25 is reached. The delivery time from the supplier has a mean of 5 days, with a standard deviation of 0.25 days. The time between customer demands is exponential with a mean of 2 hours. The store is open 10 hours per day, 7 days per week. Suppose that the order cost is $50 per order, and the cost of holding items in inventory is $0.40 per day. The store manager wants to minimize the total supply chain costs. Develop a *SimQuick* model for this supply chain, run the simulation for a two-month period, and experiment with the model to find an order quantity that minimizes total cost (keeping the reorder point fixed at 25).

16. An emergency room consists of four stations. Arrivals to the emergency room occur at a rate of 4 patients per hour and are assumed to follow an exponential distribution. Incoming patients are initially screened to determine their level of severity. Past data indicate that 5% of incoming patients require hospital admission and leave the emergency room. Thirty percent of incoming patients require ambulatory care, after which they are released. Twenty percent of incoming patients are sent to the X-ray unit, and the last 45% are sent to the laboratory unit. Of those going to the X-ray unit, 30% require admission to the hospital system, 10% are sent to the laboratory unit for additional testing, and 60% have no need of additional care and are thus released. Of patients entering the laboratory unit, 10% require hospitalization and 90% are released. Current facilities are capable of keeping up with average traffic, although there is some concern that the existing laboratory facilities can become a bottleneck on particularly busy nights, especially as the community grows. The system is shown in Figure 12.24 and consists of the following activities, durations, and flows:

ACTIVITY	DURATION	ROUTING
Arrivals	4/hour, exponential	Initial desk
Front desk	0.05 hour (constant)	0.30 ambulatory
		0.20 X-ray
		0.45 lab
		0.05 hospital
Ambulatory care	Normal (0.25 hour, 0.1)	Released
X-ray	Normal (0.25 hour, 0.05)	0.1 lab
		0.3 hospital
		0.6 released
Laboratory testing	Normal (0.5 hour, 0.1)	0.1 hospital
		0.9 released

Develop a *SimQuick* model for this system to evaluate performance at each of the stations.

17. Consider the continuous system dynamics model of medical rates in this chapter. A proposal has been made to improve the system by limiting medical rate and/or insurance rate increases to a maximum of 5% per year.

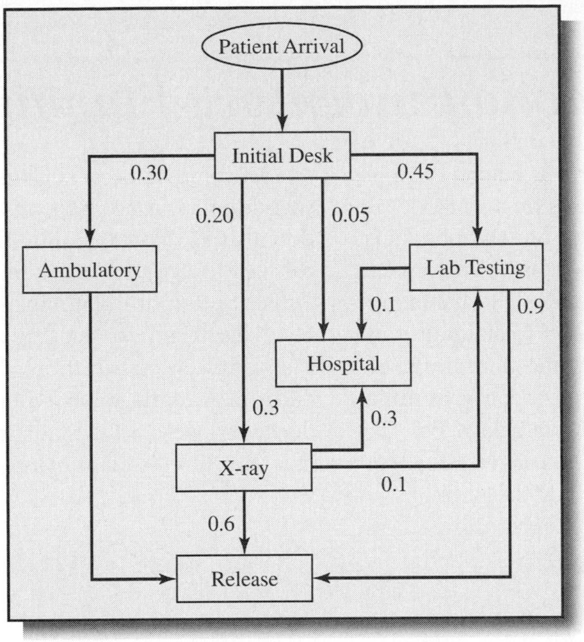

Figure 12.24 Emergency Room System

Modify the spreadsheet to simulate each of the following scenarios and discuss the results:

a. limit medical rate increases to 5% per year only

b. limit insurance rate increases to 5% per year only

c. limit both medical rate and insurance rate increases to 5% per year

18. The "cobweb" model in economics assumes that the quantity demanded of a particular product in a specified time period depends on the price in that period. The quantity supplied depends on the price in the preceding period. Also, the market is assumed to be cleared at the end of each period. These assumptions can be expressed in the following equations:

$$S(t) = c + dP(t - 1) + v(t)$$

$$D(t) = a - bP(t) + w(t)$$

$$P(t) = \frac{a - c - dP(t - 1) - v(t) + w(t)}{b} + u(t)$$

where $P(t)$ is the price in period t, $D(t)$ is the demand in period t, and $S(t)$ is the quantity supplied in period t. The variables $u(t)$, $v(t)$, and $w(t)$ are random variables with mean zero and some variance.

a. Draw an influence diagram.

b. Suppose that $a = 10,000$, $b = 2$, $c = 0$, $d = 0.1$, $u(t)$ is normal with variance 1, $v(t)$ is normal with variance 0.5, $w(t)$ is normal with variance 0.2, and $P(0) = 4,738$. Simulate this model for 50 time periods. (Note that prices are not allowed to be less than zero.)

c. Examine the effect on $P(t)$ of increasing the variance of $v(t)$ from 0.5 to 10 and from 10 to 15.

d. Compare the results from part (b) with the assumptions that u, v, and w are fixed at zero.

Production/Inventory Planning

A manufacturing plant supplies various engine components to manufacturers of motorcycles on a just-in-time basis. Planned production capacity for one component is 100 units per shift. Because of fluctuations in customers' assembly operations, demand fluctuates and is historically between 80 and 130 units per day. To maintain sufficient inventory to meet its just-in-time commitments, Tracway's management is considering a policy to run a second shift if inventory falls to 50 or below. For the annual budget planning process, managers need to know how many additional shifts will be needed.

The fundamental equation that governs this process each day is:

$$\text{Ending inventory} = \text{Beginning inventory} + \text{Production} - \text{Demand}$$

Develop an Excel spreadsheet model to simulate 260 working days (one year), and count the number of additional shifts that are required. Use this variable as a forecast cell for a *Crystal Ball* simulation of the model to find a distribution of the number of shifts required over the next year. Then use the *Crystal Ball Bootstrap* tool to find the sampling distribution of the maximum number of additional shifts required, and compute a 95% confidence interval for this statistic. Summarize your findings in a report to the plant manager.

Appendix

SimQuick *Reference Manual*

This appendix contains a concise description of all the features of *SimQuick* for reference.[2] The key methods for using *SimQuick* for modeling processes are introduced in the chapter.

Installing and Running *SimQuick*
Please refer to the *SimQuick* Note in the chapter.

SimQuick **Statistical Distributions**
A key feature of simulation software is a way to choose numbers randomly from statistical distributions. *SimQuick* provides the following formulas for doing this.

- Nor(*m*, *s*)—Numbers are chosen from a normal distribution with mean = *m*, standard deviation = *s*.

- Uni(a, b)—Numbers are chosen from a uniform distribution between *a* and *b*, with *a* < *b*.

- Exp(*m*)—Numbers are chosen from an exponential distribution with mean *m*.

- Dis(*i*)—Numbers are chosen randomly from a discrete distribution described in table *i* of the worksheet *Discrete Distributions*. (Click on *Other Features* then *Discrete Distributions* to see the tables.) The numbers in the first column of the table are chosen according to the corresponding probabilities in the second column. The numbers in the second column must be between 0 and 100 and must sum to 100.

The numbers used in the above distributions (*m*, *s*, *a*, and *b*) must lie between 10^{-100} and 10^{100}. *SimQuick* is case sensitive; hence, the three-letter abbreviations for the distributions must start with a capital letter.

These distributions are used for describing the working times at Work Stations, the arrival/departure schedules at Entrances/Exits, and the initial number of objects at Buffers. *SimQuick* also allows the arrival/departure schedules to be described with Custom Schedules. With this feature, schedules described by virtually any fixed numbers can be used.

Rules for All Elements
For each element in your model, you must fill in a table that describes the characteristics of that element.

---

[2] Adapted from David Hartvigsen, *SimQuick: Process Simulation with Excel*, 2nd ed. (Upper Saddle River, NJ: Prentice Hall, 2004).

SimQuick is case sensitive; hence, whenever a name for an element is entered into two or more places, it must be entered with the same upper- and lowercase letters. Each type of element has its own type of table. Blank tables can be found by clicking on the buttons on the Control Panel with the element names. When filling in tables, be sure to fill in table 1, then table 2, and so on, without leaving any blank tables between two filled-in tables. Additional tables may be added (see *Additional Model Limits* towards the end of this appendix).

Entrances and Exits

Entrances are where objects enter the model, and Exits are where objects leave the model. If your model has several types of objects entering (perhaps they follow different paths through a job shop or are combined on an assembly line), then you should have a separate Entrance for each type. The same holds true if your model has several types of objects that leave (perhaps your model produces several different finished goods).

When filling in a table for an Entrance or an Exit, you must fill in the following cells:

- *Name:* This must be distinct from the names of other elements.
- *Time between arrivals (departures)* and *Num. objects per arrival (departure):* These two cells define the arrival/departure schedule for the objects that arrive/depart at this Entrance/Exit. The choices for both are any one of the four *SimQuick* distributions or a number between 10^{-100} and 10^{100}.

An additional choice is to write the word Cus(*i*) into both cells, where *i* is the number of a Custom Schedule table. See the example in the chapter for the details.

When filling in a table for an Entrance, you must also fill in the following:

- At least one *Output destination* (in the top row): This is a list of the elements to which objects at this Entrance are sent. Don't leave any blanks in the middle of the list.

If objects arrive at an Entrance during the simulation, then as many as will fit enter the model. If there is not enough space for all of them, then those not entering are rejected from the model (they cannot enter at a later time).

If objects are scheduled to depart from an Exit during a simulation, then as many as are available at that time at any of the inputs to the Exit will depart up to the amount specified in *Num. objects per departure.*

The following statistics are reported on the Results sheet for Entrances:

- *Objects entering process:* This is the number of objects that arrived at the Entrance during the simulation and moved to one of its output destinations.
- *Objects unable to enter:* This is the number of objects that arrived at the Entrance during the simulation but were unable to move to one of its output destinations (perhaps a Buffer was full or a Work Station was working at the time of arrival).
- *Service level:* This is equal to

$$Objects\ entering\ process/(Objects\ entering\ process + Objects\ unable\ to\ enter)$$

The following statistics are reported on the Results sheet for Exits:

- *Objects leaving process:* This is the number of objects that leave the model at the Exit during the simulation (this typically represents filled demand).
- *Object departures missed:* This is the number of objects that could have left the model at the Exit during the simulation but did not because they were not available at the inputs to the Exit (this typically represents unfilled demand).
- *Service level:* This is equal to

$$Objects\ leaving\ process/(Objects\ leaving\ process + Objects\ departures\ missed)$$

Buffers

Buffers simply hold objects. When filling in a table for a Buffer, you must fill in the following cells:

- *Name:* This must be distinct from the names of other elements.
- *Capacity:* This is the maximum number of objects that can be held at one time. It can be either an integer between 1 and 10^{100} or the word *Unlimited* (which is actually equal to 10^{100}).
- *Initial # of objects:* This is the number of objects in the Buffer at the beginning of each simulation. It can be an integer between 0 and the capacity or one of the four *SimQuick* distributions.

The following is optional:

- *Output destination(s):* This is a list of the elements to which objects at this Buffer are sent. If you fill in a row, you must fill in the top row. Don't leave any blanks in the middle of the list.

- The *Output group size* (an integer between 1 and 10^{100}) must be filled in for each output destination. This means that the Buffer groups together a specified number of objects before outputting them as a single object (e.g., if objects are put into boxes or sent to a machine as a batch). Hence, a Buffer will not send output to an element until it contains a number of objects equal to or greater than its group size.

If an output destination is not specified for a Buffer, then objects that enter the Buffer simply remain at the Buffer. Note that all objects entering a Buffer become indistinguishable. Hence, if you have two objects that represent different products, you probably shouldn't put them into the same Buffer.

The following statistics are reported on the Results sheet for Buffers:

- *Objects leaving:* This is the number of objects that leave the Buffer and go to one of its output destinations during the simulation.

- *Final inventory:* This is the number of objects in the Buffer at the end of the simulation.

- *Minimum inventory:* This is the minimum number of objects in the Buffer during the simulation.

- *Maximum inventory:* This is the maximum number of objects in the Buffer during the simulation.

- *Mean inventory:* This is the mean number of objects in the Buffer during the simulation.

- *Mean cycle time:* This is the mean amount of time that an object spends in the Buffer during a simulation. If no objects leave the Buffer during a simulation, then the mean cycle time is "Infinite."

Work Stations

Work is performed on the objects at Work Stations. When filling in a table for a Work Station, the following cells must be filled in:

- *Name:* This must be distinct from the names of other elements.

- *Working time:* This is the amount of work time per cycle. The choices are any one of the four *SimQuick* distributions or a number between 0 and 10^{100}.

- *Output destination(s):* This is a list of the elements to which objects at this Work Station are sent. A Work Station must have at least one output destination. You must fill in the top row of the list; don't leave any blanks in the middle of the list. For each output destination, you must

specify the *# of output objects* (an integer between 1 and 10^{100}). (Use this, for example, when a Work Station breaks or disassembles an input into a number of identical outputs; typically, this number is 1.)

The following is optional:

- *Resource names:* This is a list of the resources that this Work Station needs in order to work. For each name entered, you must also enter the *# of units* of this resource that is needed (an integer between 1 and 10^{100}). Each resource used must also be entered in the table accessed by clicking on the *Other Features* and *Resource(s)* buttons.

When a Work Station finishes working on an object, it creates as many objects as designated for each output. Objects going to different outputs typically represent different things in the real process. If one type of output object is to be sent to any of several elements, then this one object type should be output to a single Buffer that then sends its outputs to these elements.

Work Stations retain output objects (in an internal buffer) until they have all been passed to subsequent elements in the model. A Work Station is *blocked* (i.e., it can perform no further work) until the created objects have all left the Work Station. Hence, for a Work Station to begin working, the following conditions must hold: The Work Station is not working; there is no finished inventory in the Work Station's internal buffer; there is one of each input object at each element providing the inputs; and, if resources are assigned to the Work Station, then the resources are available. When these conditions hold, the Work Station acquires one of each type of input object and the needed resources, determines the working time, and begins work. Resources are retained at the Work Station for the duration of the work; hence, they become unavailable to other Work Stations. When a work cycle is finished at a Work Station, its resources become available for reassignment.

Note that the requirement for a Work Station to have one of each input object before it starts working allows a Work Station to model (among other things) a machine whose job is to combine or assemble two or more objects.

The following statistics are reported on the Results sheet for Work Stations:

- *Final status:* If the Work Station is working on an object at the end of the simulation, then its final status is "Working." Otherwise, its final status is "Not Working."

- *Final inventory (int. buff):* This is the number of finished objects in the internal buffer at the end of the simulation.

- *Mean inventory (int. buff):* This is the mean number of objects in the internal buffer during the simulation.

- *Mean cycle time (int. buff):* This is the mean amount of time an object spends in the internal buffer during a simulation. It does not include time spent working on the objects.

- *Work cycles started:* This is the number of times during the simulation that the Work Station has begun working on a new set of inputs.

- *Fraction time working:* This is the fraction of the time during the simulation that the Work Station is working on an object.

- *Fraction time blocked:* This is the fraction of the time during the simulation that the Work Station is blocked.

Decision Points

Decision Points route objects to two or more (up to ten) outputs. When filling in a table for a Decision Point, the following cells must be filled in:

- *Name:* This must be distinct from the names of other elements.

- At least two *output destinations:* These are the elements to which objects at this Decision Point can be sent.

- An *output percentage* for each output destination: These numbers must be between 0 and 100 and must add up to 100.

When an object enters a Decision Point, it is randomly sent to one of the output destinations based on the percentages. A Decision Point requires zero time and has a capacity of one object. Hence, a Decision Point can be *blocked* (i.e., it will not route any additional objects) if the unit it holds cannot go to its output destination.

The following statistics are reported on the Results sheet for Decision Points:

- *Objects leaving:* This is the number of objects that leave the Decision Point and go to one of its output destinations during the simulation.

- *Final inventory (int. buff):* This is the number of objects at the Decision Point (in its internal buffer) at the end of the simulation. The number can be 0 or 1.

- *Mean inventory (int. buff):* This is the mean number of objects in the internal buffer during a simulation.

- *Mean cycle time (int. buff):* This is the mean amount of time an object spends in the internal buffer during a simulation.

Simulation Control Parameters

To run a simulation, you need to specify two numbers on the Control Panel. You must enter the *Number of time units per simulation* (a number between 10^{-100} and 10^{100}). This determines how long each simulation runs. A good strategy is to first choose what time units in the model represent in the real world so the arrival/departure schedules and the working times are relatively small numbers compared with the *Number of time units per simulation*. You also need to specify the *Number of simulations* (an integer between 1 and 1,000). Because the numbers used in the simulation are generated randomly, you can get different results each time you run the simulations. Thus, you may want to run the simulations more than once and perform some statistical analysis of the results.

To Perform the Simulation(s)

Click the *Run Simulation(s)* button.

Running Time

SimQuick can take a few minutes or longer to run if the number of time units is large, the number of simulations is large, and the number of elements is large. Running times will also depend on the speed of your computer. *SimQuick* updates the user on its progress in the lower right region of the Control Panel (and beeps when it finishes successfully). If *SimQuick* is not finished running after 30 seconds, a window appears and informs the user of *SimQuick's* progress. The user is given the option of stopping *SimQuick* or letting it continue. If the user elects to have *SimQuick* continue, then the user can specify a length of time. If *SimQuick* is not finished by this time, then this window reappears. *SimQuick* can be stopped while running at any time by hitting the Esc key.

Resources

Each resource that is used must be listed in the resources table, which can be accessed by clicking on the *Other Features* button on the Control Panel, followed by the *Resources* button. It is also necessary to enter the number (or amount) of each resource that is available (an integer between 1 and 10^{100}). Resources are assigned to Work Stations in the corresponding tables (the number needed must also be specified: an integer between 1 and 10^{100}). Resources are useful for modeling situations where one person is operating several machines, where a machine has several setup configurations, and so on. The *Mean number in use* of each resource during each simulation is provided on the Results sheet. This can be used to assess the *utilization* of resources.

Custom Schedules

For each Entrance or Exit, arrival/departure schedules may be explicitly defined as Custom Schedules. When using this option for an Entrance or Exit, the word Cus(i) must appear in the cells labeled "Time between arrivals/departures" and "# objects per arrival/departure," where i is the number of a table on the worksheet obtained by clicking *Other Features* followed by *Custom Schedules.* In these tables "Times" are the specific times during the simulation when objects arrive/depart at the corresponding element. They should be numbered between 0 and the *Number of time units per simulation,* and should increase as you move down in the table. Each time must be accompanied by a "Quantity arriving/departing," which represents the number of objects arriving/departing at that time. These should be integers between 1 and 10^{100}. A Custom Schedule can contain up to 1,000 rows.

Viewing the Model

Click the *View Model* button to see copies of all the simulation control and element tables on a single worksheet. This sheet can be used to check the logic of the model and for printing the model. The model cannot be edited from this sheet.

Clearing the Model

Clicking on the *Clear Model* button on the Control Panel clears all the *SimQuick* tables and the simulation control parameters.

Results

Results of any simulation run can be obtained by clicking on the *View Results* button on the Control Panel. The first two columns contain the element types and names in the *SimQuick* model. In the third column, the types of statistics collected during the simulations appear.

Slightly different statistics are collected for each type of element. The statistics collected for each simulation are in the columns labeled 1, 2, 3 (assuming the "Show Results Details" option has been chosen under "Other Features"; see *Results Details* in the next column). The final location of every object that arrived at the model (or started at a Buffer) can be determined, as well as how many objects passed through each element. Other summary statistics are also available, including statistics for Resources. Each number in the column labeled "Overall means" is the mean of the numbers to its right.

Results Details

Click on *Other Features* and look under *Results Details.* There are two choices: *Show Results Details* and *Hide Results Details.* If the *Show* option is chosen (and *Number of simulations* is less than or equal to 200), then the Results worksheet displays the statistics for each individual simulation. If the *Hide* option is chosen, the Results page displays only the overall means of the statistics. If the *Number of simulations* is greater than 200, then the *Hide* option must be chosen. In general, *SimQuick* will run a bit faster if the *Hide* option is chosen.

Additional Model Limits

At most, a model can have 250 elements, 10 outputs from an element, and 20 resources. (There is no additional restriction on the number of inputs to an element.) Initially, *SimQuick* provides 20 tables for each type of element. To add more elements of a certain type, simply paste copies of the given tables to the right.

How Each Simulation Is Performed

When a *SimQuick* simulation begins, a simulation clock starts in the computer and runs for the designated duration of the simulation. While this clock is running, a series of *events* sequentially takes place. There are three types of events in *SimQuick:* the arrival of a shipment of objects at an Entrance, the departure of a shipment of objects from an Exit, and the completion of work at a Work Station. Whenever an event occurs, the elements are scanned a number of times in the order of their *priority:* entrances before Work Stations before Buffers before Decision Points before Exits and, subject to this, in increasing order of table numbers.

A special purpose of the first scan is to consider the resources: If a Work Station has just finished working, then its resources become available and its newly finished objects become available to subsequent elements. Resources are not reassigned until the second scan (hence, a high-priority Work Station may reacquire its resources immediately after completing work). After the first scan, elements (except Entrances) attempt to "pull in" objects from their input elements. If a Buffer, Decision Point, or Exit has more than one input element, then it will attempt to pull in objects from its input elements in priority order.

Chapter **13**

Linear Optimization

INTRODUCTION

*T*hroughout this book, we have explored the role of data and analysis tools in managerial decisions. While many decisions involve only a limited number of alternatives and can be addressed using statistical analysis, simple decision models, or simulation, others have a very large or even an infinite number of possibilities. We introduced **optimization**—the process of selecting values of decision variables that *minimize* or *maximize* some quantity of interest—in Chapter 9. Recall that we developed a simple optimization model for setting the best price to maximize revenue:

$$Maximize\ Revenue = -2.794 \times Price^2 + 3149 \times Price$$

The quantity we seek to minimize or maximize is called the **objective function,** and any set of decision variable values that maximize or minimize the objective function is called an **optimal solution.** We showed that we could estimate the optimal price rather easily by developing a simple data table in a spreadsheet that evaluates this function for various values of the decision variable, price, or use Excel's *Solver* to find the best price exactly.

Optimization models have been used extensively in operations and supply chains, finance, marketing, and other disciplines for over 50 years to help managers allocate resources efficiently and make more cost-effective or profitable decisions. Optimization is a very broad and complex topic; in this chapter we introduce you to the most common class of optimization models—linear optimization models. In the next chapter, we will discuss more complex types of optimization models, called integer and nonlinear optimization models. The key concepts we will discuss are:

➤ *Developing mathematical models for linear optimization problems*

➤ *Implementing and solving linear optimization models on spreadsheets using Excel* Solver

➤ *Interpreting* Solver *solution reports, with an emphasis on understanding and using the managerial information contained in the output*

BUILDING LINEAR OPTIMIZATION MODELS

Most practical optimization problems consist of many decision variables and numerous **constraints**—limitations or requirements that decision variables must satisfy. The presence of constraints along with a large number of variables usually makes identifying an optimal solution considerably more difficult, and necessitates the use of powerful software tools.

Some examples of constraints are:

* The amount of material used to produce a set of products cannot exceed the available amount of 850 square feet.
* The amount of money spent on research and development projects cannot exceed the assigned budget of $300,000.
* Contractual requirements specify that at least 500 units of product must be produced.
* A mixture of fertilizer must contain exactly 30% nitrogen.
* We cannot produce a negative amount of product (this is called a *nonnegativity* constraint).

The essence of building optimization models is to translate constraints into mathematical expressions. Constraints are generally expressed mathematically as inequalities or equations. Note that the phrase "cannot exceed" specifies a "≤" inequality, "at least" specifies a "≥" inequality, and "must contain exactly" specifies an "=" relationship. All constraints in optimization models must be one of these three forms. Thus, for the examples previously provided, we would write:

- Amount of material used ≤ 850 square feet
- Amount spent on research and development ≤ $300,000
- Number of units of product produced ≥ 500
- Amount of nitrogen in mixture/total amount in mixture = 0.30
- Amount of product produced ≥ 0

The left-hand side of each of these expressions is called a **constraint function**. A constraint function is a function of the decision variables in the problem. For example, suppose that in the first case, we are producing three products. Further assume that the material requirements of these three products are 3.0, 3.5, and 2.3 square feet per unit, respectively. If A, B, and C represent the number of units of each product to produce, then $3.0A$ represents the amount of material used to produce A units of product A, $3.5B$ represents the amount of material used to produce B units of product B, and $2.3C$ represents the amount of material used to produce C units of product C. Note that dimensions of these terms are (square feet/unit)(units) = square feet. Hence, "amount of material used" can be expressed mathematically as the constraint function $3.0A + 3.5B + 2.3C$. Therefore, the constraint that limits the amount of material that can be used is written as:

$$3.0A + 3.5B + 2.3C \leq 850$$

As another example, if two ingredients contain 20% and 33% nitrogen, respectively, then the fraction of nitrogen in a mixture of x pounds of the first ingredient and y pounds of the second ingredient is expressed by the constraint function:

$$(0.20x + 0.33y)/(x + y)$$

If the fraction of nitrogen in the mixture must be 0.30, then we would have:

$$(0.20x + 0.33y)/(x + y) = 0.3$$

This can be rewritten as:

$$(0.20x + 0.33y) = 0.3(x + y)$$

and simplified as:

$$-0.1x + 0.03y = 0$$

Developing an optimization model consists of four basic steps:

1. Define the decision variables.
2. Identify the objective function.
3. Identify all appropriate constraints.
4. Write the objective function and constraints as mathematical expressions.

To see this process in action, let us examine a typical decision scenario:

Sklenka Ski Company (SSC) is a small manufacturer of two types of popular all-terrain snow skis, the Jordanelle and Deercrest models. The manufacturing process consists of two principal departments: fabrication and finishing. The fabrication department has 12 skilled workers, each of whom works 7 hours per day. The finishing department has three workers, who also work a seven-hour shift. Each pair of Jordanelle skis requires 3.5

labor hours in the fabricating department and one labor hour in finishing. The Deercrest model requires four labor hours in fabricating and 1.5 labor hours in finishing. The company operates five days per week. SSC makes a net profit of $50 on the Jordanelle model, and $65 on the Deercrest model. In anticipation of the next ski sale season, SSC must plan its production of these two models. Because of the popularity of its products and limited production capacity, its products are in high demand and SSC can sell all it can produce each season. The company anticipates selling at least twice as many Deercrest models as Jordanelle models. The company wants to determine how many of each model should be produced on a daily basis to maximize net profit.

Step 1: ***Define the decision variables.*** SSC wishes to determine how many of each model ski to produce. Thus, we may define

Jordanelle = number of pairs of Jordanelle skis produced/day

Deercrest = number of pairs of Deercrest skis produced/day

We usually represent decision variables by short, descriptive names, abbreviations, or subscripted letters such as X_1 and X_2. For many mathematical formulations involving many variables, subscripted letters are often more convenient; however, in spreadsheet models we recommend using more descriptive names to make the models and solutions easier to understand. Also, it is very important to clearly specify the dimensions of the variables, for example "pairs/day" rather than simply "Jordanelle skis."

Step 2: ***Identify the objective function.*** The problem states that SSC wishes to maximize profit. In some problems, the objective is not explicitly stated, and you must use logic and business experience to identify the appropriate objective.

Step 3: ***Identify the constraints.*** From the information provided, we see that labor hours are limited in both the fabrication department and finishing department. Therefore, we have the constraints:

Fabrication: Total labor used in fabrication cannot exceed
the amount of labor available.

Finishing: Total labor used in finishing cannot exceed
the amount of labor available.

In addition, the problem notes that the company anticipates selling at least twice as many Deercrest models as Jordanelle models. Thus, we need a constraint that states:

Number of pairs of Deercrest skis must be at least twice
the number of Jordanelle skis.

Finally, we must ensure that negative values of the decision variables cannot occur. Nonnegativity constraints are assumed in nearly all optimization models.

Step 4: ***Write the objective function and constraints as mathematical expressions.*** Because SSC makes a net profit of $50 on the Jordanelle model, and $65 on the Deercrest model, the objective function is:

Maximize *Total Profit* = 50 *Jordanelle* + 65 *Deercrest*

For the constraints, we will use the approach described earlier in this chapter. First, consider the fabrication and finishing constraints. Write these as:

Fabrication: Total labor used in fabrication
$\leq$ the amount of labor available

Finishing: Total labor used in finishing
$\leq$ the amount of labor available

Now translate both the constraint functions on the left and the limitations on the right into mathematical or numerical terms. Note that the amount of labor available in fabrication is (12 workers) (7 hours/day) = 84 hours/day, while in finishing we have (3 workers)(7 hours/day) = 21 hours/day. Because each pair of Jordanelle skis requires 3.5 labor hours and Deercrest skis require 4 labor hours in the fabricating department, the total labor used in fabrication is 3.5 *Jordanelle* + 4 *Deercrest*. Note that the dimensions of these terms are (hours/pair of skis)(number of skis produced) = hours. Similarly, for the finishing department, the total labor used is 1 *Jordanelle* + 1.5 *Deercrest*. Therefore, the appropriate constraints are:

Fabrication: 3.5 *Jordanelle* + 4 *Deercrest* $\leq$ 84

Finishing: 1 *Jordanelle* + 1.5 *Deercrest* $\leq$ 21

For the market mixture constraint, "Number of pairs of Deercrest skis must be at least twice the number of Jordanelle skis," we have:

Deercrest $\geq$ 2 *Jordanelle*

It is customary to write all the variables on the left hand side of the constraint. Thus, an alternative expression for this constraint is:

Deercrest − 2 *Jordanelle* $\geq$ 0

The difference between the number of Deercrest skis and twice the number of Jordanelle skis can be thought of as the excess number of Deercrest skis produced over the minimum market mixture requirement. Finally, nonnegativity constraints are written as:

Deercrest $\geq$ 0

Jordanelle $\geq$ 0

The complete optimization model is:

Maximize *Total Profit* = 50 *Jordanelle* + 65 *Deercrest*

3.5 *Jordanelle* + 4 *Deercrest* $\leq$ 84

1 *Jordanelle* + 1.5 *Deercrest* $\leq$ 21

Deercrest − 2 *Jordanelle* $\geq$ 0

Deercrest $\geq$ 0

Jordanelle $\geq$ 0

Characteristics of Linear Optimization Models

A linear optimization model (often called a **linear program**, or **LP**) has two basic properties. First, the objective function and all constraints are *linear functions* of the decision variables. This means that each function is simply a sum of terms, each of which is some constant multiplied by a decision variable. The SSC model has this property. Recall the constraint example that we developed earlier for the nitrogen requirement. Notice that the constraint function on the left-hand side of the constraint:

$$(0.20x + 0.33y)/(x + y) = 0.3$$

as originally written is not linear. However, we were able to convert it to a linear form using simple algebra. This is advantageous, as special, highly efficient solution algorithms are used for linear optimization problems.

The second property of a linear optimization problem is that all variables are *continuous,* meaning that they may assume any real value (typically, nonnegative). Of course, this assumption may not be realistic for a practical business problem (you cannot produce half a refrigerator!). However, because this assumption simplifies the solution method and analysis, we often apply it in many situations where the solution would not be seriously affected. In the next chapter, we will discuss situations where it is necessary to force variables to be whole numbers (integers). For all examples and problems in this chapter, we will assume continuity of the variables.

IMPLEMENTING LINEAR OPTIMIZATION MODELS ON SPREADSHEETS

We will learn how to solve optimization models using an Excel tool called *Solver.* To facilitate the use of *Solver,* we suggest the following guidelines for designing spreadsheet models for optimization problems:

- Put the objective function coefficients, constraint coefficients, and right-hand values in a logical format in the spreadsheet. For example, you might assign the decision variables to columns and the constraints to rows, much like the mathematical formulation of the model, and input the model parameters in a matrix. If you have many more variables than constraints, it might make sense to use rows for the variables and columns for the constraints.
- Define a set of cells (either rows or columns) for the values of the decision variables. In some models, it may be necessary to define a matrix to represent the decision variables. The names of the decision variables should be listed directly above the decision variable cells. Use shading or other formatting to distinguish these cells.
- Define separate cells for the objective function and *each* constraint function (the left-hand side of a constraint). Use descriptive labels directly above these cells.

We will illustrate these principles for the Sklenka Ski example. Figure 13.1 shows a spreadsheet model for the product mix example (Excel file *Sklenka Skis*). The *Data* portion of the spreadsheet provides the objective function coefficients, constraint coefficients, and right-hand sides of the model. Such data should be kept separate from the actual model so that if any data are changed, the model will automatically be updated. In the *Model* section, the number of each product to make is given in cells B14 and C14. Also in the *Model* section are calculations for the constraint functions:

3.5 *Jordanelle* + 4 *Deercrest* (hours used in fabrication, cell D15)

1 *Jordanelle* + 1.5 *Deercrest* (hours used in finishing, cell D16)

Deercrest − 2 *Jordanelle* (market mixture, cell D19)

and the objective function, 50 *Jordanelle* + 65 *Deercrest* (cell D22).

Figure 13.1 Sklenka Skis Model Spreadsheet Implementation

	A	B	C	D
1	Sklenka Skis			
2				
3	Data			
4			Product	
5	Department	Jordanelle	Deercrest	Limitation (hours)
6	Fabrication	3.5	4	84
7	Finishing	1	1.5	21
8				
9	Profit/unit	$ 50.00	$ 65.00	
10				
11				
12	Model			
13		Jordanelle	Deercrest	
14	Quantity Produced	0	0	Hours Used
15	Fabrication	0	0	0
16	Finishing	0	0	0
17				
18				Excess Deercrest
19	Market mixture			0
20				
21				Total Profit
22	Profit Contribution	$ -	$ -	$ -

	A	B	C	D
1	Sklenka Skis			
2				
3	Data			
4			Product	
5	Department	Jordanelle	Deercrest	Limitation (hours)
6	Fabrication	3.5	4	84
7	Finishing	1	1.5	21
8				
9	Profit/unit	50	65	
10				
11				
12	Model			
13		Jordanelle	Deercrest	
14	Quantity Produced	0	0	Hours Used
15	Fabrication	=B6*$B$14	=C6*$C$14	=B15+C15
16	Finishing	=B7*$B$14	=C7*$C$14	=B16+C16
17				
18				Excess Deercrest
19	Market mixture			=C14-2*B14
20				
21				Total Profit
22	Profit Contribution	=B9*$B$14	=C9*$C$14	=B22+C22

To show the correspondence between the mathematical model and the spreadsheet model more clearly, we will write the model in terms of the spreadsheet cells:

$$\text{Maximize Profit} = D22 = B9*B14 + C9*C14$$

subject to the constraints:

$$D15 = B6*B14 + C6*C14 \leq D6 \quad \text{(fabrication)}$$
$$D16 = B7*B14 + C7*C14 \leq D7 \quad \text{(finishing)}$$
$$D19 = C14 - 2*B14 \geq 0 \quad\quad \text{(market mixture)}$$
$$B14 \geq 0, C14 \geq 0 \quad\quad \text{(nonnegativity)}$$

Therefore, if you can formulate the model on a spreadsheet, you can easily write down the mathematical model and vice versa.

In Excel, the pair-wise sum of products of terms can easily be computed using the SUMPRODUCT function. This often simplifies the model-building process, particularly when many variables are involved. For example, the objective function formula could have been written as:

$$B9*B14 + C9*C14 = \text{SUMPRODUCT}(B9:C9,B14:C14)$$

Similarly, the labor limitation constraints could have been expressed as:

$$B6*B14 + C6*C14 = \text{SUMPRODUCT}(B6:C6,B14:C14)$$
$$B7*B14 + C7*C14 = \text{SUMPRODUCT}(B7:C7,B14:C14)$$

Excel Functions to Avoid in Modeling Linear Programs

Several common functions in Excel can cause difficulties when attempting to solve linear programs using *Solver* because they are discontinuous at some point and introduce nonlinearities into what might appear to be a linear model. For instance, in the formula

IF(A12 < 45, 0, 1), the cell value jumps from 0 to 1 when the value of cell A12 crosses 45. In such situations, the correct solution may not be identified. Common Excel functions to avoid are ABS, MIN, MAX, INT, ROUND, IF, and COUNT. While these are useful in general modeling tasks with spreadsheets, you should avoid them in linear optimization models.

SOLVING LINEAR OPTIMIZATION MODELS

To solve an optimization problem, we seek values of the decision variables that maximize or minimize the objective function and also satisfy all constraints. Any solution that satisfies all constraints of a problem is called a **feasible solution**. Finding an optimal solution among the infinite number of possible feasible solutions to a given problem is not an easy task. A simple approach is to try to manipulate the decision variables in the spreadsheet models to find the best solution possible; however, for many problems, it might be very difficult to find a feasible solution, let alone an optimal solution. You might wish to try to find the best solution you can for the Sklenka Ski problem by using the spreadsheet model. With a little experimentation and perhaps a bit of luck, you might be able to zero in on the optimal solution or something close to it. However, to guarantee finding an optimal solution, some type of systematic mathematical solution procedure is necessary. Fortunately, such a procedure is provided by the Excel *Solver* tool, which we discuss next.

Solver is an add-in packaged with Excel that was developed by Frontline Systems Inc. (www.solver.com) and can be used to solve many different types of optimization problems. The educational version of *Premium Solver* that accompanies this book is an improved alternative to the standard Excel-supplied *Solver*. Both are limited to 200 decision variables and 100 constraints (not including simple bounds on the variables); however, *Premium Solver for Education* includes better functionality, numerical accuracy, reporting, and user interface. See *Excel Note: Using Standard vs. Premium Solver* for further information about the differences between these two versions. We highly recommend using the premium version and we will use it in the remainder of this chapter. In *Premium Solver,* decision variables are called *changing cells;* and the objective function cell is called the *target cell* or *set cell. Solver* identifies values of the changing cells that minimize or maximize the target cell value. Constraints are entered by referencing constraint functions and data in spreadsheet cells using a special dialog box.

EXCEL NOTE
Using Standard vs. *Premium Solver*

In Excel 2007, *Solver* is launched by selecting *Solver* from the *Analysis* menu in the *Data* tab; after *Premium Solver* is installed, it may be launched from the *Menu Commands* menu under the *Add-ins* tab. The *Solver Parameters* dialog box shown in Figure 13.2 will then be displayed. Figure 13.3 shows the *Premium Solver* dialog. The principal difference with the *Premium Solver Parameters* dialog is the drop-down box in the middle of the right side. This allows you to select the proper

solution procedure for the type of optimization problem you wish to solve. Three options are available:

1. *Standard GRG Nonlinear*—used for solving nonlinear optimization problems
2. *Standard LP Simplex*—used for solving linear and linear integer optimization problems
3. *Standard Evolutionary*—used for solving complex nonlinear and nonlinear integer problems

Figure 13.2 Standard *Solver* Dialog Box

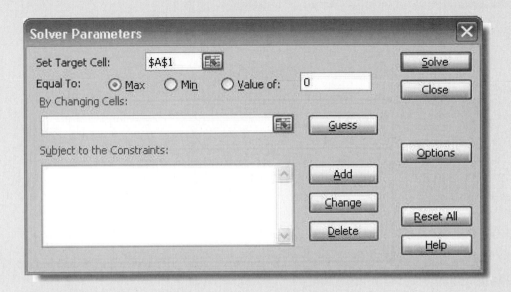

Figure 13.3 *Premium Solver* Dialog Box (Standard LP Simplex chosen)

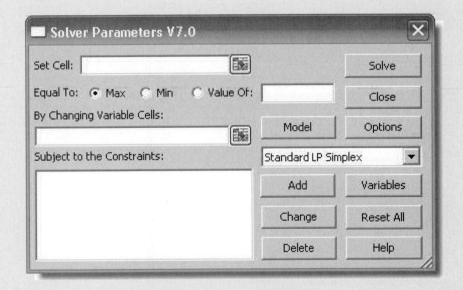

For linear optimization models, always choose *Standard Simplex LP*; the other options will be discussed in Chapter 14.

The other important difference between the standard and premium versions is in the *Options* dialog. Figure 13.4 shows the *Solver Options* dialog for standard *Solver*, and Figure 13.5 shows it for *Premium Solver*. In both versions, be sure to check the *Assume Non-Negative* box to guarantee nonnegativity of the variables. The standard version also has a check box entitled *Assume Linear Model*. This should be checked if you are solving a linear or an integer linear model. In *Premium Solver*, the choice of the solution procedure automatically incorporates the correct assumption, so this check box need not be considered and thus does not appear as an option. We suggest you consult the User Manual that accompanies the *Premium Solver for Education* installation.

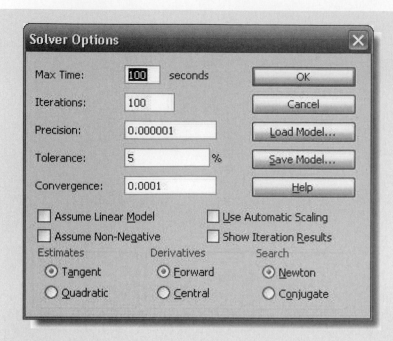

Figure 13.4 Standard *Solver* Options

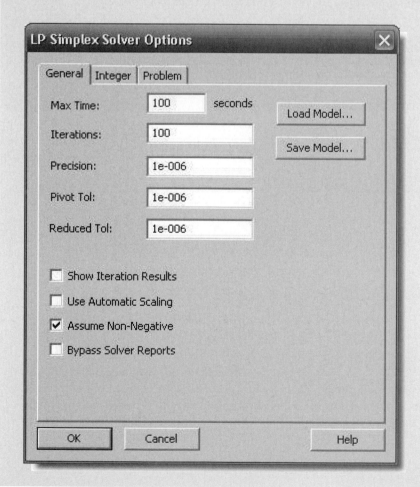

Figure 13.5 *Premium Solver* Options

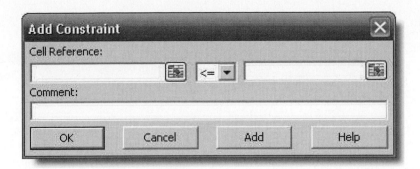

Figure 13.6
Add Constraint
Dialog Box

Solving the SSC Model

We will now describe how to solve the SSC model in Figure 13.1 using *Premium Solver*. Start by inputting the objective function cell in the *Set Cell* field by either typing in D22 or clicking on cell D22 in the spreadsheet. Next, select the type of optimization objective option (*Max* or *Min*) by choosing the appropriate radio button; *Max* is the default. Finally, define the *Changing Variable Cells* range by either entering B14:C14 in the *By Changing Variable Cells* field, or by clicking in the field and then highlighting the cell range in the spreadsheet.

To add constraints, click the *Add* button. The *Add Constraint* dialog box (Figure 13.6) will appear. *Cell Reference:* refers to the cell defining the constraint function on the left-hand side of a constraint. The drop-down menu in the center of the dialog box allows you to choose the type of constraint: $\leq$, $\geq$, $=$. The other options are discussed in the next chapter. Then input the spreadsheet cell containing the right-hand side of the constraint value in the field on the right.

You may enter one constraint at a time or define a group of constraints that all have the same algebraic form (either all $\leq$, all $\geq$, or all $=$). For example, the department resource limitation constraints are expressed within the spreadsheet model as:

$$D15 \leq D6$$
$$D16 \leq D7$$

For the first constraint, we enter (or highlight) D15 in the *Cell Reference* field, choose $\leq$ from the drop-down list, and enter (or highlight) D6 in the field on the right. Then click *Add* to add the next constraint. Because both constraints are $\leq$ types, we could have defined them as a group by entering the range D15:D16 in the *Cell Reference* field and D6:D7 in the right-hand field to simplify the input process. Finally, we add the market mixture constraint, D19 $\geq$ 0. When all constraints are added, click *OK* to return to the *Solver Parameters* dialog box. The constraints will be displayed in the *Solver Parameters* dialog as shown in Figure 13.7 You may add, change, or delete these as necessary by clicking the appropriate buttons.

For linear models, be sure to choose "Standard LP Simplex" in the drop-down box! Also, be sure to check "Assume Non-Negative" in the *Options* dialog. Unless you are an advanced user, you may generally leave the other options at their default values.

To solve the model, click the *Solve* button. The *Solver Results* dialog box will appear, as shown in Figure 13.8, with the message "Solver found a solution." If a

Figure 13.7 Final *Solver* Model

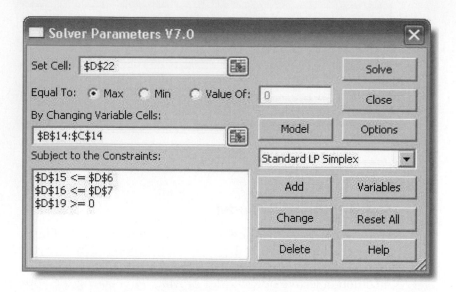

Figure 13.8 *Solver Results* Dialog Box

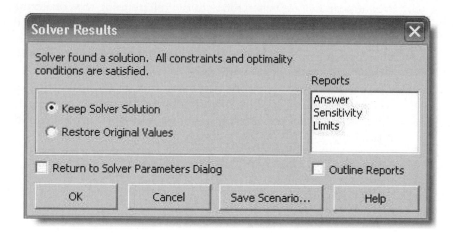

solution could not be found, *Solver* will notify you with a message to this effect. This generally means that you have an error in your model or you have included conflicting constraints that no solution can satisfy. In such cases, you will need to reexamine your model.

Solver generates three reports as shown in Figure 13.8: Answer, Sensitivity, and Limits. To add them to your Excel workbook, hold the Ctrl key down, click on the ones you want, and then click *OK*. The optimal solution will be shown in the spreadsheet as in Figure 13.9. The maximum profit is $945, obtained by producing 5.25 pairs of Jordanelle skis and 10.5 pairs of Deercrest skis per day (remember that linear models allow fractional values for the decision variables!).

	A	B	C	D
1	**Sklenka Skis**			
2				
3	**Data**			
4		**Product**		
5	**Department**	Jordanelle	Deercrest	Limitation (hours)
6	Fabrication	3.5	4	84
7	Finishing	1	1.5	21
8				
9	Profit/unit	$ 50.00	$ 65.00	
10				
11				
12	**Model**			
13		Jordanelle	Deercrest	
14	Quantity Produced	5.25	10.5	Hours Used
15	Fabrication	18.375	42	60.375
16	Finishing	5.25	15.75	21
17				
18				Excess Deercrest
19	Market mixture			0
20				
21				Total Profit
22	Profit Contribution	$ 262.50	$ 682.50	$ 945.00

Figure 13.9 Optimal Solution to the Product Mix Model

Solver Outcomes and Solution Messages

Solving a linear optimization model can result in four possible outcomes:

1. Unique optimal solution
2. Alternate optimal solutions
3. Unboundedness
4. Infeasibilty

When a model has a *unique optimal solution,* it means that there is exactly one solution that will result in the maximum (or minimum) objective. The solution to the SSC model is unique. If a model has *alternate optimal solutions,* the objective is maximized (or minimized) by more than one combination of decision variables, all of which have the same objective function value. *Solver* does not tell you when alternate solutions exist and only reports one of the many possible alternate optimal solutions. However, you can use the sensitivity report information to identify the existence of alternate optimal solutions. When any of the Allowable Increase or Allowable Decrease values for changing cells are zero, then alternate optimal solutions exist, although *Solver* does not provide an easy way to find them.

A problem is *unbounded* if the objective can be increased or decreased without bound (i.e., to infinity or negative infinity) while the solution remains feasible. A model is unbounded if *Solver* reports "The Set Cell values do not converge." This generally indicates an incorrect model, usually when some constraint or set of constraints have been left out.

Finally, an *infeasible* model is one for which no feasible solution exists; that is, when there is no solution that satisfies all constraints together. When a problem is infeasible,

Solver will report "Solver could not find a feasible solution." Infeasible problems *can* occur in practice, for example, when a demand requirement is higher than available capacity, or when managers in different departments have conflicting requirements or limitations. In such cases, the model must be reexamined and modified. Sometimes infeasibility or unboundedness is simply a result of a misplaced decimal or other error in the model or spreadsheet implementation, so accuracy checks should be made.

Interpreting *Solver* Reports

The Answer Report (Figure 13.10) provides basic information about the solution, including the values of the optimal objective function (in the *Target Cell* section) and decision variables (in the *Adjustable Cells* section). In the *Constraints* section, *Cell Value* refers to the value of the constraint function using the optimal values of the decision variables. In other words, we used 60.375 minutes in the fabrication department and 21 minutes in the finishing department by producing 5.25 pairs of Jordanelle skis and 10.5 pairs of Deercrest skis. The *Status* column tells whether each constraint is binding or not binding. A **binding constraint** is one for which the *Cell Value* is equal to the right-hand side of the value of the constraint. In this example, the constraint for fabrication is

Figure 13.10 *Solver* Answer Report

Microsoft Excel 12.0 Answer Report				
Worksheet: [Sklenka Skisl solved.xlsx]Sheet1				
Report Created: 06/04/2008 14:36:59 PM				
Result: Solver found a solution. All constraints and optimality conditions are satisfied.				
Engine: Standard LP Simplex				
Solution Time: 00 Seconds				
Iterations: 2				
Subproblems: 0				
Incumbent Solutions: 0				

Target Cell (Max)

Cell	Name	Original Value	Final Value
$D$22 Profit Contribution Total Profit		0	945

Adjustable Cells

Cell	Name	Original Value	Final Value
$B$14 Quantity Produced Jordanelle		0	5.25
$C$14 Quantity Produced Deercrest		0	10.5

Constraints

Cell	Name	Cell Value	Formula	Status	Slack
$D$15 Fabrication Hours Used		60.375	$D$15<=$D$6	Not Binding	23.625
$D$16 Finishing Hours Used		21	$D$16<=$D$7	Binding	0
$D$19 Market mixture Excess Deercrest		0	$D$19>=0	Binding	0

not binding, while the constraints for finishing and market mixture are binding. This means that there is excess time that is not used in fabrication; this value is shown in the *Slack* column as 23.626 hours. For finishing, we used all the time available, and hence, the slack value is zero. Because we produced exactly twice the number of Deercrest skis as Jordanelle skis, the market mixture constraint is binding. It would have been not binding if we had produced more than twice the number of Deercrest skis as Jordanelle.

In general, the **slack** is the difference between the right- and left-hand sides of a constraint. Examine the fabrication constraint:

$$3.5 \, Jordanelle + 4 \, Deercrest \leq 84$$

We interpret this as:

$$\text{Number of fabrication hours used} \leq \text{Hours available}$$

Note that if the amount used is strictly less than the availability, we have slack, which represents the amount unused; thus,

Number of fabrication hours used + Number of fabrication hours unused

$$= \text{Hours available}$$

or

$$\text{Slack} = \text{Number of hours unused}$$
$$= \text{Hours Available} - \text{Number of fabrication hours used}$$
$$= 84 - (3.5 \times 5.25 + 4 \times 10.5) = 23.625$$

Slack variables are always nonnegative, so for ≥ constraints, slack represents the difference between the left-hand side of the constraint function and the right-hand side of the requirement. The slack on a binding constraint will always be zero.

The Sensitivity Report (Figure 13.11) provides a variety of useful information for managerial interpretation of the solution. Specifically, it allows us to understand how

Figure 13.11 *Solver* Sensitivity Report

	A	B	C	D	E	F	G	H
1		Microsoft Excel 12.0 Sensitivity Report						
2		Worksheet: [Sklenka Skisl solved.xlsx]Sheet1						
3		Report Created: 06/04/2008 14:36:59 PM						
4								
5		Target Cell (Max)						
6		Cell	Name	Final Value				
7		$D$22	Profit Contribution Total Profit	945				
8								
9		Adjustable Cells						
10				Final	Reduced	Objective	Allowable	Allowable
11		Cell	Name	Value	Cost	Coefficient	Increase	Decrease
12		$B$14	Quantity Produced Jordanelle	5.25	0	50	1E+30	6.666666667
13		$C$14	Quantity Produced Deercrest	10.5	0	65	10	90
14								
15		Constraints						
16				Final	Shadow	Constraint	Allowable	Allowable
17		Cell	Name	Value	Price	R.H. Side	Increase	Decrease
18		$D$15	Fabrication Hours Used	60.375	0	84	1E+30	23.625
19		$D$16	Finishing Hours Used	21	45	21	8.217391304	21
20		$D$19	Market mixture Excess Deercrest	0	-2.5	0	14	42

the optimal objective value and optimal decision variables are affected by changes in the objective function coefficients, the impact of forced changes in certain decision variables, or the impact of changes in the constraint resource limitations or requirements. In the *Adjustable Cells* section, the final value for each decision variable is given, along with its reduced cost, objective coefficient, and allowable increase and decrease. The **reduced cost** tells *how much the objective coefficient needs to be reduced in order for a nonnegative variable that is zero in the optimal solution to become positive.* If a variable is positive in the optimal solution, as it is for both variables in this example, its reduced cost is always zero. We will see an example later in this chapter that will help you to understand reduced costs.

The Allowable Increase and Allowable Decrease values tell how much an individual objective function coefficient can change before the optimal values of the decision variables will change (a value listed as "1E + 30" is interpreted as infinity). For example, if the unit profit for Deercrest skis either increases by more than 10 or decreases by more than 90, then the optimal values of the decision variables will change (as long as the other objective coefficient stays the same). For instance, if we increase the unit profit by 11 (to 76) and re-solve the model, the new optimal solution will be to produce 14 pairs of Deercrest skis and no Jordanelle skis. However, any increase less than 10 will keep the current solution optimal. For Jordanelle skis, we can increase the unit profit as much as we wish without affecting the current optimal solution; however, a decrease of at least 6.67 will cause the solution to change.

Note that if the objective coefficient of any one variable that has positive value in the current solution changes but stays within the range specified by the Allowable Increase and Allowable Decrease, the optimal decision variables will stay the same; however, the objective function value will change. For example, if the unit profit of Jordanelle skis were changed to $46 (a decrease of 4, within the allowable increase), then we are guaranteed that the optimal solution will still be to produce 5.25 pairs of Jordanelle and 10.5 pairs of Deercrest. However, each of the 5.25 pairs of Jordanelle skis produced and sold would realize $4 less profit—a total decrease of 5.25($4) = $21. Thus, the new value of the objective function would be $945 − $21 = $924. If an objective coefficient changes beyond the Allowable Increase or Allowable Decrease, then we must re-solve the problem with the new value to find the new optimal solution and profit.

The range within which the objective function coefficients will not change the optimal solution provides a manager with some confidence about the stability of the solution in the face of uncertainty. If the allowable ranges are large, then reasonable errors in estimating the coefficients will have no effect on the optimal policy (although they will affect the value of the objective function). Tight ranges suggest that more effort might be spent in ensuring that accurate data or estimates are used in the model.

The *Constraints* section of the Sensitivity Report lists the final value of the constraint function (the left-hand side), the shadow price, the original constraint right-hand side, and an Allowable Increase and Allowable Decrease. The **shadow price** tells *how much the value of the objective function will change as the right-hand side of a constraint is increased by 1.* Whenever a constraint has positive slack, the shadow price is zero. For example, in the fabrication constraint, we are not using all of the available hours in the optimal solution. Thus, having one more hour available will not help us to improve the solution. However, if a constraint is binding, then any change in the right-hand side will cause the optimal values of the decision variables as well as the objective function value to change.

Let us illustrate this with the finishing constraint. The shadow price of 45 states that if an additional hour of finishing time is available, then the total profit will change by $45. To see this, change the limitation of the number of finishing hours available to 22 and re-solve the problem. The new solution is to produce 5.5 pairs of Jordanelle and 11.0 pairs of Deercrest, yielding a profit of $990. We see that the total profit increases by $45 as predicted. Thus, the shadow price represents the economic value of having an additional unit of a resource.

The shadow price is a valid predictor of the change in the objective function value for each unit of increase in the constraint right-hand side up to the value of the Allowable Increase. Thus, if up to about 8.2 additional hours of finishing time were available, profit would increase by $45 for each additional hour (but we would have to resolve the problem to actually find the optimal values of the decision variables). Similarly, the negative of the shadow price predicts the change in the objective function value for each unit the constraint right-hand side is decreased, up to the value of the Allowable Decrease. For example, if one person was ill or injured, resulting in only 14 hours of finishing time available, then profit would decrease by $7($45) = 315. This can be predicted because a decrease of 7 hours is within the Allowable Decrease of 21. Beyond these ranges, the shadow price does not predict what will happen, and the problem must be re-solved.

Another way of understanding the shadow price is to break down the impact of a change in the right-hand side of the value. How was the extra hour of finishing time used? After solving the model with 22 hours of finishing time, we see that we were able to produce an additional 0.25 pairs of Jordanelle and 0.5 pairs of Deercrest skis as compared to the original solution. Therefore, the profit increased by $0.25($50) + 0.5(65) = $12.50 + 32.50 = 45. In essence, a change in a binding constraint causes a reallocation of how the resources are used.

Interpreting the shadow price associated with the market mixture constraint is a bit more difficult. If you examine the constraint, *Deercrest* − 2 *Jordanelle* ≥ 0, closely, an increase in the right-hand side from 0 to 1 results in a change of the constraint to:

$$(Deercrest − 1) − 2\, Jordanelle ≥ 0$$

This means that the number of pairs of Deercrest skis produced would be one short of the requirement that it be at least twice the number of Jordanelle skis. If the problem is re-solved with this constraint, we find the new optimal solution to be 4.875 Jordanelle, 10.75 Deercrest, and profit = 942.50. The profit changed by the value of the shadow price and we see that 2 × Jordanelle = 9.75, one short of the requirement.

Why are shadow prices useful to a manager? They provide guidance on how to reallocate resources or change values over which the manager may have control. In linear optimization models, the parameters of some constraints cannot be controlled. For instance, the amount of time available for production or physical limitations on machine capacities would clearly be uncontrollable. Other constraints represent policy decisions, which, in essence, are arbitrary. Although it is correct to state that having an additional hour of finishing time will improve profit by $45, does this necessarily mean that the company should spend up to this amount for additional hours? This depends on whether the relevant costs have been included in the objective function coefficients. If the cost of labor *has not* been included in the objective function unit profit coefficients, then the company will benefit by paying less than $45 for additional hours. However, if the cost of labor *has* been included in the profit calculations, the company should be willing to pay up to an *additional* $45 over and above the labor costs that have already been included in the unit profit calculations.

The Limits Report (Figure 13.12) shows the lower limit and upper limit that each variable can assume while satisfying all constraints and holding all of the other

Figure 13.12 *Solver* Limits Report

	A	B	C	D	E	F	G	H	I	J
1	**Microsoft Excel 12.0Limits Report**									
2	**Worksheet: [Sklenka Skis! solved.xlsx]Sheet1**									
3	**Report Created: 06/04/2008 14:36:59 PM**									
4										
5										
6			**Target**							
7		**Cell**	**Value**							
8		$D$22	Profit Contribution Total Profit	$945.00						
9										
10										
11			**Adjustable**			**Lower**	**Target**		**Upper**	**Target**
12		**Cell**	**Name**	**Value**		**Limit**	**Result**		**Limit**	**Result**
13		$B$14	Quantity Produced Jordanelle	5.25		0	$682.50		5.25	$945.00
14		$C$14	Quantity Produced Deercrest	10.5		10.5	$945.00		10.5	$945.00

variables constant. Generally, this report provides little useful information for decision making and can be effectively ignored.

How *Solver* Creates Names in Reports

How you design your spreadsheet model will affect on how *Solver* creates the names used in the output reports. Poor spreadsheet design can make it difficult or confusing to interpret the Answer and Sensitivity reports. Thus, it is important to understand how to do this properly.

Solver assigns names to target cells, changing cells, and constraint function cells by concatenating the text in the first cell containing text to the left of the cell with the first cell containing text above it. For example, in the SSC model, the target cell is D22. The first cell containing text to the left of D15 is "Profit Contribution" in A22, and the first cell containing text above D22 is "Total Profit" in cell D21. Concatenating these text strings yields the target cell name "Profit Contribution Total Profit," which is found in the *Solver* reports. The constraint functions are calculated in cells D15 and D16. Note that their names are "Fabrication Hours Used" and "Finishing Hours Used." Similarly, the changing cells in B14 and C14 have the names "Quantity Produced Jordanelle" and "Quantity Produced Deercrest." These names make it easy to interpret the information in the Answer and Sensitivity reports. We encourage you to examine each of the target cells, changing variable cells, and constraint function cells in your models carefully so that names are properly established.

Difficulties with *Solver*

A poorly scaled model—one in which the parameters of the objective and constraint functions differ by several orders of magnitude (as we have in the transportation example where costs are in tens and supplies/demands in thousands) may cause round-off errors in internal computations or error messages such as "The conditions for Assume Linear Model are not satisfied." This does not happen often (but may in older versions of *Solver*); if it does, you should consult the Frontline Systems' Web site for additional information. Usually, all you need to do is to keep the solution that *Solver* found and run *Solver* again starting from that solution. Experts often suggest that the values of the coefficients in the objective function and constraints, as well as the right-hand sides, should not differ from each other by a factor of more than 1,000 or 10,000. *Solver* also has a checkbox for *Use Automatic Scaling* that can be used, especially if *Solver* gives an error message that linearity is not satisfied.

APPLICATIONS OF LINEAR OPTIMIZATION

Linear optimization models are the most ubiquitous of optimization models used in organizations today. Applications abound in operations, finance, marketing, engineering, and many other disciplines. Table 13.1 summarizes some common types of generic linear optimization models. We already saw an example of a product mix model with the Sklenka Ski problem. This list represents but a very small sample of the many practical types of linear optimization models that are used.

Building optimization models is more of an art than a science, as there often are several ways of formulating a particular problem. Learning how to build optimization models requires logical thought but can be facilitated by studying examples of different models and observing their characteristics.

Table 13.1   Generic Examples of Linear Optimization Models

TYPE OF MODEL	DECISION VARIABLES	OBJECTIVE FUNCTION	TYPICAL CONSTRAINTS
Product mix	Quantities of product to produce and sell	Maximize contribution to profit	Resource limitations (e.g., production time, labor, material); minimum sales requirements; maximum sales potential
Process selection	Quantities of product to make using alternative processes	Minimize cost	Demand requirements; resource limitations
Blending	Quantity of materials to mix to produce one unit of output	Minimize cost	Specifications on acceptable mixture
Portfolio selection	Proportions to invest in different financial instruments	Maximize future return or minimize risk exposure	Limit on available funds; sector requirements or restrictions; proportional relationships on investment mix
Transportation	Amount to ship between sources of supply and destinations	Minimize total transportation cost	Limited availability at sources; required demands met at destinations
Multiperiod production planning	Quantities of product to produce in each of several time periods; amount of inventory to hold between periods	Minimize total production and inventory costs	Limited production rates; material balance equations
Multiperiod financial management	Amounts to invest in short-term instruments	Maximize cash on hand	Cash balance equations; required cash obligations
Production/ marketing	Allocation of advertising expenditures; production quantities	Maximize profit	Budget limitation; production limitations; demand requirements

The most challenging aspect of model formulation is identifying constraints. Understanding the different types of constraints can help in proper identification and modeling. Constraints generally fall into one of the following categories:

- *Simple Bounds.* Simple bounds constrain the value of a single variable. You can recognize simple bounds in problem statements like "no more than $10,000 may be invested in stock XYZ" or "we must produce at least 350 units of product Y to meet customer commitments this month." The mathematical forms for these examples are:

$$XYZ \leq 10,000$$
$$Y \geq 350$$

- *Limitations.* Limitations usually involve the allocation of scarce resources. Problem statements such as "the amount of material used in production cannot exceed the amount available in inventory," "minutes used in assembly cannot exceed the available labor hours," or "the amount shipped from the Austin plant in July cannot exceed the plant's capacity" are typical of these types of constraints.
- *Requirements.* Requirements involve the specification of minimum levels of performance. Such statements as "enough cash must be available in February to meet financial obligations," "production must be sufficient to meet promised customer orders," or "the marketing plan should ensure that at least 400 customers are contacted each month" are some examples.
- *Proportional Relationships.* Proportional relationships are often found in problems involving mixtures or blends of materials or strategies. Examples include "the amount invested in aggressive growth stocks cannot be more than twice the amount invested in equity-income funds," or "the octane rating of gasoline obtained from mixing different crude blends must be at least 89."
- *Balance Constraints.* Balance constraints essentially state that "input = output" and ensure that the flow of material or money is accounted for at locations or between time periods. Examples include "production in June plus any available inventory must equal June's demand plus inventory held to July," "the total amount shipped to a distribution center from all plants must equal the amount shipped from the distribution center to all customers," or "the total amount of money invested or saved in March must equal the amount of money available at the end of February."

Constraints in linear optimization models are generally some combination of constraints from these categories. Problem data or verbal clues in a problem statement often help you identify the appropriate constraint. In some situations, all constraints may not be explicitly stated, but are required for the model to represent the real problem accurately. An example of implicit constraints is nonnegativity of the decision variables.

In the following sections, we present examples of different types of linear optimization applications. Each of these models has different characteristics, and by studying how they are developed, you will improve your ability to model other problems. We encourage you to use the four-step process that we illustrated with the Sklenka Ski problem; however, to conserve space in this book, we will go directly to the mathematical model instead of first conceptualizing the constraints and objective functions in verbal terms. We will also use these models to illustrate specific issues associated with formulation, implementation on spreadsheets, and using *Solver*.

FABRIC	DEMAND (YARDS)	DOBBIE LOOM CAPACITY (YARDS/HOUR)	REGULAR LOOM CAPACITY (YARDS/HOUR)	MILL COST ($/YARD)	OUTSOURCING COST ($/YARD)
1	45,000	4.7	0.00	$0.65	$0.85
2	76,500	5.2	5.2	$0.61	$0.75
3	10,000	4.4	4.4	$0.50	$0.65

Table 13.2
Textile Production Data

Process Selection

Process selection models generally involve choosing among different types of processes to produce a good. Make-or-buy decisions are examples of process selection models whereby one must choose whether to make one or more products in-house or subcontract them out to another firm. The following example illustrates these concepts.

Camm Textiles has a mill that produces three types of fabrics on a make-to-order basis. The mill operates on a 24/7 basis. The key decision facing the plant manager is on what type of loom to process each fabric during the coming quarter (13 weeks). Two types of looms are used: dobbie and regular. Dobbie looms can be used to make all fabrics and are the only looms that can weave certain fabrics, such as plaids. Demands, variable costs for each fabric, and production rates on the looms are given in Table 13.2. The mill has 15 regular looms and 3 dobbie looms. After weaving, fabrics are sent to the finishing department and then sold. Any fabrics that cannot be woven in the mill because of limited capacity will be purchased from an external supplier, finished at the mill, and sold at the selling price. In addition to determining which looms to process the fabrics, the manager also needs to determine which fabrics to buy externally.

To formulate a linear programming model, define:

D_i = number of yards of fabric i to produce on dobbie looms, $i = 1, \ldots, 3$
R_i = number of yards of fabric i to produce on regular looms, $i = 1, \ldots, 3$
P_i = number of yards of fabric i to purchase from an outside supplier,
        $i = 1, \ldots, 3$

Note that we are using *subscripted variables* to simplify their definition, rather than defining nine individual variables with unique names. The objective function is to minimize total cost:

$$\text{Min } 0.65D_1 + 0.61D_2 + 0.50D_3 + 0.61R_2 + 0.50R_3 + 0.85P_1 + 0.75P_2 + 0.65P_3$$

Constraints to ensure meeting production requirements are:

$$D_1 + P_1 = 45,000$$
$$D_2 + R_2 + P_2 = 76,500$$
$$D_3 + R_3 + P_3 = 10,000$$

To specify the constraints on loom capacity, we must convert yards per hour into hours per yard. For example, for fabric 1 on a dobbie loom, 4.7 yards/hour = 0.213 hours/yard. Therefore, the term $0.213D_1$ represents the total time required to produce D_1 yards of fabric 1 on a dobbie loom. The total capacity for dobbie looms is (24 hours/day)(7 days/week)(13 weeks)(3 looms) = 6,552 hours. Thus, the constraint on available production time on dobbie looms is:

$$0.213D_1 + 0.192D_2 + 0.227D_3 \le 6,552$$

For regular looms we have:

$$0.192R_2 + 0.227R_3 \leq 32{,}760$$

Finally, all variables must be nonnegative. An exercise at the end of this chapter asks you to implement and solve this model using Excel.

Blending

Blending problems involve mixing several raw materials that have different characteristics to make a product that meets certain specifications. Dietary planning, gasoline and oil refining, coal and fertilizer production, and the production of many other types of bulk commodities involve blending. We typically see proportional constraints in blending problems.

To illustrate this type of model, consider the BG Seed Company, which specializes in food products for birds and other household pets. In developing a new birdseed mix, company nutritionists have specified that the mixture must contain at least 13% protein and 15% fat, and no more than 14% fiber. The percentages of each of these nutrients in eight types of ingredients that can be used in the mix are given in Table 13.3 along with the wholesale cost per pound. What is the minimum cost mixture that meets the stated nutritional requirements?

In this example, the decisions are the amount of each ingredient to include in a given quantity—for example, one pound—of mix. Define X_i = number of pounds of ingredient i to include in one pound of the mix, for $i = 1, \ldots, 8$. The use of subscripted variables like this simplifies modeling large problems. The objective is to minimize total cost:

$$\text{Minimize } 0.22X_1 + 0.19X_2 + 0.10X_3 + 0.10X_4 + 0.07X_5 + 0.05X_6 + 0.26X_7 + 0.11X_8$$

To ensure that the mix contains the appropriate proportion of ingredients, observe that multiplying the number of pounds of each ingredient by the percentage of nutrient in that ingredient (a dimensionless quantity) specifies the number of pounds of nutrient provided. For example, $0.169X_1$ represents the number of pounds of protein in sunflower seeds. Therefore, the total number of pounds of protein provided by all ingredients is:

$$0.169X_1 + 0.12X_2 + 0.085X_3 + 0.154X_4 + 0.085X_5 + 0.12X_6 + 0.18X_7 + 0.119X_8$$

Because the total number of pounds of ingredients that are mixed together equals $X_1 + X_2 + X_3 + X_4 + X_5 + X_6 + X_7 + X_8$, the proportion of protein in the mix is:

$$(0.169X_1 + 0.12X_2 + 0.085X_3 + 0.154X_4 + 0.085X_5 + 0.12X_6 + 0.18X_7 + 0.119X_8)/(X_1 + X_2 + X_3 + X_4 + X_5 + X_6 + X_7 + X_8)$$

Table 13.3 Birdseed Nutrition Data

INGREDIENT	PROTEIN %	FAT %	FIBER %	COST/LB
Sunflower seeds	16.9	26	29	$0.22
White millet	12	4.1	8.3	$0.19
Kibble corn	8.5	3.8	2.7	$0.10
Oats	15.4	6.3	2.4	$0.10
Cracked corn	8.5	3.8	2.7	$0.07
Wheat	12	1.7	2.3	$0.05
Safflower	18	17.9	28.8	$0.26
Canary grass seed	11.9	4	10.9	$0.11

This proportion must be at least 0.13. However, we wish to determine the best amount of ingredients to include in *one pound* of mix; therefore, we add the constraint:

$$X_1 + X_2 + X_3 + X_4 + X_5 + X_6 + X_7 + X_8 = 1$$

Now we can substitute 1 for the denominator in the proportion of protein, yielding the constraint:

$$0.169X_1 + 0.12X_2 + 0.085X_3 + 0.154X_4 + 0.085X_5 + 0.12X_6 + 0.18X_7 + 0.119X_8 \geq 0.13$$

This ensures that at least 13% of the mixture will be protein. In a similar fashion, the constraints for the fat and fiber requirements are:

$$0.26X_1 + 0.014X_2 + 0.038X_3 + 0.063X_4 + 0.038X_5 + 0.017X_6 + 0.179X_7 + 0.04X_8 \geq 0.15$$

$$0.29X_1 + 0.083X_2 + 0.027X_3 + 0.024X_4 + 0.027X_5 + 0.023X_6 + 0.288X_7 + 0.109X_8 \leq 0.14$$

Finally, we have nonnegative constraints:

$$X_i \geq 0, \text{ for } i = 1, 2, \ldots, 8$$

The complete model is:

Minimize $0.22X_1 + 0.19X_2 + 0.10X_3 + 0.10X_4 + 0.07X_5 + 0.05X_6 + 0.26X_7 + 0.11X_8$

$$X_1 + X_2 + X_3 + X_4 + X_5 + X_6 + X_7 + X_8 = 1$$

$$0.169X_1 + 0.12X_2 + 0.085X_3 + 0.154X_4 + 0.085X_5 + 0.12X_6$$
$$+ 0.18X_7 + 0.119X_8 \geq 0.13$$

$$0.26X_1 + 0.041X_2 + 0.038X_3 + 0.063X_4 + 0.038X_5 + 0.017X_6$$
$$+ 0.179X_7 + 0.04X_8 \geq 0.15$$

$$0.29X_1 + 0.083X_2 + 0.027X_3 + 0.024X_4 + 0.027X_5 + 0.023X_6$$
$$+ 0.288X_7 + 0.109X_8 \leq 0.14$$

$$X_i \geq 0, \quad \text{for } i = 1, 2, \ldots, 8$$

We will ask you to implement and solve this model as an exercise at the end of the chapter.

Portfolio Investment

Many types of financial investment problems are modeled and solved using linear optimization. Such problems have the basic characteristics of blending models. Innis Investments has a client that has acquired $500,000 from an inheritance. Innis Investments manages six mutual funds:

	FUND	EXPECTED ANNUAL RETURN	STANDARD DEVIATION
1.	Innis Low Priced Stock Fund	8.13%	10.57
2.	Innis Multinational Fund	9.02%	13.22
3.	Innis Mid-cap Stock Fund	7.56%	14.02
4.	Innis Mortgage Fund	3.62%	2.39
5.	Innis Income Equity Fund	7.79%	9.30
6.	Innis Balanced Fund	4.40%	7.61

The standard deviation measures the historical variation of returns from the average return, and therefore is a measure of risk. The greater the variability, the larger the standard deviation and hence the risk. The greater the standard deviation, the greater the probability of realizing a return less that the average return. Of course, the upside potential is also larger. Innis Investments recommends that no more than $200,000 be invested in any individual fund, that at least $50,000 be invested in each of the multi-national and balanced funds, and that at least 40% be invested in the Income Equity and Balanced funds. The client would like to have an average return of at least 5% but would like to minimize risk. What portfolio would achieve this?

Let X_1 through X_6 represent the amount invested in funds 1 through 6, respectively. The total risk would be measured by the weighted standard deviation of the portfolio, where the weights are the proportion of the total investment in any fund $(X_j/500,000)$. Thus, the objective function is:

$$\text{Minimize Total Risk} = (10.57X_1 + 13.22X_2 + 14.02X_3 + 2.39X_4 + 9.30X_5 + 7.61X_6)/500,000$$

The first constraint ensures that $500,000 is invested:

$$X_1 + X_2 + X_3 + X_4 + X_5 + X_6 = 500,000$$

The next constraint ensures that the weighted return is at least 5%:

$$(8.13X_1 + 9.02X_2 + 7.56X_3 + 3.62X_4 + 7.79X_5 + 4.40X_6)/500,000 \geq 5.00$$

The next constraint ensures that at least 40% be invested in the Income Equity and Balanced funds:

$$X_5 + X_6 \geq 0.4(500,000)$$

The following specify that at least $50,000 be invested in each of the multinational and balanced funds:

$$X_2 \geq 50,000$$
$$X_6 \geq 50,000$$

Finally, we restrict each investment to a maximum of $200,000 and include nonnegativity:

$$X_j \leq 200,000 \quad \text{for } j = 1, \ldots, 6$$
$$X_j \geq 0 \quad \text{for } j = 1, \ldots, 6$$

We ask you to solve this model at the end of the chapter.

Transportation Problem

Many practical models in supply chain optimization stem from a very simple model called the transportation problem. This involves determining how much to ship from a set of sources of supply (factories, warehouses, etc.) to a set of demand locations (warehouses, customers, etc.) at minimum cost. We illustrate this with the following scenario.

General Appliance Corporation (GAC) produces refrigerators at two plants: Marietta, Georgia, and Minneapolis, Minnesota. They ship them to major distribution centers in Cleveland, Baltimore, Chicago, and Phoenix. The Accounting, Production, and Marketing departments have provided the information in Table 13.4, which shows the unit cost of shipping between any plant and distribution center, plant capacities over the next planning period, and distribution center demands. GAC's supply chain manager faces the problem of determining how much to ship between each plant and distribution center to minimize the total transportation cost, not exceed available capacity, and meet customer demand.

To develop a linear optimization model, we first define the decision variables as the amount to ship between each plant and distribution center. In this model, we will use *double-subscripted variables* to simplify the formulation. Define X_{ij} = amount

Table 13.4 Cost, Capacity, and Demand Data

	DISTRIBUTION CENTER				
PLANT	CLEVELAND	BALTIMORE	CHICAGO	PHOENIX	CAPACITY
Marietta	$12.60	$14.35	$11.52	$17.58	1,200
Minneapolis	$9.75	$16.26	$8.11	$17.92	800
Demand	150	350	500	1,000	

shipped from plant i to distribution center j, where $i = 1$ represents Marietta, $i = 2$ represents Minneapolis, $j = 1$ represents Cleveland, and so on. Using the unit cost data in Table 13.4, the total cost of shipping is equal to the unit cost times amount shipped, summed over all combinations of plants and distribution centers. Therefore, the objective function is to minimize total cost:

$$\text{Minimize } 12.60X_{11} + 14.35X_{12} + 11.52X_{13} + 17.58X_{14} + 9.75X_{21}$$
$$+ 16.26X_{22} + 8.11X_{23} + 17.92X_{24}$$

Because capacity is limited, the amount shipped from each plant cannot exceed its capacity. The total amount shipped from Marietta, for example, is $X_{11} + X_{12} + X_{13} + X_{14}$. Therefore, we have the constraint:

$$X_{11} + X_{12} + X_{13} + X_{14} \leq 1,200$$

Similarly, the capacity limitation at Minneapolis leads to the constraint:

$$X_{21} + X_{22} + X_{23} + X_{24} \leq 800$$

Next, we must ensure that the demand at each distribution center is met. This means that the total amount shipped to any distribution center from both plants must equal the demand. For instance, at Cleveland, we must have:

$$X_{11} + X_{21} = 150$$

For the remaining three distribution centers, the constraints are:

$$X_{12} + X_{22} = 350$$
$$X_{13} + X_{23} = 500$$
$$X_{14} + X_{24} = 1,000$$

Last, we need nonnegativity, $X_{ij} \geq 0$, for all i and j. The complete model is:

$$\text{Minimize } 12.60X_{11} + 14.35X_{12} + 11.52X_{13} + 17.58X_{14} + 9.75X_{21}$$
$$+ 12.63X_{22} + 8.11X_{23} + 17.92X_{24}$$
$$X_{11} + X_{12} + X_{13} + X_{14} \leq 1200$$
$$X_{21} + X_{22} + X_{23} + X_{24} \leq 800$$
$$X_{11} + X_{21} = 150$$
$$X_{12} + X_{22} = 350$$
$$X_{13} + X_{23} = 500$$
$$X_{14} + X_{24} = 1000$$
$$X_{ij} \geq 0, \quad \text{for all } i \text{ and } j$$

Figure 13.13 shows a spreadsheet implementation for the GAC transportation problem (Excel file *Transportation Model*). In the *Model* section, the decision variables are

Figure 13.13 *Transportation Model* Spreadsheet

	A	B	C	D	E	F
1	**Transportation Model**					
2						
3	**Data**					
4			**Distribution Center**			
5	**Plant**	Cleveland	Baltimore	Chicago	Phoenix	**Capacity**
6	Marietta	$ 12.60	$ 14.35	$ 11.52	$ 17.58	1200
7	Minneapolis	$ 9.75	$ 16.26	$ 8.11	$ 17.92	800
8	**Demand**	150	350	500	1000	
9						
10	**Model**					
11			**Distribution Center**			
12	**Plant**	Cleveland	Baltimore	Chicago	Phoenix	**Total shipped**
13	Marietta	0	0	0	0	0
14	Minneapolis	0	0	0	0	0
15	**Demand met**	0	0	0	0	
16						
17	**Total cost**					
18	$ -					

stored in the plant-distribution center matrix. The objective function is computed in cell A18 as:

Total Cost = B6 × B13 + C6 × C13 + D6 × D13 + E6 × E13 + B7 × B14 + C7 × C14 + D7 × D14 + E7 × E14

The SUMPRODUCT function is particularly useful for such large expressions; we could write the total cost as:

SUMPRODUCT(B6:E7;B13:E14)

Note that the SUMPRODUCT function applies to a matrix of values as long as the dimensions are the same.

To ensure that we do not exceed the capacity of any plant, the total shipped from each plant (cells F13:F14) cannot be greater than the plant capacities (cells F6:F7). For example,

Total Shipped from Marietta (cell F13) = B13 + C13 + D13 + E13 = SUM(B13:E13) ≤ F6

The constraint for Minneapolis is similar. Can you write it?

To ensure that demands are met, the total shipped to each distribution center (cells B15:E15) must equal or exceed the demands (cells B8:E8). Thus, for Cleveland,

Total Shipped to Cleveland (cell B15) = B13 + B14 = SUM(B13:B14) = B8

As you become more proficient in using spreadsheets, you should consider creating range names for the decision variables and constraint functions. This allows you to locate and manipulate elements of the model more easily. For example, in the Sklenka Skis product mix model, you might define the range B14:C14 as *Decisions* and the range B9:C9 as *Profit*. The total profit can then be computed easily as SUMPRODUCT

Figure 13.14 Original Sensitivity Report for GAC Transportation Model

Microsoft Excel 12.0 Sensitivity Report		
Worksheet: [Transportation Model solved.xlsx]Sheet1		
Report Created: 06/04/2008 16:58:42 PM		

Target Cell (Min)

Cell	Name	Final Value
$A$18	Total cost	28171

Adjustable Cells

Cell	Name	Final Value	Reduced Cost	Objective Coefficient	Allowable Increase	Allowable Decrease
$B$13	Marietta Cleveland	0	3	12.6	1E+30	3.19
$C$13	Marietta Baltimore	350	0	14.35	1.57	1E+30
$D$13	Marietta Chicago	0	4	11.52	1E+30	3.75
$E$13	Marietta Phoenix	850	0	17.58	0.34	1.57
$B$14	Minneapolis Cleveland	150	0	9.75	3.19	1E+30
$C$14	Minneapolis Baltimore	0	2	16.26	1E+30	1.57
$D$14	Minneapolis Chicago	500	0	8.11	3.75	1E+30
$E$14	Minneapolis Phoenix	150	0	17.92	1.57	0.34

Constraints

Cell	Name	Final Value	Shadow Price	Constraint R.H. Side	Allowable Increase	Allowable Decrease
$F$13	Marietta Total shipped	1200	0	1200	150	0
$F$14	Minneapolis Total shipped	800	0	800	1E+30	0
$B$15	Demand met Cleveland	150	10	150	0	150
$C$15	Demand met Baltimore	350	15	350	0	150
$D$15	Demand met Chicago	500	8	500	0	500
$E$15	Demand met Phoenix	1000	18	1000	0	150

(Decisions, Profit). In this book, however, we will stick with using cell references in all formulas to keep it simple.

A very important note on formatting the Sensitivity Report: Depending on how cells in your spreadsheet model are formatted, the Sensitivity Report produced by *Solver* may not reflect the accurate values of reduced costs or shadow prices because an insufficient number of decimal places may be displayed. For example, Figure 13.14 shows the Sensitivity Report created by *Solver*. Note that the data in columns headed Reduced Cost and Shadow Price are formatted as whole numbers. The correct values are shown in Figure 13.15 (obtained by simply formatting the data to have two decimal places)! Thus, we *highly recommend* that after you save the Sensitivity Report to your workbook, you select the reduced cost and shadow price ranges and format them to have at least two or three decimal places.

Interpreting Reduced Costs

The transportation model is a good example to use to discuss the interpretation of reduced costs. First, note that the reduced costs are zero for all variables that are

Figure 13.15 Accurate Sensitivity Report for GAC Transportation Model

	A	B	C	D	E	F	G	H
1	**Microsoft Excel 12.0 Sensitivity Report**							
2	**Worksheet: [Transportation Model solved.xlsx]Sheet1**							
3	**Report Created: 06/04/2008 16:58:42 PM**							
4								
5	Target Cell (Min)							
6		Cell	Name	Final Value				
7		$A$18	Total cost	28171				
8								
9	Adjustable Cells							
10				Final	Reduced	Objective	Allowable	Allowable
11		Cell	Name	Value	Cost	Coefficient	Increase	Decrease
12		$B$13	Marietta Cleveland	0	3.19	12.6	1E+30	3.19
13		$C$13	Marietta Baltimore	350	0.00	14.35	1.57	1E+30
14		$D$13	Marietta Chicago	0	3.75	11.52	1E+30	3.75
15		$E$13	Marietta Phoenix	850	0.00	17.58	0.34	1.57
16		$B$14	Minneapolis Cleveland	150	0.00	9.75	3.19	1E+30
17		$C$14	Minneapolis Baltimore	0	1.57	16.26	1E+30	1.57
18		$D$14	Minneapolis Chicago	500	0.00	8.11	3.75	1E+30
19		$E$14	Minneapolis Phoenix	150	0.00	17.92	1.57	0.34
20								
21	Constraints							
22				Final	Shadow	Constraint	Allowable	Allowable
23		Cell	Name	Value	Price	R.H. Side	Increase	Decrease
24		$F$13	Marietta Total shipped	1200	-0.34	1200	150	0
25		$F$14	Minneapolis Total shipped	800	0.00	800	1E+30	0
26		$B$15	Demand met Cleveland	150	9.75	150	0	150
27		$C$15	Demand met Baltimore	350	14.69	350	0	150
28		$D$15	Demand met Chicago	500	8.11	500	0	500
29		$E$15	Demand met Phoenix	1000	17.92	1000	0	150

positive in the solution. Now examine the reduced cost, 3.19, associated with shipping from Marietta to Cleveland. A question to ask is "why does the optimal solution ship nothing between these cities?" The answer is simple: It is not economical to do so! In other words, it costs too much to ship from Marietta to Cleveland; the demand can be met less expensively by shipping from Minneapolis. The next logical question to ask is "what would the unit shipping cost have to be to make it attractive to ship from Marietta instead of Minneapolis?" The answer is given by the reduced cost. If the unit cost is reduced by at least $3.19, then the optimal solution will change and would include a positive value for the Marietta-Cleveland variable. Again, this is only true if all other data are held constant.

Multiperiod Planning

Many linear optimization problems involve planning over multiple time periods. We will present two examples, a production-inventory model and an investment model.

Kristin's Creations is a home-based company that makes hand-painted jewelry boxes for teenage girls. Forecasts of sales for the next year are 150 in the autumn, 400 in the winter, and 50 in the spring. Plain jewelry boxes are purchased from a supplier for $20. The cost of capital is estimated to be 24% per year (or 6% per quarter); thus, the holding cost per item is 0.06($20) = $1.20 per quarter. Kristin hires art students part-time to craft her designs during the autumn, and they earn $5.50 per hour. Because of the high demand for part-time help during the winter holiday season, labor rates are higher in the winter, and workers earn $7.00 per hour. In the spring, labor is more difficult to keep, and the owner must pay $6.25 per hour to retain qualified help. Each jewelry box takes two hours to complete. How should production be planned over the three quarters to minimize the combined production and inventory holding costs?

The principal decision variables are the number of jewelry boxes to produce during each of the three quarters. While it might seem obvious to simply produce to the anticipated level of sales, it may be advantageous to produce more during some quarter and carry the items in inventory, thereby letting lower labor rates offset the carrying costs. Therefore, we must also define decision variables for the number of units to hold in inventory at the end of each quarter. The decision variables are:

$$P_A = \text{amount to produce in autumn}$$
$$P_W = \text{amount to produce in winter}$$
$$P_S = \text{amount to produce in spring}$$
$$I_A = \text{inventory held at the end of autumn}$$
$$I_W = \text{inventory held at the end of winter}$$
$$I_S = \text{inventory held at the end of spring}$$

The production cost per unit is computed by multiplying the labor rate by the number of hours required to produce one. Thus, the unit cost in the autumn is ($5.50)(2) = $11.00; in the winter, ($7.00)(2) = $14.00; and in the spring, ($6.25)(2) = $12.50. The objective function is to minimize the total cost of production and inventory. (Because the cost of the boxes themselves is constant, it is not relevant to the problem we are addressing.) The objective function is therefore:

$$\text{Minimize } 11P_A + 14P_W + 12.50P_S + 1.20I_A + 1.20I_W + 1.20I_S$$

The only explicit constraint is that demand must be satisfied. Note that both the production in a quarter as well as the inventory held from the *previous* time quarter can be used to satisfy demand. In addition, any amount in excess of the demand is held to the next quarter. Therefore, the constraints take the form of *inventory balance equations* that essentially say "what is available in any time period must be accounted for somewhere." More formally,

Production + Inventory from the Previous Quarter = Demand
+ Inventory Held to the Next Quarter

This can be represented visually using the diagram in Figure 13.16. For each quarter, the sum of the variables coming in must equal the sum of the variables going out. Drawing such a figure is very useful for any type of multiple time period planning model. This results in the constraint set:

$$P_A + 0 = 150 + I_A$$
$$P_W + I_A = 400 + I_W$$
$$P_S + I_W = 50 + I_S$$

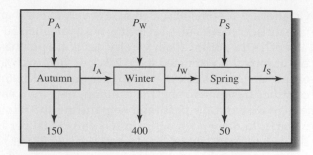

Figure 13.16 Material Balance Constraint Structure

Moving all variables to the left side results in the model:

$$\text{Minimize } 11P_A + 14P_W + 12.50P_S + 1.20I_A + 1.20I_W + 1.20I_S$$

Subject to

$$P_A - I_A = 150$$
$$P_W + I_A - I_W = 400$$
$$P_s + I_w - I_s = 50$$
$$P_i \geq 0, \quad \text{for all } i$$
$$I_j \geq 0, \quad \text{for all } j$$

As we have noted, developing models is more of an art than a science; consequently, there is often more than one way to model a particular problem. Using the ideas presented in this example, we may construct an alternative model involving only the production variables. We simply have to make sure that demand is satisfied. We can do this by ensuring that the cumulative production in each quarter is at least as great as the cumulative demand. This is expressed by the following constraints:

$$P_A \geq 150$$
$$P_A + P_W \geq 550$$
$$P_A + P_W + P_S \geq 600$$
$$P_A, P_W, P_S \geq 0$$

The differences between the left- and right-hand sides of these constraints are the ending inventories for each period (and we need to keep track of these amounts because inventory has a cost associated with it). Thus, we use the following objective function:

$$\text{Minimize } 11P_A + 14P_W + 12.50P_S + 1.20(P_A - 150) + 1.20(P_A + P_W - 550)$$
$$+ 1.20(P_A + P_W + P_S - 600)$$

Of course, this function can be simplified algebraically by combining like terms. Although these two models look very different, they are mathematically equivalent and will produce the same solution. An exercise at the end of this chapter will ask you to verify this.

Financial planning often occurs over an extended time horizon and can be formulated as multiperiod optimization models. For example, a financial manager at D.A. Branch & Sons must ensure that funds are available to pay company expenditures in the future, but would also like to maximize investment income. Three short-term investment options are available over the next six months: A, a one-month CD that pays 0.5%, available each month; B, a three-month CD that pays 1.75%, available at the beginning of the first four months; and C, a six-month CD that pays 2.3%, available in the first month. The net expenditures for the next six months are forecast as $50,000, ($12,000), $23,000, ($20,000), $41,000, ($13,000). Amounts in parentheses indicate a net inflow of cash. The company must maintain a cash balance of at least $10,000 at the end of each month. The company currently has $200,000 in cash.

Figure 13.17 Cash Balance Constraint Structure

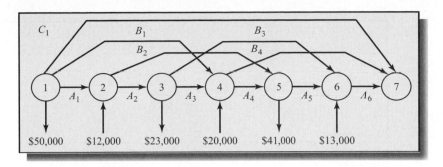

At the beginning of each month, the manager must decide how much to invest in each alternative that may be available. Define:

A_i = amount (\$) to invest in a one-month CD at the start of month i

B_i = amount (\$) to invest in a three-month CD at the start of month i

C_i = amount (\$) to invest in a six-month CD at the start of month i

Because the time horizons on these alternatives vary, it is helpful to draw a picture to represent the investments and returns for each year as shown in Figure 13.17. Each circle represents the beginning of a month. Arrows represent the investments and cash flows. For example, investing in a three-month CD at the start of month 1 (B_1) matures at the beginning of month 4. It is reasonable to assume that all funds available would be invested.

From Figure 13.17, we see that investments A_6, B_4, and C_1 will mature at the end of month 6; that is, at the beginning of month 7. To maximize the amount of cash on hand at the end of the planning period, we have the objective function:

$$\text{Maximize } 1.005A_6 + 1.0175B_4 + 1.023C_1$$

The only constraints necessary are minimum cash balance equations. For each month, the net cash available, which is equal to the cash in less cash out, must be at least \$100,000. These follow directly from Figure 13.17. The complete model is:

$$\text{Maximize } 1.005A_6 + 1.0175B_4 + 1.023C_1$$

Subject to

$$200,000 - (A_1 + B_1 + C_1 + 50,000) \geq 10,000 \quad \text{(Month 1)}$$
$$1.005A_1 + 12,000 - (A_2 + B_2) \geq 10,000 \quad \text{(Month 2)}$$
$$1.005A_2 - (A_3 + B_3 + 23,000) \geq 10,000 \quad \text{(Month 3)}$$
$$1.005A_3 + 1.0175B_1 + 20,000 - (A_4 + B_4) \geq 10,000 \quad \text{(Month 4)}$$
$$1.005A_4 + 1.0175B_2 - (A_5 + 41,000) \geq 10,000 \quad \text{(Month 5)}$$
$$1.005A_5 + 1.0175B_3 + 13,000 - A_6 \geq 10,000 \quad \text{(Month 6)}$$
$$A_i, B_i, C_i \geq 0, \quad \text{for all } i$$

A Model with Bounded Variables

Product mix problems involve determining the best mix of products to produce with limited resources and other requirements (the Sklenka Ski Company problem is a simple example). Suppose that J&M Manufacturing makes four models of gas grills, *A*, *B*, *C*, and *D*. Each grill must flow through five departments, stamping, painting, assembly,

Table 13.5   J&M Manufacturing Data

GRILL MODEL	SELLING PRICE/UNIT	VARIABLE COST/UNIT	MINIMUM MONTHLY SALES REQUIREMENTS	MAXIMUM MONTHLY SALES POTENTIAL
A	$250	$210	0	4,000
B	$300	$240	0	3,000
C	$400	$300	500	2,000
D	$650	$520	500	1,000

DEPARTMENT	A	B	C	D	HOURS AVAILABLE
Stamping	40	30	10	10	320
Painting		20	10	10	320
Assembly	25	15	15	12	320
Inspection	20	20	25	15	320
Packaging	50	40	40	30	320

inspection, and packaging. Table 13.5 shows the relevant data. Production rates are shown in units/hour. (Grill A uses imported parts and does not require painting).

J&M wants to determine how many grills to make to maximize monthly profit. To formulate this as a linear optimization model, let:

A, B, C, and D = number of units of model A, B, C, and D to produce, respectively

The objective function is to maximize the total net profit:

$$\text{Maximize } (250 - 210)A + (300 - 240)B + (400 - 300)C + (650 - 520)D$$
$$= 40A + 60B + 100C + 130D$$

The constraints include limitations on the amount of production hours available in each department, the minimum sales requirements, and maximum sales potential limits. Here is an example of where you must carefully look at the dimensions of the data. The production rates are given in units/hour, so if you multiply these values by the number of units produced, you will have an expression that makes no sense. Therefore, you must divide the decision variables by units per hour or equivalently, convert these data to hours/unit, and then multiply by the decision variables:

$$A/40 + B/30 + C/10 + D/10 \le 320 \quad \text{(Stamping)}$$
$$B/20 + C/10 + D/10 \le 320 \quad \text{(Painting)}$$
$$A/25 + B/15 + C/15 + D/12 \le 320 \quad \text{(Assembly)}$$
$$A/20 + B/20 + C/25 + D/15 \le 320 \quad \text{(Inspection)}$$
$$A/50 + B/40 + C/40 + D/30 \le 320 \quad \text{(Packaging)}$$

The sales constraints are simple upper and lower bounds on the variables:

$$A \ge 0$$
$$B \ge 0$$
$$C \ge 500$$
$$D \ge 500$$
$$A \le 4,000$$
$$B \le 3,000$$
$$C \le 2,000$$
$$D \le 1,000$$

Nonnegativity constraints are implied by the lower bounds on the variables and, therefore, do not need to be explicitly stated. Figure 13.18 shows a spreadsheet implementation (Excel file *J&M Manufacturing*) with the optimal solution and the *Solver* model used to find it.

Solver handles simple lower bounds (e.g., $x \geq 10$) and upper bounds (e.g., $x \leq 150$) quite differently from ordinary constraints in the Sensitivity Report. To see this, look at the Answer and Sensitivity reports for the J&M Manufacturing model in Figures 13.19 and 13.20 (the spreadsheet model is available on the CD-ROM). In the Answer Report, all constraints are listed along with their status. For example, we see that the lower bound on A, lower bound on D, and upper bound on B are all binding. However, none of the bound constraints appear in the Constraints section of the Sensitivity Report.

In *Solver*, nonzero lower and upper bounds are treated in a manner similar to non-negativity constraints, which also do not appear explicitly as constraints in the model. *Solver* does this to increase the efficiency of the solution procedure used; for large models this can represent significant savings in computer-processing time. However, it makes it more difficult to interpret the sensitivity information, because we no longer have the shadow prices and allowable increases and decreases associated with these constraints. Actually, this isn't quite true; the shadow prices are there, but in a different form.

Figure 13.18 Spreadsheet Implementation and *Solver* Model for J&M Manufacturing

	A	B	C	D	E	F
1	J&M Manufacturing					
2						
3	Data					
4	Grill model	Selling price	Variable cost	Min Sales	Max Sales	
5	A	$ 250.00	$ 210.00	0	4000	
6	B	$ 300.00	$ 240.00	0	3000	
7	C	$ 400.00	$ 300.00	500	2000	
8	D	$ 650.00	$ 520.00	500	1000	
9						
10	Production rates (hours/unit)	A	B	C	D	Hours Available
11	Stamping	40	30	10	10	320
12	Painting		20	10	10	320
13	Assembly	25	15	15	12	320
14	Inspection	20	20	25	15	320
15	Packaging	50	40	40	30	320
16						
17	Model					
18	Department	A	B	C	D	Hours Used
19	Stamping	0.000	100.000	117.500	50.000	267.500
20	Painting		150.000	117.500	50.000	317.500
21	Assembly	0.000	200.000	78.333	41.667	320.000
22	Inspection	0.000	150.000	47.000	33.333	230.333
23	Packaging	0.000	75.000	29.375	16.667	121.042
24						
25	Number produced	0	3000	1175	500	
26	Net profit/unit	$ 40.00	$ 60.00	$ 100.00	$ 130.00	Total Profit
27	Profit contribution	$ -	$ 390,000.00	$ 117,500.00	$ 20,000.00	$ 527,500.00

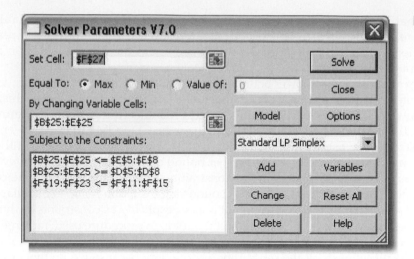

Figure 13.18 *continued*

Figure 13.19 J&M Manufacturing *Solver* Answer Report

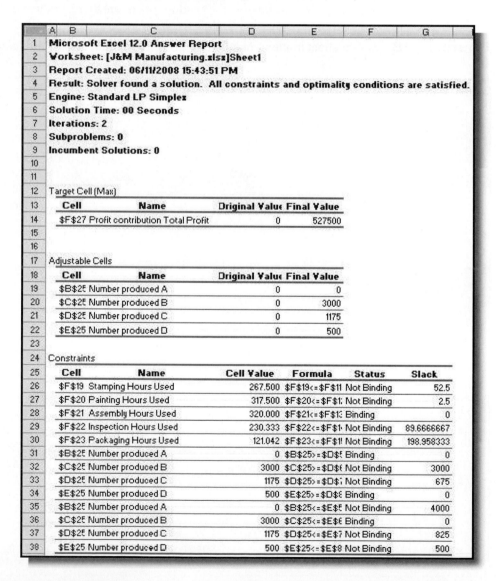

Figure 13.20 J&M Manufacturing *Solver* Sensitivity Report

	A	B	C	D	E	F	G	H
1	Microsoft Excel 12.0 Sensitivity Report							
2	Worksheet: [J&M Manufacturing.xlsx]Sheet1							
3	Report Created: 06/11/2008 15:43:51 PM							
4								
5	Target Cell (Max)							
6		Cell	Name	Final Value				
7		$F$27	Profit contribution Total Profit	527500				
8								
9	Adjustable Cells							
10				Final	Reduced	Objective	Allowable	Allowable
11		Cell	Name	Value	Cost	Coefficient	Increase	Decrease
12		$B$25	Number produced A	0	0	60	0	1E+30
13		$C$25	Number produced B	3000	30	130	1E+30	30
14		$D$25	Number produced C	1175	0	100	30	0
15		$E$25	Number produced D	500	-85	40	85	1E+30
16								
17	Constraints							
18				Final	Shadow	Constraint	Allowable	Allowable
19		Cell	Name	Value	Price	R.H. Side	Increase	Decrease
20		$F$19	Stamping Hours Used	267.500	0.000	320	1E+30	52.5
21		$F$20	Painting Hours Used	317.500	0.000	320	1E+30	2.5
22		$F$21	Assembly Hours Used	320.000	1500.000	320	1.666666667	45
23		$F$22	Inspection Hours Used	230.333	0.000	320	1E+30	89.66666667
24		$F$23	Packaging Hours Used	121.042	0.000	320	1E+30	198.9583333

First, let us interpret the reduced costs. Recall that in an ordinary model with only nonnegativity constraints and no other simple bounds, the reduced cost tells how much the objective coefficient needs to be reduced in order for a variable to become positive in an optimal solution. A nonzero lower bound is very similar to nonnegativity; it represents the minimum value for that variable. For product D, we have $D \geq 500$. Note that the optimal solution specifies that we produce only the minimum amount of D required. Why? It is simply not economical to produce more because the profit contribution of D is too low relative to the other products. How much more would the profit on D have to be in order for it to be economical to produce anything other than the minimum amount required? As we saw earlier for the transportation problem, the answer is given by the reduced cost. The unit profit on D would have to be reduced by at least $-\$85$ (that is, *increased* by at least \$85).

For product B, the reduced cost is \$30. Note that B is at its upper bound, 3,000. We want to produce as much of B as possible because it generates a large profit. How much would the unit profit have to be lowered before it is no longer economical to produce the maximum amount? Again, the answer is the reduced cost, or \$30.

Now, let's ask these questions in a different way. For product D, what would the effect be of increasing the right-hand side value of the bound constraint, $D \geq 500$, by one unit? If we increase the right-hand side of the lower bound constraint by 1, we are essentially forcing the solution to produce more than the minimum requirement. How would the objective function change if we do this? It would have to decrease because we would lose money by producing an extra unit of a nonprofitable product. How much? The answer again is the reduced cost! Producing an additional unit of product D will result in a profit reduction of $85. Similarly, increasing the right-hand side of the constraint $B \leq 3,000$ by one will increase the profit by $30. *The reduced cost associated with a bounded variable is the same as the shadow price of the bound constraint.* However, we no longer have the allowable range over which we can change the constraint values. (Important: The Allowable Increase and Allowable Decrease values in the Sensitivity Report refer to the objective coefficients, not the reduced costs.)

Interpreting reduced costs as shadow prices for bounded variables can be a bit confusing. Fortunately, there is a neat little trick that you can use to eliminate this issue. In your spreadsheet model, define a new set of cells for any decision variables that have upper or lower bound constraints by referencing (not copying) the original changing cells. This is shown in row 29 of Figure 13.21.

In the *Solver* model, use these auxiliary variable cells—*not* the changing variable cells as defined—to define the bound constraints. Thus, the *Solver* constraints would be as shown in Figure 13.22. The Sensitivity Report for this model is shown in Figure 13.23. We now see that the *Constraints* section has rows corresponding to the bound constraints and that the shadow prices are the same as the reduced costs in the previous sensitivity report. Moreover, we now know the allowable increases and decreases for each shadow price, which we did not have in the previous sensitivity

Figure 13.21 Auxiliary Variable Cells in J&M Manufacturing Model

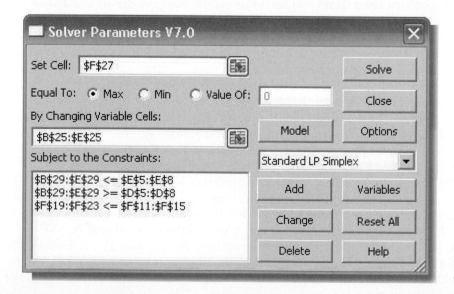

	A	B	C	D	E	F
24						
25	Number produced	0	3000	1175	500	
26	Net profit/unit	=B5-C5	=B6-C6	=B7-C7	=B8-C8	Total Profit
27	Profit contribution	=B25*C26	=C25*E26	=D25*D26	=E25*B26	=SUM(B27:E27)
28						
29	Auxiliary variable	=B25	=C25	=D25	=E25	

Solver Parameters V7.0

Set Cell: $F$27

Equal To: ● Max   ○ Min   ○ Value Of: 0

By Changing Variable Cells:

$B$25:$E$25

Subject to the Constraints:

$B$29:$E$29 <= $E$5:$E$8
$B$29:$E$29 >= $D$5:$D$8
$F$19:$F$23 <= $F$11:$F$15

Standard LP Simplex

Solve
Close
Model
Options
Add
Variables
Change
Reset All
Delete
Help

Figure 13.22 J&M Manufacturing *Solver* Model with Auxiliary Variables

	A B	C	D	E	F	G	H
1	Microsoft Excel 12.0 Sensitivity Report						
2	Worksheet: [J&M Manufacturing.xlsx]Sheet1						
3	Report Created: 06/11/2008 16:22:16 PM						
4							
5	Target Cell (Max)						
6	Cell	Name	Final Value				
7	$F$27 Profit contribution Total Profit		527500				
8							
9	Adjustable Cells						
10			Final	Reduced	Objective	Allowable	Allowable
11	Cell	Name	Value	Cost	Coefficient	Increase	Decrease
12	$B$25 Number produced A		0	0	60	0	1E+30
13	$C$25 Number produced B		3000	0	130	1E+30	30
14	$D$25 Number produced C		1175	0	100	30	0
15	$E$25 Number produced D		500	0	40	85	1E+30
16							
17	Constraints						
18			Final	Shadow	Constraint	Allowable	Allowable
19	Cell	Name	Value	Price	R.H. Side	Increase	Decrease
20	$F$19 Stamping Hours Used		267.500	0.000	320	1E+30	52.5
21	$F$20 Painting Hours Used		317.500	0.000	320	1E+30	2.5
22	$F$21 Assembly Hours Used		320.000	1500.000	320	1.6666667	45
23	$F$22 Inspection Hours Used		230.333	0.000	320	1E+30	89.666667
24	$F$23 Packaging Hours Used		121.042	0.000	320	1E+30	198.95833
25	$B$29 Auxiliary variable A		0	0	0	1125	0
26	$C$29 Auxiliary variable B		3000	0	0	3000	1E+30
27	$D$29 Auxiliary variable C		1175	0	500	675	1E+30
28	$E$29 Auxiliary variable D		500	-85	500	500	100
29	$B$29 Auxiliary variable A		0	0	4000	1E+30	4000
30	$C$29 Auxiliary variable B		3000	30	3000	675	50
31	$D$29 Auxiliary variable C		1175	0	2000	1E+30	825
32	$E$29 Auxiliary variable D		500	0	1000	1E+30	500

report. Thus, we recommend that you use this approach unless solution efficiency is an important issue.

A Production/Marketing Allocation Model

Many problems involve allocation of marketing effort, such as advertising dollars. The following is an example of combining elements of a product mix model with marketing budget allocation decisions based on demand elasticity.[1] This example also illustrates some important issues of properly interpreting sensitivity results and the influence that modeling approaches can have.

A small winery, Walker Wines, buys grapes from local growers and blends the pressings to make two types of wine: Shiraz and merlot. It costs $1.60 to purchase the grapes needed to make a bottle of Shiraz, and $1.40 to purchase the grapes needed to

---

[1] Adapted from an example in Roger D. Eck, *Operations Research for Business* (Belmont, CA: Wadsworth, 1976), 129–131.

make a bottle of merlot. The contract requires that they provide at least 40% but not more than 70% Shiraz. Based on market research related to it, it is estimated that the base demand for Shiraz is 1,000 bottles, but increases by 5 bottles for each $1 spent on advertising while the base demand for merlot is 2,000 bottles and increases by 8 bottles for each $1 spent on advertising. Production should not exceed demand. Shiraz sells to retail stores for $6.25 per bottle while merlot is sold for $5.25 per bottle. Walker Wines has $50,000 available to purchase grapes and advertise its products, with an objective of maximizing profit contribution.

To formulate this model, let:

S = number of bottles of Shiraz produced
M = number of bottles of merlot produced
A_s = dollar amount spent on advertising Shiraz
A_m = dollar amount spent on advertising merlot

The objective is to maximize profit (revenue minus costs) = ($6.25S + $5.25M) − ($1.60S + $1.40M + $A_s + A_m$) = 4.65S + 4.85M − A_s − A_m$
Constraints are defined as follows:

1. Budget cannot be exceeded:

$$\$1.60S + \$1.40M + A_s + A_m \leq \$50,000$$

2. Contractual requirements must be met:

$$0.4 \leq S/(S + M) \leq 0.7$$

or, expressed in linear form:

$$0.6S - 0.4M \geq 0 \text{ and } 0.3S - 0.7M \leq 0$$

3. Production must not exceed demand:

$$S \leq 1,000 + 5A_s$$
$$M \leq 2,000 + 8A_m$$

4. Nonnegativity

Figure 13.24 shows a spreadsheet implementation of this model (Excel file *Walker Wines*) along with the *Solver* solution. Figure 13.25 shows the *Solver* Sensitivity Report. A variety of practical questions can be posed around the sensitivity report. For example, suppose that the accountant noticed a small error in computing the profit contribution for Shiraz. The cost of Shiraz grapes should have been $1.65 instead of $1.60. How will this affect the solution?

In the model formulation, you can see that a $0.05 increase in cost results in a drop in the unit profit of Shiraz from $4.65 to $4.60. In the Sensitivity Report, the change in the profit coefficient is within the allowable decrease of 0.05328, thus concluding that no change in the optimal solution will result. However, this is *not* the correct interpretation! If the model is resolved using the new cost parameter, the solution changes dramatically as shown in Figure 13.26.

Why did this happen? One crucial assumption in interpreting sensitivity analysis information for changes in model parameters is that *all other model parameters are held constant*. It is easy to fall into a trap of ignoring this assumption and blindly crunching through the numbers. This is particularly true when using spreadsheet models. In this case, the unit cost is also reflected in the binding budget constraint. When we change the cost parameter, the constraint also changes. This violates the assumption. The change causes the budget constraint to become infeasible, and the solution must be adjusted to maintain feasibility.

Figure 13.24 Walker Wines Spreadsheet Model

	A	B	C	D	E	F
1		Walker Wines Product Mix				
2						
3		Data				
4			Shiraz	Merlot		
5		Cost/bottle	$ 1.60	$ 1.40		
6		Price/bottle	$ 6.25	$ 5.25		
7						
8		Base demand	1,000.00	2,000.00		
9		Increase/$1 Adv.	5	8		
10		Min. percent requirement	40%			
11		Max. percent limitation	70%			
12						
13		Total Budget	$ 50,000.00			
14						
15		Model				
16						
17		Total profit				
18		$ 124,775.84				
19			Shiraz	Merlot	Total	
20		Unit profit	$ 4.65	$ 3.85		
21		Advertising dollars	$ 3,912.37	$ 851.53	$ 4,763.90	
22		Demand	20,561.86	8,812.23	29,374.09	
23		Quantity produced	20,561.86	8,812.23	29,374.09	
24						
25		Min. percent requirement	8812.227074	>=	0	
26		Max. percent limitation	0	<=	0	
27						
28					Used	Unused
29		Budget	$ 36,811.35	$ 13,188.65	$ 50,000.00	$ -

	A	B	C	D	E	F
1		Walker Wines Product Mix				
2						
3		Data				
4			Shiraz	Merlot		
5		Cost/bottle	1.6	1.4		
6		Price/bottle	6.25	5.25		
7						
8		Base demand	1000	2000		
9		Increase/$1 Adv.	5	8		
10		Min. percent requirement	0.4			
11		Max. percent limitation	0.7			
12						
13		Total Budget	50000			
14						
15		Model				
16						
17		Total profit				
18		=(C20*C23)+(D20*D23)-C21-D21				
19			Shiraz	Merlot	Total	
20		Unit profit	=C6-C5	=D6-D5		
21		Advertising dollars	0	0	=SUM(C21:D21)	
22		Demand	=C8+(C9*C21)	=D8+(D9*D21)	=SUM(C22:D22)	
23		Quantity produced	0	0	=SUM(C23:D23)	
24						
25		Min. percent requirement	=(1-C10)*C23-C10*D23	>=	0	
26		Max. percent limitation	=(1-C11)*C23-C11*D23	<=	0	
27						
28					Used	Unused
29		Budget	=C21+(C23*C5)	=D21+(D23*D5)	=SUM(C29:D29)	=C13-E29

Figure 13.25 Walker Wines *Solver* Sensitivity Report

	A B C	D	E	F	G	H
1	Microsoft Excel 12.0 Sensitivity Report					
2	Worksheet: [Walker Wines Revised.xls]Model					
3	Report Created: 06/12/2008 11:01:42 AM					
4						
5	Target Cell (Max)					
6	Cell Name	Final Value				
7	$B$18 Total profit	124775.837				
8						
9	Adjustable Cells					

			Final	Reduced	Objective	Allowable	Allowable
10	Cell	Name	Value	Cost	Coefficient	Increase	Decrease
11							
12	$C$21	Advertising dollars Shiraz	$ 3,912.37 $ -		-1	3.771790808	0.266393443
13	$D$21	Advertising dollars Merlot	$ 851.53 $ -		-1	0.361111111	112.8666667
14	$C$23	Quantity produced Shiraz	20,561.86	0.00	4.65	1E+30	0.053278689
15	$D$23	Quantity produced Merlot	8,812.23	0.00	3.85	0.045138889	14.10833333
16							
17	Constraints						

			Final	Shadow	Constraint	Allowable	Allowable
18			Value	Price	R.H. Side	Increase	Decrease
19	Cell	Name					
20	$C$23	Quantity produced Shiraz	20,561.86	0.69	0	21297.93978	195000
21	$C$25	Min. percent requirement Shiraz	8812.227074	0	0	8812.227074	1E+30
22	$C$26	Max. percent limitation Shiraz	0	0.047307132	0	6500	9256.880734
23	$D$23	Quantity produced Merlot	8,812.23	0.43	0	6964.285714	383971.4286
24	$E$29	Budget Used	$ 50,000.00 $	2.46	50000	1E+30	39000

Figure 13.26 Walker Wines *Solver* Solution After Cost Increase

	A	B	C	D	E	F
1	**Walker Wines Product Mix**					
2						
3	**Data**					
4			Shiraz	Merlot		
5		Cost/bottle	$ 1.65	$ 1.40		
6		Price/bottle	$ 6.25	$ 5.25		
7						
8		Base demand	1,000.00	2,000.00		
9		Increase/$1 Adv.	5	8		
10		Min. percent requirement	40%			
11		Max. percent limitation	70%			
12						
13		Total Budget	$ 50,000.00			
14						
15	**Model**					
16						
17		Total profit				
18		$ 122,231.12				
19			Shiraz	Merlot	Total	
20		Unit profit	$ 4.60	$ 3.85		
21		Advertising dollars	$ 2,238.67	$ 2,036.25	$ 4,274.92	
22		Demand	12,193.35	18,290.03	30,483.38	
23		Quantity produced	12,193.35	18,290.03	30,483.38	
24						
25		Min. percent requirement	0	>=	0	
26		Max. percent limitation	-9145.015106	<=	0	
27						
28					Used	Unused
29		Budget	$ 22,357.70	$ 27,642.30	$ 50,000.00	$ -

This example points out the importance of fully understanding the mathematical model when analyzing sensitivity information. One suggestion to ensure that sensitivity analysis information is interpreted properly in spreadsheet models is to use Excel's formula auditing capability. If you select the cost of Shiraz (cell C5) and apply the "Trace Dependents" command from the *Formula Auditing* menu, you will see that the unit cost influences both the unit profit (cell C20) *and* the budget constraint function (cell C29).

HOW *SOLVER* WORKS

Solver uses a mathematical algorithm called the "simplex method." This was developed in 1947 by the late Dr. George Dantzig. The simplex method characterizes feasible solutions algebraically by solving systems of linear equations (obtained by adding slack variables to the constraints as we described earlier in this chapter). It moves systematically from one solution to another to improve the objective function until an optimal solution is found (or until the problem is deemed infeasible or unbounded). Because of the linearity of the constraints and objective function, the simplex method is guaranteed to find an optimal solution if one exists, and usually does so quickly and efficiently.

To gain some intuition into the logic of *Solver*, consider the following example. Crebo Manufacturing produces four types of structural support fittings—plugs, rails, rivets, and clips—that are machined on two CNC machining centers. The machining centers have a capacity of 280,000 minutes per year. The gross margin per unit and machining requirements are provided below:

PRODUCT	PLUGS	RAILS	RIVETS	CLIPS
Gross margin/unit	$0.30	$1.30	$0.75	$1.20
Minutes/unit	1	2.5	1.5	2

How many of each product should be made to maximize gross profit margin?

To formulate this as a linear optimization model, define $X1$, $X2$, $X3$, and $X4$ to be the number of plugs, rails, rivets, and clips to produce. The problem is to:

$$\text{Maximize Gross Margin} = 3X1 + 1.3X2 + .75X3 + 1.2X4$$

subject to the constraint that limits the machining capacity and nonnegativity of the variables:

$$1X1 + 2.5X2 + 1.5X3 + 2X4 \leq 280,000$$

$$X1, X2, X3, X4 \geq 0$$

To solve this problem, your first thought might be to choose the variable with the highest marginal profit. Because $X2$ has the highest marginal profit, you might try producing as many rails as possible. Since each rail requires 2.5 minutes, the maximum number that can be produced is $280,000/2.5 = 112,000$, for a total profit of $\$1.3(112,000) = \$145,600$. However, notice that each rail uses a lot more machining time than the other products. The best solution isn't necessarily the one with the highest marginal profit, but the one that provides the highest *total* profit. Therefore, more profit might be realized by producing a proportionately larger quantity of a different product having a smaller marginal profit. This is the key insight. What the simplex method does is essentially evaluate the impact of constraints in terms of their contribution to the objective function for each variable. For the simple case of only one constraint, the optimal (maximum) solution is found by simply choosing the variable with the highest ratio of the objective coefficient to the constraint coefficient; in this example, the gross margin/unit to the minutes per unit of machining capacity used. The highest ratio occurs for $X4$, $1.2/2 = 0.6$. This can be interpreted as the marginal profit per unit of resource consumed. If we produce the maximum number of clips, $280,000/2 = 140,000$, the total profit is $1.2(140,000) = \$168,000$, considerably more than the profit for rails alone. The mathematics gets complicated with more constraints and requires multiple iterations to systematically improve the solution, but that's the basic idea!

The simplex method allows many real business problems involving thousands or even millions of variables, and often hundreds or thousands of constraints, to be solved in reasonable computational time, and is the basis for advanced optimization algorithms involving integer variables that we describe in the next chapter.

BASIC CONCEPTS REVIEW QUESTIONS

1. Explain the following terms: optimization, objective function, optimal solution, constraint, constraint function, feasible solution, binding constraint, slack.
2. Explain how *Solver* identifies a unique optimal solution, alternative optimal solutions, an unbounded problem, and an infeasible problem.
3. List the important guidelines to follow for modeling optimization problems on spreadsheets.
4. What Excel functions should you avoid when implementing linear optimization models on spreadsheets?
5. Explain what reduced cost and shadow price signify in *Solver* Sensitivity Reports.
6. For each of the examples in this chapter, classify the constraints into the following categories. Are there any other types of constraints that do not fall into these categories?
   a. Simple bounds
   b. Limitations
   c. Requirements
   d. Proportional relationships
   e. Balance constraints
7. Explain the advantages and disadvantages of the way *Solver* handles simple bound constraints. How can complete shadow price information for bound constraints be generated using *Solver*?
8. Explain the shadow prices for the GAC transportation model in Figure 13.15 in simple managerial terms.

SKILL-BUILDING EXERCISES

1. Make the following changes to the Sklenka Ski Company model, re-solve using *Solver*, and answer the following questions:
   a. Increase the unit profit on Jordanelle skis by $10. What happens to the solution? Could you have predicted this from the Sensitivity Report (Figure 13.11)?
   b. Decrease the unit profit on Jordanelle skis by $10. What happens to the solution? Could you have predicted this from the Sensitivity Report (Figure 13.11)?
   c. Increase the number of finishing hours available by 10. What happens to the solution? Could you have predicted this from the Sensitivity Report (Figure 13.11)?
   d. Decrease the number of finishing hours available by 10. What happens to the solution? Could you have predicted this from the Sensitivity Report (Figure 13.11)?
   e. Change the unit profit for Deercrest skis to $75. What solution do you get? Do alternate optimal solutions exist? Verify that producing 0 pairs of Jordanelle and 14 pairs of Deercrest skis is an alternate optimal solution for this scenario.
   f. Change the finishing and fabrication constraints to be ≥ instead of ≤ type of constraints. What happens? Why did this occur?
   g. Change the finishing and fabrication constraints to = instead of ≤ type of constraints. What happens? Why did this occur?
2. Implement the Camm Textiles process selection model on a spreadsheet and solve it using *Solver*.
3. Implement the BG Seed Company blending model on a spreadsheet and solve it using *Solver*. Explain the *Solver* results.
4. Implement the Innis Investments portfolio investment model on a spreadsheet and solve it using *Solver*.
5. Implement both the original and alternative models developed for Kristin's Creations multiperiod planning model to verify that they result in the same optimal solution. In the alternative model, how can you determine the values of the inventory variables?
6. Implement the D.A. Branch & Sons multiperiod investment model on a spreadsheet and solve it using *Solver*.

PROBLEMS AND APPLICATIONS

Data for most of these problems can be found in the Excel file *Chapter 13 Problem Data* to facilitate model development.

1. Burger Office Equipment produces two types of desks, standard and deluxe. Deluxe desks have oak tops and more expensive hardware and require additional time for finishing and polishing. Standard desks require 80 square feet of pine and 12 hours of labor, while deluxe desks require 62 square feet of pine, 18 square feet of oak, and 18 hours of labor. For the next week, the company has 5000 square feet of pine, 700 square feet of oak, and 400 hours of labor available. Standard desks net a profit of $75, while deluxe desks net a profit of $160. All desks can be sold to national chains such as Staples or Office Depot.
   a. Develop a linear optimization model to determine how many of each the company should make next week to maximize profit contribution.
   b. Implement your model on a spreadsheet and find an optimal solution.
   c. Explain the reduced cost associated with standard desks.
   d. What constraints are binding? Explain how the shadow price can be used to make decisions that will improve profitability.
   e. If 25% of the oak is deemed to be cosmetically defective, how will the optimal solution be affected?
   f. The shop foreman is suggesting that his workforce be allowed to work an additional 100 hours at an overtime premium of $6/hour. Is this a good suggestion? Why or why not?

2. Sandford Tile Company makes ceramic and porcelain tile for residential and commercial use. They produce three different grades of tile (for walls, residential flooring, and commercial flooring), each of which requires different amounts of materials and production time, and generates different contribution to profit. The information below shows the percentage of materials needed for each grade and the profit per square foot.

	GRADE I	GRADE II	GRADE III
Profit/square foot	$2.50	$4.00	$5.00
Clay	50%	30%	25%
Silica	5%	15%	10%
Sand	20%	15%	15%
Feldspar	25%	40%	50%

Each week, Sanford Tile receives raw material shipments and the operations manager must schedule the plant to efficiently use the materials to maximize profitability. Currently, inventory consists of 6,000 pounds of clay, 3,000 pounds of silica, 5,000 pounds of sand, and 8,000 pounds of feldspar. Because demand varies for the different grades, marketing estimates that at most 8,000 square feet of Grade III tile should be produced, and that at least 1,500 square feet of Grade I tiles are required. Each square foot of tile weighs approximately two pounds.
   a. Develop a linear optimization model to determine how many of each grade of tile the company should make next week to maximize profit contribution.
   b. Implement your model on a spreadsheet and find an optimal solution.
   c. Explain the sensitivity information for the objective coefficients. What happens if the profit on Grade I is increased by $0.05?
   d. If an additional 500 pounds of feldspar is available, how will the optimal solution be affected?

e. Suppose that 1,000 pounds of clay are found to be of inferior quality. What should the company do?

f. Use the auxiliary variable cells technique to handle the bound constraints and generate all shadow prices.

3. Worley Fluid Supplies produces three types of fluid-handling equipment: control valves, metering pumps, and hydraulic cylinders. All three products require assembly and testing before they can be shipped to customers.

	CONTROL VALVE	METERING PUMP	HYDRAULIC CYLINDER
Assembly time (min)	45	20	30
Testing time (min)	20	15	25
Profit/unit	$372	$174	$288
Maximum sales	20	50	45
Minimum sales	5	12	22

A total of 3,060 minutes of assembly time and 2,100 minutes of testing time are available next week.

a. Develop a linear optimization model to determine how many pieces of equipment the company should make next week to maximize profit contribution.

b. Implement your model on a spreadsheet and find an optimal solution.

c. Explain the sensitivity information for the objective coefficients. What happens if the profit on hydraulic cylinders is decreased by $10?

d. Due to scheduled maintenance, the assembly time is expected to be only 3,000 minutes. How will this affect the solution?

e. A worker in the testing department has to take a personal leave because of a death in the family and will miss 3 days (24 hours). How will this affect the optimal solution?

f. Use the auxiliary variable technique to handle the bound constraints and generate all shadow prices.

4. Rosenberg Land Development (RLD) is a developer of condominium properties in the Southwest United States. RLD has recently acquired a 40.625 acre site outside of Phoenix, Arizona. Zoning restrictions allow at most 8 units per acre. Three types of condominiums are planned: one, two, and three bedroom units. The average construction costs for each type of unit are $450,000, $600,000, and $750,000. These units will generate a net profit of 10%. The company has equity and loans totaling $180 million dollars for this project. From prior development projects, senior managers have determined that there must be a minimum of 15% one bedroom units, 25% two bedroom units, and 25% three bedroom units.

a. Develop a linear optimization model to determine how many of each type of unit the developer should build.

b. Implement your model on a spreadsheet and find an optimal solution.

c. Explain the value of increasing the budget for the project.

5. Marketing managers have various media alternatives, such as radio, TV, magazines, etc., in which to advertise and must determine which to use, the number of insertions in each, and the timing of insertions to maximize advertising effectiveness within a limited budget. Suppose that three media options are available to Kernan Services Corporation: radio, TV, and magazine. The following table provides some information about costs, exposure values, and bounds on the permissible number of ads in each medium desired by the firm. The exposure value is a measure of the number of people exposed to the advertisement and is derived from market research studies, and the client's objective is to

maximize the total exposure value. The company would like to achieve a total exposure value of at least 90,000.

	MEDIUM COST/AD	EXPOSURE VALUE/AD	MIN UNITS	MAX UNITS
Radio	$500	2,000	0	15
TV	$2,000	4,000	10	
Magazine	$200	2,700	6	12

How many of each type of ad should be placed in order to minimize the cost of achieving the minimum required total exposure? Use the auxiliary variable approach to model this problem, and write a short memo to the marketing manager explaining the solution and sensitivity information.

6. Klein Industries manufactures three types of air compressors: small, medium, and large, which have unit profits of $20.46, $34.28, and $16.22, respectively. The projected monthly sales are:

	SMALL	MEDIUM	LARGE
Minimum	14,000	6,200	2,600
Maximum	21,000	12,500	4,200

The production process consists of three primary activities: bending and forming, welding, and painting. The amount of time in minutes needed to process each product in each department is shown below:

	SMALL	MEDIUM	LARGE	AVAILABLE TIME
Bending/forming	0.4	0.7	0.8	23,400
Welding	0.6	1.0	1.2	23,400
Painting	1.4	2.6	3.1	46,800

How many of each type of air compressor should the company produce to maximize profit?

a. Formulate and solve a linear optimization model using the auxiliary variable cells method and write a short memo to the production manager explaining the sensitivity information.

b. Solve the model without the auxiliary variables and explain the relationship between the reduced costs and the shadow prices found in part a.

7. The International Chef, Inc. markets three blends of oriental tea: premium, Duke Grey, and breakfast. The firm uses tea leaves from India, China, and new domestic California sources.

	TEA LEAVES (PERCENT)		
QUALITY	INDIAN	CHINESE	CALIFORNIA
Premium	40	20	20
Duke Grey	20	30	40
Breakfast	40	40	40

Net profit per pound for each blend is $0.50 for premium, $0.30 for Duke Grey, and $0.20 for breakfast. The firm's regular weekly supplies are 20,000 pounds of

Indian tea leaves, 22,000 pounds of Chinese tea leaves, and 16,000 pounds of California tea leaves. Develop and solve a linear optimization model to determine the optimal mix to maximize profit, and write a short memo to the president, Kathy Chung, explaining the sensitivity information in language that she can understand.

8. Young Energy operates a power plant that includes a coal-fired boiler to produce steam to drive a generator. The company can purchase different types of coals and blend them to meet the requirements for burning in the boiler. The table below shows the characteristics of the different types of coals:

TYPE	BTU/LB.	% ASH	% MOISTURE	COST ($/LB)
A	11,500	13%	10%	$2.49
B	11,800	10%	8%	$3.04
C	12,200	12%	8%	$2.99
D	12,100	12%	8%	$2.61

The required BTU/lb must be at least 11,900. In addition, the ash content can be at most 12.2% and the moisture content at most 9.4%. Develop and solve a linear optimization model to find the best coal blend for Young Energy. Explain how the company might reduce its costs by changing the blending restrictions.

9. The Hansel Corporation, located in Bangalore, India, makes plastics materials that are mixed with various additives and reinforcing materials before being melted, extruded, and cut into small pellets for sale to other manufacturers. Four grades of plastic are made, each of which might include up to four different additives. The table below shows the number of pounds of additive per pound of each grade of final product, the weekly availability of the additives, and cost and profitability information.

	GRADE 1	GRADE 2	GRADE 3	GRADE 4	AVAILABILITY
Additive A	0.40	0.37	0.34	0.90	100,000
Additive B	0.30	0.33	0.33		90,000
Additive C	0.20	0.25	0.33		40,000
Additive D	0.10	0.05		0.10	10,000
Profit/lb	$2.00	$1.70	$1.50	$2.80	

Because of marketing considerations, the total amount of grades 1 and 2 should not exceed 60% of the total of all grades produced, and at least 30% of the total product mix should be grade 4.

a. How much of each grade should be produced to maximize profit? Develop and solve a linear optimization model.

b. A labor strike in India leads to a shortage of 20,000 units of additive C. What should the production manager do?

c. Management is considering raising the price on grade 2 to $2.00 per pound. How will the solution be changed?

10. Janette Douglas is coordinating a bake sale for a nonprofit organization. The organization has acquired $2,000 in donations to hold the sale. The table

below shows the amounts and costs of ingredients used per batch of each baked good:

Ingredient	Brownies	Cupcakes	Peanut Butter Cups	Shortbread Cookies	Cost/Unit
Butter (cups)	0.67	0.33	1	0.75	$1.44
Flour (cups)	1.5	1.5	1.25	2	$0.09
Sugar (cups)	1.75	1	2	0.25	$0.16
Vanilla (tsp)	2	0.5	0	0	$0.06
Eggs	3	2	1	0	$0.12
Walnuts (cups)	2	0	0	0	$0.31
Milk (cups)	0.5	1	2	0	$0.05
Chocolate (oz)	8	2.5	9	0	$0.10
Baking soda (tsp)	2	1	0	0	$0.07
Frosting (cups)	0.5	1.5	0	1	$2.74
Peanut butter (cups)	0	0	2.5	0	$2.04

One batch results in 10 brownies, 12 cupcakes, 8 peanut butter cups, and 12 shortbread cookies. Each batch of brownies can be sold for $6.00, cupcakes for $10.00, peanut butter cups for $12.00, and shortbread cookies for $7.50. The organization anticipates that a total of at least 4,000 baked goods must be made. For adequate variety, at least 30 batches of each baked good are required, except for the popular brownies, which require at least 100 batches. In addition, no more than 40 batches of shortbread cookies should be made. How can the organization best use its budget and make the largest amount of money?

11. Liquid Gold, Inc. transports radioactive waste from nuclear power plants to disposal sites around and about the country. Each plant has an amount of material that must be moved each period. Each site has a limited capacity per period. The cost of transporting between sites is given here (some combinations of plants and storage sites are not to be used, and no figure is given). Develop and solve a transportation model for this problem.

Plant	Material	S1	S2	S3	S4	Site	Capacity
P1	20,876	105	86	—	23	S1	285,922
P2	50,870	86	58	41	—	S2	308,578
P3	38,652	93	46	65	38	S3	111,955
P4	28,951	116	27	94	—	S4	208,555
P5	87,423	88	56	82	89		
P6	76,190	111	36	72	—		
P7	58,237	169	65	48	—		

12. Jason Wright is a part-time MBA student who would like to optimize his financial decisions. Currently, he has $15,000 in his savings account. Based on an analysis of his take-home pay, expected bonuses, and anticipated tax refund, he has estimated his income for each month over the next year. In addition, he has estimated his monthly expenses, which vary because of scheduled payments for

insurance, utilities, tuition and books, and so on. The table below summarizes his estimates:

MONTH	INCOME	EXPENSES
1 January	$3,400	$3,360
2 February	$3,400	$2,900
3 March	$3,400	$6,600
4 April	$9,500	$2,750
5 May	$3,400	$2,800
6 June	$5,000	$6,800
7 July	$4,600	$3,200
8 August	$3,400	$3,600
9 September	$3,400	$6,550
10 October	$3,400	$2,800
11 November	$3,400	$2,900
12 December	$5,000	$6,650

Jason has identified several short-term investment opportunities:
- A 3-month CD yielding 0.74% at maturity
- A 6-month CD yielding 1.72% at maturity
- An 11-month CD yielding 4.08% at maturity
- His savings account yields 0.0375% per month

To ensure enough cash for emergencies, he would like to maintain at least $2,000 in the savings account. Jason's objective is to maximize his cash balance at the end of the year. Develop a linear optimization model to find the best investment strategy.

13. Shannon Board manages a nonprofit children's theater company. The theater performs in two venues: Beverly Hall and the Lauren Elizabeth Theater. For the upcoming season, seven shows have been chosen. The question Shannon faces is how many performances of each of the seven shows should be scheduled. A financial analysis has estimated revenues for each performance of the seven shows, and Shannon has set the minimum number of performances of each show based upon union agreements with Actor's Equity Association and the popularity of the shows in other markets:

SHOW	REVENUE	COST	MINIMUM NUMBER OF PERFORMANCES
1	$2,217	$968	32
2	$2,330	$1,568	13
3	$1,993	$755	23
4	$3,364	$1,148	34
5	$2,868	$1,180	35
6	$3,851	$1,541	16
7	$1,836	$1,359	21

Beverly Hall is available for 60 performances during the season, while Lauren Elizabeth Theater is available for 150 performances. Shows 3 and 7 must be performed in Beverly Hall, while the other shows are performed in the Lauren Elizabeth Theater. The company wants to achieve revenues of at least $550,000 while minimizing its production costs. Develop and solve a linear optimization model to determine the best way to schedule the shows. Is it possible to achieve revenues of $600,000? What is the highest amount of revenue that can be achieved?

14. Bratcher Cosmetics has created two new perfumes: Lavender Breeze and Lavender Passion. It costs $3.00 to purchase the lavender needed for each bottle of Lavender Passion, and $2.60 for each bottle of Lavender Breeze. Marketing executives have set a requirement that at least 30% but no more than 60% of the product mix be Lavender Passion. The average monthly demand for Lavender Passion is 5,000 bottles, and is estimated to increase by 10 bottles for each $1 spent on advertising. For Lavender Breeze, the monthly demand is estimated to be 9000 bottles, but is expected to increase by 50 bottles for each $1 spent on advertising. Lavender Passion sells for $12.00 per bottle, and Lavender Breeze for $10.50 per bottle. A monthly budget of $100,000 is available for both advertising and purchase of the lavender. How many of each type of perfume should be produced to maximize the net profit?

15. An international graduate student will receive a $28,000 foundation scholarship and reduced tuition. He must pay $1,500 in tuition for each of the autumn, winter, and spring quarters, and $500 in the summer. Payments are due on the first day of September, December, March, and May, respectively. Living expenses are estimated to be $1,500 per month, payable on the first day of the month. The foundation will pay him $18,000 on August 1, and the remainder on May 1. To earn as much interest as possible, the student wishes to invest the money. Three types of investments are available at his bank: a 3-month CD, earning 0.75% (net 3-month rate); a 6-month CD, earning 1.9%; and a 12-month CD, earning 4.2%. Develop a linear optimization model to determine how he can best invest the money and meet his financial obligations.

16. A recent MBA graduate, Dara, has gained control over custodial accounts that her parents had established. Currently, her money is invested in four funds but she has identified several other funds as options for investment. She has $100,000 to invest with the following restrictions:
   - Keep at least $5,000 in savings.
   - Invest at least 14% in the money market fund.
   - Invest at least 16% in international funds.
   - Keep 35% of funds in current holdings.
   - Do not allocate more than 20% of funds to any one investment except for the money market and savings account.
   - Allocate at least 30% into new investments.

	AVERAGE RETURN	EXPENSES	
1 Large cap blend	17.2%	0.93%	(current holding)
2 Small cap growth	20.4%	0.56%	(current holding)
3 Green fund	26.3%	0.70%	(current holding)
4 Growth & income	15.6%	0.92%	(current holding)
5 Multi-cap growth	19.8%	0.92%	
6 Mid-cap index	22.1%	0.22%	
7 Multi-cap core	27.9%	0.98%	
8 Small cap international	35.0%	0.54%	
9 Emerging international	36.1%	1.17%	
10 Money market fund	4.75%	0	
11 Savings account	1.0%	0	

Develop a linear optimization model to determine the best investment strategy.

17. Holcomb Candles, Inc. manufactures decorative candles and has contracted with a national retailer to supply a set of special holiday candles to its 8,500 stores. These include large jars, small jars, large pillars, small pillars, and a package of

four votive candles. In negotiating the contract for the display, the manufacturer and retailer agreed that 8 feet would be designated for the display in each store, but that at least 2 feet be dedicated to large jars and large pillars, and at least one foot to the votive candle packages. At least as many jars as pillars must be provided. The manufacturer has obtained 200,000 pounds of wax, 250,000 feet of wick, and 100,000 ounces of holiday fragrance. The amount of materials and display size required for each product is shown in the table below:

	LARGE JAR	SMALL JAR	LARGE PILLAR	SMALL PILLAR	VOTIVE PACK
Wax	0.5	0.25	0.5	0.25	0.3125
Fragrance	0.24	0.12	0.24	0.12	0.15
Wick	0.43	0.22	0.58	0.33	0.8
Display feet	0.48	0.24	0.23	0.23	0.26
Profit/unit	$0.25	$0.20	$0.24	$0.21	$0.16

How many of each product should be made to maximize the profit? Interpret the shadow prices in the sensitivity report.

18. A department store chain is planning to open a new store. It needs to decide how to allocate the 100,000 square feet of available floor space among seven departments. Data on expected performance of each department per month, in terms of square feet (sf), are shown below.

DEPARTMENT	INVESTMENT/ SF	RISK AS A % OF $ INVESTED	MINIMUM SF	MAXIMUM SF	EXPECTED PROFIT PER SF
Electronics	$100	24	6,000	30,000	$12.00
Furniture	50	12	10,000	30,000	6.00
Men's Clothing	30	5	2,000	5,000	2.00
Clothing	600	10	3,000	40,000	30.00
Jewelry	900	14	1,000	10,000	20.00
Books	50	2	1,000	5,000	1.00
Appliances	400	3	12,000	40,000	13.00

The company has gathered $20 million to invest in floor stock. The risk column is a measure of risk associated with investment in floor stock based on past data from other stores and accounts for outdated inventory, pilferage, breakage, etc.. For instance, electronics loses 24% of its total investment, furniture loses 12% of its total investment, etc.

a. Develop a linear optimization model to maximize profit.
b. If the chain obtains another $1 million of investment capital for stock, what would the new solution be?

19. A South American honey farm makes five types of honey: cream, filtered, pasteurized, mélange (a mixture of several types), and strained, which are sold in 1 or 0.5 kilogram glass containers, 1 kg and 0.75 kg plastic containers, or in bulk. Key data are shown in the following table.

SELLING PRICES (CHILEAN PESOS)					
	0.75 KG PLASTIC	1 KG PLASTIC	0.5 KG GLASS	1 KG GLASS	BULK/KG
cream	744	880	760	990	616
filtered	635	744	678	840	521
pasteurized	696	821	711	930	575
mélange	669	787	683	890	551
strained	683	804	697	910	563

MINIMUM DEMAND				
	0.75 KG PLASTIC	1 KG PLASTIC	0.5 KG GLASS	1 KG GLASS
cream	300	250	350	200
filtered	250	240	300	180
pasteurized	230	230	350	300
mélange	350	300	250	350
strained	360	350	250	380

MAXIMUM DEMAND				
	0.75 KG PLASTIC	1 KG PLASTIC	0.5 KG GLASS	1 KG GLASS
cream	550	350	470	310
filtered	400	380	440	300
pasteurized	360	390	490	400
mélange	530	410	390	430
strained	480	420	380	500

PACKAGE COSTS (CHILEAN PESOS)			
0.75 KG PLASTIC	1 KG PLASTIC	0.5 KG GLASS	1 KG GLASS
91	112	276	351

Harvesting and production costs for each product per kilogram in pesos are:
  Cream: 322
  Filtered: 255
  Pasteurized: 305
  Mélange: 272
  Strained: 287
Develop a linear optimization model to maximize profit if a total of 10,000 kg of honey is available.

20. Mirza Manufacturing makes four electronic products, each of which is comprised of three main materials: magnet, wiring, and casing. The products are shipped to three distribution centers in North America, Europe, and Asia. Marketing has specified that no location should receive more than the maximum demand and should receive at least the minimum demand. The material costs/unit are: magnet—$0.59, wire—$0.29, and casing—$0.31. The table below shows the number of units of each material required in each unit of end product and the production cost per unit.

PRODUCT	PRODUCTION COST/UNIT	MAGNETS	WIRE	CASING
A	$0.25	4	2	2
B	$0.35	3	1	3
C	$0.15	2	2	1
D	$0.10	8	3	2

Additional information is provided below.

MIN DEMAND			
PRODUCT	NA	EU	ASIA
A	850	900	100
B	700	200	500
C	1,100	800	600
D	1,500	3,500	2,000

MAX DEMAND			
PRODUCT	NA	EU	ASIA
A	2,550	2,700	300
B	2,100	600	1,500
C	3,300	2,400	1,800
D	4,500	10,500	6,000

SHIPPING COST/UNIT			
PRODUCT	NA	EU	ASIA
A	0.2	0.25	0.35
B	0.2	0.2	0.3
C	0.2	0.2	0.3
D	0.2	0.2	0.25

UNIT SALES REVENUE			
PRODUCT	NA	EU	ASIA
A	4.0	4.5	4.55
B	3.7	3.9	3.95
C	2.7	2.9	2.4
D	6.8	6.5	6.9

AVAILABLE RAW MATERIAL	
Magnet	120,000
Wire	50,000
Casing	40,000

Develop an appropriate linear optimization model to maximize net profit.

CASE

Haller's Pub & Brewery

Jeremy Haller of Haller's Pub & Brewery has compiled data describing the amount of different ingredients and labor resources needed to brew six different types of beers that the brewery makes. He also gathered financial information and estimated demand over a 26-week forecast horizon. These data are provided in the *Haller's Pub* tab of the Excel file *Chapter 13 Problem Data*. The profits for each batch of each type of beer are:

Light Ale: $3,925.78
Golden Ale: $4,062.75

Freedom Wheat: $3,732.34
Berry Wheat: $3,704.49
Dark Ale: $3,905.79
Hearty Stout: $3,490.22

These values incorporate fixed overhead costs of $7,500 per batch. Use the data to validate the profit figures and develop a linear optimization model to maximize profit. Write a report to Mr. Haller explaining the results and sensitivity analysis information in language that he (a nonquantitative manager) can understand.

Chapter 14

Integer and Nonlinear Optimization

INTRODUCTION

*T*his chapter extends the concepts developed in Chapter 13 to integer and nonlinear optimization models. Both integer and nonlinear models have many important applications in areas such as scheduling, supply chains, and portfolio management.

In an **integer linear optimization model (integer program)**, some or all of the variables are restricted to being *whole numbers*. If only a subset of variables is restricted to being integer while others are continuous, we call this a **mixed integer linear optimization model**. A special type of integer problem is one in which variables can only be 0 or 1. These *binary variables* help us to model logical "yes or no" decisions. Integer linear optimization models are generally more difficult to solve than pure linear optimization models.

In a **nonlinear optimization model (nonlinear program)**, the objective function and/or constraint functions are *nonlinear functions* of the decision variables; that is, terms cannot be written as a constant times a variable. Some examples of nonlinear terms are $3x^2, 4/y$, and $6xy$. Nonlinear optimization models are considerably more difficult to solve than either linear or integer models.

In this chapter, we will discuss:

➤ *Developing mathematical models for integer and nonlinear optimization problems*

➤ *Implementing integer and nonlinear optimization models on spreadsheets*

➤ *Using Solver to find optimal solutions to integer and nonlinear optimization models*

➤ *Using Evolutionary Solver for obtaining good solutions to difficult optimization problems*

➤ *Applying simulation and risk analysis to better understand the implications of solutions to optimization models*

➤ *Using OptQuest for optimization models with stochastic conditions*

INTEGER OPTIMIZATION MODELS

An *integer (linear) optimization model* is a linear model in which some or all of the decision variables are restricted to integer (whole-number) values. For many practical applications, we need not worry about forcing the decision variables to be integers. For example, in deciding on the optimal number of cases of diapers to produce next month, we could use a linear model, since rounding a value like 5,621.63 would have little impact on the results. However, in a production planning decision involving low-volume, high-cost items such as airplanes, an optimal value of 10.42 would make little sense, and a difference of one unit (rounded up or down) could have significant economic consequences. Decision variables that we force to be integers are called *general integer variables.* An example for which general integer variables are needed follows.

A Cutting Stock Problem

The paper industry uses integer optimization to find the best mix of cutting patterns to meet demand for various sizes of paper rolls. In a similar fashion, sheet steel producers cut strips of different sizes from rolled coils of thin steel. Suppose that a company makes standard 110-inch-wide rolls of thin sheet metal, and slits them into smaller rolls to meet customer orders for widths of 12, 15, and 30 inches. The demands for these widths vary from week to week.

From a 110-inch roll, there are many different ways to slit 12-, 15-, and 30-inch pieces. A *cutting pattern* is a configuration of the number of smaller rolls of each type that are cut from the raw stock. Of course, one would want to use as much of the roll as possible to avoid costly scrap. For example, one could cut seven 15-inch rolls, leaving a 5-inch piece of scrap. Finding good cutting patterns for a large set of end products is in itself a challenging problem. Suppose that the company has proposed the following cutting patterns:

	Size of End Item			
Pattern	12″	15″	30″	Scrap
1	0	7	0	5″
2	0	1	3	5″
3	1	0	3	8″
4	9	0	0	2″
5	2	1	2	11″
6	7	1	0	11″

Demands this week are 500 12-inch rolls, 715 15-inch rolls, and 630 30-inch rolls. The problem is to develop a model that will determine how many 110-inch rolls to cut into each of the six patterns in order to meet demand and scrap.

Define X_i to be the number of 110-inch rolls to cut using cutting pattern i, for $i = 1, \ldots, 6$. Note that X_i needs to be a whole number because each roll that is cut generates a different number of end items. Thus, X_i will be modeled using general integer variables. Because the objective is to minimize scrap, the objective function is:

$$\text{Min } 5X_1 + 5X_2 + 8X_3 + 2X_4 + 11X_5 + 11X_6$$

The only constraints are that end item demand must be met; that is, we must produce at least 500 12-inch rolls, 715 15-inch rolls, and 630 30-inch rolls. The number of end item rolls produced is found by multiplying the number of end item rolls produced by each cutting pattern by the number of 110-inch rolls cut using that pattern. Therefore, the constraints are:

$$0X_1 + 0X_2 + 1X_3 + 9X_4 + 2X_5 + 7X_6 \geq 500 \quad \text{(12-inch rolls)}$$
$$7X_1 + 1X_2 + 0X_3 + 0X_4 + 1X_5 + 1X_6 \geq 715 \quad \text{(15-inch rolls)}$$
$$0X_1 + 3X_2 + 3X_3 + 0X_4 + 2X_5 + 0X_6 \geq 630 \quad \text{(30-inch rolls)}$$

Finally, we include nonnegativity and integer restrictions:

$$X_i \geq 0 \text{ and integer}$$

Solving Integer Optimization Models

Integer optimization models are set up in the same manner as linear models in a spreadsheet. Figure 14.1 shows the cutting stock model implementation on a spreadsheet (Excel file *Cutting Stock Model*) and the solution that results if we ignore the integer restrictions on the variables (that is, the solution to the linear optimization model). The *Solver* model is shown in Figure 14.2. Note that the optimal solution results in fractional values of the variables.

To enforce integer restrictions on variables using *Solver*, add a constraint by selecting the variable range and choosing *int* from the drop-down box in the *Add Constraint* dialog

Figure 14.1 *Cutting Stock Model* Spreadsheet and Linear
Programming Solution

	A	B	C	D	E
1	**Cutting Stock Problem**				
2					
3	**Data**				
4	Pattern	12-in rolls	15-in rolls	30-in rolls	Scrap
5	1	0	7	0	5
6	2	0	1	3	5
7	3	1	0	3	8
8	4	9	0	0	2
9	5	2	1	2	11
10	6	7	1	0	11
11	Demand	500	715	630	
12					
13	**Model**				
14		No. of rolls			
15	Pattern 1	72.14			
16	Pattern 2	210.00			
17	Pattern 3	0.00			
18	Pattern 4	55.56			
19	Pattern 5	0.00			
20	Pattern 6	0.00			
21					
22		12-in rolls	15-in rolls	30-in rolls	
23	Number produced	500	715	630	
24					
25		Total			
26	Scrap	1521.8254			

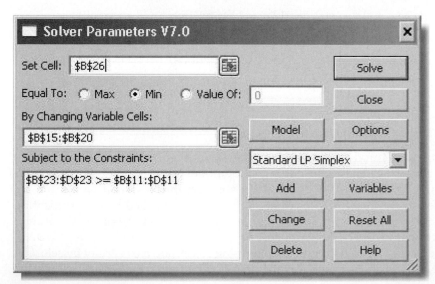

Figure 14.2 *Solver* Model for
Cutting Stock Problem

Figure 14.3
Enforcing Integer
Restrictions in
Solver

as shown in Figure 14.3. Figure 14.4 shows the resulting solution. Notice that the minimum value of the objective function (total scrap) is larger than the linear programming solution. This is expected because we have added an additional constraint (the integer restrictions). Whenever you add a constraint to a model, the value of the objective function can never improve and usually worsens. Also note that simply rounding the linear programming solution would not have provided the optimal integer solution.

There is one important thing to remember when using Solver for integer optimization. Solver sets a default tolerance of 0.05, meaning that the solution procedure will stop if a solution is found that can be proven to be within 5% of the optimal value. To

	A	B	C	D	E
1	**Cutting Stock Problem**				
2					
3	**Data**				
4	Pattern	12-in rolls	15-in rolls	30-in rolls	Scrap
5	1	0	7	0	5
6	2	0	1	3	5
7	3	1	0	3	8
8	4	9	0	0	2
9	5	2	1	2	11
10	6	7	1	0	11
11	Demand	500	715	630	
12					
13	**Model**				
14		No. of rolls			
15	Pattern 1	72.00			
16	Pattern 2	211.00			
17	Pattern 3	0.00			
18	Pattern 4	56.00			
19	Pattern 5	0.00			
20	Pattern 6	0.00			
21					
22		12-in rolls	15-in rolls	30-in rolls	
23	Number produced	504	715	633	
24					
25		Total			
26	Scrap	1527			

Figure 14.4 Optimal
Integer Solution to the Cutting
Stock Problem

guarantee that the optimum value will be found by Solver, this tolerance should be set to 0. This is done by clicking the Options button and changing the tolerance under the Integer tab. For very large and complex problems, setting the tolerance to 0 may result in a very long computational time; in these cases, we often leave the tolerance at the default value.

If the optimal linear programming solution had turned out to have all integer values, then it clearly would have solved the integer model. In fact, the algorithm used to solve integer optimization models begins by solving the associated linear model without the integer restrictions and proceeds to enforce the integer restrictions using a systematic search process that involves solving a series of modified linear programs.

Because integer models are discontinuous by their very nature, sensitivity information cannot be generated in the same manner as for linear models, and therefore no sensitivity report is provided by *Solver*; only the Answer Report is available. To investigate changes in model parameters, it is necessary to re-solve the model.

INTEGER OPTIMIZATION MODELS WITH BINARY VARIABLES

Many optimization models require *binary variables,* which are variables that are restricted to being either 0 or 1. Mathematically, a binary variable x is simply a general integer variable that is restricted to being between 0 and 1:

$$0 \leq x \leq 1 \text{ and integer}$$

However, we usually just write this as:

$$x = 0 \text{ or } 1$$

Binary variables enable us to model logical decisions in optimization models. For example, binary variables can be used to model decisions such as whether ($x = 1$) or not ($x = 0$) to place a facility at a certain location, whether or not to run a production line, or whether or not to invest in a certain stock. One common example we present next is project selection or capital budgeting, in which a subset of potential projects must be selected with limited resource constraints.

Project Selection

Hahn Engineering's research and development group has identified five potential new engineering and development projects; however, the firm is constrained by its available budget and human resources. Each project is expected to generate a return (given by the net present value) but requires a fixed amount of cash and personnel. Because the resources are limited, all projects cannot be selected. Projects cannot be partially completed; thus, either the project must be undertaken completely or not at all. The data are given in Table 14.1. If a project is selected, it generates the full value of the expected return and requires the full amount of cash and personnel shown in Table 14.1. For example, if we select projects 1 and 3, the total

Table 14.1   Project Selection Data

	PROJECT 1	PROJECT 2	PROJECT 3	PROJECT 4	PROJECT 5	AVAILABLE RESOURCES
Expected return (NPV)	$180,000	$220,000	$150,000	$140,000	$200,000	
Cash requirements	$55,000	$83,000	$24,000	$49,000	$61,000	$150,000
Personnel requirements	5	3	2	5	3	12

return is $180,000 + $150,000 = $330,000$, and these projects require cash totaling $55,000 + $24,000 = $79,000$ and $5 + 2 = 7$ personnel.

To model this situation, we define the decision variables to be binary, corresponding to either not selecting or selecting each project, respectively. Define $x_i = 1$ if project i is selected, and 0 otherwise. By multiplying these binary variables by the expected returns, the objective function is:

$$\text{Maximize } \$180,000x_1 + \$220,000x_2 + \$150,000x_3 + \$140,000x_4 + \$200,000x_5$$

Because cash and personnel are limited, we have the constraints:

$$\$55,000x_1 + \$83,000x_2 + \$24,000x_3 + \$49,000x_4$$
$$+ \$61,000x_5 \leq \$150,000 \quad \text{(cash limitation)}$$
$$5x_1 + 3x_2 + 2x_3 + 5x_4 + 3x_5 \leq 12 \quad \text{(personnel limitation)}$$

Note that if projects 1 and 3 are selected, then $x_1 = 1$ and $x_3 = 1$ and the objective and constraint functions equal:

$$\text{Return} = \$180,000(1) + \$220,000(0) + \$150,000(1) + \$140,000(0)$$
$$+ \$200,000(0) = \$330,000$$
$$\text{Cash Required} = \$55,000(1) + \$83,000(0) + \$24,000(1) + \$49,000(0)$$
$$+ \$61,000(0) = \$79,000$$
$$\text{Personnel Required} = 5(1) + 3(0) + 2(1) + 5(0) + 3(0) = 7$$

This model is easy to implement on a spreadsheet, as shown in Figure 14.5 (Excel file *Project Selection Model*). The decision variables are defined in cells B11:F11. The objective function, computed in cell G14, is the total return, which can be expressed as the sum of the product of the return from each project and the binary decision variable:

$$\text{Total Return} = \text{B5} \times \text{B11} + \text{C5} \times \text{C11} + \text{D5} \times \text{D11} + \text{E5} \times \text{E11} + \text{F5} \times \text{F11}$$

These constraints can be written as:

$$\text{Cash Used} = \text{B6} \times \text{B11} + \text{C6} \times \text{C11} + \text{D6} \times \text{D11} + \text{E6} \times \text{E11} + \text{F6} \times \text{F11} \leq \text{G6}$$
$$\text{Personnel Used} = \text{B7} \times \text{B11} + \text{C7} \times \text{C11} + \text{D7} \times \text{D11} + \text{E7} \times \text{E11} + \text{F7} \times \text{F11} \leq \text{G7}$$

The left-hand sides of these functions can be found in cells G12 and G13.

Figure 14.5 *Project Selection Model* Spreadsheet

	A	B	C	D	E	F	G
1	**Project Selection Model**						
2							
3	**Data**						
4		**Project 1**	**Project 2**	**Project 3**	**Project 4**	**Project 5**	**Available**
5	**Expected Return (NPV)**	$ 180,000	$ 220,000	$ 150,000	$ 140,000	$ 200,000	**Resources**
6	Cash requirements	$ 55,000	$ 83,000	$ 24,000	$ 49,000	$ 61,000	$ 150,000
7	Personnel requirements	5	3	2	5	3	12
8							
9	**Model**						
10							
11	**Project selection decisions**	1	0	1	0	1	**Total**
12	Cash Used	$ 55,000	$ -	$ 24,000	$ -	$ 61,000	$ 140,000
13	Personnel Used	5	0	2	0	3	10
14	Return	$ 180,000	$ -	$ 150,000	$ -	$ 200,000	$ 530,000

Figure 14.6 *Solver* Model for Project Selection

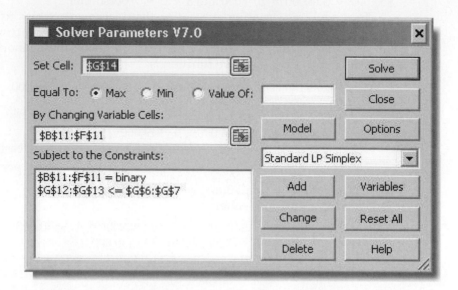

The *Solver* model is shown in Figure 14.6. To invoke the binary constraints on the variables, select the range of the variables in the *Add Constraint* dialog, and choose *bin* in the drop-down box. The resulting constraint is $B$11:$F$11 = binary as shown in the *Solver* model. The optimal solution, shown in Figure 14.5, is to select projects 1, 3, and 5 for a total return of $530,000.

Site Location Model

Integer optimization models have wide applications in locating facilities. The following is an example of a "covering problem," one in which we seek to choose a subset of locations that serve, or cover, all locations in a service area. Suppose that an unincorporated township wishes to find the best locations for fire stations. Assume that the township is divided into smaller districts or neighborhoods, and that transportation studies have estimated the response time for emergency vehicles to travel between each pair of districts. The township wants to locate the fire stations so that all districts can be reached within an eight-minute response time. The table below shows the estimated response time in minutes between each pair of districts:

FROM/TO	1	2	3	4	5	6	7
1	0	2	10	6	12	5	8
2	2	0	6	9	11	7	10
3	10	6	0	5	5	12	6
4	6	9	5	0	9	4	3
5	12	11	5	9	0	10	8
6	5	7	12	4	10	0	6
7	8	10	6	3	8	6	0

Define $X_j = 1$ if a fire station is located in district j, and 0 if not. The objective is to minimize the number of fire stations that need to be built:

$$\text{Min } X_1 + X_2 + X_3 + X_4 + X_5 + X_6 + X_7$$

Each district must be reachable within 8 minutes by some fire station. Thus, from the table, for example, we see that in order to be able to respond to district 1 in 8 minutes or less, a station must be located in either district 1, 2, 4, 6, or 7. Therefore, we must have the constraint:

$$X_1 + X_2 + X_4 + X_6 + X_7 \geq 1$$

Similar constraints may be formulated for each of the other districts:

$$
\begin{aligned}
X_1 + X_2 + X_3 + X_4 + X_5 + X_6 + X_7 &\geq 1 \\
X_1 + X_2 + X_3 + X_6 &\geq 1 \\
X_2 + X_3 + X_4 + X_5 + X_7 &\geq 1 \\
X_1 + X_3 + X_4 + X_6 + X_7 &\geq 1 \\
X_1 + X_2 + X_4 + X_6 + X_7 &\geq 1 \\
X_1 + X_3 + X_4 + X_5 + X_6 + X_7 &\geq 1
\end{aligned}
$$

Figure 14.7 shows a spreadsheet model for this problem (Excel file *Fire Station Location Model*). To develop the constraints in the model, we construct a matrix by converting all response times that are within 8 minutes to 1s, and those that exceed 8 minutes to 0s. Then the constraint functions for each district are simply the

Figure 14.7   *Fire Station Location Model* Spreadsheet

	A	B	C	D	E	F	G	H	I	J
1	**Fire Station Location Model**									
2										
3	**Data**									
4										
5	**Response time**		8							
6										
7	Response Times									
8	From/To	1	2	3	4	5	6	7		
9	1	0	2	10	6	12	5	8		
10	2	2	0	6	9	11	7	10		
11	3	10	6	0	5	5	12	6		
12	4	6	9	5	0	9	4	3		
13	5	12	11	5	9	0	10	8		
14	6	5	7	12	4	10	0	6		
15	7	8	10	6	3	8	6	0		
16										
17	**Model**									
18										
19	From/To	1	2	3	4	5	6	7	Covered?	Requirement
20	1	1	1	0	1	0	1	1	1	1
21	2	1	1	1	0	0	1	0	1	1
22	3	0	1	1	1	1	0	1	2	1
23	4	1	0	1	1	0	1	1	2	1
24	5	0	0	1	0	1	0	1	2	1
25	6	1	1	0	1	0	1	1	1	1
26	7	1	0	1	1	1	1	1	2	1
27									Total	
28	Location	0	0	1	0	0	0	1	2	

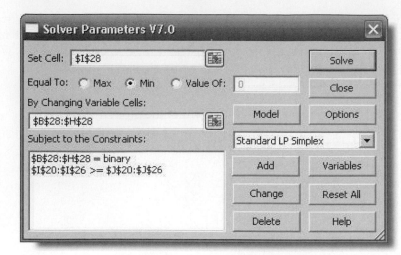

Figure 14.8 *Solver* Model for Fire Station Location

SUMPRODUCT of the decision variables and the rows of this matrix, making the *Solver* model, shown in Figure 14.8, easy to define. For this example, the solution is to site fire stations in districts 3 and 7.

As noted, sensitivity analysis can only be conducted for integer optimization by re-solving the model. Suppose that the township's Board of Trustees wants to better understand the trade-offs between the response time and minimum number of fire stations needed. We could change the value of the response time in cell C5 and resolve the model, obtaining the results:

RESPONSE TIME	MIN. # OF SITES
5	3
6	2
7	2
8	2
9	1
10	1

These results show the maximum response time can be reduced to six minutes while still using only two fire stations (the model solution yields districts 1 and 3). This would clearly be a better alternative. Also, if the response time is increased by only one minute from its original target, the township could save the cost of building a second facility. Of course, such decisions need to be evaluated carefully.

The next example illustrates a more complicated example involving the use of binary variables to configuring a personal computer.

Computer Configuration

Consumers and business customers have the opportunity to select the features and options of many products when ordering. For example, companies like Dell allow customers to choose the configuration of computers. Suppose that a customer wants to buy a new notebook computer with a limited budget of $1,800. Many options are generally available. For example, suppose that the base price of a notebook is $1,000, and includes 1-year warranty, 1 GB RAM, an 80 GB hard drive; CD-ROM/DVD module, and a 4-cell lithium ion battery The following options are available, along with price changes:

- Processor: slower (subtract \$179), faster (add \$100), fastest (add \$300)
- Warranty: 2 year (add \$129), 3 year (add \$269)
- Memory: 2 GB (add \$50), 4 GB (add \$500)
- Hard drive: 60 GB (subtract \$29), 120 GB (add \$39), 160 GB (add \$79)
- Modular bay: CD-ROM only (subtract \$39): CD/RW (add \$79), DVD/RW (add \$179)
- Battery: 8-cell lithium ion (add \$59)
- Enhanced video and photo software (add \$79)

If the customer wants to configure the top-of-the-line system, it would require an additional \$1,465 over the base price, exceeding the budget. Thus, the customer must choose the options carefully. We can build a model for doing this by defining binary variables corresponding to each possible choice. Let:

X_{p1} = 1 if slower processor is selected; X_{p2} = 1 if faster processor is selected, and X_{p3} = 1 if fastest processor is selected

X_{w1} = 1 if 2-year warranty is chosen; X_{w2} = 1 if 3-year warranty is chosen

X_{m1} = 1 if 2 GB memory is chosen; X_{m2} = 1 if 4 GB is chosen

X_{d1} = 1 if 60 GB hard drive is chosen; X_{d2} = 1 if 120 GB hard drive is chosen; X_{d3} = 1 if 160 GB hard drive is chosen

X_{c1} = 1 if CD only is chosen; X_{c2} = 1 if CD/RW is chosen, and X_{c3} = 1 if DVD/RW is chosen

X_b = 1 if 8-cell battery is chosen

X_s = 1 if enhanced video and photo software is chosen

One way of specifying an appropriate objective function is to develop preferences for each option. This can be done using the notion of utility that we described in Chapter 10. For example, suppose that the customer is intending to purchase the machine for music and photo storage; in this case, a large hard drive might be the most important feature. If the computer is to be used for video editing, then a fast processor, large hard drive, and large RAM memory would also be most important. For business travel, then a more minimal configuration with a longer warranty might be preferred. Suppose that the consumer has ranked the options and assigned a utility to each as summarized as follows:

OPTION	VARIABLE	UTILITY
Slower processor	X_{p1}	0.20
Faster processor	X_{p2}	0.70
Fastest processor	X_{p3}	0.90
2-year warranty	X_{w1}	0.50
3-year warranty	X_{w2}	0.55
2 GB memory	X_{m1}	0.70
4 GB memory	X_{m2}	0.80
60 GB hard drive	X_{d1}	0.10
120 GB hard drive	X_{d2}	0.30
160 GB hard drive	X_{d3}	1.00
CD only	X_{c1}	0.00
CD/RW	X_{c2}	0.95
DVD/RW	X_{c3}	0.45
8-cell battery	X_b	0.15
Video and photo software	X_s	0.85

The objective function would be to maximize utility:

Maximize $0.20X_{p1} + 0.70X_{p2} + 0.90X_{p3} + 0.50X_{w1} + 0.55X_{w2} + 0.70X_{m1} + 0.80X_{m2}$
$+ 0.10X_{d1} + 0.30X_{d2} + 1.0X_{d3} + 0.0X_{c1} + 0.95X_{c2} + 0.45X_{c3} + 0.15X_b + 0.85X_s$

The first constraint is that the budget not be exceeded:

$-179X_{p1} + 100X_{p2} + 300X_{p3} + 129X_{w1} + 269X_{w2} + 50X_{m1} + 500X_{m2}$
$-29X_{d1} + 39X_{d2} + 79X_{d3} - 39X_{c1} + 79X_{c2} + 179X_{c3} + 59X_b + 79X_s \leq 800$

Because we defined the variables as choices over and above the base configuration, the budget constraint reflects the amount available over the base price. Next, for each group of options, at most one can be chosen:

$$X_{p1} + X_{p2} + X_{p3} \leq 1 \quad \text{(processor)}$$
$$X_{w1} + X_{w2} \leq 1 \quad \text{(warranty)}$$
$$X_{m1} + X_{m2} \leq 1 \quad \text{(memory)}$$
$$X_{d1} + X_{d2} + X_{d3} \leq 1 \quad \text{(hard drive)}$$
$$X_{c1} + X_{c2} + X_{c3} \leq 1 \quad \text{(modular bay)}$$
$$X_b \leq 1 \quad \text{(battery)}$$
$$X_s \leq 1 \quad \text{(software)}$$

Finally, the variables must be binary.

$$X_{ij} = 0 \text{ or } 1 \text{ for each } i \text{ and } j$$

Binary variables allow us to model a wide variety of logical constraints. For example, in the computer configuration example, there are often technical restrictions or recommendations that require a specific option if another one is chosen. For example, suppose that choosing a DVD/RW requires at least 2 GB of memory. This means that if $X_{c3} = 1$ then either X_{m1} or X_{m2} must be equal to 1. We can model this with the constraint:

$$X_{m1} + X_{m2} \geq X_{c3}$$

Note that if $X_{c3} = 1$, then we must have $X_{m1} + X_{m2} \geq 1$, which will force either X_{m1} or X_{m2} to be 1. However, if $X_{c3} = 0$, then X_{m1} or X_{m2} can assume any value.

As another example, suppose that if we choose the slowest processor, then we cannot choose the DVD/RW. Thus, if $X_{p1} = 1$, then X_{c3} must be 0. This can be modeled by the constraint:

$$X_{c3} \leq 1 - X_{p1}$$

As a third example, suppose that the customer wants to ensure that if the enhanced video and photo software is chosen, then both the 4 GB memory and 160 GB hard drive should be chosen. In other words, if $X_s = 1$, we want to force $X_{d3} = 1$ and $X_{m2} = 1$. We can model this in one of two ways. First, use two constraints:

$$X_{d3} \geq X_s$$
$$X_{m2} \geq X_s$$

An alternative way of doing this is to use one constraint that is the sum of both of these:

$$X_{d3} + X_{m2} \geq 2X_s$$

Table 14.2 summarizes common types of logical conditions and how they can be modeled using binary variables. An exercise at the end of this chapter will ask you to solve this problem.

LOGICAL CONDITION	CONSTRAINT MODEL FORM
If A, then B	$B \geq A$ or $B - A \geq 0$
If not A, then B	$B \geq 1 - A$ or $A + B \geq 1$
If A, then not B	$B \leq 1 - A$ or $B + A \leq 1$
At most one of A and B	$A + B \leq 1$
If A, then B and C	$(B \geq A$ and $C \geq A)$ or $B + C \geq 2A$
If A and B, then C	$C \geq A + B - 1$ or $A + B - C \leq 1$

Table 14.2
Modeling Logical
Conditions Using Binary
Variables

A Supply Chain Facility Location Model

In 1993, Procter & Gamble began an effort entitled Strengthening Global Effectiveness (SGE) to streamline work processes, drive out non-value-added costs, and eliminate duplication.[1] A principal component of SGE was the North American Product Supply Study, designed to reexamine and reengineer P&G's product-sourcing and distribution system for its North American operations, with an emphasis on plant consolidation. Prior to the study, the North American supply chain consisted of hundreds of suppliers, over 50 product categories, over 60 plants, 15 distribution centers (DCs), and over 1,000 customers. The need to consolidate plants was driven by the move to global brands and common packaging, and the need to reduce manufacturing expense, improve speed to market, avoid major capital investments, and deliver better consumer value.

One of the key submodels in the overall optimization effort was an integer optimization model to identify optimal distribution center (DC) locations in the supply chain and to assign customers to the DCs. Customers were aggregated into 150 zones. P&G had a policy of single sourcing—that is, each customer should be served by only one DC. The optimization model used in the analysis was:

$$\text{Min } \Sigma C_{ij}X_{ij}$$
$$\Sigma X_{ij} = 1, \text{for every } j$$
$$\Sigma Y_i = k, \text{for every } i$$
$$X_{ij} \leq Y_i, \text{for every } i \text{ and } j$$

In this model, $X_{ij} = 1$ if customer zone j is assigned to DC i, and 0 if not, and $Y_i = 1$ if DC i is chosen from among a set of k potential locations. In the objective function, C_{ij} is the total cost of satisfying the demand in customer zone j from DC i. The first constraint ensures that each customer zone is assigned to exactly one DC. The next constraint limits the number of DCs to be selected from among the candidate set. The parameter k was varied by the analysis team to examine the effects of choosing different numbers of locations. The final constraint ensures that customer zone j cannot be assigned to DC i unless DC i is selected in the supply chain.

This model was used in conjunction with a simple transportation model for each of 30 product categories. Product-strategy teams used these models to specify plant locations and capacity options, and optimize the flow of product from plants to DCs and customers. In reconfiguring the supply chain, P&G realized annual cost savings of over $250 million.

In the next example, we will develop a different optimization model without the single sourcing restriction for a similar type of supply chain design problem.

---

[1] Jeffrey D. Camm, Thomas E. Chorman, Franz A. Dill, James R. Evans, Dennis J. Sweeney, and Glenn W. Wegryn, "Blending OR/MS, Judgment, and GIS: Restructuring P&G's Supply Chain," *Interfaces*, 27, no. 1 (January–February, 1997), 128–142.

Mixed Integer Optimization Models

Many practical applications of optimization involve a combination of continuous variables and binary variables. This provides the flexibility to model many different types of complex decision problems.

Plant Location Model

Suppose that in the transportation model example discussed in Chapter 13, demand forecasts exceed the existing capacity and the company is considering adding a new plant from among two choices: Fayetteville, Arkansas, or Chico, California. Both plants would have a capacity of 1,500 units but only one can be built. Table 14.3 shows the revised data.

The company now faces two decisions. It must decide which plant to build, and then how to best ship the product from the plant to the distribution centers. Of course, one approach would be to solve two separate transportation models, one that includes the Fayetteville plant, and the other that includes the Chico plant. However, we will demonstrate how to answer both questions simultaneously, as this provides the most efficient approach, especially if the number of alternatives and combinations is larger than for this example.

Define a binary variable for the decision of which plant to build: $Y_1 = 1$ if the Fayetteville plant is built, and $Y_2 = 1$ if the Chico plant is built. The objective function now includes terms for the proposed plant locations:

$$\text{Minimize } 12.60X_{11} + 14.35X_{12} + 11.52X_{13} + 17.58X_{14} + 9.75X_{21} + 16.26X_{22}$$
$$+ 8.11X_{23} + 17.92X_{24} + 10.41X_{31} + 11.54X_{32} + 9.87X_{33} + 11.64X_{34}$$
$$+ 13.88X_{41} + 16.95X_{42} + 12.51X_{43} + 8.32X_{44}$$

Capacity constraints for the Marietta and Minneapolis plants remain as before. However, for Fayetteville and Chico, we can only allow shipping from those locations if a plant is built there. In other words, if we do not build a plant in Fayetteville (if $Y_1 = 0$), for example, then we must ensure that the amount shipped from Fayetteville to any distribution center must be zero, or $X_{3j} = 0$ for $j = 1$ to 4. To do this, we multiply the capacity by the binary variable corresponding to the location:

$$X_{11} + X_{12} + X_{13} + X_{14} \leq 1,200$$
$$X_{21} + X_{22} + X_{23} + X_{24} \leq 800$$
$$X_{31} + X_{32} + X_{33} + X_{34} \leq 1,500Y_1$$
$$X_{41} + X_{42} + X_{43} + X_{44} \leq 1,500Y_2$$

Table 14.3   Plant Location Data

PLANT	CLEVELAND	BALTIMORE	CHICAGO	PHOENIX	CAPACITY
Marietta	$12.60	$14.35	$11.52	$17.58	1,200
Minneapolis	$9.75	$16.26	$8.11	$17.92	800
Fayetteville	$10.41	$11.54	$9.87	$11.64	1,500
Chico	$13.88	$16.95	$12.51	$8.32	1,500
Demand	300	500	700	1,800	

(column header spanning CLEVELAND, BALTIMORE, CHICAGO, PHOENIX: DISTRIBUTION CENTER)

Figure 14.9 *Plant Location Model* Spreadsheet

	A	B	C	D	E	F	G	H	I	J
1	**Plant Location Model**									
2										
3	**Data**									
4			**Distribution Center**							
5	**Plant**	Cleveland	Baltimore	Chicago	Phoenix	**Capacity**				
6	Marietta	$ 12.60	$ 14.35	$ 11.52	$ 17.58	1200				
7	Minneapolis	$ 9.75	$ 16.26	$ 8.11	$ 17.92	800				
8	Fayetteville	$ 10.41	$ 11.54	$ 9.87	$ 11.64	1500				
9	Chico	$ 13.88	$ 16.95	$ 12.51	$ 8.32	1500				
10	**Demand**	300	500	700	1800					
11										
12	**Model**									
13										
14	Amount Shipped		**Distribution Center**							
15	**Plant**	Cleveland	Baltimore	Chicago	Phoenix	**Total shipped**			New Plant Chosen	Surplus Capacity
16	Marietta	200	500	0	300	1000		Fayetteville	0	0
17	Minneapolis	100	0	700	0	800		Chico	1	0
18	Fayetteville	0	0	0	0	0		Total	1	
19	Chico	0	0	0	1500	1500				
20	**Demand met**	300	500	700	1800					
21										
22	**Total cost**									
23	$	34,101								

Note that if the binary variable is zero, then the right-hand side of the constraint is zero, forcing all shipment variables to be zero also. If, however, a particular Y variable is 1, then shipping up to the plant capacity is allowed. The demand constraints are the same as before, except that additional variables corresponding to the possible plant locations are added and new demand values are used:

$$X_{11} + X_{21} + X_{31} + X_{41} = 300$$
$$X_{12} + X_{22} + X_{32} + X_{42} = 500$$
$$X_{13} + X_{23} + X_{33} + X_{43} = 700$$
$$X_{14} + X_{24} + X_{34} + X_{44} = 1,800$$

To guarantee that only one new plant is built, we must have:

$$Y_1 + Y_2 = 1$$

Finally, we have nonnegativity for the continuous variables. $X_{ij} \geq 0$, for all i and j.

Figure 14.9 shows the spreadsheet model (Excel file *Plant Location Model*) and optimal solution. Note that in addition to the continuous variables X_{ij}, in the range B16:E19, we defined binary variables Yi in cells I16 and I17. Cells J16 and J17 represent the constraint functions $1,500Y_1 - X_{31} - X_{32} - X_{33} - X_{34}$ and $1500Y_2 - X_{41} - X_{42} - X_{43} - X_{44}$, respectively. These are restricted to be ≥ 0 to enforce the capacity constraints at the potential locations in the *Solver* model (Figure 14.10). You should closely examine the other constraints in the *Solver* model to verify that they are correct. The solution specifies selecting the Chico location.

Models of this type are commonly used in supply chain design and other facility location applications.

A Model with Fixed Costs

Many business problems involve fixed costs; they are either incurred in full, or not at all. Binary variables can be used to model such problems. To illustrate this, consider

Figure 14.10 *Solver* Model for Plant Location

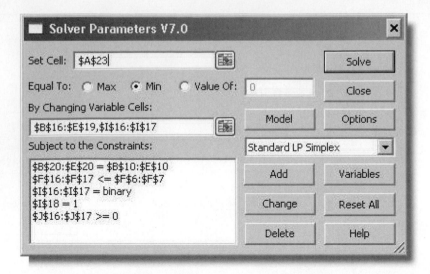

the multi-period production-inventory planning model for Kristin's Creations that we developed in Chapter 13. Suppose that Kristin must rent some equipment to produce her products, which costs $65 for three months. The equipment can be rented or returned each quarter, so if nothing is produced in a quarter, it makes no sense to incur the rental cost.

The fixed costs can be incorporated into the model by defining an additional set of variables:

$$Y_A = 1 \text{ if production occurs during the Autumn, and 0 if not}$$
$$Y_W = 1 \text{ if production occurs during the Winter, and 0 if not}$$
$$Y_S = 1 \text{ if production occurs during the Spring, and 0 if not}$$

Then the objective function becomes:

$$\text{Minimize } 11P_A + 14P_W + 12.50P_S + 1.20I_A + 1.20I_W + 1.20I_S \\ + 65(Y_A + Y_W + Y_S)$$

The basic material balance equations are the same:

$$P_A - I_A = 150$$
$$P_W + I_A - I_W = 400$$
$$P_S + I_W - I_S = 50$$

However, we must ensure that whenever a production variable, P, is positive, that the corresponding Y variable is equal to 1; and conversely, if the Y variable is 0 (you don't rent the equipment), then the corresponding production variable must also be 0. This can be accomplished with the following constraints:

$$P_A \le 600Y_A$$
$$P_W \le 600Y_W$$
$$P_S \le 600Y_S$$

Note that if any Y is 0 in a solution, then P is forced to be zero, and if P is positive, then Y must be 1. Because we don't know how much the value of any production variable will be, we use 600, which is the sum of the demands over the time horizon to multiply by Y. So when Y is 1, any amount up to 600 units can be produced. Actually any large number can be used, so long as it doesn't restrict the possible values of P. Generally, the smallest value should be used for efficiency. Finally, P_A, P_w, and P_s must be nonnegative, and Y_A, Y_W, and Y_S are binary.

You might observe that this model does not preclude feasible solutions in which a production variable is 0 while its corresponding Y variable is 1. This implies that we incur the fixed cost even though no production is incurred during that time period. While such a solution is feasible, it can never be optimal, as a lower cost could be obtained by setting the Y variable to 0 without affecting the value of the production variable, and the solution algorithm will always ensure this.

Logical Conditions and Spreadsheet Implementation

In the fixed-cost problem, we needed binary variables in order to incorporate the fixed costs into the objective function and model the logical conditions that ensure that the fixed cost will only be incurred if the production in a period is positive. We did this for an important reason—to preserve linearity of the constraints so that we could use the simplex method incorporated into *Solver* to find an optimal solution. Recall in Chapter 13 that we described certain Excel functions, such as IF, that should not be used to implement linear optimization models on spreadsheets. However, it might have been much simpler to use an IF function in the spreadsheet to model the fixed costs. For example, we could include the fixed cost in the objective function by using an IF function in the following way: IF(cell corresponding to $P_A > 0, 65, 0$). In this way, there would be no need for the binary variables and the additional constraints. However, doing so would violate the linearity conditions required for the simplex method. Therefore, to use *Solver*, it is important to first formulate the appropriate mathematical model and implement that on the spreadsheet. Later in this chapter, we will discuss some alternatives that can be used for spreadsheet models that violate linearity requirements.

NONLINEAR OPTIMIZATION

In many situations, the relationship among variables in a model is not linear. Whenever either the objective function or a constraint is not linear, the model becomes a *nonlinear optimization problem,* requiring different solution techniques. Nonlinear models do not have a common structure as do linear models, making it more difficult to develop appropriate models. We present two examples of nonlinear optimization models in business.

Hotel Pricing

The Marquis Hotel is considering a major remodeling effort and needs to determine the best combination of rates and room sizes to maximize revenues. Currently, the hotel has 450 rooms with the following history:

ROOM TYPE	RATE	DAILY AVG. NO. SOLD	REVENUE
Standard	$85	250	$21,250
Gold	$98	100	$9,800
Platinum	$139	50	$6,950
		Total Revenue	$38,000

Each market segment has its own price/demand elasticity. Estimates are:

ROOM TYPE	PRICE ELASTICITY OF DEMAND
Standard	−1.5
Gold	−2.0
Platinum	−1.0

This means, for example, that a *1% decrease* in the price of a standard room will *increase* the number of rooms sold by *1.5%*. Similarly, a 1% increase in the price will decrease the number of rooms sold by 1.5%. For any pricing structure (in $), the projected number of rooms of a given type sold (we will allow continuous values for this example) can be found using the formula:

(Historical Average Number of Rooms Sold) + (Elasticity)(New Price − Current Price)(Historical Average Number of Rooms Sold)/(Current Price)

The hotel owners want to keep the price of a standard room between $70 and $90; a gold room between $90 and $110; and a platinum room between $120 and $149. Define S = price of a standard room, G = price of a gold room, and P = price of a platinum room. Thus, for standard rooms, the projected number of rooms sold is $250 - 1.5(S - 85)(250)/85 = 625 - 4.41176S$. The objective is to set the room prices to maximize total revenue. Total revenue would equal the price times the projected number of rooms sold, summed over all three types of rooms. Therefore, total revenue would be:

$$\text{Total Revenue} = S(625 - 4.41176S) + G(300 - 2.04082G) + P(100 - 0.35971P)$$
$$= 625S + 300G + 100P - 4.41176S^2 - 2.04082G^2 - 0.35971P^2$$

To keep prices within the stated ranges, we need constraints:

$$70 \leq S \leq 90$$
$$90 \leq G \leq 110$$
$$120 \leq P \leq 149$$

Finally, although the rooms may be renovated, there are no plans to expand beyond the current 450-room capacity. Thus, the projected number of total rooms sold cannot exceed 450:

$$(625 - 4.41176S) + (300 - 2.04082G) + (100 - 0.35971P) \leq 450$$

or simplified as:

$$1025 - 4.41176S - 2.04082G - 0.35971P \leq 450$$

The complete model is:

$$\text{Maximize } 625S + 300G + 100P - 4.41176S^2 - 2.04082G^2 - 0.35971P^2$$
$$70 \leq S \leq 90$$
$$90 \leq G \leq 110$$
$$120 \leq P \leq 149$$
$$1025 - 4.41176S - 2.04082G - 0.35971P \leq 450$$

Figure 14.11 shows a spreadsheet model (Excel file *Hotel Pricing Model*) for this example showing the optimal solution. The decision variables, the new prices to

Figure 14.11 *Hotel Pricing Model* Spreadsheet and Optimal Solution

	A	B	C	D	E	F
1	Marquis Hotel					
2						
3	Data					
4		Current	Average		Total Room	
5	Room type	Rate	Daily Sold	Elasticity	Capacity	
6	Standard	$ 85.00	250	-1.5	450	
7	Gold	$ 98.00	100	-2		
8	Platinum	$ 139.00	50	-1		
9						
10	Model				Projected	
11					Rooms	Projected
12	Room type	New Price	Price Range		Sold	Revenue
13	Standard	$ 76.87	$ 70.00	$ 90.00	286	$ 21,974.39
14	Gold	$ 90.00	$ 90.00	$ 110.00	116	$ 10,469.39
15	Platinum	$ 145.04	$ 120.00	$ 149.00	48	$ 6,936.87
16				Totals	450	$ 39,380.65

charge, are given in cells B13:B15. The projected numbers of rooms sold are computed in cells E13:E15 using the preceding formula. By multiplying the number of rooms sold by the new price for each room type, the projected revenue is calculated, as given in cells F13:F15. The total revenue in cell F16 represents the objective function. The constraints—(1) the new price must fall within the allowable price range and (2) the total projected number of rooms sold must not exceed 450—can be expressed within the spreadsheet model as:

$$B13:B15 \geq C13:C15$$
$$B13:B15 \leq D13:D15$$

and

$$E16 \leq E6$$

Note that it is easier to formulate this model more as a financial spreadsheet than to enter the analytical formulas as they were developed. The optimal prices predict a demand for all 450 rooms with a total revenue of $39,380.65.

Solving Nonlinear Optimization Models

Nonlinear optimization models are formulated with *Solver* in the same fashion as linear or integer models. In *Premium Solver*, you should select *Standard GRG Nonlinear* as the solution procedure. "GRG" stands for "generalized reduced gradient," which is the name of the algorithm used to solve nonlinear models in *Solver*. Figure 14.12 shows the premium version *Solver Parameters* dialog box for the hotel pricing model. The optimal solution was shown in Figure 14.11.

The information contained in the Answer Report (Figure 14.13) is the same as for linear models. Because nonlinear models are continuous (as long as they do not contain any integer variables), *Solver* produces a sensitivity report. However, for nonlinear models, the Sensitivity Report (Figure 14.14) is quite different. In the *Adjustable Cells* section, the *Reduced Gradient* is analogous to the *Reduced Cost* in linear models.

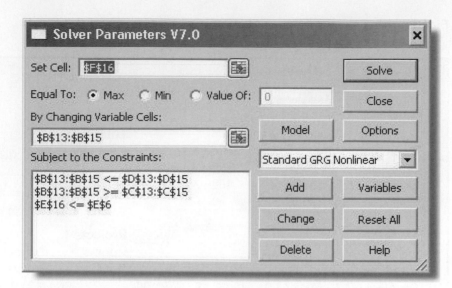

Figure 14.12
Solver Model for Hotel Pricing Example

Figure 14.13 Hotel Pricing Example Answer Report

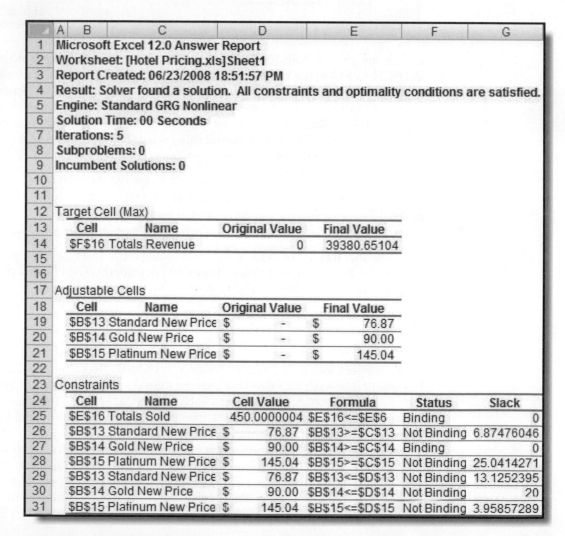

Figure 14.14 Hotel Pricing Example Sensitivity Report

	A	B	C	D	E
1	**Microsoft Excel 12.0 Sensitivity Report**				
2	**Worksheet: [Hotel Pricing.xls]Sheet1**				
3	**Report Created: 06/23/2008 18:51:58 PM**				
4					
5	Target Cell (Max)				
6		Cell	Name	Final Value	
7		$F$16	Totals Revenue	39380.65104	
8					
9	Adjustable Cells				
10				Final	Reduced
11		Cell	Name	Value	Gradient
12		$B$13	Standard New Price	$ 76.87	$ -
13		$B$14	Gold New Price	$ 90.00	$ (42.69)
14		$B$15	Platinum New Price	$ 145.04	$ -
15					
16	Constraints				
17				Final	Lagrange
18		Cell	Name	Value	Multiplier
19		$E$16	Totals Sold	450.0000004	12.08293216

For this problem, however, the objective function coefficient of each price depends on many parameters, and therefore, the reduced gradient is more difficult to interpret in relation to the problem data. *Lagrange Multipliers* in the *Constraints* section are similar to shadow prices for linear models. For nonlinear models, the Lagrange multipliers give the *approximate* rate of change in the objective function as the right-hand side of a binding constraint is increased by one unit. Thus, for the hotel pricing problem, if the number of available rooms in increased by 1 to 451, the total revenue would increase by *approximately* $12.08. For linear models, shadow prices give the *exact* rate of change within the Allowable Increase and Decrease limits. Thus, you should be somewhat cautious when interpreting these values and will need to re-solve the models to find the true effect of changes to constraints.

Many nonlinear problems are notoriously difficult to solve. *Solver* cannot guarantee finding the absolute best solution (called a *global optimal solution*) for all problems. A *local optimum solution* is one for which all points close by are no better than the solution (think of the analogy of being at the top of a hill when the highest peak is on another mountain). The solution found often depends much on the starting solution in your spreadsheet. For complex problems, it is wise to run *Solver* from different starting points. You should also look carefully at the *Solver* results dialog box when the model has completed running. If it indicates "Solver has found a solution. All constraints and optimality conditions are satisfied," then at least a local optimal solution has been found. If you get the message: "Solver has converged to the current solution. All constraints are satisfied." then you should run *Solver* again from the current solution to try to find a better solution.

Markowitz Portfolio Model

The Markowitz portfolio model[2] is a classic optimization model in finance that seeks to minimize the risk of a portfolio of stocks subject to a constraint on the portfolio's expected return. For example, suppose an investor is considering three stocks. The expected return for stock 1 is 10%; for stock 2, 12%; and for stock 3, 7%; and she would like an expected return of at least 10%. Clearly one option is to invest everything in stock 1; however, this may not be a good idea as the risk might be too high. Recall from Chapter 10 that we can measure risk by the standard deviation, or equivalently, the variance. Research has found the variance–covariance matrix of the individual stocks to be:

	Stock 1	Stock 2	Stock 3
Stock 1	0.025	0.015	−0.002
Stock 2		0.030	0.005
Stock 3			0.004

Thus, the decision variables are the percent of each stock to allocate to the portfolio. (You might be familiar with the term "asset allocation model" that many financial investment companies suggest to their clients; for example, "maintain 60% equities, 30% bonds, and 10% cash.") Define x_j to be the fraction of the portfolio to invest in stock j.

The objective function is to minimize the risk of the portfolio as measured by its variance. Because stock prices are correlated with one another, the variance of the portfolio must reflect not only variances of the stocks in the portfolio but also the covariance between stocks. The variance of a portfolio is the weighted sum of the variances and covariances:

$$\text{Variance of Portfolio} = \sum_{i=1}^{k} s_i^2 x_i^2 + \sum_{i=1}^{k}\sum_{j>1} 2s_{ij}x_i x_j$$

where

$s_i^2 =$ the sample variance in the return of stock i
$s_{ij} =$ the sample covariance between stocks i and j

Using the preceding data, the objective function is:

$$\text{Minimize Variance} = 0.025x_1^2 + 0.030x_2^2 + 0.004x_3^2 + 2(0.015)x_1 x_2$$
$$+ 2(-0.002)x_1 x_3 + 2(0.005)x_2 x_3$$

The constraints must first ensure that we invest 100% of our budget. Because the variables are defined as fractions, we must have:

$$x_1 + x_2 + x_3 = 1$$

Second, the portfolio must have an expected return of at least 10%. The return on a portfolio is simply the weighted sum of the returns of the stocks in the portfolio. This results in the constraint:

$$10x_1 + 12x_2 + 7x_3 \geq 10$$

---

[2] H.M. Markowitz, *Portfolio Selection, Efficient Diversification of Investments* (New York: John Wiley & Sons, 1959).

Finally, we cannot invest negative amounts:

$$x_1, x_2, x_3 \geq 0$$

The complete model is:

$$\text{Minimize Variance} = 0.025x_1^2 + 0.030x_2^2 + 0.004x_3^2 + 0.03x_1x_2 - 0.004x_1x_3 + 0.010x_2x_3$$

$$x_1 + x_2 + x_3 = 1$$
$$10x_1 + 12x_2 + 7x_3 \geq 10$$
$$x_1, x_2, x_3 \geq 0$$

Figure 14.15 shows a spreadsheet model for this example (Excel file *Markowitz Model*). The decision variables (fraction of each stock in the portfolio) are entered in cells B14:B16. The expected return and variance of the portfolio are computed in cells B20 and C20. The variance of the optimal portfolio is 0.012.

The *Solver* Sensitivity Report is shown in Figure 14.16. As we noted, for nonlinear models, the Lagrange multipliers are only approximate indicators of shadow prices. For this solution, the Lagrange multiplier predicts that the minimum variance will increase by 63.2% if the target return is increased from 10% to 11%. If you re-solve the model, you will find that the minimum variance increases to 0.020, a 66.67% increase.

Optimization models can and should provide decision makers with valuable insight beyond simply finding an optimal solution. Using spreadsheet models and *Solver*, it is easy to systematically vary a parameter of a model and investigate its

Figure 14.15   *Markowitz Model* Spreadsheet

	A	B	C	D	E	F	G
1	**Markowitz Model**						
2							
3	**Data**						
4		Expected				**Variance-Covariance Matrix**	
5		Return			Stock 1	Stock 2	Stock 3
6	Stock 1	10%		Stock 1	0.025	0.015	-0.002
7	Stock 2	12%		Stock 2		0.03	0.005
8	Stock 3	7%		Stock 3			0.004
9	Target Return	10%					
10							
11	**Model**						
12					Variance Calculations		
13		Allocation			Squared Terms	Cross-Products	
14	Stock 1	0.25			0.001579256	0.003387001	
15	Stock 2	0.45			0.006053362	-0.000301067	
16	Stock 3	0.30			0.000358718	0.001345191	
17	Total	1					
18							
19		Return	Variance				
20	Portfolio	10.0%	0.012				

Figure 14.16 *Markowitz Model* Sensitivity Report

	A	B	C	D	E
1	**Microsoft Excel 12.0 Sensitivity Report**				
2	**Worksheet: [Markowitz Model.xls]Markowitz Model**				
3	**Report Created: 06/23/2008 19:23:47 PM**				
4					
5	Target Cell (Min)				
6		**Cell**	**Name**	**Final Value**	
7		$C$20	Portfolio Variance	0.012422461	
8					
9	Adjustable Cells				
10				**Final**	**Reduced**
11		**Cell**	**Name**	**Value**	**Gradient**
12		$B$14	Stock 1 Allocation	0.25	0.00
13		$B$15	Stock 2 Allocation	0.45	0.00
14		$B$16	Stock 3 Allocation	0.30	0.00
15					
16	Constraints				
17				**Final**	**Lagrange**
18		**Cell**	**Name**	**Value**	**Multiplier**
19		$B$17	Total Allocation	1	-0.038363698
20		$B$20	Portfolio Return	10.0%	63.2%

impact on the solution. For example, we might be interested in understanding the relationship between the minimum risk and the target return. By changing the target return, and re-solving the model, we obtain the chart shown in Figure 14.17 (advanced users of Excel can program a macro to automate this process). This clearly shows that the minimum variance increases at a faster rate as the target return increases, indicating that the investor faces a risk premium in seeking higher returns.

An alternative modeling approach would be to maximize the return subject to a constraint on risk. For example, suppose the investor wants to maximize expected return subject to a risk (variance) no greater than 1%. This form of the model would be:

$$\text{Maximize } 10x_1 + 12x_2 + 7x_3$$

$$x_1 + x_2 + x_3 = 1$$

$$0.025x_1^2 + 0.030x_2^2 + 0.004x_3^2 + 0.03x_1x_2 - 0.004x_1x_3 + 0.010x_2x_3 \leq 0.01$$

$$x_1, x_2, x_3 \geq 0$$

In this case, we would have a linear objective function and a mixture of linear and nonlinear constraints. An exercise will ask you to solve this by modifying the example spreadsheet.

Evolutionary Solver for Nonlinear Optimization

Solver is not guaranteed to find an optimal solution for every nonlinear optimization problem, particularly those with many local optimum solutions. Other

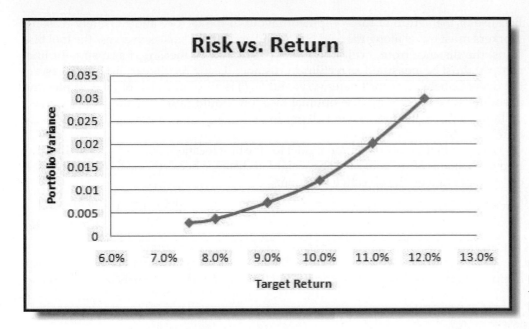

Figure 14.17
Risk vs. Return Profile
for Markowitz
Portfolio Example

problems that involve both nonlinear functions and integer variables, or discontinuous spreadsheet functions such as IF, ABS, ROUND, and so on that we cautioned you not to use for linear or integer optimization, are "non-smooth," making them difficult to solve using conventional techniques. To overcome the limitations of conventional optimization procedures, new approaches called *metaheuristics* have been proposed. These approaches have some exotic names, including genetic algorithms, neural networks, and tabu search. Such approaches use heuristics—intelligent rules for systematically searching among solutions—that remember the best solutions they find, then modifying or combining them in attempting to find better solutions.

The *Premium Solver Standard Evolutionary* algorithm uses such an approach. *Evolutionary Solver* can also be used for spreadsheet models that contain discontinuous functions. For many practical situations, however, it may be more convenient to build the spreadsheet model using these types of functions and then apply the evolutionary algorithm rather than the standard simplex LP or GRG algorithms. To illustrate *Evolutionary Solver*, consider the following location problem.

Edwards Manufacturing is studying where to locate a tool bin on the factory floor. The locations of five production cells are expressed as *x*- and *y*-coordinates on a rectangular grid of the factory layout. The daily demand for tools (measured as the number of trips to the tool bin) at each production cell is also known. The relevant data are:

CELL	X-COORDINATE	Y-COORDINATE	DEMAND
Fabrication	1	4	12
Paint	1	2	24
Subassembly 1	2.5	2	13
Subassembly 2	3	5	7
Assembly	4	4	17

Because of the nature of the equipment layout in the factory and for safety reasons, workers must travel along marked horizontal and vertical aisles to access the tool bin. Thus, the distance from a cell to the tool bin cannot be measured as a straight line; rather it must be measured as rectilinear distance. Using rectilinear distance measure, the distance between coordinates (x, y) and (a, b) is absolute value of $(x - a)$ plus the absolute value of $(y - b)$. The optimal location should minimize the total weighted distance between the tool bin and all production cells, where the weights are the daily number of trips to the tool bin.

To formulate an optimization model for the best location, define (X, Y) as the location coordinates of the tool bin. The weighted distance between the tool bin and each cell is expressed by the objective function:

$$\text{Minimize } 12(|X - 1| + |Y - 4|) + 24(|X - 1| + |Y - 2|) + 13(|X - 2.5| + |Y - 2|) + 7(|X - 3| + |Y - 5|) + 17(|X - 4| + |Y - 4|)$$

The absolute value functions create a non-smooth set of solutions.

To use *Evolutionary Solver*, the decision variables must have both upper and lower bounds. Figure 14.18 shows a spreadsheet model for the Edwards Manufacturing

Figure 14.18 *Edwards Manufacturing* Spreadsheet

	A	B	C	D
1	**Edwards Manufacturing**			
2				
3	**Data**			
4				
5	Cell	x-coordinate	y-coordinate	Demand
6	Fabrication	1	4	12
7	Paint	1	2	24
8	Subassembly 1	2.5	2	13
9	Subassembly 2	3	5	7
10	Assembly	4	4	17
11	Maximum	4	5	
12				
13	**Model**			
14	Tool bin location	2.499876246	2.000005905	
15				
16	Cell	Weighted Distance		
17	Fabrication	41.9984441		
18	Paint	35.99717162		
19	Subassembly 1	0.001685555		
20	Subassembly 2	24.50082494		
21	Assembly	59.50200343		
22	Total	162.0001297		

Figure 14.19 *Solver* Model for Edwards Manufacturing

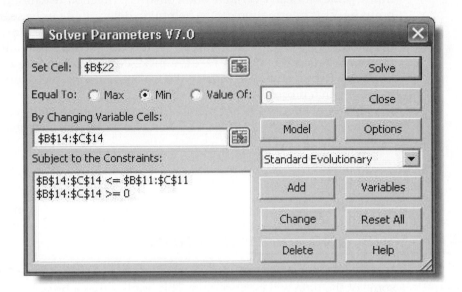

example (Excel file *Edwards Manufacturing Model*). The upper bounds are chosen as the maximum coordinate values and the lower bounds are zero. The *Solver* model is shown in Figure 14.19.

Evolutionary Solver produces an Answer Report similar to the ones we have already seen, and a new report, called a Population Report, shown in Figure 14.20. This report provides statistics on the solutions encountered during the search

Figure 14.20 *Evolutionary Solver* Population Report

	A	B	C	D	E	F	G	H
1	**Microsoft Excel 12.0 Population Report**							
2	**Worksheet: [Book2]Sheet1**							
3	**Report Created: 06/28/2008 09:50:05 AM**							
4								
5								
6	**Adjustable Cells**							
7				**Best**	**Mean**	**Standard**	**Maximum**	**Minimum**
8		**Cell**	**Name**	**Value**	**Value**	**Deviation**	**Value**	**Value**
9		$B$14	x-coordinate	2.499999686	2.22804413	0.767961989	2.717950279	0
10		$C$14	y-coordinate	2.000001358	1.919093264	0.534385873	2.801005462	0
11								
12	**Constraints**							
13		None						

process, showing the best value, mean, standard deviation, maximum, and minimum values for each variable and constraint function (excluding upper and lower bounds). This information can be helpful in determining if the search process should be continued further. The results depend heavily on the starting values of the decision variables; different values can produce different results for complicated models. Thus, it is wise to run the procedure from different starting points. A Solutions Report is also provided, which gives details of the solutions generated during the search process.

As a final note, *Solver* has many nuances and options. "Power users" should consult the *Premium Solver Platform User Guide* for additional information.

RISK ANALYSIS AND OPTIMIZATION

It is rare that any optimization model is completely deterministic; in most cases, some of the data will be uncertain. This implies that inherent risk exists in using the optimal solution obtained from a model. Using the capabilities of risk analysis software such as *Crystal Ball*, these risks can be better understood and mitigated. To illustrate this, we will use the hotel pricing problem.

In this problem, the price–demand elasticities of demand are only estimates and most likely are quite uncertain. Because we probably will not know anything about their distributions, let us conservatively assume that the true values might vary from the estimates by plus or minus 25%. Thus, we model the elasticities by uniform distributions. Using the optimal prices identified by *Solver* earlier in this chapter, let us see what happens to the forecast of the number of rooms sold under this assumption using *Crystal Ball*.

In the spreadsheet model, select cells D6:D8 as assumption cells with uniform distributions having minimum and maximum values equal to 75% and 125% of the estimated values, respectively. The item of total rooms sold (E16) is defined as a forecast cell. The model was replicated 5,000 times, creating the report in Figure 14.21. We see that the mean number of rooms sold under these prices is 450, which should be expected, since the mean values of the elasticities were used to derive the optimal prices. However, because of the uncertainty associated with the elasticities, the probability that *more* than 450 rooms will be sold (demanded) is approximately 0.5! This suggests that if the assumptions of the uncertain elasticities are true, the hotel might anticipate that demand will exceed its room capacity about half the time, resulting in many unhappy customers.

We could use these results, however, to identify the appropriate hotel capacity to ensure, for example, only a 10% chance exists that demand will exceed capacity. Figure 14.22 shows the forecast chart when the certainty level is set at 90% and the left grabber is anchored. We could interpret this as stating that if the hotel capacity were about 457 or 458 rooms, then demand will exceed capacity at most 10% of the time. So if we shift the capacity constraint down by 7 rooms to 443 and find the optimal prices associated with this constraint, we would expect demand to exceed 450 at most 10% of the time. The *Solver* results for this case are shown in Figure 14.23, and Figure 14.24 shows the results of a *Crystal Ball* run confirming that with these prices, demand will exceed 450 less than 10% of the time.

Figure 14.21 *Crystal Ball* Results for Hotel Pricing Example

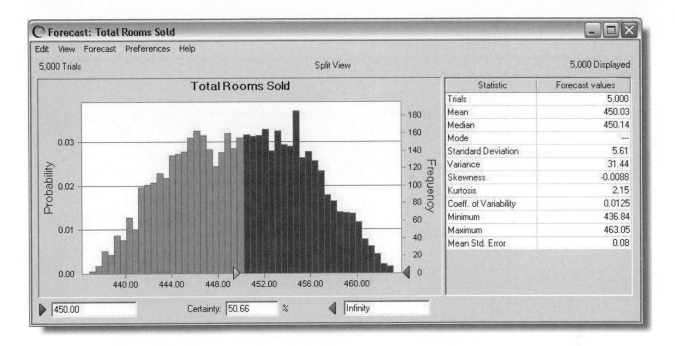

Figure 14.22 Forecast Chart for a 10% Risk of Exceeding Capacity

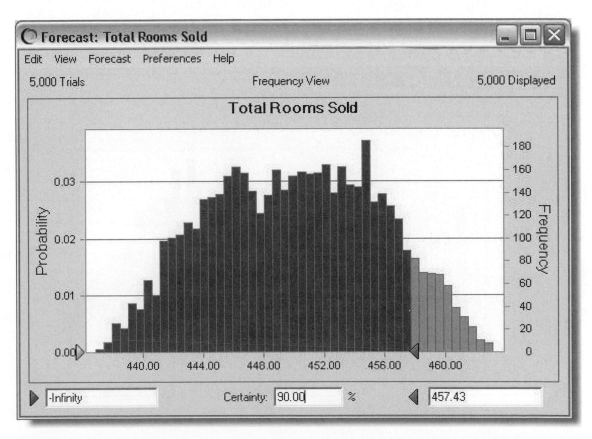

Figure 14.23 *Solver Solution* for 443-Room Capacity

	A	B	C		D		E	F
1	**Marquis Hotel**							
2								
3	**Data**							
4		**Current**	**Average**				**Total Room**	
5	**Room type**	**Rate**	**Daily Sold**	**Elasticity**			**Capacity**	
6	Standard	$ 85.00	250		-1.5		443	
7	Gold	$ 98.00	100		-2			
8	Platinum	$ 139.00	50		-1			
9								
10	**Model**						**Projected**	
11							**Rooms**	**Projected**
12	**Room type**	**New Price**	**Price Range**				**Sold**	**Revenue**
13	Standard	$ 78.34	$ 70.00	$	90.00		279	$21,886.69
14	Gold	$ 90.00	$ 90.00	$	110.00		116	$10,469.39
15	Platinum	$ 146.51	$ 120.00	$	149.00		47	$ 6,929.72
16					Totals		443	$39,285.80

Figure 14.24 *Crystal Ball* Confirmation Run Results

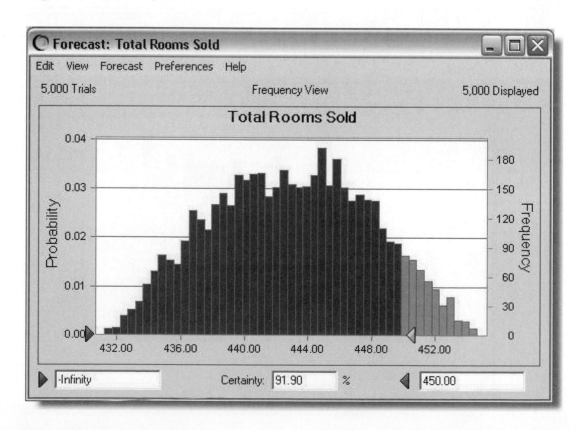

COMBINING OPTIMIZATION AND SIMULATION

To find an optimal set of decision variables for any simulation-based model, you generally need to search in a heuristic or ad hoc fashion. This usually involves running a simulation for an initial set of variables, analyzing the results, changing one or more variables, rerunning the simulation, and repeating this process until a satisfactory solution is obtained. This process can be very tedious and time-consuming, and often how to adjust the variables from one iteration to the next is not clear. While the *Crystal Ball Decision Table* tool is available, its functionality is somewhat limited.

OptQuest enhances the analysis capabilities of *Crystal Ball* by automatically searching for optimal solutions within *Crystal Ball* simulation model spreadsheets. Within *OptQuest*, you describe your optimization problem and search for values of decision variables that maximize or minimize a predefined objective. Additionally, *OptQuest* is designed to find solutions that satisfy a wide variety of constraints or a set of goals that you may define.

A Portfolio Allocation Model

We will use a portfolio allocation model to illustrate the steps of setting up and running an optimization problem using *Crystal Ball* and *OptQuest*. An investor has $100,000 to invest in four assets. The expected annual returns and minimum and maximum amounts with which the investor will be comfortable allocating to each investment follow:

INVESTMENT	ANNUAL RETURN	MINIMUM	MAXIMUM
1. Life insurance	5%	$ 2,500	$5,000
2. Bond mutual funds	7%	$30,000	none
3. Stock mutual funds	11%	$15,000	none
4. Savings account	4%	none	none

The major source of uncertainty in this problem is the annual return of each asset. In addition, the decision maker faces other risks, for example, unanticipated changes in inflation or industrial production, the spread between high- and low-grade bonds, and the spread between long- and short-term interest rates. One approach to incorporating such risk factors in a decision model is arbitrate pricing theory (APT).[3] APT provides estimates of the sensitivity of a particular asset to these types of risk factors. Let us assume that the risk factors per dollar allocated to each asset have been determined as follows:

INVESTMENT	RISK FACTOR/DOLLAR INVESTED
1. Life insurance	−0.5
2. Bond mutual funds	1.8
3. Stock mutual funds	2.1
4. Savings account	−0.3

---

[3] See Schniederjans, M., T. Zorn, and R. Johnson, "Allocating Total Wealth: A Goal Programming Approach," *Computers and Operations Research*, 20, no. 7(1993): 679–685.

The investor may specify a target level for the weighted risk factor, leading to a constraint that limits the risk to the desired level. For example, suppose that our investor will tolerate a weighted risk per dollar invested of at most 1.0. Thus, the weighted risk for a $100,000 total investment will be limited to 100,000. If our investor allocates $5,000 in life insurance, $50,000 in bond mutual funds, $15,000 in stock mutual funds, and $30,000 in a savings account (which fall within the minimum and maximum amounts specified), the total expected annual return would be:

$$0.05(\$5,000) + 0.07(\$50,000) + 0.11(\$15,000) + 0.04(\$30,000) = \$6,600$$

However, the total weighted risk associated with this solution is:

$$-0.5(5,000) + 1.8(50,000) + 2.1(15,000) - 0.3(30,000) = 110,000$$

Because this is greater than the limit of 100,000, this solution could not be chosen.

The decision problem, then, is to determine how much to invest in each asset to maximize the total expected annual return, remain within the minimum and maximum limits for each investment, and meet the limitation on the weighted risk.

Using *OptQuest*

The basic process for using *OptQuest* is described as follows:

1. Create a *Crystal Ball* model of the decision problem.
2. Define the decision variables within *Crystal Ball*.
3. Invoke *OptQuest* from the *Crystal Ball* toolbar.
4. Create a new optimization file.
5. Specify objectives.
6. Specify decision variables.
7. Specify any constraints.
8. Specify *OptQuest* run options.
9. Solve the optimization problem.

Create the Crystal Ball *Spreadsheet Model*

An important task in using *OptQuest* is to create a useful spreadsheet model. A spreadsheet for this problem is shown in Figure 14.25 (Excel file *Portfolio Allocation Model*). Problem data are specified in rows 4 through 8. On the bottom half of the spreadsheet, we specify the model outputs, namely, the values of the decision variables, objective function, and constraints (the total weighted risk and total amount invested). You can see that this particular solution is not feasible because the total weighted risk exceeds the limit of 100,000.

Now that the basic model is developed, we define the assumptions and forecast cells in *Crystal Ball*. We will assume that the annual returns for life insurance and mutual funds are uncertain, but that the rate for the savings account is constant. We will make the following assumptions in the *Crystal Ball* model:

- Cell B4: uniform distribution with minimum 4% and maximum 6%
- Cell B5: normal distribution with mean 7% and standard deviation 1%
- Cell B6: lognormal distribution with mean 11% and standard deviation 4%

We define the forecast cell to be the total expected return, cell E16. As would be the case with any *Crystal Ball* application, you would select *Run Preferences* from the *Run* menu and choose appropriate settings. For this example, we set the number of trials per simulation to 2,000.

Figure 14.25 *Portfolio Allocation Model* Spreadsheet

	A	B	C	D	E
1	Portfolio Allocation Model				
2		Annual			Risk factor
3	Investment	return	Minimum	Maximum	per dollar
4	Life Insurance	5.0%	$ 2,500.00	$ 5,000.00	-0.5
5	Bond mutual funds	7.0%	$ 30,000.00	none	1.8
6	Stock mutual funds	11.0%	$ 15,000.00	none	2.1
7	Savings Account	4.0%	none	none	-0.3
8	Total amount available	$100,000		Limit	100,000
9					
10		Amount			Total weighted
11	Decision variables	invested			risk
12	Life Insurance	$ 5,000.00			146,000.00
13	Bond mutual funds	$ 50,000.00			
14	Stock mutual funds	$ 30,000.00			Total expected
15	Savings Account	$ 15,000.00			return
16	Total amount invested	$ 100,000.00			$ 7,650.00

Define Decision Variables

The next step is to identify the decision variables in the model. This is accomplished using the *Define Decision* option in the *Define* group. Position the cursor on cell B12 and click *Define Decision*. Set the minimum and maximum values according to the problem data (i.e., columns C and D in the spreadsheet), as shown in Figure 14.26.

Next, we repeat the process of defining decision variables for cells B13, B14, and B15. When the maximum limit is "none," you may use a value of $100,000 because this is the total amount available. You are now ready to run *OptQuest* by clicking on *OptQuest* on the from the *Tools* menu.

Figure 14.26 *Define Decision Variable* Dialog

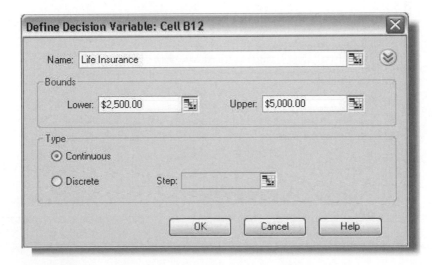

Create a New Optimization File and Specify Objectives

In the *Welcome* screen in *OptQuest*, click *Next* to get started. This will step you through the process of setting up your optimization model. You will then see the screen shown in Figure 14.27 to specify your objectives. Click on *Add Objective*. For the portfolio example, the window will display the objective *"Maximize the mean of Total expected return"* which you may customize as appropriate. We will discuss adding requirements after completing this example.

Specify Decision Variables

After clicking *Next* (or choosing *Decision Variables* from the left panel), the *Decision Variables* screen is displayed (Figure 14.28). The decision variables defined in your *Crystal Ball* model will be shown. You may unselect (freeze) any of them as appropriate. The *Base Case* values are the values currently in the spreadsheet. The *Type* column indicates whether a variable is discrete or continuous. The variable type can be changed in this window or in the *Define Decision* dialog of *Crystal Ball*. A step size is associated with discrete variables. A variable of the type Discrete_2, for example, has a step size of 2. Therefore, if the lower and upper bounds for this variable are 0 and 7, respectively, the only feasible values are 0, 2, 4, and 6. Any values or type may be changed by clicking on the appropriate value or name. Click *Next* to continue.

Specify Constraints

The *Constraints* screen displayed allows you to specify any constraints. A constraint is any limitation or requirement that restricts the possible solutions to the problem.

Figure 14.27   *OptQuest* Objectives Input Screen

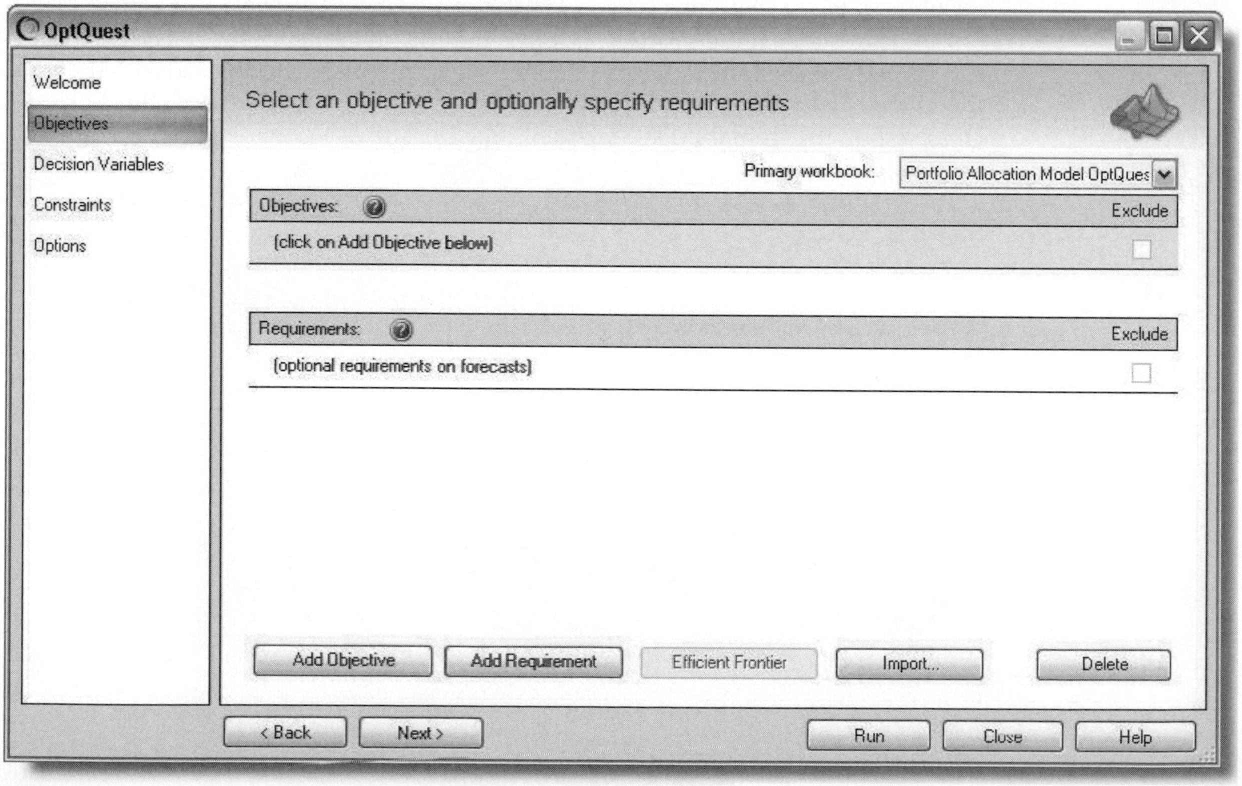

Figure 14.28 *OptQuest Decision Variables Input Screen*

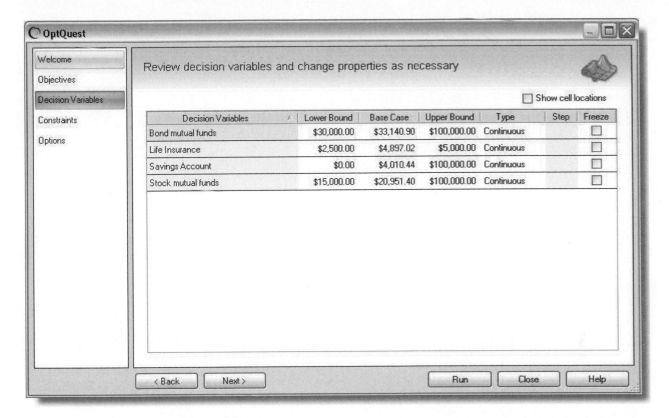

In our example, we have two constraints. The first constraint limits the total weighted risk to 100,000, and the second ensures that we do not allocate more than $100,000 in total to all assets. Constraints are entered in a manner similar to the way they are defined in *Solver*; that is, by specifying the cells corresponding to the left and right hand side of the constraint. In this example, the risk constraint is expressed as E12 ≤ E8, and the allocation constraint as B16 ≤ B8. This is shown in Figure 14.29.

Specify OptQuest *Run Options*

Next, the *Options* window allows you to select various options for running the simulation (see Figure 14.30). *Optimization control* allows you to specify the total time that the system is allowed to search for the best values for the optimization variables. You may enter either the total number of minutes or number of simulations. Performance will depend on the speed of your microprocessor. The default optimization time is 10 minutes; however, you are able to choose any time limit you desire. Selecting a very long time limit does not present a problem because you are always able to terminate the search in the control panel. You may click the Run Preferences button to change the number of trials and other *Crystal Ball* preferences (in this example, we set the number of trials to 2,000). Additionally, you will be given the option to extend the search and carry the optimization process farther once the selected time has expired. Under *Type of optimization,* you can set the optimization type as stochastic—that is, with *Crystal Ball* assumptions or deterministic without assumptions. You may also specify what windows are displayed while the simulation is running and whether to set the decision variable cells to the best solution upon completion.

Figure 14.29 *OptQuest Constraints* Input Screen

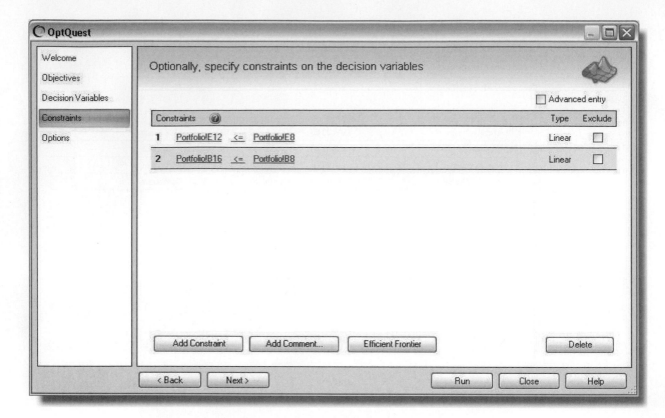

Solve the Optimization Problem

The optimization process is initiated by clicking the *Run* button. As the simulation is running, you will see a *Performance Chart* that shows a plot of the value of the objective as a function of the number of simulations evaluated, and details of the best solutions generated during the search. Figure 14.31 shows the *OptQuest Results* screen upon completion of the optimization.

Interpreting OptQuest Results

You should note that the "best" *OptQuest* solution identified may not be the true optimal solution to the problem, but will, it is hoped, be close to the actual optimal solution. The accuracy of the results depends on the time limit you select for searching, the number of decision variables, the number of trials per simulation, and the complexity of the problem. With more decision variables, you need a larger number of trials.

After solving an optimization problem with *OptQuest,* you probably would want to examine the *Crystal Ball* simulation using the optimal values of the decision variables in order to assess the risks associated with the recommended solution. Figure 14.32 shows the *Crystal Ball* forecast chart associated with the best solution. Although the mean value was optimized, we see that a high amount of variability exists in the actual return because of the uncertainty in the returns of the individual investments. Also note that the mean value for the simulation is different than that found in the *OptQuest* solution because we used a larger number of trials. With *OptQuest*, we used only 1,000 trials per simulation and resulted in more sampling error. Had we used a larger number of trials,

Figure 14.30 *OptQuest Options* Input Screen

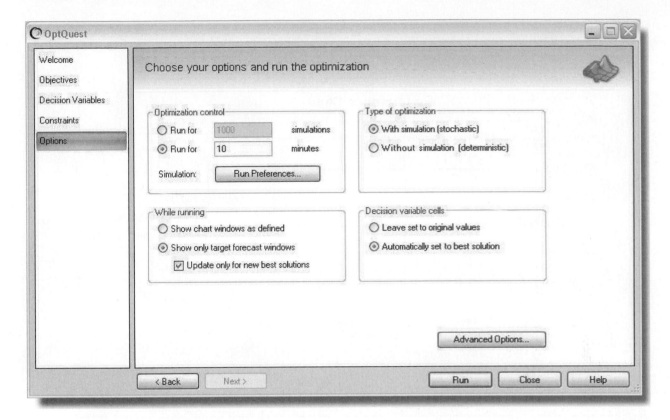

the search would have uncovered fewer solutions in the time allotted. Thus, a trade-off exists between precision and time.

Adding a Requirement

A *requirement* is a forecast statistic that is restricted to fall within a specified lower and upper bound. The forecast statistic may be one of the following:

- Mean
- Median
- Mode
- Standard deviation
- Variance
- Percentile (as specified by the user)
- Skewness
- Kurtosis
- Coefficient of variation
- Range (minimum, maximum, and width)
- Standard error

For example, to reduce the uncertainty of returns in the portfolio while also attempting to maximize the expected return, we might want to restrict the standard deviation to be no greater than 1,000. To add such a requirement in *OptQuest,* click the *Add Requirement* button in the *Objectives* input screen. You will need to change the

Figure 14.31 *OptQuest Results* Summary

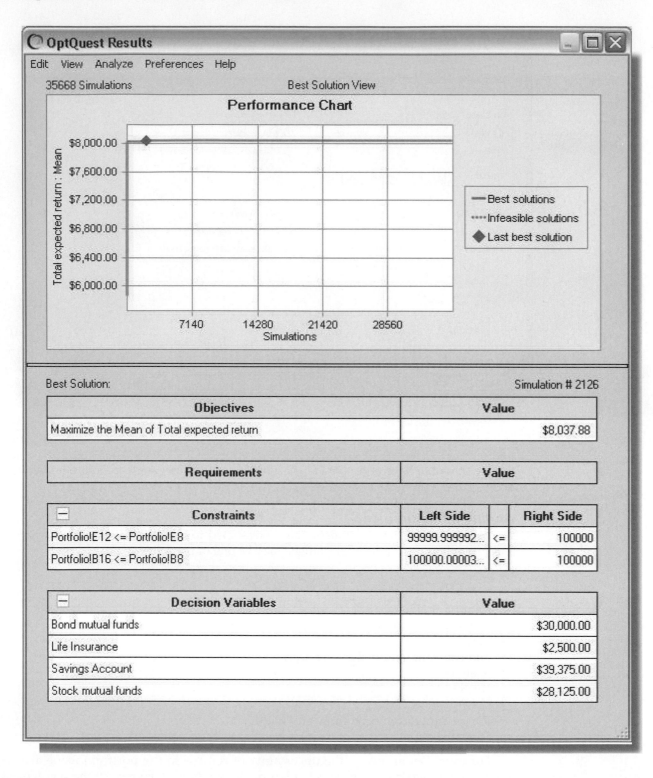

Figure 14.32 *Crystal Ball* Simulation Results for Best *OptQuest* Solution

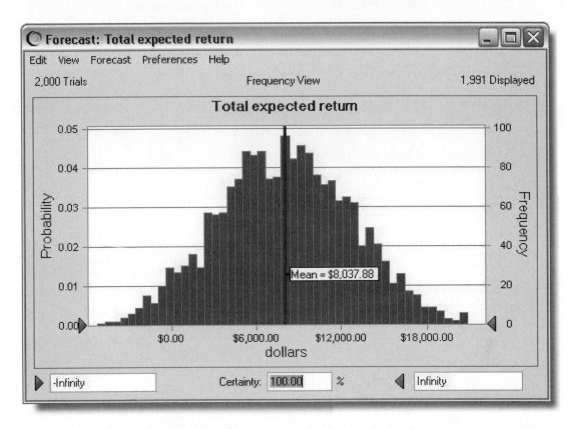

Figure 14.33 *Forecast Selection* Window with Standard Deviation Requirement

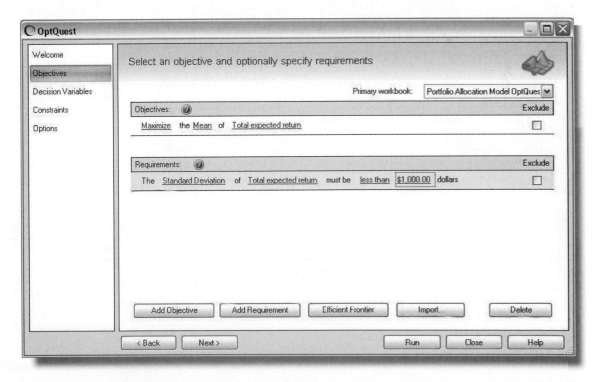

default requirement to reflect the correct values. This is shown in Figure 14.33. You may now run the new model. The results (starting from the solution from the previous optimization run) are shown in Figure 14.34. The best solution among those with standard deviations less than or equal to 1,000 is identified.

Figure 14.34 *OptQuest* Results with Standard Deviation Requirement

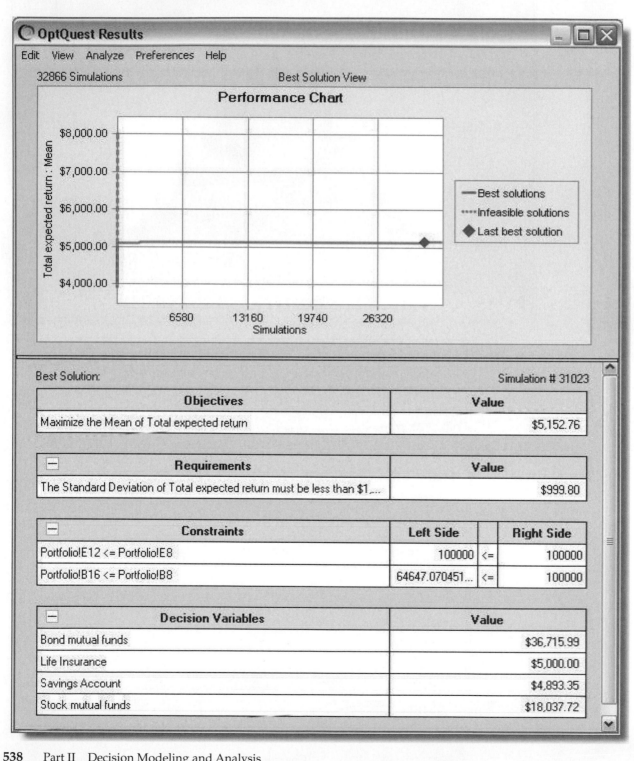

Basic Concepts Review Questions

1. Explain the difference between general integer variables and binary variables.
2. What is the difference between linear and nonlinear optimization models?
3. How are integer variables defined in *Solver*?
4. How can binary variables be used to model logical conditions? Provide several examples.
5. In both the plant location and fixed-cost models presented in this chapter, we used "linking" constraints that established a relationship between the binary variables and other continuous variables in the model. Why are such constraints necessary in any mixed integer model?
6. How do Sensitivity Reports differ between linear and nonlinear optimization models? How do you interpret reduced gradients and Lagrange multipliers?
7. Explain why *Evolutionary Solver* is needed for certain types of optimization problems.
8. How can *Crystal Ball* and *OptQuest* enhance the analysis of optimization problems?

Skill-Building Exercises

1. Solve the fire station location model without the binary restrictions. Were the binary restrictions needed?
2. Find other feasible cutting patterns for the cutting stock model and include them in the model. Does adding the additional patterns improve the solution?
3. In the fire station location model, suppose that the township wants to be able to reach each district from at least two alternative fire stations. Modify the model to find a solution.
4. Implement and solve the computer configuration model on a spreadsheet.
5. Modify the spreadsheet for Kristin's Creations in Chapter 13 to include the fixed costs as described in this chapter and find an optimal solution.
6. Redesign the spreadsheet model for the fixed-cost scenario by eliminating the binary variables and using IF functions as described in the chapter. Try to solve your model using the standard simplex LP procedure in *Solver*? What happens?
7. For the Markowitz portfolio example, solve models for required returns from 7% to 12% and develop a stacked bar chart showing the proportion of each stock in the portfolio. How would you explain this to an investor?
8. Solve the hotel pricing problem using *Evolutionary Solver* from several different starting points. Do you get the optimal solution each time?
9. Solve the Edwards Manufacturing tool bin location model using conventional nonlinear optimization with *Solver* and compare the solution found with *Evolutionary Solver*.
10. Solve the portfolio allocation model with *OptQuest* using 100 trials per simulation, and again using 2,500 trials per simulation. How do the results differ?

Problems and Applications

Note: Data for most of these problems are provided in the Excel file *Chapter 14 Problem Data* to facilitate model building.

1. Thermal transfer printing transfers ink from a ribbon onto paper through a combination of heat and pressure. Different types of printers use different sizes of ribbons. A ribbon manufacturer has forecasted demands for seven different ribbon sizes:

Size	Forecast (Rolls)
60 mm	1,620
83 mm	512
102 mm	840
110 mm	2,640
120 mm	502
130 mm	755
165 mm	680

The rolls from which ribbons are cut are 880 mm in length. Scrap is valued at $0.05 per mm. Generate 10 different cutting patterns so that each size can be cut from at least one pattern. Use your data to construct and solve an optimization model for finding the number of patterns to cut to meet demand and minimize trim loss.

2. For the cutting stock example in the chapter, suppose that the scrap generated from each pattern costs $2 per inch. The company also incurs a fixed cost for setting up the machine to cut different patterns. These costs are:

PATTERN	FIXED COST
1	$70
2	$40
3	$40
4	$90
5	$50
6	$80

Modify the model developed in this chapter to include the scrap and set up costs, and solve the model with *Solver*.

3. The personnel director of a company that recently absorbed another firm and is now downsizing and must relocate five information systems analysts from recently closed locations. Unfortunately, there are only three positions available for five people. Salaries are fairly uniform among this group (those with higher pay were already given the opportunity to begin anew). Moving expenses will be used as the means of determining who will be sent where. Estimated moving expenses are:

	MOVING COST TO		
ANALYST	GARY	SALT LAKE CITY	FRESNO
Arlene	$8,000	$7,000	$5,000
Bobby	$5,000	$8,000	$12,000
Charlene	$9,000	$15,000	$16,000
Douglas	$4,000	$8,000	$13,000
Emory	$7,000	$3,000	$4,000

Model this as an integer optimization model to minimize cost and determine which analysts to relocate to the three locations.

4. A medical device company is allocating next year's budget among its divisions. As a result, the R&D Division needs to determine which research and development projects to fund. Each project requires various software and hardware and consulting expenses, along with internal human resources. A budget allocation of $1,250,000 has been approved, and 32 engineers are available to work on the projects. The R&D group has determined that at most one of projects 1 and 2 should be pursued, and that if project 4 is chosen, then project 2 must also be chosen. Develop a model to select the best projects within the budget.

PROJECT	NPV	INTERNAL ENGINEERS	ADDITIONAL COSTS
1	$600,000	9	$196,000
2	680,000	12	400,000
3	550,000	7	70,000
4	400,000	4	180,000
5	350,000	8	225,000
6	725,000	10	275,000
7	340,000	8	130,000

5. A software support division of a major corporation has eight projects that can be performed. Each project requires different amounts of development time and testing time. In the coming planning period, 1,190 hours of development time and 1,000 hours of testing time are available, based on the skill mix of the staff. The internal transfer price (revenue to the support division) and the times required for each project are shown in the table. Which projects should be selected to maximize revenue?

PROJECT	DEVELOPMENT TIME	TESTING TIME	TRANSFER PRICE
1	80	67	$23,520
2	248	208	$72,912
3	41	34	$12,054
4	10	92	$32,340
5	240	202	$70,560
6	195	164	$57,232
7	269	226	$79,184
8	110	92	$32,340

6. Soapbox is a local band that plays classic and contemporary rock. The band members charge $600 for a three-hour gig. They would like to play at least 30 gigs per year but need to determine the best way to promote themselves. The most they are willing to spend on promotion is $2,500. The possible promotion options are as follows:
   - Playing free gigs
   - Making a demo CD
   - Hiring an agent
   - Handing out fliers
   - Creating a Web site

   Each free gig costs them about $250 for travel and equipment, but generates about 3 paying gigs. A high-quality studio demo CD should help the band book about 20 gigs, but will cost about $1,000. A demo CD made on home recording equipment will cost only $400 but may result in only 10 bookings. A good agent will get the band about 15 gigs, but will charge $1,500. The band can create a Web site for $400 and would expect to generate 6 gigs from this exposure. They also estimate that they may book 1 gig for every 500 fliers they hand out, which would cost $0.08 each. They don't want to play more than 10 free gigs or send out more than 2,500 fliers. Develop and solve an optimization model to find the best promotion strategy to maximize their revenue.

7. Premier Paints supplies to major contractors. One of their contracts for a specialty paint requires them to supply 750, 500, 400, and 950 gallons over the next 4 months. To produce this paint requires a shutdown and cleaning of one of their manufacturing departments at a cost of $1,000. The entire contract requirement can be produced during the first month in one production run; however, the inventory that must be held until delivery costs $0.75 per gallon per month. If the paint is produced in other months, then the cleaning costs are incurred during each month of production. Formulate and solve an integer optimization model to determine the best monthly production schedule to meet delivery contracts and minimize total costs.

8. Dannenfelser Design works with clients in three major project categories: architecture, interior design, and combined. Each type of project requires an estimated number of hours for different categories of employees as shown in the table below.

	ARCHITECTURE	INTERIOR DESIGN	COMBINED	HOURLY RATE
Principal	15	5	18	$115
Sr. Designer	25	35	40	$85
Draftsman	40	30	60	$60
Administrator	5	5	8	$40

In the coming planning period, 184 hours of Principal time, 414 hours of Sr. Designer time, 588 hours of draftsman time, and 72 hours of administrator time are available. Profit per project averages $1,290 for architecture, $1,110 for interior design, and $1,878 for combined projects. The firm would like to work on at least one of each type of project for exposure among clients. Assuming that the firm has more demand than they can possibly handle, find the best mix of projects to maximize profit.

9. Anya is a part-time MBA student who works full-time and is constantly on the run. She recognized the challenge of eating a balanced diet and wants to minimize cost while meeting some basic nutritional requirements. Based on some research, she found that a very active woman should consume 2,250 calories per day. According to one author's guidelines, the following daily nutritional requirements are recommended:

SOURCE	RECOMMENDED INTAKE (GRAMS)
Fat	maximum 75
Carbohydrates	maximum 225
Fiber	maximum 30
Protein	at least 168.75

She chose a sample of meals that could be obtained from healthier quick-service restaurants around town as well as some that could be purchased at the grocery store.

FOOD	COST/SERVING	CALORIES	FAT	CARBS	FIBER	PROTEIN
Turkey sandwich	$4.69	530	14	73	4	28
Baked potato soup	$3.39	260	16	23	1	6
Whole grain chicken sandwich	$6.39	750	28	83	10	44
Bacon turkey sandwich	$5.99	770	28	84	5	47
Southwestern refrigerated chicken wrap	$3.69	220	8	29	15	21
Sesame chicken refrigerated chicken wrap	$3.69	250	10	26	15	26
Yogurt	$0.75	110	2	19	0	5
Raisin bran with skim milk	$0.40	270	1	58	8	12
Cereal bar	$0.43	110	2	22	0	1
1 cup broccoli	$0.50	25	0.3	4.6	2.6	2.6
1 cup carrots	$0.50	55	0.25	13	3.8	1.3
1 scoop protein powder	$1.29	120	4	5	0	17

Anya does not want to eat the same entrée (first six foods) more than once each day but does not mind eating breakfast or side items (last five foods) twice a day and protein powder-based drinks up to four times a day for convenience. Develop an integer linear optimization model to find the number of servings of each food choice in a daily diet to minimize cost and meet the nutritional targets.

10. Brewer Services contracts with outsourcing partners to handle various customer service functions. The customer service department is open Monday through Friday from 8 A.M. to 5 P.M. Calls vary over the course of a typical day. Based on a study of call volumes provided by one of the firm's partners, the minimum number of staff needed each hour of the day are:

HOUR	MINIMUM STAFF REQUIRED
8–9	5
9–10	12
10–11	15
11–noon	12
Noon–1	11
1–2	18
2–3	17
3–4	19
4–5	14

Mr. Brewer wants to hire some permanent employees and staff the remaining requirements using part-time employees who work four-hour shifts (four consecutive hours starting as early as 8 A.M. or as late as 1 P.M.). Develop and solve an integer optimization model to answer the following questions.
a. Suppose Mr. Brewer uses five permanent employees. What is the minimum number of part-time employees he will need for each four-hour shift to ensure meeting the staffing requirements?
b. What if he uses 11 permanent employees?
c. Investigate other possible numbers of permanent employees between 5 and 15, compare the differences, and make a recommendation.

11. Josh Steele manages a professional choir in a major city. His marketing plan is focused on generating additional local demand for concerts and increasing ticket revenue, and also gaining attention at the national level to build awareness of the ensemble across the country. He has $20,000 to spend on media advertising. The goal of the advertisement campaign is to generate as much local recognition as possible while reaching at least 4,000 units of national exposure. He has set a limit of 100 total ads. Additional information is shown below.

MEDIA	PRICE	LOCAL EXPOSURE	NATIONAL EXPOSURE	LIMIT
FM radio spot	$80.00	110	40	30
AM radio spot	$65.00	55	20	30
Cityscape ad	$250.00	80	5	24
MetroWeekly ad	$225.00	65	8	24
Hometown paper ad	$500.00	400	70	10
Neighborhood paper ad	$300.00	220	40	10
Downtown magazine ad	$55.00	35	0	15
Choir journal ad	$350.00	10	75	12
Professional organization magazine ad	$300.00	20	65	12

The last column sets limits on the number of ads to ensure that the advertising markets do not become saturated.

a. Find the optimal number of ads of each type to run to meet the choir's goals by developing and solving an integer optimization model.

b. What if he decides to use no more than six different types of ads? Modify the model in part (a) to answer this question.

12. Chris Corry has a company-sponsored retirement plan at a major brokerage firm. He has the following funds available:

FUND	RISK	TYPE	RETURN
1	High	Stock	11.98%
2	High	Stock	13.18%
3	High	Stock	9.40%
4	High	Stock	7.71%
5	High	Stock	8.35%
6	High	Stock	16.38%
7	Medium	Blend	4.10%
8	Medium	Blend	12.52%
9	Medium	Blend	8.62%
10	Medium	Blend	11.14%
11	Medium	Blend	8.78%
12	Low	Blend	9.44%
13	Low	Blend	8.38%
14	Low	Bond	7.65%
15	Low	Bond	6.90%
16	Low	Bond	5.53%
17	Low	Bond	6.30%

His financial advisor has suggested that at most 60% of the portfolio should be composed of high-risk funds. At least 20% should be invested in bond funds, and at most 40% can be invested in any single fund. At least 5 funds should be selected, and if a fund is selected, it should be funded with at least 5% of the total contribution.

Develop and solve an integer optimization model to determine which funds should be selected and what percentage of his total investment should be allocated to each fund.

13. Mark Haynes is interested in buying a new car. He decided on a particular model, which has lots of options from which to choose. The base price of the car is $16,510 and he allotted a budget of $19,250 to purchase it. The table below shows the possible options over and above the base model that he could choose, their cost, and the utility that he assigned to each option:

Option	Utility	Cost
Slower engine	−0.10	$(300.00)
Faster engine	0.40	$600.00
Fastest engine	0.50	$1,000.00
No warranty	0.00	$(250.00)
3-year warranty	0.30	$450.00
5-year warranty	0.70	$750.00
Automatic transmission	0.80	$800.00
15-inch wheels	−0.15	$(150.00)
16-inch wheels	0.25	$300.00
Alloy wheels	0.35	$500.00
AM/FM radio	0.10	$200.00
AM/FM/CD	0.30	$300.00
AM FM/CD/DVD	0.50	$400.00
AM/BM/CD/DVD 6 speaker	0.65	$750.00
Sunroof	0.25	$50.00
Moonroof	0.40	$150.00
2-wheel disc brakes	−0.10	$(250.00)
4-wheel disc brakes with ABS	0.35	$250.00

All of these are options to replace the base model configuration. Thus, for example, he may choose one of the three engine options or keep the base model engine, choose one of these three warranty options or keep the base model warranty, choose automatic transmission or keep the base model manual transmission, and so on. In addition, if the automatic transmission is chosen, then a three- or five-year warranty must be chosen, and if the fastest engine is chosen, then either the 16-inch or alloy wheels must be chosen.

a. Develop and solve an integer optimization model to maximize his utility and stay within budget.

b. If he has an additional $500 to spend, how would his choices change?

c. In a magazine, Mark found a car stereo system for $300. If he decides to replace the base model radio with this one, how would the model and his decisions change?

14. The Spurling Group is considering using magazine outlets to advertise their online Web site. The company has identified seven publishers. Each publisher

breaks down their subscriber base into a number of groups based on demographics and location. These data are shown in the table below:

PUBLISHER	GROUPS	SUBSCRIBERS/GROUP	COST/GROUP
A	5	460,000	$1,560
B	10	50,000	$290
C	4	225,000	$1,200
D	20	24,000	$130
E	5	1,120,000	$2,500
F	1	1,700,000	$7,000
G	2	406,000	$1,700

The company has set a budget of $25,000 for advertising and wants to maximize the number of subscribers exposed to their ads. However, publishers B and D are competitors and only one of these may be chosen. A similar situation exists with publishers C and G. Formulate and solve an integer optimization model to determine which publishers to select and how many groups to purchase for each publisher.

15. A young entrepreneur has invented a new air-adjustable basketball shoe with pump, similar to those advertised widely by more expensive brand names. He contacted a supplier of Victor basketball shoes, a little-known brand with low advertising. This supplier would provide shoes at the nominal price of $6 per pair of shoes. He needs to know the best price at which to sell these shoes. As a business student with strong economics training, he remembered that the volume sold is affected by the product's price—the higher the price, the lower the volume. He asked his friends and acquaintances what they would pay for a premium pair of basketball shoes that were a "little off-brand." Based on this information, he developed the formula:

$$\text{Volume} = 1,000 - 20\,\text{Price}$$

There are some minor expenses involved, including a $50 fee for selling shoes in the neighborhood (a fixed cost), as well as his purchase price of $6 per shoe. Develop an appropriate objective function and find the optimal price level using *Solver*.

16. The entrepreneur in the previous problem did very well selling Victor shoes. His shoe supplier told him of a new product, Top Notch, that was entering the market. This shoe would be a product substitute for Victors, so that the higher the price of either shoe, the greater the demand for the other. He interviewed more potential clients to determine price response and cross elasticities. This yielded the following relationships:

$$\text{Volume of Victors} = 1,000 - 20P_v + 1P_n$$
$$\text{Volume of Top Notch} = 800 + 2P_v - 18P_n$$

where P_v = price of Victors and P_n = price of Top Notch. Develop a model to maximize the total revenue and find the optimal prices using *Solver*.

17. The Hal Chase Investment Planning Agency is in business to help investors optimize their return from investment. Hal deals with three investment mediums: a stock fund, a bond fund, and his own Sports and Casino Investment Plan (SCIP). The stock fund is a mutual fund investing in openly traded stocks. The bond fund focuses on the bond market, which has a more stable but lower expected return. SCIP is a high-risk scheme, often resulting in heavy losses but occasionally

coming through with spectacular gains. Average returns, their variances, and covariances are:

	STOCK	BOND	SCIP
Average return	0.148	0.060	0.152
Variance	0.014697	0.000155	0.160791
Covariance with stock		0.000468	−0.002222
Covariance with bond			−0.000227

Develop and solve a portfolio optimization model for this situation for a target return of 12%.

18. Stout Investments wishes to design a minimum variance portfolio of index funds. The funds selected for consideration and their variance-covariance matrix and average returns are given below:

	BOND	S&P 500	SMALL CAP	MID CAP	LARGE CAP	EMERGING MARKET	COMMODITY
Bond	0.002%						
S&P 500	−0.001%	0.020%					
Small cap	−0.001%	0.027%	0.047%				
Mid cap	−0.001%	0.024%	0.039%	0.033%			
Large cap	−0.001%	0.019%	0.027%	0.023%	0.027%		
Emerging market	0.000%	0.032%	0.050%	0.043%	0.041%	0.085%	
Commodity	0.000%	0.000%	0.005%	0.005%	0.009%	0.015%	0.054%
Average weekly return	0.044%	0.118%	0.256%	0.226%	0.242%	0.447%	0.053%

Stout Investments would like to achieve an average weekly return of 0.19%, or roughly a 10% annual return.

a. Formulate and solve a Markowitz portfolio optimization model for this situation.

b. Suppose the company wants to restrict the percentage of investments in each fund as follows:

Bond: between 10% and 50%
S&P 500: between 30% and 50%
Small cap: no more than 20%
Mid cap: no more than 20%
Large cap: no more than 20%
Emerging market: no more than 10%
Commodity: no more than 20%

How would the optimal portfolio change? Compare the solutions obtained using the GRG and Evolutionary algorithms provided by *Solver*.

19. Tejeda Investment Management, LLC manages 401K retirement plans. A client has asked them to recommend a portfolio. In discussing options with the client, 15 mutual funds were selected. The variance-covariance matrix and other relevant information are provided in the *Problem 19* tab of the *Chapter 14 Problem Data* Excel file (using the quotation symbols for these funds). The client would

like to achieve a minimum return of at least 10%. The company also recommends the following:

- At least five funds be selected,
- No more than three funds in the portfolio should have a Morningstar rating of three or less,
- At most two funds should have an above average risk rating, and
- If a fund is selected, then at least 5% but no more than 25% should be invested in that fund.

a. Formulate and solve an integer nonlinear optimization model to minimize the portfolio variance and meet the recommended restrictions. Use *Evolutionary Solver*.

b. Suppose the client wants to maximize the expected return with the restriction that the portfolio variance be no higher than 15%. How does the solution change?

20. For the project selection example, suppose that the returns in the objective function are normally distributed with means as given by the expected returns and standard deviation equal to 10% of the mean. However, also assume that each project has a success rate modeled as a Bernoulli distribution. That is, the return will be realized only if the project is successful (use the "Yes–No" distribution in *Crystal Ball*). Success rates for the five projects are 0.80, 0.70, 0.90, 0.40, and 0.60, respectively. Modify the spreadsheet model, and use *OptQuest* to find a solution that maximizes the mean return.

CASE

Tindall Bookstores

Tindall Bookstores[4] is a major national retail chain with stores located principally in shopping malls. For many years, the company has published a Christmas catalog that was sent to current customers on file. This strategy generated additional mail-order business while also attracting customers to the stores. However the cost-effectiveness of this strategy was never determined. In 2008, John Harris, vice president of marketing, conducted a major study on the effectiveness of direct-mail delivery of Tindall's Christmas catalog. The results were favorable: Patrons who were catalog recipients spent more, on average, than did comparable nonrecipients. These revenue gains more than compensated for the costs of production, handling, and mailing, which had been substantially reduced by cooperative allowances from suppliers.

With the continuing interest in direct mail as a vehicle for delivering holiday catalogs, Harris continued to investigate how new customers could most effectively be reached. One of these ideas involved purchasing mailing lists of magazine subscribers through a list broker. In order to determine which magazines might be more appropriate, a mail questionnaire was administered to a sample of current customers to ascertain which magazines they regularly read. Ten magazines were selected for the survey. The assumption behind this strategy is that subscribers of magazines that a high proportion of current customers read would be viable targets for future purchases at Tindall stores. The question is which magazine lists should be purchased to maximize reaching of potential customers in the presence of a limited budget for the purchase of lists.

Data from the customer survey have begun to trickle in. The information about the 10 magazines a customer subscribes to is provided on the returned questionnaire. Harris has asked you to develop a prototype model, which later can be used to decide which lists to purchase. So far only 53 surveys have been returned. To keep the prototype model manageable Harris has instructed you to go ahead with the model development using the data from the 53

---

[4] This case is based on the case by the same name developed by James R. Evans of the University of Cincinnati, and was sponsored by the Direct Marketing Policy Center.

returned surveys. These data are shown in Table 14.4. The costs of the first ten lists are given below and your budget is $3,000.

List	1	2	3	4	5	6	7	8	9	10
Cost(000)	$1	$1	$1	$1.5	$1.5	$1.5	$1	$1.2	$.5	$1.1

DATA FOR TINDALL BOOKSTORES SURVEY

What magazines should be chosen to maximize overall exposure? Conduct a budget sensitivity analysis on the Tindall magazine list selection problem. Solve the problem for a variety of budgets and graph percentage of total reach (number reached /53) versus budget amount. In your opinion, when is an increment in budget no longer warranted (based on this limited data)?

Table 14.4 Survey Results

CUSTOMER	MAGAZINES	CUSTOMER	MAGAZINES
1	10	28	4, 7
2	1, 4	29	6
3	1	30	3, 4, 5, 10
4	5, 6	31	4
5	5	32	8
6	10	33	1, 3, 10
7	2, 9	34	4, 5
8	5, 8	35	1, 5, 6
9	1, 5, 10	36	1, 3
10	4, 6, 8, 10	37	3, 5, 8
11	6	38	3
12	3	39	2, 7
13	5	40	2, 7
14	2, 6	41	7
15	8	42	4, 5, 6
16	6	43	NONE
17	4, 5	44	5, 10
18	7	45	1, 2
19	5, 6	46	7
50	2, 8	47	1, 5, 10
21	7, 9	48	3
22	6	49	1, 3, 4
23	3, 6, 10	50	NONE
24	NONE	51	2, 6
25	5, 8	52	NONE
26	3, 10	53	2, 5, 8, 9, 10
27	2, 8		

APPENDIX

Table A.1   The Cumulative Standard Normal Distribution

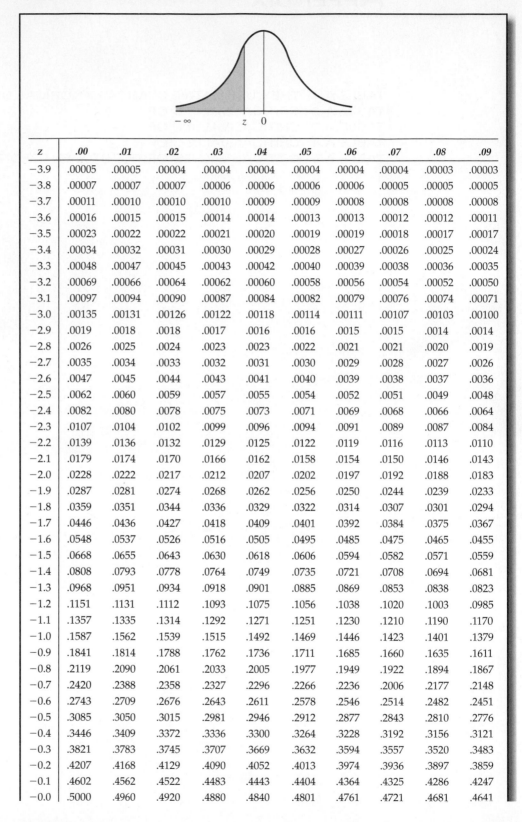

z	.00	.01	.02	.03	.04	.05	.06	.07	.08	.09
−3.9	.00005	.00005	.00004	.00004	.00004	.00004	.00004	.00004	.00003	.00003
−3.8	.00007	.00007	.00007	.00006	.00006	.00006	.00006	.00005	.00005	.00005
−3.7	.00011	.00010	.00010	.00010	.00009	.00009	.00008	.00008	.00008	.00008
−3.6	.00016	.00015	.00015	.00014	.00014	.00013	.00013	.00012	.00012	.00011
−3.5	.00023	.00022	.00022	.00021	.00020	.00019	.00019	.00018	.00017	.00017
−3.4	.00034	.00032	.00031	.00030	.00029	.00028	.00027	.00026	.00025	.00024
−3.3	.00048	.00047	.00045	.00043	.00042	.00040	.00039	.00038	.00036	.00035
−3.2	.00069	.00066	.00064	.00062	.00060	.00058	.00056	.00054	.00052	.00050
−3.1	.00097	.00094	.00090	.00087	.00084	.00082	.00079	.00076	.00074	.00071
−3.0	.00135	.00131	.00126	.00122	.00118	.00114	.00111	.00107	.00103	.00100
−2.9	.0019	.0018	.0018	.0017	.0016	.0016	.0015	.0015	.0014	.0014
−2.8	.0026	.0025	.0024	.0023	.0023	.0022	.0021	.0021	.0020	.0019
−2.7	.0035	.0034	.0033	.0032	.0031	.0030	.0029	.0028	.0027	.0026
−2.6	.0047	.0045	.0044	.0043	.0041	.0040	.0039	.0038	.0037	.0036
−2.5	.0062	.0060	.0059	.0057	.0055	.0054	.0052	.0051	.0049	.0048
−2.4	.0082	.0080	.0078	.0075	.0073	.0071	.0069	.0068	.0066	.0064
−2.3	.0107	.0104	.0102	.0099	.0096	.0094	.0091	.0089	.0087	.0084
−2.2	.0139	.0136	.0132	.0129	.0125	.0122	.0119	.0116	.0113	.0110
−2.1	.0179	.0174	.0170	.0166	.0162	.0158	.0154	.0150	.0146	.0143
−2.0	.0228	.0222	.0217	.0212	.0207	.0202	.0197	.0192	.0188	.0183
−1.9	.0287	.0281	.0274	.0268	.0262	.0256	.0250	.0244	.0239	.0233
−1.8	.0359	.0351	.0344	.0336	.0329	.0322	.0314	.0307	.0301	.0294
−1.7	.0446	.0436	.0427	.0418	.0409	.0401	.0392	.0384	.0375	.0367
−1.6	.0548	.0537	.0526	.0516	.0505	.0495	.0485	.0475	.0465	.0455
−1.5	.0668	.0655	.0643	.0630	.0618	.0606	.0594	.0582	.0571	.0559
−1.4	.0808	.0793	.0778	.0764	.0749	.0735	.0721	.0708	.0694	.0681
−1.3	.0968	.0951	.0934	.0918	.0901	.0885	.0869	.0853	.0838	.0823
−1.2	.1151	.1131	.1112	.1093	.1075	.1056	.1038	.1020	.1003	.0985
−1.1	.1357	.1335	.1314	.1292	.1271	.1251	.1230	.1210	.1190	.1170
−1.0	.1587	.1562	.1539	.1515	.1492	.1469	.1446	.1423	.1401	.1379
−0.9	.1841	.1814	.1788	.1762	.1736	.1711	.1685	.1660	.1635	.1611
−0.8	.2119	.2090	.2061	.2033	.2005	.1977	.1949	.1922	.1894	.1867
−0.7	.2420	.2388	.2358	.2327	.2296	.2266	.2236	.2006	.2177	.2148
−0.6	.2743	.2709	.2676	.2643	.2611	.2578	.2546	.2514	.2482	.2451
−0.5	.3085	.3050	.3015	.2981	.2946	.2912	.2877	.2843	.2810	.2776
−0.4	.3446	.3409	.3372	.3336	.3300	.3264	.3228	.3192	.3156	.3121
−0.3	.3821	.3783	.3745	.3707	.3669	.3632	.3594	.3557	.3520	.3483
−0.2	.4207	.4168	.4129	.4090	.4052	.4013	.3974	.3936	.3897	.3859
−0.1	.4602	.4562	.4522	.4483	.4443	.4404	.4364	.4325	.4286	.4247
−0.0	.5000	.4960	.4920	.4880	.4840	.4801	.4761	.4721	.4681	.4641

Table A.1 (*Continued*)

z	.00	.01	.02	.03	.04	.05	.06	.07	.08	.09
0.0	.5000	.5040	.5080	.5120	.5160	.5199	.5239	.5279	.5319	.5359
0.1	.5398	.5438	.5478	.5517	.5557	.5596	.5636	.5675	.5714	.5753
0.2	.5793	.5832	.5871	.5910	.5948	.5987	.6026	.6064	.6103	.6141
0.3	.6179	.6217	.6255	.6293	.6331	.6368	.6406	.6443	.6480	.6517
0.4	.6554	.6591	.6628	.6664	.6700	.6736	.6772	.6808	.6844	.6879
0.5	.6915	.6950	.6985	.7019	.7054	.7088	.7123	.7157	.7190	.7224
0.6	.7257	.7291	.7324	.7357	.7389	.7422	.7454	.7486	.7518	.7549
0.7	.7580	.7612	.7642	.7673	.7704	.7734	.7764	.7794	.7823	.7852
0.8	.7881	.7910	.7939	.7967	.7995	.8023	.8051	.8078	.8106	.8133
0.9	.8159	.8186	.8212	.8238	.8264	.8289	.8315	.8340	.8365	.8389
1.0	.8413	.8438	.8461	.8485	.8508	.8531	.8554	.8577	.8599	.8621
1.1	.8643	.8665	.8686	.8708	.8729	.8749	.8770	.8790	.8810	.8830
1.2	.8849	.8869	.8888	.8907	.8925	.8944	.8962	.8980	.8997	.9015
1.3	.9032	.9089	.9066	.9082	.9099	.9115	.9131	.9147	.9162	.9177
1.4	.9192	.9207	.9222	.9236	.9251	.9265	.9279	.9292	.9306	.9319
1.5	.9332	.9345	.9357	.9370	.9382	.9394	.9406	.9418	.9429	.9441
1.6	.9452	.9463	.9474	.9484	.9495	.9505	.9515	.9525	.9535	.9545
1.7	.9554	.9564	.9573	.9582	.9591	.9599	.9608	.9616	.9625	.9633
1.8	.9641	.9649	.9656	.9664	.9671	.9678	.9686	.9693	.9699	.9706
1.9	.9713	.9719	.9726	.9732	.9738	.9744	.9750	.9756	.9761	.9767
2.0	.9772	.9778	.9783	.9788	.9793	.9798	.9803	.9808	.9812	.9817
2.1	.9821	.9826	.9830	.9834	.9838	.9842	.9846	.9850	.9854	.9857
2.2	.9861	.9864	.9868	.9871	.9875	.9878	.9881	.9884	.9887	.9890
2.3	.9893	.9896	.9898	.9901	.9904	.9906	.9909	.9911	.9913	.9916
2.4	.9918	.9920	.9922	.9925	.9927	.9929	.9931	.9932	.9934	.9936
2.5	.9938	.9940	.9941	.9943	.9945	.9946	.9948	.9949	.9951	.9952
2.6	.9953	.9955	.9956	.9957	.9959	.9960	.9961	.9962	.9963	.9964
2.7	.9965	.9966	.9967	.9968	.9969	.9970	.9971	.9972	.9973	.9974
2.8	.9974	.9975	.9976	.9977	.9977	.9978	.9979	.9979	.9980	.9981
2.9	.9981	.9982	.9982	.9983	.9984	.9984	.9985	.9985	.9986	.9986
3.0	.99865	.99869	.99874	.99878	.99882	.99886	.99889	.99893	.99897	.99900
3.1	.99903	.99906	.99910	.99913	.99916	.99918	.99921	.99924	.99926	.99929
3.2	.99931	.99934	.99936	.99938	.99940	.99942	.99944	.99946	.99948	.99950
3.3	.99952	.99953	.99955	.99957	.99958	.99960	.99961	.99962	.99964	.99965
3.4	.99966	.99968	.99969	.99970	.99971	.99972	.99973	.99974	.99975	.99976
3.5	.99977	.99978	.99978	.99979	.99980	.99981	.99981	.99982	.99983	.99983
3.6	.99984	.99985	.99985	.99986	.99986	.99987	.99987	.99988	.99988	.99989
3.7	.99989	.99990	.99990	.99990	.99991	.99991	.99992	.99992	.99992	.99992
3.8	.99993	.99993	.99993	.99994	.99994	.99994	.99994	.99995	.99995	.99995
3.9	.99995	.99995	.99996	.99996	.99996	.99996	.99996	.99996	.99997	.99997

Entry represents area under the cumulative standardized normal distribution from $-\infty$ to z.

Table A.2 Critical Values of *t*

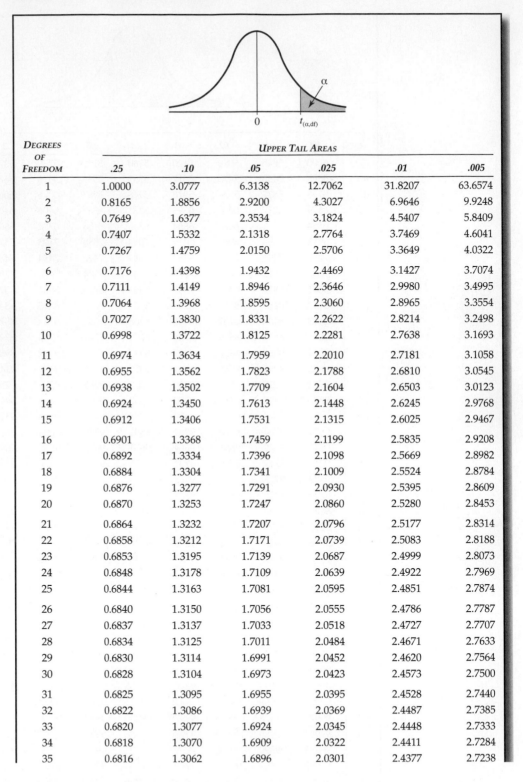

DEGREES OF FREEDOM	UPPER TAIL AREAS					
	.25	.10	.05	.025	.01	.005
1	1.0000	3.0777	6.3138	12.7062	31.8207	63.6574
2	0.8165	1.8856	2.9200	4.3027	6.9646	9.9248
3	0.7649	1.6377	2.3534	3.1824	4.5407	5.8409
4	0.7407	1.5332	2.1318	2.7764	3.7469	4.6041
5	0.7267	1.4759	2.0150	2.5706	3.3649	4.0322
6	0.7176	1.4398	1.9432	2.4469	3.1427	3.7074
7	0.7111	1.4149	1.8946	2.3646	2.9980	3.4995
8	0.7064	1.3968	1.8595	2.3060	2.8965	3.3554
9	0.7027	1.3830	1.8331	2.2622	2.8214	3.2498
10	0.6998	1.3722	1.8125	2.2281	2.7638	3.1693
11	0.6974	1.3634	1.7959	2.2010	2.7181	3.1058
12	0.6955	1.3562	1.7823	2.1788	2.6810	3.0545
13	0.6938	1.3502	1.7709	2.1604	2.6503	3.0123
14	0.6924	1.3450	1.7613	2.1448	2.6245	2.9768
15	0.6912	1.3406	1.7531	2.1315	2.6025	2.9467
16	0.6901	1.3368	1.7459	2.1199	2.5835	2.9208
17	0.6892	1.3334	1.7396	2.1098	2.5669	2.8982
18	0.6884	1.3304	1.7341	2.1009	2.5524	2.8784
19	0.6876	1.3277	1.7291	2.0930	2.5395	2.8609
20	0.6870	1.3253	1.7247	2.0860	2.5280	2.8453
21	0.6864	1.3232	1.7207	2.0796	2.5177	2.8314
22	0.6858	1.3212	1.7171	2.0739	2.5083	2.8188
23	0.6853	1.3195	1.7139	2.0687	2.4999	2.8073
24	0.6848	1.3178	1.7109	2.0639	2.4922	2.7969
25	0.6844	1.3163	1.7081	2.0595	2.4851	2.7874
26	0.6840	1.3150	1.7056	2.0555	2.4786	2.7787
27	0.6837	1.3137	1.7033	2.0518	2.4727	2.7707
28	0.6834	1.3125	1.7011	2.0484	2.4671	2.7633
29	0.6830	1.3114	1.6991	2.0452	2.4620	2.7564
30	0.6828	1.3104	1.6973	2.0423	2.4573	2.7500
31	0.6825	1.3095	1.6955	2.0395	2.4528	2.7440
32	0.6822	1.3086	1.6939	2.0369	2.4487	2.7385
33	0.6820	1.3077	1.6924	2.0345	2.4448	2.7333
34	0.6818	1.3070	1.6909	2.0322	2.4411	2.7284
35	0.6816	1.3062	1.6896	2.0301	2.4377	2.7238

Table A.2 *(Continued)*

DEGREES OF FREEDOM	UPPER TAIL AREAS					
	.25	.10	.05	.025	.01	.005
36	0.6814	1.3055	1.6883	2.0281	2.4345	2.7195
37	0.6812	1.3049	1.6871	2.0262	2.4314	2.7154
38	0.6810	1.3042	1.6860	2.0244	2.4286	2.7116
39	0.6808	1.3036	1.6849	2.0227	2.4258	2.7079
40	0.6807	1.3031	1.6839	2.0211	2.4233	2.7045
41	0.6805	1.3025	1.6829	2.0195	2.4208	2.7012
42	0.6804	1.3020	1.6820	2.0181	2.4185	2.6981
43	0.6802	1.3016	1.6811	2.0167	2.4163	2.6951
44	0.6801	1.3011	1.6802	2.0154	2.4141	2.6923
45	0.6800	1.3006	1.6794	2.0141	2.4121	2.6896
46	0.6799	1.3002	1.6787	2.0129	2.4102	2.6870
47	0.6797	1.2998	1.6779	2.0117	2.4083	2.6846
48	0.6796	1.2994	1.6772	2.0106	2.4066	2.6822
49	0.6795	1.2991	1.6766	2.0096	2.4049	2.6800
50	0.6794	1.2987	1.6759	2.0086	2.4033	2.6778
51	0.6793	1.2984	1.6753	2.0076	2.4017	2.6757
52	0.6792	1.2980	1.6747	2.0066	2.4002	2.6737
53	0.6791	1.2977	1.6741	2.0057	2.3988	2.6718
54	0.6791	1.2974	1.6736	2.0049	2.3974	2.6700
55	0.6790	1.2971	1.6730	2.0040	2.3961	2.6682
56	0.6789	1.2969	1.6725	2.0032	2.3948	2.6665
57	0.6788	1.2966	1.6720	2.0025	2.3936	2.6649
58	0.6787	1.2963	1.6716	2.0017	2.3924	2.6633
59	0.6787	1.2961	1.6711	2.0010	2.3912	2.6618
60	0.6786	1.2958	1.6706	2.0003	2.3901	2.6603
61	0.6785	1.2956	1.6702	1.9996	2.3890	2.6589
62	0.6785	1.2954	1.6698	1.9990	2.3880	2.6575
63	0.6784	1.2951	1.6694	1.9983	2.3870	2.6561
64	0.6783	1.2949	1.6690	1.9977	2.3860	2.6549
65	0.6783	1.2947	1.6686	1.9971	2.3851	2.6536
66	0.6782	1.2945	1.6683	1.9966	2.3842	2.6524
67	0.6782	1.2943	1.6679	1.9960	2.3833	2.6512
68	0.6781	1.2941	1.6676	1.9955	2.3824	2.6501
69	0.6781	1.2939	1.6672	1.9949	2.3816	2.6490
70	0.6780	1.2938	1.6669	1.9944	2.3808	2.6479
71	0.6780	1.2936	1.6666	1.9939	2.3800	2.6469
72	0.6779	1.2934	1.6663	1.9935	2.3793	2.6459
73	0.6779	1.2933	1.6660	1.9930	2.3785	2.6449
74	0.6778	1.2931	1.6657	1.9925	2.3778	2.6439
75	0.6778	1.2929	1.6654	1.9921	2.3771	2.6430
76	0.6777	1.2928	1.6652	1.9917	2.3764	2.6421
77	0.6777	1.2926	1.6649	1.9913	2.3758	2.6412
78	0.6776	1.2925	1.6646	1.9908	2.3751	2.6403
79	0.6776	1.2924	1.6644	1.9905	2.3745	2.6395
80	0.6776	1.2922	1.6641	1.9901	2.3739	2.6387

(Continued)

Table A.2 (Continued)

DEGREES OF FREEDOM	UPPER TAIL AREAS					
	.25	.10	.05	.025	.01	.005
81	0.6775	1.2921	1.6639	1.9897	2.3733	2.6379
82	0.6775	1.2920	1.6636	1.9893	2.3727	2.6371
83	0.6775	1.2918	1.6634	1.9890	2.3721	2.6364
84	0.6774	1.2917	1.6632	1.9886	2.3716	2.6356
85	0.6774	1.2916	1.6630	1.9883	2.3710	2.6349
86	0.6774	1.2915	1.6628	1.9879	2.3705	2.6342
87	0.6773	1.2914	1.6626	1.9876	2.3700	2.6335
88	0.6773	1.2912	1.6624	1.9873	2.3695	2.6329
89	0.6773	1.2911	1.6622	1.9870	2.3690	2.6322
90	0.6772	1.2910	1.6620	1.9867	2.3685	2.6316
91	0.6772	1.2909	1.6618	1.9864	2.3680	2.6309
92	0.6772	1.2908	1.6616	1.9861	2.3676	2.6303
93	0.6771	1.2907	1.6614	1.9858	2.3671	2.6297
94	0.6771	1.2906	1.6612	1.9855	2.3667	2.6291
95	0.6771	1.2905	1.6611	1.9853	2.3662	2.6286
96	0.6771	1.2904	1.6609	1.9850	2.3658	2.6280
97	0.6770	1.2903	1.6607	1.9847	2.3654	2.6275
98	0.6770	1.2902	1.6606	1.9845	2.3650	2.6269
99	0.6770	1.2902	1.6604	1.9842	2.3646	2.6264
100	0.6770	1.2901	1.6602	1.9840	2.3642	2.6259
110	0.6767	1.2893	1.6588	1.9818	2.3607	2.6213
120	0.6765	1.2886	1.6577	1.9799	2.3578	2.6174
∞	0.6745	1.2816	1.6449	1.9600	2.3263	2.5758

For particular number of degrees of freedom, entry represents the critical value of t corresponding to a specified upper tail area (α).

Table A.3   Critical Values of χ^2

DEGREES OF FREEDOM	.995	.99	.975	.95	.90	.75	.25	.10	.05	.025	.01	.005
1			0.001	0.004	0.016	0.102	1.323	2.706	3.841	5.024	6.635	7.879
2	0.010	0.020	0.051	0.103	0.211	0.575	2.773	4.605	5.991	7.378	9.210	10.597
3	0.072	0.115	0.216	0.352	0.584	1.213	4.108	6.251	7.815	9.348	11.345	12.838
4	0.207	0.297	0.484	0.711	1.064	1.923	5.385	7.779	9.488	11.143	13.277	14.860
5	0.412	0.554	0.831	1.145	1.610	2.675	6.626	9.236	11.071	12.833	15.086	16.750
6	0.676	0.872	1.237	1.635	2.204	3.455	7.841	10.645	12.592	14.449	16.812	18.548
7	0.989	1.239	1.690	2.167	2.833	4.255	9.037	12.017	14.067	16.013	18.475	20.278
8	1.344	1.646	2.180	2.733	3.490	5.071	10.219	13.362	15.507	17.535	20.090	21.955
9	1.735	2.088	2.700	3.325	4.168	5.899	11.389	14.684	16.919	19.023	21.666	23.589
10	2.156	2.558	3.247	3.940	4.865	6.737	12.549	15.987	18.307	20.483	23.209	25.188
11	2.603	3.053	3.816	4.575	5.578	7.584	13.701	17.275	19.675	21.920	24.725	26.757
12	3.074	3.571	4.404	5.226	6.304	8.438	14.845	18.549	21.026	23.337	26.217	28.299
13	3.565	4.107	5.009	5.892	7.042	9.299	15.984	19.812	22.362	24.736	27.688	29.819
14	4.075	4.660	5.629	6.571	7.790	10.165	17.117	21.064	23.685	26.119	29.141	31.319
15	4.601	5.229	6.262	7.261	8.547	11.037	18.245	22.307	24.996	27.488	30.578	32.801
16	5.142	5.812	6.908	7.962	9.312	11.912	19.369	23.542	26.296	28.845	32.000	34.267
17	5.697	6.408	7.564	8.672	10.085	12.792	20.489	24.769	27.587	30.191	33.409	35.718
18	6.265	7.015	8.231	9.390	10.865	13.675	21.605	25.989	28.869	31.526	34.805	37.156
19	6.844	7.633	8.907	10.117	11.651	14.562	22.718	27.204	30.144	32.852	36.191	38.582
20	7.434	8.260	9.591	10.851	12.443	15.452	23.828	28.412	31.410	34.170	37.566	39.997
21	8.034	8.897	10.283	11.591	13.240	16.344	24.935	29.615	32.671	35.479	38.932	41.401
22	8.643	9.542	10.982	12.338	14.042	17.240	26.039	30.813	33.924	36.781	40.289	42.796
23	9.260	10.196	11.689	13.091	14.848	18.137	27.141	32.007	35.172	38.076	41.638	44.181
24	9.886	10.856	12.401	13.848	15.659	19.037	28.241	33.196	36.415	39.364	42.980	45.559
25	10.520	11.524	13.120	14.611	16.473	19.939	29.339	34.382	37.652	40.646	44.314	46.928
26	11.160	12.198	13.844	15.379	17.292	20.843	30.435	35.563	38.885	41.923	45.642	48.290
27	11.808	12.879	14.573	16.151	18.114	21.749	31.528	36.741	40.113	43.194	46.963	49.645
28	12.461	13.565	15.308	16.928	18.939	22.657	32.620	37.916	41.337	44.461	48.278	50.993
29	13.121	14.257	16.047	17.708	19.768	23.567	33.711	39.087	42.557	45.722	49.588	52.336
30	13.787	14.954	16.791	18.493	20.599	24.478	34.800	40.256	43.773	46.979	50.892	53.672

UPPER TAIL AREAS (α)

For a particular number of degrees of freedom, entry represents the critical value of χ^2 corresponding to a specified upper tail area (α).

For larger values of degrees of freedom (df) the expression $Z = \sqrt{2\chi^2} - \sqrt{2(df) - 1}$ may be used, and the resulting upper tail area can be obtained from the table of the standard normal distribution (Table A.1a).

Table A.4 Critical Values of F

$\alpha = .05$

$F_{U(\alpha, df_1, df_2)}$

NUMERATOR df_1

DENOMINATOR df_2	1	2	3	4	5	6	7	8	9	10	12	15	20	24	30	40	60	120	∞
1	161.4	199.5	215.7	224.6	230.2	234.0	236.8	238.9	240.5	241.9	243.9	245.9	248.0	249.1	250.1	251.1	252.2	253.3	254.3
2	18.51	19.00	19.16	19.25	19.30	19.33	19.35	19.37	19.38	19.40	19.41	19.43	19.45	19.45	19.46	19.47	19.48	19.49	19.50
3	10.13	9.55	9.28	9.12	9.01	8.94	8.89	8.85	8.81	8.79	8.74	8.70	8.66	8.64	8.62	8.59	8.57	8.55	8.53
4	7.71	6.94	6.59	6.39	6.26	6.16	6.09	6.04	6.00	5.96	5.91	5.86	5.80	5.77	5.75	5.72	5.69	5.66	5.63
5	6.61	5.79	5.41	5.19	5.05	4.95	4.88	4.82	4.77	4.74	4.68	4.62	4.56	4.53	4.50	4.46	4.43	4.40	4.36
6	5.99	5.14	4.76	4.53	4.39	4.28	4.21	4.15	4.10	4.06	4.00	3.94	3.87	3.84	3.81	3.77	3.74	3.70	3.67
7	5.59	4.74	4.35	4.12	3.97	3.87	3.79	3.73	3.68	3.64	3.57	3.51	3.44	3.41	3.38	3.34	3.30	3.27	3.23
8	5.32	4.46	4.07	3.84	3.69	3.58	3.50	3.44	3.39	3.35	3.28	3.22	3.15	3.12	3.08	3.04	3.01	2.97	2.93
9	5.12	4.26	3.86	3.63	3.48	3.37	3.29	3.23	3.18	3.14	3.07	3.01	2.94	2.90	2.86	2.83	2.79	2.75	2.71
10	4.96	4.10	3.71	3.48	3.33	3.22	3.14	3.07	3.02	2.98	2.91	2.85	2.77	2.74	2.70	2.66	2.62	2.58	2.54
11	4.84	3.98	3.59	3.36	3.20	3.09	3.01	2.95	2.90	2.85	2.79	2.72	2.65	2.61	2.57	2.53	2.49	2.45	2.40
12	4.75	3.89	3.49	3.26	3.11	3.00	2.91	2.85	2.80	2.75	2.69	2.62	2.54	2.51	2.47	2.43	2.38	2.34	2.30
13	4.67	3.81	3.41	3.18	3.03	2.92	2.83	2.77	2.71	2.67	2.60	2.53	2.46	2.42	2.38	2.34	2.30	2.25	2.21
14	4.60	3.74	3.34	3.11	2.96	2.85	2.76	2.70	2.65	2.60	2.53	2.46	2.39	2.35	2.31	2.27	2.22	2.18	2.13
15	4.54	3.68	3.29	3.06	2.90	2.79	2.71	2.64	2.59	2.54	2.48	2.40	2.33	2.29	2.25	2.20	2.16	2.11	2.07
16	4.49	3.63	3.24	3.01	2.85	2.74	2.66	2.59	2.54	2.49	2.42	2.35	2.28	2.24	2.19	2.15	2.11	2.06	2.01
17	4.45	3.59	3.20	2.96	2.81	2.70	2.61	2.55	2.49	2.45	2.38	2.31	2.23	2.19	2.15	2.10	2.06	2.01	1.96
18	4.41	3.55	3.16	2.93	2.77	2.66	2.58	2.51	2.46	2.41	2.34	2.27	2.19	2.15	2.11	2.06	2.02	1.97	1.92
19	4.38	3.52	3.13	2.90	2.74	2.63	2.54	2.48	2.42	2.38	2.31	2.23	2.16	2.11	2.07	2.03	1.98	1.93	1.88
20	4.35	3.49	3.10	2.87	2.71	2.60	2.51	2.45	2.39	2.35	2.28	2.20	2.12	2.08	2.04	1.99	1.95	1.90	1.84
21	4.32	3.47	3.07	2.84	2.68	2.57	2.49	2.42	2.37	2.32	2.25	2.18	2.10	2.05	2.01	1.96	1.92	1.87	1.81
22	4.30	3.44	3.05	2.82	2.66	2.55	2.46	2.40	2.34	2.30	2.23	2.15	2.07	2.03	1.98	1.94	1.89	1.84	1.78
23	4.28	3.42	3.03	2.80	2.64	2.53	2.44	2.37	2.32	2.27	2.20	2.13	2.05	2.01	1.96	1.91	1.86	1.81	1.76
24	4.26	3.40	3.01	2.78	2.62	2.51	2.42	2.36	2.30	2.25	2.18	2.11	2.03	1.98	1.94	1.89	1.84	1.79	1.73
25	4.24	3.39	2.99	2.76	2.60	2.49	2.40	2.34	2.28	2.24	2.16	2.09	2.01	1.96	1.92	1.87	1.82	1.77	1.71
26	4.23	3.37	2.98	2.74	2.59	2.47	2.39	2.32	2.27	2.22	2.15	2.07	1.99	1.95	1.90	1.85	1.80	1.75	1.69
27	4.21	3.35	2.96	2.73	2.57	2.46	2.37	2.31	2.25	2.20	2.13	2.06	1.97	1.93	1.88	1.84	1.79	1.73	1.67
28	4.20	3.34	2.95	2.71	2.56	2.45	2.36	2.29	2.24	2.19	2.12	2.04	1.96	1.91	1.87	1.82	1.77	1.71	1.65
29	4.18	3.33	2.93	2.70	2.55	2.43	2.35	2.28	2.22	2.18	2.10	2.03	1.94	1.90	1.85	1.81	1.75	1.70	1.64
30	4.17	3.32	2.92	2.69	2.53	2.42	2.33	2.27	2.21	2.16	2.09	2.01	1.93	1.89	1.84	1.79	1.74	1.68	1.62
40	4.08	3.23	2.84	2.61	2.45	2.34	2.25	2.18	2.12	2.08	2.00	1.92	1.84	1.79	1.74	1.69	1.64	1.58	1.51
60	4.00	3.15	2.76	2.53	2.37	2.25	2.17	2.10	2.04	1.99	1.92	1.84	1.75	1.70	1.65	1.59	1.53	1.47	1.39
120	3.92	3.07	2.68	2.45	2.29	2.17	2.09	2.02	1.96	1.91	1.83	1.75	1.66	1.61	1.55	1.50	1.43	1.35	1.25
∞	3.84	3.00	2.60	2.37	2.21	2.10	2.01	1.94	1.88	1.83	1.75	1.67	1.57	1.52	1.46	1.39	1.32	1.22	1.00
1	647.8	799.5	864.2	899.6	921.8	937.1	948.2	956.7	963.3	968.6	976.7	984.9	993.1	997.2	1001	1006	1010	1014	1018

Table A.4 (Continued)

$\alpha = .025$

$F_{U(\alpha, df_1, df_2)}$

NUMERATOR df_1

DENOMINATOR df_2	1	2	3	4	5	6	7	8	9	10	12	15	20	24	30	40	60	120	∞
2	38.51	39.00	39.17	39.25	39.30	39.33	39.36	39.37	39.39	39.40	39.41	39.43	39.45	39.46	39.46	39.47	39.48	39.49	39.50
3	17.44	16.04	15.44	15.10	14.88	14.73	14.62	14.54	14.47	14.42	14.34	14.25	14.17	14.12	14.08	14.04	13.99	13.95	13.90
4	12.22	10.65	9.98	9.60	9.36	9.20	9.07	8.98	8.90	8.84	8.75	8.66	8.56	8.51	8.46	8.41	8.36	8.31	8.26
5	10.01	8.43	7.76	7.39	7.15	6.98	6.85	6.76	6.68	6.62	6.52	6.43	6.33	6.28	6.23	6.18	6.12	6.07	6.02
6	8.81	7.26	6.60	6.23	5.99	5.82	5.70	5.60	5.52	5.46	5.37	5.27	5.17	5.12	5.07	5.01	4.96	4.90	4.85
7	8.07	6.54	5.89	5.52	5.29	5.12	4.99	4.90	4.82	4.76	4.67	4.57	4.47	4.42	4.36	4.31	4.25	4.20	4.14
8	7.57	6.06	5.42	5.05	4.82	4.65	4.53	4.43	4.36	4.30	4.20	4.10	4.00	3.95	3.89	3.84	3.78	3.73	3.67
9	7.21	5.71	5.08	4.72	4.48	4.32	4.20	4.10	4.03	3.96	3.87	3.77	3.67	3.61	3.56	3.51	3.45	3.39	3.33
10	6.94	5.46	4.83	4.47	4.24	4.07	3.95	3.85	3.78	3.72	3.62	3.52	3.42	3.37	3.31	3.26	3.20	3.14	3.08
11	6.72	5.26	4.63	4.28	4.04	3.88	3.76	3.66	3.59	3.53	3.43	3.33	3.23	3.17	3.12	3.06	3.00	2.94	2.88
12	6.55	5.10	4.47	4.12	3.89	3.73	3.61	3.51	3.44	3.37	3.28	3.18	3.07	3.02	2.96	2.91	2.85	2.79	2.72
13	6.41	4.97	4.36	4.00	3.77	3.60	3.48	3.39	3.31	3.25	3.15	3.05	2.95	2.89	2.84	2.78	2.72	2.66	2.60
14	6.30	4.86	4.24	3.89	3.66	3.50	3.38	3.29	3.21	3.15	3.05	2.95	2.84	2.79	2.73	2.67	2.61	2.55	2.49
15	6.20	4.77	4.15	3.80	3.58	3.41	3.29	3.20	3.12	3.06	2.96	2.86	2.76	2.70	2.64	2.59	2.52	2.46	2.40
16	6.12	4.69	4.08	3.73	3.50	3.34	3.22	3.12	3.05	2.99	2.89	2.79	2.68	2.63	2.57	2.51	2.45	2.38	2.32
17	6.04	4.62	4.01	3.66	3.44	3.28	3.16	3.06	2.98	2.92	2.82	2.72	2.62	2.56	2.50	2.44	2.38	2.32	2.25
18	5.98	4.56	3.95	3.61	3.38	3.22	3.10	3.01	2.93	2.87	2.77	2.67	2.56	2.50	2.44	2.38	2.32	2.26	2.19
19	5.92	4.51	3.90	3.56	3.33	3.17	3.05	2.96	2.88	2.82	2.72	2.62	2.51	2.45	2.39	2.33	2.27	2.20	2.13
20	5.87	4.46	3.86	3.51	3.29	3.13	3.01	2.91	2.84	2.77	2.68	2.57	2.46	2.41	2.35	2.29	2.22	2.16	2.09
21	5.83	4.42	3.82	3.48	3.25	3.09	2.97	2.87	2.80	2.73	2.64	2.53	2.42	2.37	2.31	2.25	2.18	2.11	2.04
22	5.79	4.38	3.78	3.44	3.22	3.05	2.93	2.84	2.76	2.70	2.60	2.50	2.39	2.33	2.27	2.21	2.14	2.08	2.00
23	5.75	4.35	3.75	3.41	3.18	3.02	2.90	2.81	2.73	2.67	2.57	2.47	2.36	2.30	2.24	2.18	2.11	2.04	1.97
24	5.72	4.32	3.72	3.38	3.15	2.99	2.87	2.78	2.70	2.64	2.54	2.44	2.33	2.27	2.21	2.15	2.08	2.01	1.94
25	5.69	4.29	3.69	3.35	3.13	2.97	2.85	2.75	2.68	2.61	2.51	2.41	2.30	2.24	2.18	2.12	2.05	1.98	1.91
26	5.66	4.27	3.67	3.33	3.10	2.94	2.82	2.73	2.65	2.59	2.49	2.39	2.28	2.22	2.16	2.09	2.03	1.95	1.88
27	5.63	4.24	3.65	3.31	3.08	2.92	2.80	2.71	2.63	2.57	2.47	2.36	2.25	2.19	2.13	2.07	2.00	1.93	1.85
28	5.61	4.22	3.63	3.29	3.06	2.90	2.78	2.69	2.61	2.55	2.45	2.34	2.23	2.17	2.11	2.05	1.98	1.91	1.83
29	5.59	4.20	3.61	3.27	3.04	2.88	2.76	2.67	2.59	2.53	2.43	2.32	2.21	2.15	2.09	2.03	1.96	1.89	1.81
30	5.57	4.18	3.59	3.25	3.03	2.87	2.75	2.65	2.57	2.51	2.41	2.31	2.20	2.14	2.07	2.01	1.94	1.87	1.79
40	5.42	4.05	3.46	3.13	2.90	2.74	2.62	2.53	2.45	2.39	2.29	2.18	2.07	2.01	1.94	1.88	1.80	1.72	1.64
60	5.29	3.93	3.34	3.01	2.79	2.63	2.51	2.41	2.33	2.27	2.17	2.06	1.94	1.88	1.82	1.74	1.67	1.53	1.48
120	5.15	3.80	3.23	2.89	2.67	2.52	2.39	2.30	2.22	2.16	2.05	1.94	1.82	1.76	1.69	1.61	1.53	1.43	1.31
∞	5.02	3.69	3.12	2.79	2.57	2.41	2.29	2.19	2.11	2.05	1.94	1.83	1.71	1.64	1.57	1.48	1.39	1.27	1.00

(Continued)

Table A.4 (Continued)

$\alpha = .005$

$F_{U(\alpha, df_1, df_2)}$

NUMERATOR df_1

DENOMINATOR df_2	1	2	3	4	5	6	7	8	9	10	12	15	20	24	30	40	60	120	∞
1	16211	20000	21615	22500	23056	23437	23715	23925	24091	24224	24426	24630	24836	24940	25044	25148	25253	25359	25465
2	198.5	199.0	199.2	199.2	199.3	199.3	199.4	199.4	199.4	199.4	199.4	199.4	199.4	199.5	199.5	199.5	199.5	199.5	199.5
3	55.55	49.80	47.47	46.19	45.39	44.84	44.43	44.13	43.88	43.69	43.39	43.08	42.78	42.62	42.47	42.31	42.15	41.99	41.83
4	31.33	26.28	24.26	23.15	22.46	21.97	21.62	21.35	21.14	20.97	20.70	20.44	20.17	20.03	19.89	19.75	19.61	19.47	19.32
5	22.78	18.31	16.53	15.56	14.94	14.51	14.20	13.96	13.77	13.62	13.38	13.15	12.90	12.78	12.66	12.53	12.40	12.27	12.14
6	18.63	14.54	12.92	12.03	11.46	11.07	10.79	10.57	10.39	10.25	10.03	9.81	9.59	9.47	9.36	9.24	9.12	9.00	8.88
7	16.24	12.40	10.88	10.05	9.52	9.16	8.89	8.68	8.51	8.38	8.18	7.97	7.75	7.65	7.53	7.42	7.31	7.19	7.08
8	14.69	11.04	9.60	8.81	8.30	7.95	7.69	7.50	7.34	7.21	7.01	6.81	6.61	6.50	6.40	6.29	6.18	6.06	5.95
9	13.61	10.11	8.72	7.96	7.47	7.13	6.88	6.69	6.54	6.42	6.23	6.03	5.83	5.73	5.62	5.52	5.41	5.30	5.19
10	12.83	9.43	8.08	7.34	6.87	6.54	6.30	6.12	5.97	5.85	5.66	5.47	5.27	5.17	5.07	4.97	4.86	4.75	4.64
11	12.23	8.91	7.60	6.88	6.42	6.10	5.86	5.68	5.54	5.42	5.24	5.05	4.86	4.76	4.65	4.55	4.44	4.34	4.23
12	11.75	8.51	7.23	6.52	6.07	5.76	5.52	5.35	5.20	5.09	4.91	4.72	4.53	4.43	4.33	4.23	4.12	4.01	3.90
13	11.37	8.19	6.93	6.23	5.79	5.48	5.25	5.08	4.94	4.82	4.64	4.46	4.27	4.17	4.07	3.97	3.87	3.76	3.65
14	11.06	7.92	6.68	6.00	5.56	5.26	5.03	4.86	4.72	4.60	4.43	4.25	4.06	3.96	3.86	3.76	3.66	3.55	3.44
15	10.80	7.70	6.48	5.80	5.37	5.07	4.85	4.67	4.54	4.42	4.25	4.07	3.88	3.79	3.69	3.58	3.48	3.37	3.26
16	10.58	7.51	6.30	5.64	5.21	4.91	4.69	4.52	4.38	4.27	4.10	3.92	3.73	3.64	3.54	3.44	3.33	3.22	3.11
17	10.38	7.35	6.16	5.50	5.07	4.78	4.56	4.39	4.25	4.14	3.97	3.79	3.61	3.51	3.41	3.31	3.21	3.10	2.98
18	10.22	7.21	6.03	5.37	4.96	4.66	4.44	4.28	4.14	4.03	3.86	3.68	3.50	3.40	3.30	3.20	3.10	2.99	2.87
19	10.07	7.09	5.92	5.27	4.85	4.56	4.34	4.18	4.04	3.93	3.76	3.59	3.40	3.31	3.21	3.11	3.00	2.89	2.78
20	9.94	6.99	5.82	5.17	4.76	4.47	4.26	4.09	3.96	3.85	3.68	3.50	3.32	3.22	3.12	3.02	2.92	2.81	2.69
21	9.83	6.89	5.73	5.09	4.68	4.39	4.18	4.01	3.88	3.77	3.60	3.43	3.24	3.15	3.05	2.95	2.84	2.73	2.61
22	9.73	6.81	5.65	5.02	4.61	4.32	4.11	3.94	3.81	3.70	3.54	3.36	3.18	3.08	2.98	2.88	2.77	2.66	2.55
23	9.63	6.73	5.58	4.95	4.54	4.26	4.05	3.88	3.75	3.64	3.47	3.30	3.12	3.02	2.92	2.82	2.71	2.60	2.48
24	9.55	6.66	5.52	4.89	4.49	4.20	3.99	3.83	3.69	3.59	3.42	3.25	3.06	2.97	2.87	2.77	2.66	2.55	2.43
25	9.48	6.60	5.46	4.84	4.43	4.15	3.94	3.78	3.64	3.54	3.37	3.20	3.01	2.92	2.82	2.72	2.61	2.50	2.38
26	9.41	6.54	5.41	4.79	4.38	4.10	3.89	3.73	3.60	3.49	3.33	3.15	2.97	2.87	2.77	2.67	2.56	2.45	2.33
27	9.34	6.49	5.36	4.74	4.34	4.06	3.85	3.69	3.56	3.45	3.28	3.11	2.93	2.83	2.73	2.63	2.52	2.41	2.29
28	9.28	6.44	5.32	4.70	4.30	4.02	3.81	3.65	3.52	3.41	3.25	3.07	2.89	2.79	2.69	2.59	2.48	2.37	2.25
29	9.23	6.40	5.28	4.66	4.26	3.98	3.77	3.61	3.48	3.38	3.21	3.04	2.86	2.76	2.66	2.56	2.45	2.33	2.21
30	9.18	6.35	5.24	4.62	4.23	3.95	3.74	3.58	3.45	3.34	3.18	3.01	2.82	2.73	2.63	2.52	2.42	2.30	2.18
40	8.83	6.07	4.98	4.37	3.99	3.71	3.51	3.35	3.22	3.12	2.95	2.78	2.60	2.50	2.40	2.30	2.18	2.06	1.93
60	8.49	5.79	4.73	4.14	3.76	3.49	3.29	3.13	3.01	2.90	2.74	2.57	2.39	2.29	2.19	2.08	1.96	1.83	1.69
120	8.18	5.54	4.50	3.92	3.55	3.28	3.09	2.93	2.81	2.71	2.54	2.37	2.19	2.09	1.98	1.87	1.75	1.61	1.43
∞	7.88	5.30	4.28	3.72	3.35	3.09	2.90	2.74	2.62	2.52	2.36	2.19	2.00	1.90	1.79	1.67	1.53	1.36	1.00

For a particular combination of numerator and denominator degrees of freedom, entry represents the critical values of F corresponding to a specified upper tail area (α).

Source: Reprinted from E. S. Pearson and H. O. Hartley, eds., Biometrika Tables for Statisticians, 3d ed., 1966, by permission of the Biometrika Trustees.

Table A.5 Critical Values[a] of the Studentized Range Q

UPPER 5% POINTS ($\alpha = 0.05$)

$\nu \backslash \eta$	2	3	4	5	6	7	8	9	10	11	12	13	14	15	16	17	18	19	20
1	18.00	27.00	32.80	37.10	40.40	43.10	45.40	47.40	49.10	50.60	52.00	53.20	54.30	55.40	56.30	57.20	58.00	58.80	59.60
2	6.09	8.30	9.80	10.90	11.70	12.40	13.00	13.50	14.00	14.40	14.70	15.10	15.40	15.70	15.90	16.10	16.40	16.60	16.80
3	4.50	5.91	6.82	7.50	8.04	8.48	8.85	9.18	9.46	9.72	9.95	10.15	10.35	10.52	10.69	10.84	10.98	11.11	11.24
4	3.93	5.04	5.76	6.29	6.71	7.05	7.35	7.60	7.83	8.03	8.21	8.37	8.52	8.66	8.79	8.91	9.03	9.13	9.23
5	3.64	4.60	5.22	5.67	6.03	6.33	6.58	6.80	6.99	7.17	7.32	7.47	7.60	7.72	7.83	7.93	8.03	8.12	8.21
6	3.46	4.34	4.90	5.31	5.63	5.89	6.12	6.32	6.49	6.65	6.79	6.92	7.03	7.14	7.24	7.34	7.43	7.51	7.59
7	3.34	4.16	4.68	5.06	5.36	5.61	5.82	6.00	6.16	6.30	6.43	6.55	6.66	6.76	6.85	6.94	7.02	7.09	7.17
8	3.26	4.04	4.53	4.89	5.17	5.40	5.60	5.77	5.92	6.05	6.18	6.29	6.39	6.48	6.57	6.65	6.73	6.80	6.87
9	3.20	3.95	4.42	4.76	5.02	5.24	5.43	5.60	5.74	5.87	5.98	6.09	6.19	6.28	6.36	6.44	6.51	6.58	6.64
10	3.15	3.88	4.33	4.65	4.91	5.12	5.30	5.46	5.60	5.72	5.83	5.93	6.03	6.11	6.20	6.27	6.34	6.40	6.47
11	3.11	3.82	4.26	4.57	4.82	5.03	5.20	5.35	5.49	5.61	5.71	5.81	5.90	5.99	6.06	6.14	6.20	6.26	6.33
12	3.08	3.77	4.20	4.51	4.75	4.95	5.12	5.27	5.40	5.51	5.62	5.71	5.80	5.88	5.95	6.03	6.09	6.15	6.21
13	3.06	3.73	4.15	4.45	4.69	4.88	5.05	5.19	5.32	5.43	5.53	5.63	5.71	5.79	5.86	5.93	6.00	6.05	6.11
14	3.03	3.70	4.11	4.41	4.64	4.83	4.99	5.13	5.25	5.36	5.46	5.55	5.64	5.71	5.79	5.85	5.92	5.97	6.03
15	3.01	3.67	4.08	4.37	4.60	4.78	4.94	5.08	5.20	5.31	5.40	5.49	5.58	5.65	5.72	5.79	5.85	5.90	5.96
16	3.00	3.65	4.05	4.33	4.56	4.74	4.90	5.03	5.15	5.26	5.35	5.44	5.52	5.59	5.66	5.72	5.79	5.84	5.90
17	2.98	3.63	4.02	4.30	4.52	4.71	4.86	4.99	5.11	5.21	5.31	5.39	5.47	5.55	5.61	5.68	5.74	5.79	5.84
18	2.97	3.61	4.00	4.28	4.49	4.67	4.82	4.96	5.07	5.17	5.27	5.35	5.43	5.50	5.57	5.63	5.69	5.74	5.79
19	2.96	3.59	3.98	4.25	4.47	4.65	4.79	4.92	5.04	5.14	5.23	5.32	5.39	5.46	5.53	5.59	5.65	5.70	5.75
20	2.95	3.58	3.96	4.23	4.45	4.62	4.77	4.90	5.01	5.11	5.20	5.28	5.36	5.43	5.49	5.55	5.61	5.66	5.71
24	2.92	3.53	3.90	4.17	4.37	4.54	4.68	4.81	4.92	5.01	5.10	5.18	5.25	5.32	5.38	5.44	5.50	5.54	5.59
30	2.89	3.49	3.84	4.10	4.30	4.46	4.60	4.72	4.83	4.92	5.00	5.08	5.15	5.21	5.27	5.33	5.38	5.43	5.48
40	2.86	3.44	3.79	4.04	4.23	4.39	4.52	4.63	4.74	4.82	4.91	4.98	5.05	5.11	5.16	5.22	5.27	5.31	5.36
60	2.83	3.40	3.74	3.98	4.16	4.31	4.44	4.55	4.65	4.73	4.81	4.88	4.94	5.00	5.06	5.11	5.16	5.20	5.24
120	2.80	3.36	3.69	3.92	4.10	4.24	4.36	4.48	4.56	4.64	4.72	4.78	4.84	4.90	4.95	5.00	5.05	5.09	5.13
∞	2.77	3.31	3.63	3.86	4.03	4.17	4.29	4.39	4.47	4.55	4.62	4.68	4.74	4.80	4.85	4.89	4.93	4.97	5.01

(Continued)

Table A.5 (Continued)

UPPER 1% POINTS ($\alpha = 0.01$)

$\nu \backslash \eta$	2	3	4	5	6	7	8	9	10	11	12	13	14	15	16	17	18	19	20
1	90.00	135.00	164.00	186.00	202.00	216.00	227.00	237.00	246.00	253.00	260.0	266.00	272.00	277.00	282.00	286.00	290.00	294.00	298.00
2	14.00	19.00	22.30	24.70	26.60	28.20	29.50	30.70	31.70	32.60	33.40	34.10	34.80	35.40	36.00	36.50	37.00	37.50	37.90
3	8.26	10.60	12.20	13.30	14.20	15.00	15.60	16.20	16.70	17.10	17.50	17.90	18.20	18.50	18.80	19.10	19.30	19.50	19.80
4	6.51	8.12	9.17	9.96	10.60	11.10	11.50	11.90	12.30	12.60	12.80	13.10	13.30	13.50	13.70	13.90	14.10	14.20	14.40
5	5.70	6.97	7.80	8.42	8.91	9.32	9.67	9.97	10.24	10.48	10.70	10.89	11.08	11.24	11.40	11.55	11.68	11.81	11.93
6	5.24	6.33	7.03	7.56	7.97	8.32	8.61	8.87	9.10	9.30	9.49	9.65	9.81	9.95	10.08	10.21	10.32	10.43	10.54
7	4.95	5.92	6.54	7.01	7.37	7.68	7.94	8.17	8.37	8.55	8.71	8.86	9.00	9.12	9.24	9.35	9.46	9.55	9.65
8	4.74	5.63	6.20	6.63	6.96	7.24	7.47	7.68	7.87	8.03	8.18	8.31	8.44	8.55	8.66	8.76	8.85	8.94	9.03
9	4.60	5.43	5.96	6.35	6.66	6.91	7.13	7.32	7.49	7.65	7.78	7.91	8.03	8.13	8.23	8.32	8.41	8.49	8.57
10	4.48	5.27	5.77	6.14	6.43	6.67	6.87	7.05	7.21	7.36	7.48	7.60	7.71	7.81	7.91	7.99	8.07	8.15	8.22
11	4.39	5.14	5.62	5.97	6.26	6.48	6.67	6.84	6.99	7.13	7.25	7.36	7.46	7.56	7.65	7.73	7.81	7.88	7.95
12	4.32	5.04	5.50	5.84	6.10	6.32	6.51	6.67	6.81	6.94	7.06	7.17	7.26	7.36	7.44	7.52	7.59	7.66	7.73
13	4.26	4.96	5.40	5.73	5.98	6.19	6.37	6.53	6.67	6.79	6.90	7.01	7.10	7.19	7.27	7.34	7.42	7.48	7.55
14	4.21	4.89	5.32	5.63	5.88	6.08	6.26	6.41	6.54	6.66	6.77	6.87	6.96	7.05	7.12	7.20	7.27	7.33	7.39
15	4.17	4.83	5.25	5.56	5.80	5.99	6.16	6.31	6.44	6.55	6.66	6.76	6.84	6.93	7.00	7.07	7.14	7.20	7.26
16	4.13	4.78	5.19	5.49	5.72	5.92	6.08	6.22	6.35	6.46	6.56	6.66	6.74	6.82	6.90	6.97	7.03	7.09	7.15
17	4.10	4.74	5.14	5.43	5.66	5.85	6.01	6.15	6.27	6.38	6.48	6.57	6.66	6.73	6.80	6.87	6.94	7.00	7.05
18	4.07	4.70	5.09	5.38	5.60	5.79	5.94	6.08	6.20	6.31	6.41	6.50	6.58	6.65	6.72	6.79	6.85	6.91	6.96
19	4.05	4.67	5.05	5.33	5.55	5.73	5.89	6.02	6.14	6.25	6.34	6.43	6.51	6.58	6.65	6.72	6.78	6.84	6.89
20	4.02	4.64	5.02	5.29	5.51	5.69	5.84	5.97	6.09	6.19	6.29	6.37	6.45	6.52	6.59	6.65	6.71	6.76	6.82
24	3.96	4.54	4.91	5.17	5.37	5.54	5.69	5.81	5.92	6.02	6.11	6.19	6.26	6.33	6.39	6.45	6.51	6.56	6.61
30	3.89	4.45	4.80	5.05	5.24	5.40	5.54	5.65	5.76	5.85	5.93	6.01	6.08	6.14	6.20	6.26	6.31	6.36	6.41
40	3.82	4.37	4.70	4.93	5.11	5.27	5.39	5.50	5.60	5.69	5.77	5.84	5.90	5.96	6.02	6.07	6.12	6.17	6.21
60	3.76	4.28	4.60	4.82	4.99	5.13	5.25	5.36	5.45	5.53	5.60	5.67	5.73	5.79	5.84	5.89	5.93	5.98	6.02
120	3.70	4.20	4.50	4.71	4.87	5.01	5.12	5.21	5.30	5.38	5.44	5.51	5.56	5.61	5.66	5.71	5.75	5.79	5.83
∞	3.64	4.12	4.40	4.60	4.76	4.88	4.99	5.08	5.16	5.23	5.29	5.35	5.40	5.45	5.49	5.54	5.57	5.61	5.65

[a] Range/$S \sim Q_{1-\bar{\alpha}:\eta,\nu}$, η is the size of the sample from which the range is obtained, and ν is the number of degrees of freedom of S.

Source: Reprinted from E. S. Pearson and H. O. Hartley, eds., Table 29 of Biometrika Tables for Statisticians, Vol. 1, 3rd ed., 1966, by permission of the Biometrika Trustees, London.

INDEX